AF517872

Cotton

History, Species, Varieties, Morphology, Breeding, Culture, Diseases, Marketing, and Uses

(Frontispiece)

A field of Upland long-staple cotton with open bolls awaiting pickers. This field made one and one-half bales per acre.

Cotton

History, Species, Varieties, Morphology, Breeding, Culture, Diseases, Marketing, and Uses

Harry Bates Brown

Second Edition

2002
BIOTECH BOOKS
Delhi - 110 035

Brown, Harry Bates (b.1876 –)
First Indian Edition 2002

ISBN 81–7622–067–1

Published by : **BIOTECH BOOKS**
1123/74, Tri Nagar,
DELHI – 110 035
Phone: 7109765
e-mail: biotechbooks@yahoo.co.in

Showroom : 4762-63/23, Ansari Road, Darya Ganj,
NEW DELHI - 110 035
Phone: 3245578, 3244987

Printed at : **CHAWLA OFFSET PRINTERS**
New Delhi – 110 052

PRINTED IN INDIA

TO

MY PARENTS

JOHN HOLTON BROWN

AND

SUE (BATES) BROWN

THIS BOOK

IS AFFECTIONATELY DEDICATED

PREFACE TO THE SECOND EDITION

The main object in producing the second edition of our book on cotton is to bring it up to date. Considerable progress has been made in cotton research since the first edition was issued. The general plan of the book has not been changed, but parts of nearly all chapters have been rewritten and new paragraphs added.

Mention should be made of other men who aided in the work of the second edition. Dr. U. R. Gore, of the Georgia Experiment Station, kindly gave us a number of unpublished drawings showing different stages in flower development and fertilization; Prof. R. W. Harned and others, of the Government Laboratory at Tallulah, La., contributed information and photographs; Dr. F. L. Davis and Dr. M. B. Sturgis, of the Crops and Soils Department, gave criticism and advice on fertilizers and soils; Dr. J. O. Ware, of the Bureau of Plant Industry, U. S. Department of Agriculture, furnished figures on the relative acreage of different varieties; a large number of agronomists in the various cotton-growing states gave information on fertilizers recommended or used and varieties and cultural methods in use. Director C. T. Dowell and others of the Experiment Station gave assistance in various ways.

H. B. Brown.

PREFACE TO THE FIRST EDITION

A book which treats of cotton in its various phases must make at least a brief survey of many fields, for the topic is broad, and information on it is widely scattered. Several books have been written on this subject, but much of the material in these is out of date on account both of commercial changes and of new information now available as a result of scientific research. It is the object of the present volume to bring together, in compact form, a considerable body of information concerning the most important aspects of cotton and the cotton industry.

In the treatment of the subject a brief historical survey is first made. This is followed by a discussion of the cotton plant in various relationships—its botanical species, agricultural varieties, morphology, physiology, reproduction, and heredity. Next there is a discussion of subjects relating to cotton production—cotton breeding, fertilizers, soils, climate, culture, diseases, harvesting, and ginning. This is followed, in turn, by a review of phases of cotton marketing—classing, cotton marketing proper, and cotton exchanges. Finally, products and uses of different parts of the cotton plant are considered.

Some mention is made of cottons in foreign countries and the conditions under which they are grown, but, in the main, the material here included applies especially to the upland cotton district of the American Cotton Belt. Some of the views advanced may have been unduly influenced by the writer's experience in Mississippi where he has spent many days in cotton fields engaged in cotton breeding.

The preparation of a work of this type is an arduous task and one that would hardly be possible of accomplishment at all without the aid and indulgence of associates and friends engaged in different phases of the industry. The writer wishes to acknowledge his indebtedness to all who have assisted in this work, and to mention especially Mr. C. A. Tate, Mr. Ide P. Trotter, Mr. H. M. Smith, and Mr. Newman Hancock, who aided in collecting data; Prof. A. W. Palmer, Prof. F. L. Lewton, Prof.

W. E. Ayres, Prof. J. M. Beal, Dr. C. H. Myers, Dr. W. F. Hand, Prof. M. J. Funchess, Dr. J. O. Morgan, Prof. D. C. Neal, Prof. George A. Malony, Mr. L. A. Willis, Mr. George Darden, Mr. R. W. Boys, Mr. D. A. McCandliss, and Prof. G. B. Walker, who read certain chapters of the manuscript; Prof. F. J. Weddell and Dr. R. Y. Winters who very painstakingly read the whole manuscript; Director J. R. Ricks who gave free use of the Mississippi Experiment Station Library; Mr. Whitman Davis for very liberal loan of library books; and Prof. W. Lawrence Balls and others for use of quotations and illustrations from their published works. Acknowledgment of the credit due is made on the page where the borrowed figure or illustration is used. The author is also greatly indebted to the officials of the Stoneville Pedigreed Seed Co. for their cooperation which made the preparation of the manuscript possible.

H. B. Brown.

CONTENTS

COTTON

CHAPTER I

HISTORY OF COTTON AND THE COTTON INDUSTRY*

The origin of the word "cotton" and the date when it was first used in its present meaning are somewhat uncertain. At first it meant merely a fine textile, and the word was broad enough to include flax. Watt,[2] an English authority on cotton, who has investigated the derivation of the word and various words in different languages expressing the same idea, has the following to say: "The Sanskrit word *kárpásá-i*, usually rendered 'cotton,' is connected with the Greek and Latin *karpasos* and *carbasus*, but meant originally Spanish flax." The English word "cotton" came from the Arabic word *qutun*, or *kutun*.

The earliest historical records show man using various fibers in the manufacture of cloth. Wool was the principal fiber used in early times in western Asia and in southern Europe, flax in northern Europe, silk in China, and cotton in India. The inhabitants of the various regions learned in some way to convert fibers into a cloth that would serve their needs.

When Cotton Was First Used.—When the cotton fiber was first woven into cloth is not known, but the time evidently antedates history. Gulati and Turner[12] report that in some recent excavations at Mohenjo-daro in the valley of the Indus in northwestern India, some fragments of cotton cloth and string were unearthed. Archaeological evidence indicates that they date back to about 3000 B.C. A microscopical examination showed typical cotton fibers of the arboreum type which is grown in India at present. Cotton fabrics have also been found in prehistoric pueblo ruins in Arizona. McGregor[13] states, "Material from prehistoric sites indicates that three types of yarn

* Numbered references throughout are to the "References" at the close of each chapter.

were spun: weaving yarn, sewing thread, and string. Three variations of simple weaves were noted, depending on the amount of twist of the weft threads, the resulting cloth grading from soft, loosely woven to tough canvas-like fabrics."

The first reference in literature to cotton, so far as is known at present, is to be found in a Hindu Rig-Veda hymn.[3] This hymn was written fifteen centuries before Christ and mentions "threads in the loom," showing that cotton was already being used in weaving. The ancient Hindu laws in the religious books of Manu record its use about 800 B.C., but it is to be inferred that the plant and fiber had both been known for generations before that date. It was specified in the laws that the sacrificial thread of the Brahman must be made of cotton (*karpasi*).[1] Herodotus (484–425 B.C.) said:

> There are trees which grow wild there [India] the fruit of which is a wool exceeding in beauty and goodness that of sheep. The Indians make their clothes of this tree wool.

Theophrastus (372–287 B.C.)[1] gives a little light on cotton cultivation in India at an early day. He says:

> The trees from which the Indians make cloth have a leaf like that of the black mulberry, but the whole plant resembles the dog rose. They set them in plains arranged in rows so as to look like vines at a distance.

Nearchus,[1] the admiral of Alexander the Great, who conducted an army along the shores of southeastern Asia about 327 B.C., reported:

> There are trees in India bearing, as it were, bunches of wool. The natives make linen garments of it, wearing a shirt which reached to the middle of the leg, a sheet folded around the shoulders, and a turban rolled round the head, and the linen made by them from this substance was fine, and whiter than any other.

From references made to cotton by various early writers, it appears that not only did the Hindus grow cotton and use it in making cloth from ancient times but that for 3,000 years India was the center of the cotton industry. Some of the cloth made by the Indians with primitive distaffs and rude looms was not equaled for delicacy and texture until the nineteenth century. Marco Polo, a Venetian, who traveled in Asia in the thirteenth century, mentions the coast of Coromandel (Madras, India) as producing "the finest and most beautiful cottons that are to be found in any part of the world." Others described the Indian calicuts (calicoes) as being

> . . . so fine you can hardly feel them in your hand, and the thread when spun is scarce discernible. . . . When muslin is laid on the grass to bleach and dew has fallen on it, it is no longer discernible.

Poetic writers of the Orient spoke of these fabrics as "webs of woven wind."

Hindu Textile Machinery.—The machinery used by the ancient Hindus in ginning, spinning, and weaving cotton was of the most primitive type. The gin was a small roller gin called "churka," consisting of two upright pieces between which were two horizontal rollers so set that they almost touched and to one of which a crank was attached. They had a few longitudinal flutes on the surface. The cotton seed with lint attached was fed between the rollers on one side. The lint was caught by the rollers and passed on through, but the seed, being larger, could not do so and were thus separated from the lint. (A gin of this type can be used successfully only with cotton that gins easily, as, for instance, Sea Island or other smooth-seeded varieties.)

The ginned lint was "bowed" before being spun. This process loosened neps or knotted masses of fibers, making the whole mass of cotton loose and fluffy. Considerable trash and foreign matter were also shaken out. Scherer[3] describes the bowing process as follows:

> The bow was an interesting contrivance of elastic wood made still more vibrant by the tension of taut cords. A workman, placing his bow in contact with a mass of lint, would strike the resounding strings with a wooden hammer, so that powerful vibrations forced open the knots of cotton, shook free the small rubbish of the fields, and produced a mass of downy fleece.

After the bowing process the lint was spun into a thread by a woman, who used either a one-thread spinning wheel or simply a distaff. The quality of the thread depended on the skill and dexterity of the spinner's fingers.

The Hindu weaver accomplished the greatest wonders of all. Scherer[3] says:

> Cotton, having been ginned and bowed and then spun into delicate yarn, was passed on to this magical master craftsman, plying his trade

under the friendly shade of a tree. A handful of reeds, with balances suspended from overhanging branches, made up his frail apparatus, the workman sitting in a pit beneath it, his great toes treadling with looped threads, his hands wielding the wide shuttle-batten, the warp being stretched out along the ground.

Myths about Cotton.—During the Middle Ages most of the knowledge of distant countries was obtained from travelers and explorers who had visited those lands. These early voyagers were much interested in producing an effect on their hearers and were not very careful about the scientific accuracy of their statements. In fact, many strange and fabulous tales were told, among them the story of the Vegetable Lamb, or Zoophyte. a being part animal and part plant.

Scherer[3] quotes the following description of the Vegetable Lamb of Scythia, or the Borametz of Scythia.

It was in form like a lamb, and from its navel grew a stem or root by which this Zoophyte, or plant-animal, was fixed, attached like a gourd to the soil below the surface of the ground, and, according to the length of its stem or root, it devoured all the herbage which it was able to reach within the circle of its tether. The hunters, who went in search of this creature were unable to capture or remove it until they had succeeded in cutting the stem by well-aimed arrows or darts, when the animal immediately fell prostrate to the earth and died. Its bones being placed with certain ceremonies and incantations in the mouth of one desiring to foretell the future, he was instantly seized with the spirit of divination, and endowed with the gift of prophecy.

Another story was to the effect that there was a little lamb within each boll and the lint within the boll was borne by this little lamb. Sir John Mandeville, a traveler who is said to have set out from England about 1322, mentions this strange phenomenon. He says:

Another passing marvelous thing may be related, which, however, I saw not myself but heard from trustworthy persons. And when these be ripe [referring to the bolls] they burst, and a little beast is found inside like a small lamb, so they have both melons and meat. And though some, peradventure, may find that hard to believe, yet it may be quite true.

Cotton lint or floss resembles fine wool, and those who have never seen a plant producing such material would naturally think that it must be an animal product. Thus it became easy for the travelers to find believers in their highly imaginative tales.

The Cotton Industry in India during Later Times.—As has been shown, India very early took front rank in the cotton industry and was the leading cotton country of the world for perhaps 3,000 years. The natives grew the plant extensively and made sufficient cloth to supply their own needs and to supply traders, who carried it to foreign lands. With their primitive equipment they were able to weave muslins of such delicate texture that they were not equaled in fineness by that of another country prior to the latter half of the nineteenth century.

Although cotton has been grown in India for ages, progress in improving the yield of varieties and quality of staple has been very slow. The yield per acre at present is less than 100 pounds of lint cotton, and the staple length is largely less than ½ inch. The grade is very low because of the large amount of dirt, leaf trash, etc., in the lint.

The area of India lying within the Cotton Belt is, according to Handy,[1] 874,880,000 acres, or an area about twice that of the cotton states of the United States. Much of this area consists of very fertile land, but the climate is not so favorable as that prevailing in the American Cotton Belt. A part of the area has scanty rainfall, whereas other parts have too much—as much as 500 inches per annum in some places. The Deccan, or Central India, is the great cotton section of the country. It is an elevated tableland, with splendid fertile soil which holds moisture well. Most of the export cotton is raised in this region and finds its outlet through the port of Bombay. As was shown above, the comparative cotton production of India is low. This might be increased considerably if there were sufficient demand. During the period of the Civil War in the United States (1861–1865), when the American supply was greatly reduced, there was a temporary increase in production. The estimated yield in 1934–1935 was 4,065,000 bales of 500 pounds each.

For many years the hand looms consumed most of the cotton produced in India. In 1854, a power mill was built at Bombay. This was followed by another the next year and by several more the next few years. In 1861, there were 12 mills in India; and in 1919, 263. Home consumption in 1933–1934 was 1,839,000 bales—about 45 per cent of the production.

More than half the Indian crop is exported. Japan gets more of this than any other country. About a third of the cotton produced is exported in the form of yarn, the bulk of which goes to China. India imports a limited amount of raw cotton from America and a large amount of cotton goods from Great Britain.

Cotton in China.—Cotton was introduced into China from India in early times, but there are no records showing the exact date. There was from early days considerable trade with India both by land caravans and by small boats moving along the coast of the two countries. Cotton seed naturally found its way into China, but for many years cotton plants were grown only as ornamentals. The Chinese people, until comparatively recent times, were very reluctant to make any change in habits, manner of dress, government, or business; so they continued to use the same fibers for making clothing that their fathers had used, even though cotton was better and cheaper. It was not until about A.D. 1300 that cotton came into general use, but since that time its use has spread extensively. At present several million bales are produced and consumed annually. Since a large part of this cotton is used in the home and does not reach the market, it is impossible to get any reliable data on the amount produced. The U. S. Bureau of Agricultural Economics estimated the crop for the year 1934–1935 at 3,125,000 bales of 478 pounds lint each.

Very primitive methods of culture are in use in China, the seed being either sown broadcast or planted in hills in patches ranging in size from a few square yards to 2 acres and cultivated by hand. Much of the land is very fertile, and good yields are obtained. Labor is abundant, and the climate satisfactory, but the great need for using the land for food crops will tend to prevent the increase of cotton growing. The main cotton-producing areas are the valleys of the Yellow River, or Hwang Ho, and of the Yangtze Kiang.

The varieties grown are mostly of the Asiatic type. They have small bolls of rather low lint percentage, with a coarse staple about ⅝ inch in length. Commonly the entire boll is pulled from the stalk, and the cotton is picked from the bolls later at the grower's convenience. The lint obtained is said to be clean and white. Recently some American upland varieties which have been grown and acclimated at the University of

Nanking have yielded well and given an excellent quality of lint. This shows that American varieties, if handled properly, can be grown there to advantage.

The Hindu churka, or hand-power roller gin, is used in ginning the cotton used in the home; the bow is used for cleaning, and the hand loom for weaving. According to Handy,[1] but little cotton is sold. What surplus there is, is made into cloth by the women of the family, and the cloth sold in the neighborhood. Oil is expressed from the seed and used for illuminating purposes. The cotton stalks are preserved for fuel. It is estimated by Garside[3] that the cotton mills of China contain about 4,680,000 spindles and use about half a million bales of native in addition to a small amount of imported cotton.

China, according to Todd,[4] exports annually about a quarter of a million bales, most of which goes to Japan, but some to America and Germany. She imports a small amount of raw cotton from India and a considerable quantity of cotton yarn from India and Japan.

Japan.—Japan proper produces very little cotton, but she has become a great manufacturing country, her mills containing more than 9,000,000 spindles and consuming annually more than 3,500,000 bales of cotton. The major portion of this yield comes from India and the United States.

Korea, or Chosen, which became a part of the Japanese Empire in 1910, is a promising new source of cotton for Japan. Cotton was introduced into Korea from China about 500 years ago, and Chinese varieties and methods of culture were used until recently. When Japan took over the country, much encouragement was given to cotton growing. American varieties were introduced, seed farms were established, and a cotton-cultivation association was formed. In 1934–1935, the production was 136,000 bales. Japan takes most of the cotton that is exported.

Indo-China and Siam.—These countries produce a limited amount of cotton, the most important of which is the Cambodia variety, grown in the province of Cambodia. Within the last few years, this variety has made a splendid showing in parts of India. During the year 1933–1934, Indo-China produced 5,649 bales; and Siam, 3,244.

East Indies.—Cotton is cultivated in all parts of Java, but the total amount produced is not great. Handy[1] estimated the annual production to be about 6,500 bales of 400 pounds each. Much of the cotton grown is used in stuffing cushions, mattresses, etc. A small amount is also grown in that part of Siam which is in the Malay States; this is made into coarse cloth with the primitive homemade machines in use in the homes there. The more attractive imported machine-made fabrics are displacing the homespun garments. A small amount of cotton is exported from the East Indies to Europe.

Western Asia.—Cotton is grown in various parts of western Asia, the area being widely scattered and small compared with those in the United States and India. The principal districts are in Russian Turkestan, Persia, and Asia Minor.

Cotton has been grown in parts of western Asia since ancient times. It was seen growing on the island of Tylos in the Persian Gulf by Pliny (A.D. 23–79) and in Arabia by Theophrastus. It was probably introduced into these regions and in other parts of western Asia from India. Until comparatively recently, cotton growing was not a principal industry even with the people that grew it. It was grown on land not needed for such food crops as wheat, rice, and barley, and consequently no great amount was raised. That which was produced was for domestic use only. The high prices brought about by the scarcity during the Civil War period in America stimulated the cotton-growing industry everywhere. Russia, after conquering Turkestan, also fostered cotton growing so as to have a greater supply of lint for her mills. The rainfall in Turkestan is so scanty that irrigation must be practiced. Water is supplied by a number of rivers flowing from the mountains to the plains districts. The soils are of clay mixture, which is not easily permeable to water, and are fertile, being enriched by alluvial deposits from the rivers.

Most of the varieties formerly grown in the Asiatic districts just mentioned are native and belong to the species *Gossypium herbaceum.* These varieties have a short, rough staple and are similar to Indian varieties. American varieties have been tried with varying degrees of success. Sea Island has naturally proved a failure, not being adapted to dry regions. Upland varieties which have been acclimated have been more successful.

Recently the Soviet Union, in an effort to make the country self-sufficient in respect to cotton supplies, has fostered an increase

in production. Varieties have been imported from various countries, cotton-breeding stations established, and new methods of culture introduced, and the cotton-growing area has been extended in European Russia and to the northward in the Asiatic areas. In 1917, according to the Bureau of the Census Report,[5] the cotton production of Russia was 605,000 bales (from Turkestan and Transcaucasia). In 1921, the production had decreased to 95,000 bales, owing to the unsettled conditions prevailing. It is the present plan of the Russian government to increase the annual production to 3,200,000 bales. The yield in 1933–1934 was 1,738,000 bales.

Russia has need for all the cotton she can grow, and probably more. Garside[14] reported the number of cotton spindles in mills in Russia in 1934 as 9,800,000. Home spinning and weaving call for an additional supply of cotton.

Persia produces annually about 140,000 bales of cotton. The varieties grown are native and similar to the Russian. The cotton grown is of an inferior type and of low grade. On account of the scanty rainfall, irrigation is badly needed. Private wells are used to some extent. Most of the crop is sold to Russia.

The provinces of Adana, Aydin, and Aleppo in Asia Minor produce a limited amount of cotton. In 1913–1914, this amounted to approximately 150,000 bales weighing 440 pounds each. The climate of Adana is comparable to that of Egypt. The native cottons grown are similar to the Russian. American cottons have been introduced and found to pay better than the native. Most of the product goes to mills in Continental Europe.

Southern Europe.—The only cotton-producing area of much consequence in southern Europe is Transcaucasia. This province produced 124,000 bales in 1912, according to Shepperson.[6] The climate, soils, cultural conditions, and varieties used in this region are similar to those of Turkestan.

Greece grows some cotton on the plains of Boeotia and in Salonika. Todd[4] reports that the average crop is about 30,000 bales. American, Egyptian, and native cottons are grown. No cotton is exported.

A small amount of cotton is grown in the southern part of Italy and on the island of Malta.

Cotton is said to have been introduced into Spain in the tenth century. Recently an attempt has been made there to revive interest in cotton growing, and 18,533 bales were produced in 1933–1934.

Egypt.—The civilization of Egypt dates back nearly fifty centuries, but no reliable evidence has been found to prove that cotton was in use there prior to 200 B.C. Clothing and fabrics found in ancient tombs were made from flax in most instances and never from cotton. It seems odd that a country with such an ancient civilization and one so active in the trade of the world should be slow in making use of a valuable plant like cotton. Balls[8] thinks it is strange that no evidence of the use of cotton by the ancient Egyptians has been found and believes it possible that it may yet be discovered. He says that it may be that some "lucky excavation or casual glance through a microscope will suddenly extend the known history of cotton in Egypt two or three thousand years."

Wild cottons are found growing in parts of Africa, especially in the Sudan. These may be indigenous, but no certainty can be attached to this point, since seed may have been distributed by Arab traders who covered the continent years ago. Definite information on the culture of cotton in various parts of Egypt prior to 1800 is lacking. It is known, however, that cotton has been grown and an excellent quality of staple produced in the upper regions of the Nile for a long period of time.

Definite descriptions of certain species of cotton being grown in Egypt in 1800 and a brief sketch of the cotton-growing industry were made by members of the French expedition that invaded Egypt about that time. Balls[9] gives the following abstract from their writings:

> Some cotton fields exist in almost all parts, yet the crop is peculiar to the most southerly portion of Upper Egypt, and to the Delta.
>
> Thebes Province: *G. arborescens.* Sown at intervals of a metre on ridges between hods, 200 hods to the feddan; bamia and molochieh on the same ground. Sown mostly in April; flowers after 5 to 6 months; first flowers take 90 days to mature. If sown in July, no bolls are open till early March of the following year. Bolls are picked bodily, placed to dry in the sun, and the *écailles* are removed by hand. Fields never bigger than three feddans. Same plant lasts 8 to 10 years. Never topped, but dry branches are broken off, maximum yield in third year being 300 rotls per feddan. Used *aux fabriques du toile du pays* and more highly esteemed than Syrian.

Delta: *G. herbaceum.* Only grown as an annual. Sown in early April after wheat. Three waterings during the 5 months that it remains in the soil, the third being at flood time. Harvest in early September. . . . Near Samanoud gives 180 to 240 rotls per feddan. Price 16 pataques per cantar when sea is free; 9 pataques in time of war. In Mansourah province it is resown every year, but not pulled up, the bolls being gathered bodily as they ripen.

It appears that at the time mentioned, or at the close of the eighteenth century, two very distinct species of cotton were being grown in Egypt, one a tree cotton grown in Upper Egypt and the other an annual grown in Lower Egypt, or the Delta district. The latter is thought by Balls[8] to have been identical with the native cotton grown in western Asia.

The year 1820 marks the beginning of a new era in cotton growing in Egypt, it being the year that Jumel's tree cotton was first grown commercially in Lower Egypt. Professor W. L. Balls, formerly botanist for the Khedivial Agricultural Society, has spent several years studying cotton in Egypt and is probably better qualified to outline its history than any other man. He says:[8]

The cultivation of this short-staple Asiatic cotton [the annual species mentioned above] died out in consequence of the economic development of Jumel's plant, and the last trace we find is a record in 1840 stating that it is almost extinct.

The tree cotton from Upper Egypt was next brought forward under aegis of Mohammed Ali, founder of the Khediviate, at the suggestion of M. Jumel, a Franco-Swiss engineer. Taken from the garden of Maho Bey, in Cairo, it was propagated rapidly from the year 1820 under a system of state control, and soon displaced the Asiatic type. The brown, long, strong lint, readily ginned from the almost naked seed, quickly made its reputation with the spinners, and this type of lint has been typical of the Egyptian product ever since.

To trace the origin of the present cultivated varieties from this stock is almost impossible. Still, the following interpretation meets all the facts at present known.

The success of Jumel's tree cotton led to the importation and trial of other cottons, notably Sea Island. Importations of this latter strain, an annual in habit, have continued to the present day. It is not very successful in Egypt, yielding lightly and suffering unduly from "shedding," but the lint is often of good quality, equal to that of Georgia's and Florida's. The state control of the seed supply became disorganized after a time, partly in consequence of Mohammed Ali's military activities, and the inevitable mixing of two seed stocks was accelerated. This mixing, combined with natural crossing, led to the formation of splitting forms, some of which were annual but brown linted, and these gave rise to the Ashmouni stock or old "Brown Egyptian," which dominated the fields up to 1887. The tree type disappeared in consequence of its greater liability to damage from insect pests, such as the boll worm (*Earias insulana*), which was definitely recorded in Egypt as early as 1876, and also on account of better cultivation obtained with plants of annual habit. The only remaining trace of its influence is the presence of abnormally tall rogues up to four metres in height in the field. The Hamouli variety was possibly an intermediate stage in this process of extinction by artificial and natural selection.

From the Ashmouni stock came the Afifi, in 1887, by selection, probably natural in part, and from this now degenerate complex of subvarieties and splitting forms other varieties have been selected. The Ashmouni stock was driven into Upper Egypt, and has there improved itself until it is now making a reputation anew.

The relatively white lint of the Sea Island stock has always been a feature of at least one Egyptian variety, such as Abyad and Gallini, both extinct, and the modern Abbassi. Gallini in particular, while possessing the bigger boll, higher yield, and "climatic suitability" of its Peruvian-type ancestor, was a very fine cotton, which controlled the fine spinning market for years, until it deteriorated through mixture and crossing and was driven into oblivion by competition with Georgia's and Florida's, its own ancestors. Gallini has been revenged of late years on its own unnatural ancestors by the modern Yannovitch, itself a single plant selection from the Afifi complex.

The apparent identity of all the modern varieties of Egyptian cotton in external appearance—for even when grown side by side they are scarcely distinguishable—is the natural result of their origin from two related stocks.

The cotton-growing area of Egypt consists of a narrow strip some 550 miles long along the Nile and the fan-shaped delta, containing about 4,000,000 acres, at the mouth of the river. The soil is alluvial and very fertile. The rainfall is so scanty (from 1 to 8 inches) that irrigation must be practiced everywhere. No land can be used unless water can be carried to it from the Nile. The area is thus limited by the water supply available during the growing season. To increase this supply is a difficult matter. Consequently, there is not much chance for increasing the cotton-growing area.

If the land is irrigated properly, the soil and the climate of Egypt are such as to produce an excellent yield of high-class cotton. The average yield for the whole acreage of the country from 1901 to 1921 inclusive was 393 pounds of lint cotton per acre. No other country equals this in yield. The total production averages about 1,800,000 bales of 500 pounds each. The grade is high, because much care is taken in picking and ginning, and the weather for picking is splendid, not being marred by storms or rains. The length, strength, and color of Egyptian cottons, as well as the uniformity of the fiber, are characteristics that add to their value. In length they rank next to Sea Island, ranging from 1⅛ to 1⅝ inches.

Ginning is done on roller gins at ginneries in the interior towns. Baling and hydraulic pressing are done at the gin, but the bale is rebaled and compressed with steam at Alexandria.

The bulk of the Egyptian crop is exported. In 1933–1934, Great Britian received about 33 per cent, the United States about 6 per cent, countries in the continent of Europe about 48 per cent, and other countries smaller amounts.

Anglo-Egyptian Sudan.—This immense area in east-central Africa contains much fertile land and gives promise of becoming in time one of the great cotton-producing areas of the world. There is sufficient rainfall in the southern part of the territory for cotton growing, and much of the rest can be irrigated. The yield for 1934–1935 was 227,474 bales.

British West Africa.—Native cottons have been grown in West Africa for several centuries, and enough cloth is woven to supply most of the local needs and allow some for commerce. In 1902, the British Cotton Growing Association was organized to foster cotton growing in British territory. Experiments have since been made in Gambia, Sierra Leone, Gold Coast, and Nigeria. According to Todd,[4] results were favorable in most of the places tried, but for lack of funds the work was discontinued in all the districts except Gold Coast and Nigeria prior to 1907.

It is estimated by Todd[4] that Nigeria has an area of 25,000,000 acres suitable for cotton. The soil is fertile; the rainfall is sufficient, being about 50 inches annually; and the supply of labor is sufficient. The hot, dry winds, which blow at periods from the interior of the continent while the plants are fruiting, do considerable damage. Sea Island and Egyptian varieties have proved failures there, but American uplands do well. There are also native varieties that are about equal to the uplands in value. The production of Nigeria in 1934–1935 was 42,000 bales. The greatest hindrance to the spread of the cotton-growing industry in this region is lack of transportation facilities.

British Provinces in East and South Africa.—Great Britian has several provinces in East and South Africa from which a limited amount of cotton is being obtained and from which larger quantities are expected later. The most important of these is Uganda in the eastern part. This province is said by Todd[4] to have 25,000 square miles suited to cotton growing. In 1934–1935, it produced 205,742 bales. The cotton grown is descended from American long-staple varieties and has good staple. Means of transportation are poor, and consequently the cost of getting the cotton to market is considerable.

Other cotton-growing districts under English influence in East and South Africa are Nyasaland, which produced 8,100 bales in 1934–1935; Rhodesia, 1,616 bales; and South Africa, 2,652 bales. English companies have done much to encourage cotton growing in these districts, but the production has been small in amount. The natives do not want to work, and carrying the product to market is as great a task as growing it. The character of the staple produced was all that could be desired.

Colonies Belonging to Other European Nations.—France, Germany, Italy, Portugal, and Belgium all have endeavored to encourage cotton growing in their African colonies during recent years. They have distributed seed of improved varieties, offered good marketing facilities, and tried to teach the natives better methods of culture. While the possibilities of future development are great in most of these regions, much remains to be done before any great amount of cotton will be available to the world from them. The greatest hindrance to the industry is lack of means of transporting the products to market.

Brazil.—Cotton has been grown in Brazil since prehistoric times and is doubtless indigenous to the country. It was being grown there very generally when the land was discovered by Europeans in 1500. Probably much of the lint used at that time was collected by the natives from plants growing wild. Several writers within the period from 1500 to 1781 mention the production or use of cotton in various parts of Brazil. Claude d'Abberville, a Capuchin missionary in Maranhão from 1612 to 1614, quoted by Handy,[1] in speaking of the natives growing cotton says:

> They gather, clean, beat, and spin cotton with much dexterity, and with it make open hammocks resembling nets, and others as well woven and full of figures as if they were the work of better weavers; also aprons in which they carry their children about their necks.

Although grown rather generally, it appears that no cotton was exported prior to the middle of the seventeenth century, and not much was exported before 1781, the year when the first cotton was shipped to England. From that date to 1800 the main supply for England came from Brazil. Handy[1] estimates that the amount of cotton shipped from five of the principal ports in 1800 was 22,116,000 pounds. About this time the United States began to export a considerable quantity of cotton and became a competitor in supplying Europe. This competition checked the growth of the industry in Brazil.

Cotton is grown in two distinct districts in Brazil—the Northeastern States, which are situated in the extreme northeastern section of the continent, and the Southern States, which are in the southeastern part of the country. The larger states in the northeastern section are Pará, Maranhão, Ceará, and Bahia. The principal cotton-producing states in the southern area are Minas Geraes, São Paulo, and Paraná.

The area of the 11 Northeastern States of Brazil is estimated at 1,133,253 square miles, an area four times the size of Texas. According to Norris,[15] this area may be divided into three districts. The first is the narrow coastal plain extending the full length of the coast; the second, west of the first, includes the foothills and a low coastal mountain range; and the third district is a high arid and semiarid plateau lying west of the coastal mountains. Much of the vast region of the Northeastern States is forest, jungle, desert, and waste land, uninhabited and practically uninhabitable. As viewed by Norris,[15] only a small percentage of this land can be classed as potential cotton land. He says, "The task of clearing the land, developing the transportation facilities, preparing the soil, planting and cultivating a crop over this large area is far beyond the ability of the present generation."

The acreage of cotton in the Northeastern States was 1,406,320 in 1933–1934, and the production was 447,543 bales. The average annual yield from 1921-1922 to 1933-1934 was 178 pounds per acre. Primitive methods of cultivation have been followed almost exclusively down to recent years. The custom has been to clear the land, plant the cotton in hills here and there, and do all the cultivating with a heavy type of hoe.

Several different varieties are grown, some of them descended from indigenous species. Two or more species are represented. One is the tree cotton, known as "Maranhão," or "Moco," which is grown on the interior plateau. Another is a herbaceous short-stapled kind known as "Herbaceo," or "matta." This is very similar to American uplands. Some new American varieties have been imported in recent years. The tree cotton attains a height of 15 to 20 feet and yields well for a few years. It produces a longer staple than the herbaceous kinds, the length ranging from 3/4 to 1 1/4 inches, but the staple is not very uniform. Varieties are badly mixed. Grades are low on account of lack of care in picking and poor ginning. The tree cotton is less productive than the herbaceous varieties and is more expensive to pick.

While there are large areas in Northeastern Brazil in which cotton can be grown, there are several hindrances to the increase in production. In the plateau region, rainfall is scanty and somewhat uncertain; in the Amazon region, the rainfall is too heavy; insects and fungous diseases are especially troublesome on account of the perennial cottons; labor is scarce in many sections, and much of it is indolent; transportation facilities are poor, the railroads being few and short, and there is a scarcity of improved wagon roads.

The Southern States of Brazil form an important cotton-producing region—one that has increased its production greatly within the last few years. The area of the four principal states is about equal to that of Arkansas, Louisiana, Oklahoma, and Texas. The estimated cotton acreage for the year 1933–1934 was 1,113,200 acres, and the production was 500,400 bales. The average yield per acre for the 10-year period ending 1933–1934 was 182 pounds. The climate of the Southern States is favorable for cotton, the rainfall being 50 to 60 inches per year and well distributed. The land, according to Norris,[16] is mostly rolling and moderately fertile. The people are largely of

Portuguese descent with a considerable mixture of Indian and negro blood. There are also numerous Italian, German, Spanish, and Japanese immigrants. Much of the cotton is grown by the Japanese. The labor supply is not adequate for increased production.

American upland varieties are grown in Southern Brazil, the best strains being selections from Texas varieties. The quality of the staple is not equal to that of American varieties. The fiber lacks uniformity, and the grade is often low because of poor ginning. Most of the gins lack cleaning equipment, and many are old and in poor repair. Plant diseases and insect pests are prevalent, the worst insect being the pink bollworm.

Southern Brazil is better supplied with railroads than other parts of the country; consequently, transportation facilities are much better. However, there are still but few roads in the sparsely populated areas.

The cotton textile industry of Brazil has increased rapidly during the last 10 years. Garside[14] gives the number of mill spindles in 1934 as 2,702,000. The mills now produce more than 90 per cent of the cotton goods used in the country.

The U. S. Department of Agriculture[17] gives the total cotton production of Brazil for the year 1933–1934 as 1,011,000 bales and the exports as 235,000 bales.

Peru.—Evidence points to the conclusion that cotton is indigenous to Peru. There are historical records that show its use there since about A.D. 1500. Pizarro, the Spaniard who invaded Peru in 1522, found the natives clad in cotton garments. Mummy wrappings and fabrics found in tombs indicate the use of cotton many years previous to that date. The Spanish invaders were looking for gold and consequently paid but little attention to the cotton industry discovered, doing nothing to foster it. Statistics on the production of cotton in the country for the next 300 years are not available, but doubtless enough cotton was produced to supply local needs, and, during the latter part of the period, some for export. In 1862, according to Handy,[1] 341,143 pounds was exported to Liverpool; in 1865, the exports had increased to 4,145,260 pounds; in 1892, to 6,000,000 pounds. For the year 1921, the total production, according to the U. S. Bureau of Commerce,[5] was 157,000 bales. The increase in production since 1892 was phenomenal.

The cotton-growing lands of Peru are in numerous small valleys that run from the Andes Mountains toward the ocean. These valleys have a rich alluvial soil. The rainfall is so scanty that irrigation is necessary, but that is fairly easily accomplished, for there is a river flowing through each valley, fed by melting snows from the mountains.

Transportation is a relatively simple matter, since the cotton-growing areas are not far from the ocean, and some of the valleys are supplied with railways. The cotton lands are in plantations varying from 500 to 5,000 acres and are owned by Peruvians. Most of the labor is done by Indians, who are considered satisfactory laborers and are paid a fair wage, about 65 cents a day. Chinese and Japanese laborers have been tried but were not found to be so desirable as the natives.

The acreage for 1934–1935, as given by the U. S. Department of Agriculture,[17] was 367,000 acres. The yield that year was 341,970 bales. This is nearly a bale to the acre and is a good yield. On some of the better plantations an average yield of 500 pounds of lint per acre is reported.

The larger part of the ginning is done by so-called "upcountry" gins, to which the seed cotton is brought in from the farms, usually on the backs of mules. Some large plantations have their own gins. Certain cotton-importing firms have local representatives, who buy cotton from the gins and plantations. The gin bales are not compressed before shipment. About 90 per cent of the Peruvian production is exported. The prospects for Peru as a cotton-producing country are bright. Although the area suited to cotton growing is somewhat limited, and all of it requires irrigation, a very much greater amount may be grown than is produced at present. Todd[4] estimates that the possibilities are as great as in Egypt, which produces annually over a million bales.

Other South American Countries.—In addition to the countries mentioned, other South American countries produce limited amounts of cotton. Several of these are apparently capable of yielding a much larger quantity than is now produced. Of these countries Argentina is the leader. In the northern part of Argentina there are extensive areas of fertile land, where the temperature and the rainfall are well suited to cotton. The chief limiting factors are shortage of labor and the production of other crops that are very profitable. The estimated area in cotton in 1933–1934 was 482,000 acres. Primitive cultural methods are in use, but splendid yields have been reported in some instances, some being as great as 500 pounds of lint cotton per acre. Paraguay, Uruguay, and Chile are also producing a small amount of cotton, the total yield not being more than a few thousand bales. Colombia, Venezuela, and Guiana also produce cotton in small amounts, but most of it is used by local mills and in the homes. Guiana formerly exported considerable cotton. Todd[4] reports 16,000 bales in 1827.

Mexico.—It was estimated by Ruiz (quoted by Handy[1]) that the annual production of cotton in Mexico at the beginning of the sixteenth century, or prior to the invasion by the Spaniards led by Cortez, was 116,000,000 pounds. Under Spanish rule, the production declined but has continued in a limited way in all parts of the country from that day until the present. About 1860, there was some increase in interest due to the stimulus of high prices brought on by the Civil War in the United States. Production continued to increase somewhat during the next three decades. In 1892, Ruiz estimated that the annual production was 50,000 bales. The U. S. Department of Agriculture[17] gives the production for Mexico for 1935–1936 as 234,569 bales.

The area of Mexico is somewhat greater than that of the Cotton Belt of the United States, and much of the land is suited to growing cotton, provided there is sufficient moisture. There is plenty of fertile land, but much of it has to be irrigated. The natives are largely Indians and mixed breeds. There is an abundant supply of labor.

Although cotton is grown on a small scale in widely scattered regions in Mexico, the Laguna district is by far the most important area, producing on the average about 60 per cent of the Mexican total. The Mexicali and Matamoras districts rank next in importance. With the exception of some long-staple cotton in the Mexicali district, most of the cotton grown in Mexico is of the American upland type. Two of our present American varieties, Acala and Durango, were imported from Mexico. Some of the Mexican varieties have doubtless arisen from indigenous wild species. According to reports made by the Mexican Ministry of Agriculture, much of the cotton produced has a staple length of an inch or more, and none is shorter than $\frac{15}{16}$ inch.

Railway transportation facilities in the main Mexican cotton-producing districts are good. The annual consumption in the Mexican textile mills for the 6-year period ending April, 1931, averaged 184,000 bales of 478 pounds net. This was 75 to 80 per cent of the production. The remainder was exported.

Central America.—Small amounts of cotton have been grown in Nicaragua, in Honduras, and in other Central American countries. It is thought that the lands of these states are adapted to growing cotton and may, in course of time, be of some importance in the world's production.

West Indies.—Cotton is indigenous to some of the West India islands and is now and has been for many years cultivated to a limited extent on most of them. In 1801, the West Indies produced 25,000 bales. Up to 1820, the production exceeded that of the United States. Since then it has decreased considerably. The principal variety grown is Sea Island. Yields are said to range from 100 to 300 pounds of lint per acre. On many of the islands the length of staple and the quality of fiber are considerably below that of American Sea Island. St. Vincent produces the best quality of staple. Accurate statistics on the production of many of the islands are lacking.

The West India Islands are capable of producing much more cotton than they do. Greater profits from other crops which require less labor militate against cotton production.

South Sea Islands.—Many of the islands of the Pacific Ocean produce cotton on a small scale or can grow cotton, but the amount actually made is too small to be of much consequence.

Australia can also grow cotton, but up to the present the industry has not flourished in that country, partly because of lack of labor.

Cotton in the United States.—American cottons are more or less of hybrid origin; like the American people, they are a product of the "melting pot." For about 300 years different varieties and some different species have been brought to the Cotton Belt from all parts of the cotton-growing world. The various cottons have been allowed to mix and cross. From the resulting complex of hybrids come the forms we have today. Although it is possible to trace to some extent the history of a few varieties, such as Sea Island, Pima and Yuma (the American-Egyptian varieties), and one or two recent importations from Mexico (Acala and Durango), the botanical history of the great mass of varieties is a puzzle that will probably never be solved.

Indigenous Cottons.—De Vica, quoted by Hammond,[11] reported in 1536 that he found wild cotton plants growing in territory now within the states of Louisiana and Texas. Similar reports have been made by others, but none of the reports has been authenticated by qualified persons; consequently, it is very doubtful if any species of cotton is really indigenous to any part of the American Cotton Belt, the land that produces the major portion of the world's crop today.

Introduction in the American Colonies.—Cotton was introduced into the English colonies very soon after the settlements were first made. Indeed, there is some evidence that cotton was grown in the Virginia colony in 1607, the year the colony was established. In 1621, cotton wool is listed as being worth 8 pence a pound. Cotton production was held in check in this colony by the extreme profitableness of tobacco culture. Later in the century, considerable cotton was grown and made into cloth for home use by the planters.

Handy[1] mentions cotton being grown by a colony of English at Cape Fear, in the Province of Carolina, in 1664, quoting:

> But they have brought with them most sorts of seeds and roots of the Barbados, which thrive in this most temperate clime. . . . They have indigo, very good tobacco, and cotton wool.

Cotton is referred to as one of the products of South Carolina in 1666 and of Louisiana, which then included Alabama, Mississippi, and Louisiana, in 1728.

> The colony of Louisiana . . . was in a flourishing condition, its fields being cultivated by more than 2,000 slaves in cotton, indigo, tobacco, and grain.

Handy[1] further says that cotton seed were planted in Georgia in 1734, being sent there by Philip Nutter, of Chelsea, England. Trench Coxe, of Philadelphia, mentions cotton being grown as a garden crop on the eastern shore of the Chesapeake Bay in Maryland in 1736.

Many other similar references to the growing of cotton in the colonies may be found in literature. It will suffice to say that the cotton-growing industry in this country advanced slowly. Seed was brought in at divers times and from various regions, among which may be mentioned the Barbados and other islands of the West Indies, Siam, Mexico, and the Levant. Most of the cotton produced was used in making homespun clothes. Thomas Jefferson, writing in 1786, said:

> The four southernmost states make a great deal of cotton. Their poor are almost entirely clothed with it in winter and summer. In winter they wear shirts of it and outer clothing of cotton and wool mixed. In summer their shirts are linen, but the outer clothing cotton. The dress of the women is almost entirely of cotton, manufactured by themselves, except the richer class, and even many of these wear a great deal of homespun cotton. It is as well manufactured as the calicoes of Europe.

Most of the early experiments with cotton culture were made with short-staple cotton. Mrs. Kinsey Burden's attempt in 1788 to grow Sea Island cotton on Burden's Island is considered by Handy[1] as being the first trial of Sea Island in South Carolina. Mrs. Burden's cotton failed to mature, a failure that was probably due to the fact that the variety (Bourbon) was not suited to the climate. A successful attempt at growing Sea Island was made near Beaufort in 1790 by William Elliott.

Exports.—Prior to 1793, there were no satisfactory cotton gins in use in the colonies. Most of the lint was picked from the seed by hand. This was a very slow, laborious process. A man could pick off only 1 or 2 pounds a day. On account of this extra labor, there was not much profit in growing cotton for sale or export. The labor necessary for producing the small amount needed for home weaving was not so noticeable. The few gins in use were very similar to the "churka" wooden roller gins used in India centuries before. Under such conditions, it is easy to understand why the cotton exports were limited. Dana, in "Cotton from Seed to Loom," mentions Samuel Auspourguer, a Swiss living in Georgia, as carrying a sample of cotton to London in 1739. This is taken as the first export. Various other shipments, made at irregular periods, are mentioned, but none of them made prior to 1787 contained more than 17 bales. In 1788, 58,500 pounds were exported to Liverpool; in 1789, 127,500 pounds; in 1791, 189,500 pounds; and in 1792, 138,328 pounds.

With the invention of Whitney's saw gin in 1793, a new era started in the cotton industry in America. The production, exports, and domestic consumption all increased with rapid strides following the introduction of this new labor-saving machine. In 1790, 889 bales were exported; in 1800, 91,716; in 1820, 449,257; in 1840, 1,313,500; in 1860, 3,127,568; in 1880, 4,589,346; in 1900, 6,538,000; in 1920, 6,362,000; in 1930, 8,579,000.

Cotton Manufactures.—The English government discouraged weaving and the making of cloth in the colonies on the theory that a development of that industry would work to the detriment of the spinners in the mother country. Spinning and weaving in the homes were not prohibited, but there were no factories during colonial days. The first cotton mill was built in 1787, at Beverly, Mass. Other mills were erected soon after in the New England and middle Atlantic states, and one at Statesburg in South Carolina. Until 1815, only carding and spinning were done with power machines, the weaving being done on hand looms. Spinning and weaving in the home continued in the South for many years after factories were built.

The number of mills increased rapidly. In 1831, there were 801 cotton mills in the country, in 1860, 1,091 mills with 5,235,727 spindles; in 1898, 17,450,000 spindles; in 1910, 28,500,000 spindles; in 1920, 34,500,000. In 1925, the peak number of 37,929,000 spindles was reached. Since that date there has been a decrease. In 1934, the number was 30,938,000. In the last decade there has been a small increase in the southern states, but a heavy decrease in the northern. In 1934, the southern states had 19,327,000 spindles, and the northern had 11,611,000.

For the past several decades the percentage of increase in spindleage in Great Britian, the greatest spinning country in the world, has been considerably less than that of the United States. The United States is now consuming 50 per cent of its production In 1900, it did not consume more than 33 per cent, and before that the percentage was still lower.

Area and Divisions of the Cotton Belt.—The American Cotton Belt contains about 700,000 square miles of land, but less than 10 per cent of this area is actually being cultivated in cotton.

The Cotton Belt contains five more or less distinct districts or divisions. The old Sea Island growing region may be designated as the first district. This covers parts of South Carolina, Georgia, Florida, and islands along the coast. This district is in a transition stage at present. Most of the Sea Island cotton has been discarded on account of heavy boll-weevil depredations, and uplands have been substituted. This region is thought to be especially well adapted to Sea Island on account of the very humid atmosphere.

A second area, which may be called the "short-staple district," includes North Carolina, western South Carolina, northern and western Georgia, Alabama, and the hill-land portions of Mississippi, Louisiana, and eastern Arkansas. This section is somewhat broken and subject to erosion. Much of the soil is red clay and rather unfertile. The big-boll, short-staple varieties seem to be best adapted to this region.

A third area is the Delta district. This includes the belt of alluvial land along the Mississippi River and other rivers emptying into it. The major portion of the district consists of a broad belt of alluvial soil between the Yazoo and Mississippi rivers. This land lies in northwest Mississippi and is usually designated as "The Delta." Parts of Arkansas, Louisiana, and Missouri also belong to the Delta section. The rainfall here averages about 50 inches a year, and the soils are deep, fertile, and level. This region seems to be especially adapted to long-staple upland cotton.

Texas, Oklahoma, southeastern New Mexico, and western Arkansas form a fourth district. This area has rather light rainfall and much hot, dry weather. Only hardy, vigorous varieties do well here. The principal cottons grown are the big-boll, medium-length staple varieties, such as Mebane Triumph, Acala, Rowden, and Lone Star.

The Imperial Valley in California and Salt River Valley in Arizona may be said to form a fifth division of the Cotton Belt. The cotton here can be grown only under irrigation. Pima and Yuma strains of Egyptian cotton and certain upland varieties are grown.

Yields.—The yields in the United States vary considerably in different years and in different sections. They are naturally low in places where boll-weevil infestation is heavy. In 1934, the average for Missouri, a weevil-free state with small acreage, was 366 pounds of lint per acre. The yield of North Carolina, a state typical of the short-staple area with slight weevil damage, was 316 pounds of lint. The yield for Georgia, in 1921, was 90 pounds. This represents a short-staple state with heavy weevil infestation. Arizona and California, in 1934, produced 410 and 556 pounds per acre, respectively. This cotton came from weevil-free irrigated districts. Texas and Oklahoma the same year produced 112 and 56 pounds. These states represent droughty areas. The average production of Sea Island in the Sea Island district in pre-boll-weevil days was 125 pounds of lint per acre.

The average yield for the whole Cotton Belt in 1920 was 170.9 pounds per acre; for the years 1923–1932, 169.9 pounds; for 1934, 170.9 pounds; for 1935, 188 pounds.

The total production in 1921 was 7,953,641 bales; in 1926, 17,978,000.

The 5-year average for the period 1928–1932 was 14,666,000 bales. After 1933, production was curtailed to reduce the cotton surplus.

Length and Character of Staple.—In 1934, 0.2 per cent of the cotton crop had a staple length 1¼ inches or above; 8.3 per cent, 1⅛ to 1⁷⁄₃₂; and 91.5 per cent, under 1⅛ inches, the last being considered as short cotton. The figures for other years are very similar. Arizona leads the states in the production of staple cotton 1¼ inches and over in length on account of its American-Egyptian varieties. Mississippi produces more cotton 1⅛ to 1¼ inches in length than any other state. Much of the cotton grown in the Delta district of Mississippi is 1⅛ and 1³⁄₁₆ inches. Texas leads in the production of short cotton by a wide margin, its production usually being more than 4,000,000 bales. Since 1928, the production of cotton shorter than ⅞ inch staple has decreased, both in number of bales and in relative proportion of the annual crop.

In the Cotton Belt, weather during the harvesting season is usually dry and favorable for picking, and as some care is taken by the pickers to avoid getting trash and leaf in the seed cotton, the grade of American cottons compares favorably with that of most of the cotton produced elsewhere. The staple in uniformity and character is also better than that found in several other countries, such as India, China, Brazil, or the Levant; yet there is still much to be desired. At present much interest is being taken in new and improved varieties, and numerous seed-breeding farms have been established. With the use of more good seed, advancements are sure to come.

References

1. Handy, R. B.: "History and General Statistics of Cotton," in "The Cotton Plant," *U. S. Dept. Agr. Bull.* 33, 1896.
2. Watt, Sir George: "The Wild and Cultivated Cotton Plants of the World," Longmans, Green & Co., London, 1907.
3. Scherer, James, A.B.: "Cotton as a World Power," Frederick A. Stokes Company, New York, 1916.
4. Todd, John A.: "The World's Cotton Crops," A. and C. Black, Ltd., London, 1923.
5. Anonymous: "Cotton Production and Distribution," *U. S. Dept. Commerce, Bur. Census Bull.* 150, 1921–1922.
6. Shepperson, A. B.: "Cotton Facts," Shepperson Publishing Company, New York, 1913.
7. Shepperson, A. B.: "Cotton Facts," Shepperson Publishing Company, New York, 1920.
8. Balls, W. L.: "The Cotton Plant in Egypt," The Macmillan Company, New York, 1919.
9. Balls, W. L.: "The Cotton Plant of Egypt," *Year Book*, Khedivial Agricultural Society, 1909.
10. Pearse, Arno S.: "Brazilian Cotton," *Rept. International Cotton Committee, Manchester*, 1921.
11. Hammond, M. R.: "The Cotton Industry," The Macmillan Company, New York, 1897.
12. Gulati, A. N. and A. J. Turner: "A Note on the Early History of Cotton," *Jour. Textile Inst.*, vol. 20, T1-T9. 1929.
13. McGregor, J. C.: "Prehistoric Cotton Fabrics of Arizona," *Museum Notes, Museum Northern Arizona*, vol. 4, No. 2, 1931.
14. Garside, Alston H.: "Cotton Year Book," New York Cotton Exchange, 1934.
15. Norris, P. K.: "Cotton Production in Northeast Brazil," *U. S. Dept. Agr. Bur. Agr. Econ.*, F. S. 64, 1935.
16. Norris, P. K.: "Cotton Production in Southern Brazil," *U. S. Dept. Agr. Bur. Agr. Econ.*, F. S. 63, 1935.
17. U. S. Department of Agriculture, Agricultural Statistics, 1936.

CHAPTER II

TAXONOMY OF THE COTTON PLANT

Cotton belongs to the mallow family, or Malvaceae, which has representatives in nearly all parts of the world, but its species are most abundant in the tropics and in warm regions. Several plants of economic importance are included in the family. In addition to cotton, or *Gossypium*, which is the most important, may be mentioned the Deccan hemp (*Hibiscus cannabinus*), okra (*H. esculentus*), mallow (*Malva sylvestris*), and marsh mallow (*Althaea officinalis*).

The genus *Gossypium*, to which cotton belongs, contains a number of species, but it is difficult to say just how many. Some authorities include many different varieties and forms under a few species, though others draw the species lines closer. Linnaeus is credited with 5 or 6 species; De Candolle recognized 13; Parlatore, an Italian who studied the genus about 1866, listed 7 ("Le specie dei cotoni"). Todaro, another Italian who monographed the genus, recognized 54 species ("Relazione sulla culta dei cotoni," 1877–1878). Watt,[1] who has made an extended study of the cottons grown in India and other countries and has studies herbarium specimens from all parts of the world, classifies all the well-known forms, both cultivated and wild, into 29 species and 16 botanical varieties.

Species of Gossypium Difficult to Classify.—The genus *Gossypium* is generally considered a difficult one taxonomically. This is due not to the lack of distinguishing marks on the plants but to several other causes: (1) The foundation laid by Linnaeus was faulty in that he based descriptions of some of the most important species on cultivated plants. (2) Linnaeus' descriptions were not full enough to define species clearly. (3) A large part of the tropics, where wild cotton plants are found most abundantly, has not been explored thoroughly by botanists. (4) For centuries cotton plants have been transferred from country to country for commercial planting. Some of these have escaped from cultivation, and it is now difficult to determine the original home of certain species or to get any insight into botanical relationship through geographical location. (5) There has been crossing and hybridizing due to the planting of different species and varieties in proximity.

Many species of *Gossypium* may be crossed by insects carrying pollen from one plant to another. It is doubtful if there are many of the American upland varieties that are not hybrids or blends. As evidence of hybridization may be noted the lack of stability in nearly all varieties and frequent reversions to unlike, perhaps ancestral, types. In almost every variety specimens with brown lint or floss occasionally appear. Brown is the color of the lint of many wild species and of some cultivated cottons. Okra-leaved cottons, with leaves resembling those of some of the Asiatic cotton varieties, are of common occurrence. Plants with seeds lacking fuzz, as are found in cottons of the Sea Island type, often appear among plants of fuzzy-seeded varieties. In nearly every variety some plants are found that are off type in length of lint, or lint percentage, shape of boll, size of boll, color of fuzz on seeds, color of stamens, color of petals, habit of branching, or size of plant or in respect to some particular characters.

Some of the aforementioned off-type forms may be due to mutations, but it is probable that most of them are hybrids reverting to ancestral forms, since it is known that several species hybridize readily and that there has been abundant opportunity for them to do so.

It is therefore impossible to trace definitely the origin of many of the cultivated species or, in some cases, even to give with a certainty the land of their origin.

The Malvaceae.—Small[2] characterizes the Malvaceae, or mallow family, as follows:

Herbs, shrubs, or rarely trees, of great economic importance, the vegetative parts destitute of any unwholesome substance. Foliage often pubescent with simple or branching hairs. Leaves alternate, stipulate, blades palmately nerved and often lobed, usually crenate or toothed. Inflorescence axillary, racemose, paniculate or corymbose, sometimes congested. Flowers regular, perfect, often subtended by an involucre resembling a second calyx. Calyx of five more or less united sepals; these valvate, mostly persistent. Corolla of five hypogenous convolute petals, alternate with the sepals, united; blades often inequilateral. Androecium of numerous stamens. Filaments monadelphous, united with the claws of the petals. Anthers reniform, one-celled. Pollen grains hispid. Gynoecium of several carpels. Ovary several-celled, or the several pistils distinct. Styles terminal. Stigmas capitate. Ovules amphitropous or semianatropous, pendulous or ascending,

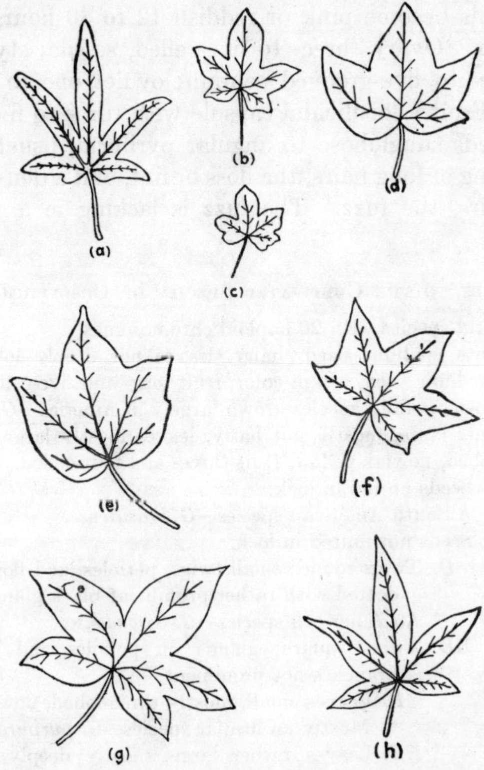

FIG. 1.—Typical leaves of the most important species of Gossypium: (*a*) *G. arboreum;* (*b*) *G. nanking;* (*c*) *G. herbaceum;* (*d*) *G. hirsutum;* (*e*) *G. purpurascens;* (*f*) *G. braziliense;* (*g*) *G. peruvianum;* (*h*) *G. barbadense.*

inserted on the inner angles of the cavities. Fruit capsular, or the carpels separate or separable. Seeds solitary or numerous, with a crustaceous, smooth or rough, sometimes pubescent testa. Endosperm scant, fleshy, mucilaginous. Embryo large, curved, or folded, with foliaceous cotyledons.

The Malvaceae contain 39 genera and some 300 species.

Gossypium, of Linnaeus.—The genus *Gossypium* may be described as follows: Annual or perennial herbs, shrubs or trees with erect branching stems. Leaves alternate; blades usually palmately lobed, three to nine lobes, rarely entire; petioled. Involucel of three large, cordate-toothed or incised bracteoles. Flowers rather large and showy, borne single. Calyx truncate or shortly five-toothed. Petals five, white, yellow, deep red, or purplish; some with purple spot on claw; the white or cream-colored petals become pink or reddish 12 to 30 hours after the flowers open. Ovary three- to five-celled, sessile; styles united into a three- to five-grooved column; ovules one to several in each cell. Fruit a loculicidal capsule with three to five leathery valves. Seeds subglobose to angular pyriform, usually clothed with a coating of long hairs, the floss or lint, and a dense covering of short hairs, the fuzz. The fuzz is lacking in a number of species.

Key to the Cultivated Species of Gossypium

A. New World species with 26 haploid chromosomes.
- *B*. Plants medium sized, hairy, leaves not deeply lobed, flowers medium sized, cream color, fruit four- and five-celled.
  A cosmopolitan species grown largely in America—*G. hirsutum*.
- *BB*. Plants large, mostly not hairy, leaves mostly large and deeply lobed, flowers yellow, fruit three- and four-celled.
  - *C*. Seeds united in lock.
    A South American species—*G. braziliense*.
  - *CC*. Seeds not united in lock.
    - *D*. Twigs round; small twigs, petioles, and flower pedicels dotted with rather prominent black glands.
      An American species—*G. barbadense*.
    - *DD*. Twigs square; glands on petioles and small stem branches not prominent.
      - *E*. Leaves small, mostly three-lobed, flowers small.
        Mostly an insular species—*G. purpurascens*.
      - *EE*. Leaves rather large, mostly deeply lobed, and densely tomentose beneath; flowers large, sulphur yellow.
        A South American species—*G. peruvianum*.

AA. Old World species with 13 haploid chromosomes.
- *B*. Leaves cleft deeply, two-thirds or more, lobes nearly linear; flowers mostly purple.
  An Asiatic species—*G. arboreum*.
- *BB*. Leaves not cleft two-thirds, lobes not linear, flowers not purple.
  - *C*. Fuzz rufous, fruit three- to four-celled, flowers rather large.
    An Asiatic species—*G. nanking*.
  - *CC*. Fuzz grayish, fruit four- to five-celled, flowers rather small.
    An Asiatic species—*G. herbaceum*.

Species Descriptions.*—The eight species described have the most important cultivated varieties. Purely wild species are not described. Sir George Watt, of England, has made an extended study of the *Gossypia* of the world and collected a large amount of data concerning their species; in his monograph[1] may be found descriptions of both wild and cultivated species.

Gossypium barbadense Linnaeus.—(*G. vitifolium* Lamarck and *G. maritimum* Todaro are here considered as belonging to

FIG. 2.—Young plants of *G. barbadense* grown on James Island, S. C. (*Courtesy C. B. Doyle, U. S. Department Agriculture.*)

the species *G. barbadense* L. Several other forms have been referred to this species by various authors.) Sea Island, Egyptian cotton.

Shrubby perennial, an annual in cultivation, 4 to 8 feet high, glabrous, dotted with rather prominent black glands. Stem erect, round, and smooth; branches wide-spreading (Fig. 2). Leaves large, deeply segmented into three to five oblong, acuminate, spreading lobes, the central one not much larger than the larger laterals; stipules linear, ovate, acute, oblique; petiole nearly as long as the leaf (Fig. 1*h*). Inflorescence solitary on an

* There is much confusion concerning the species of the genus *Gossypium*. Although considerable study has been devoted to them, the information is still unsatisfactory. Authorities do not agree. In order to get the matter cleared up, it will be necessary to make a world-wide study of types now growing, study herbarium specimens, grow various types in culture, try hybridizing different species, and study their chromosomes.

axillary leafy shoot; bracteoles rather large, erect, segments spreading at the top, not deeply cut; bracteoles attached to calyx tube, slightly united or entirely free; corolla of medium size, usually less than twice the length of the bracteoles but sometimes larger, petals pale yellow with purple claw. Calyx wide companulate with large rounded teeth. Fruit an ovate-acuminate capsule, usually three-valved, valves not reflexed; seeds, six to nine in each cell, ovate, beaked, nearly naked; lint fine creamy white, long and silky. Fibers sometimes attain a length of $2\frac{1}{2}$ inches but are commonly not longer than $1\frac{1}{2}$ inches.

Gossypium barbadense is a species that has given students of *Gossypium* considerable trouble. It has been a difficult matter to delimit its forms and to trace out their origin and distribution. Watt[1] considers that *G. barbadense* Hook. and Arn. and Parl. (in part) is equivalent to *G. vitifolium* Lamk. and that *G. barbadense* Hook. and Benth. is equivalent to *G. punctatum* Sch. and Thon. *G. barbadense* Parl. (in part) is described by Watt as *G. barbadense* L. var. *maritima* Watt.

Origin and Distribution.—Gossypium barbadense has by many been regarded as a native of the West Indies, especially Barbados, but Watt[1] in recent studies found but little evidence to substantiate this belief. He thinks that this species may have been cultivated in that region some centuries ago but that the present Sea Island stock is an improved race, probably developed from a hybrid of *G. barbadense* and *G. braziliense*. This he designates as *G. barbadense* L. var. *maritima* Watt.

Sea Island cotton is grown in the West Indies and was cultivated rather extensively in eastern South Carolina, Georgia, and northeastern Florida and on islands adjacent to those states until about 1920, when the boll weevil made its culture unprofitable. It has been tried in India, in Egypt, and in other parts of Africa and Australia and on various islands in the Pacific Ocean but found to be unsuited to all these regions except the islands. The species is not known in a wild state.

Egyptian cotton is grown in Egypt, in Central Asia, in southwestern United States, and in a limited acreage in a few other countries.

Important Cottons Derived from G. barbadense.—In addition to the splendid strains of Sea Island cotton that have been mentioned, several other valuable cottons have been developed from or probably inherit characteristics from *G. barbadense.* The American long-staple upland cottons appear to have some Sea Island traits.

Gossypium purpurascens Poiret. (Synonym, *G. racemosum* Todaro).—Bourbon, Porto Rico, or Siam cotton.

Perennial shrub or small tree; twigs angled and glabrous or nearly so. Leaves small, three-lobed, the laterals pointing outward and upward; petiole usually longer than the blade; stipules prominent, ovate oblique linear (Fig. 1*e*). Inflorescence consists of leafy lateral shoots bearing one or more flowers; bracteoles free or nearly so, ovate, acute, deeply auricled, with seven to nine long linear teeth; flowers small, yellow, often purple spot on claw of petals; calyx wide, campanulate, crenately or dentately toothed. Fruit oblong acuminate, three- to four-celled; seeds ovate, acute, not angled, with rusty fuzz around beak, otherwise smooth and naked; lint white, soft, silky, and easily removed.

Distribution.—*Gossypium purpurascens* is possibly only a cultivated form of *G. taitense* Parl. which it resembles rather closely. It is known only in cultivation and is found widely distributed on islands in the Atlantic, Indian, and Pacific oceans. It has been grown in Upper Egypt, East Africa, Madagascar, Siam, southern India, South America, and other regions.

Gossypium braziliense Macfayden. (Synonym, *G. conglomeratum* Wiesner).—Kidney, Brazilian, or Pernambuco cotton.

Perennial shrub or small tree. Stem and twigs angular above, rounded below. Leaves large, three to five palmately lobed nearly glabrous (Fig. 1*f*). Inflorescence, on axillary shoots one- to five-flowered; bracteoles large, cordate, auricled slightly, united below. Glands on upper part of pedicel and inside bracteoles; flowers large, petals pale yellow with orange or red spot on claw. Calyx large, cup-shaped, truncate, or irregularly four- to five-toothed. Fruit oblong, acuminate, beaked, when mature embraced by a large accrescent calyx and bracteoles, three-celled, dehiscing a little more than half its length, valves erect, the margins and beak rigidly reflexed. Seeds of each lock united in a fairly compact mass, medium large, many more or less flattened or angled on account of compression in the lock; naked except a little fuzz at apex; lint abundant, white, fine and silky, but fibers largely lack the twist found in most other cottons; length of staple 1 to 1½ inches. Yield of lint per acre, poor.

Distribution.—*Gossypium braziliense* is generally regarded as being indigenous to Brazil and other parts of South America. It has been found growing wild in West Africa and either wild or under cultivation in China, Japan, India, Central America, and the West Indies.

Important Cottons Derived from G. braziliense.—Watt[1] considers that Jumel's cotton, introduced in Egypt in 1820, was probably *G. braziliense* and consequently that the present-day Egyptian cotton has this species as one of its ancestors.

Fig. 3.—*Gossypium hirsutum.*

Gossypium hirsutum Linnaeus. (Synonyms, *G. latifolium* Murray; *G. barbadense*, Wight; *G. herbaceum*, Small; and others).—Upland, Saw-ginned Dharwar, New Orleans, and Georgia cotton.

Annual plants 2 to 4 feet high; lower branches ascending, upper spreading; older parts of stem and branches smooth, younger covered with moderately stiff, spreading, whitish hairs (Fig. 3). Leaves subcoriaceous, pale to dark green, varying much in shape, ovate or ovate-cordate, entire, or more commonly three to five shallow, triangular, acute lobes (Fig. 1*d*); stipules lanceolate, falcate, about ½ inch long; bracteoles rounded, upper half of the margins with about 10 falcately lanceolate acuminate teeth. Flowers varied in size; calyx campanulate, accrescent, with five triangular lobes; petals pale to deep yellow, some forms with purple spot on claws; filaments rather long, allowing the anthers to droop; stigmas long, mostly consolidated and twisted but occasionally branched at the apex. Fruit is usually four- to five-celled, spherical, ovate, short- to long-pointed; valves usually wide-spreading or reflexed; seeds mostly large, ovate, pointed at attachment end, and covered with a dense coating of grayish, rusty, or green fuzz; lint abundant and medium-fine to medium-coarse, firmly attached to seeds.

Distribution.—*Gossypium hirsutum* is not known wild, but it is widely distributed in cultivation. It is to be found all over the American Cotton Belt and has been collected in Africa, India, Java, China, Persia, and Europe.

Origin.—The origin and ancestry of *G. hirsutum* are uncertain. Watt[1] considers the species a cultivated form of *G. punctatum* Sch. and Thon. That species closely resembles *G. hirsutum*, is well known in a wild state, and has been identified in many regions in which the latter species grows.

Important Cultivated Cottons Belonging to G. hirsutum.—The number of cultivated cottons considered as arising from *G. hirsutum* depends, of course, on the conception of species. In the foregoing description the forms belonging to *G. mexicanum* Todaro, as viewed by Watt, are included. Viewed in this light, very nearly all of the American upland varieties, certain Mexican varieties, Dharwar cotton in India, Hindi cotton in Egypt, and American varieties in China and in different parts of Africa belong to the species.

Gossypium hirsutum Linnaeus, var. **religiosa,** Watt. (Synonym, *G. religiosum* of Linnaeus and others).—Nanking. Khaki, or sacred cotton. The true sacred cotton of early writers is probably *G. arboreum* or some one of its varieties grown in southern Asia. Linnaeus, however, named as *G. religiosum* a type that is very similar to *G. hirsutum.* This is made a variety of *G. hirsutum* by Watt.[1] It differs from typical *G. hirsutum* in having slender trailing branches, rust-colored fuzz, and usually rust-colored lint.

Gossypium peruvianum Cavanilles. (Synonym, *G. vitifolium* Roxburgh).—South American, Peruvian, or Andes cotton.

Perennial, bushy, twigs long, flexuous, and strongly angled. Leaves large, thick, usually densely tomentose below, especially when young, ovate-cordate; leaves entire on lower part of the bush, but three- to five-lobed on the rest, lobes broad oblong, acuminate pointed; stipules very large, broad oblong (Fig. 1*g*). Inflorescence on elongated leafy shoots with generally solitary extra axillary flowers; bracteoles with six distinct glands, cordate

FIG. 4.—*Gossypium barbadense.* Pima American-Egyptian cotton. Considered as belonging to the species *G. peruvianum* by some authorities. (*After Kearney.*)

auriculate, free from each other; nerves of bracteoles numerous, raised, and parallel; flowers sulphur yellow, petals with purple claws; calyx loose, obscurely toothed. Fruit ovate-oblong, suddenly acuminate, three-valved, scarcely exceeding the bracteoles; seeds large, free from each other, with a gray, rufous, or green fuzz; lint copious but wiry and harsh.

Distribution.—Watt[1] considers *G. peruvianum* as probably indigenous to the Andes in equatorial regions. It has been found growing in Peru and in other parts of South America, in Central America, in Africa, and in other cotton-growing countries. It is thought by some authorities to be one of the ancestors of the modern Egyptian cottons.

Gossypium herbaceum Linnaeus.—Levant, or Arabian, cotton. Annual, herbaceous, 2 to 5 feet high; stem and branches rounded, slightly zigzag owing to bending at joints, somewhat hairy when young but nearly glabrous when mature (Fig. 5). Leaves leathery, glabrous when mature, or nearly so, cordate to subreniform in outline, five to seven lobes, the three middle ones obovate and short-pointed; petiole very long, stipules awl-

Fig. 5.—*Gossypium herbaceum.* Grown at Agricultural School, Wuhn, China, 1923. (*Courtesy R. R. Childs, University of Georgia.*)

shaped (Fig. 1*c*). Pedicels usually shorter than bracteoles; bracteoles large, green, slightly united at the base; flowers hardly twice the length of the bracteoles, rather small, yellow, with purple claws; calyx large, loose about the ovary, with short, rounded teeth. Fruit ovoid, beaked, four- to five-celled; seeds large, angled, with gray fuzz and coarse grayish-white lint.

Distribution.—*Gossypium herbaceum* is widely grown in Asia and was formerly grown in Southern Europe. It is probably a native of Arabia and Asia Minor. Some authorities consider this the first species of herbaceous cotton grown. It is said to have been taken to America during early colonial days and grown more extensively than any other cotton but was later displaced by other species.

Important Cottons Derived from G. herbaceum.—This species is the parent of many Asiatic varieties of cotton. Although many of them are not choice ones, they are grown extensively. The Levant cotton, grown in Arabia, Asia Minor, and southern Europe, is one of the most important. Gammie[3] traces Broach and several other Indian varieties to this species.

Gossypium nanking Meyen. (Synonym; *G. indicum* Lamarck and others; *G. herbaceum,* Roxburgh, and others; probably classified by most authorities as *G. herbaceum*).—Chinese or Siam cotton. Watt[1] considers the strain with rufous-colored lint the "Nanking" or "Nankeen" cotton.

Annual or perennial, rather delicate, sparsely branched plant, stems and leaves somewhat hairy; leaf segments extending to middle of the leaf, three to five lobes, larger ones constricted somewhat below, bluntly acute to acuminate at apex; two lower lobes appear to be an extra addition to leaf (Fig. 1*b*). Bracteoles are large, purplish, united below, with three to four sharp teeth on the apex. Flowers rather large, yellow; petals with faint-purple claws. Fruit somewhat angled, three- to four-celled; seeds large, irregular, covered with dense coating of rufous fuzz; lint white, tending to become rufous.

Distribution.—*Gossypium nanking* is not known wild but, as determined by Watt,[1] is cultivated in China, Japan, Malaya, Siam, Burma, India, Central Asia, Arabia, Madagascar, and Africa.

Important Varieties.—A large part of the native cultivated cottons of China and Japan belong to *G. nanking*, as probably do many of the cultivated varieties in India and other parts of Asia.

Gossypium arboreum Linnaeus. (Synonyms *Xylon arboreum* Boerhaave, *G. rubrum* Forskal).—Tree cotton of India and Africa, Deo, or Nurma cotton.

Perennial, 6 to 10 feet high; branches long and slender; young branches and petioles purple (Fig. 6). Leaves thick, leathery, smooth, cordate, deeply segmented into five to seven oblong lanceolate lobes; often a secondary lobe within the lateral sinuses (Fig. 1*a*.) Bracteoles comparatively small, united below, ovate-cordate, entire or occasionally toothed. Flower large, white to purple with darker spots on petals. Fruit rounded with flattened sides, three- to four-celled, valves with margins reflexed; seeds rather large, coated with grayish-green fuzz; lint white in cultivated forms, rufous in wild.

FIG. 6.—*Gossypium arboreum.* (*After Leake.*)

Distribution.—*Gossypium arboreum* is not known in its original wild state. It was probably the first species cultivated and was observed in cultivation in India, Arabia, and Africa centuries ago. It is now found in gardens in India, Arabia, Indo-China, Japan, Java, and Malaya.

Cultivated Races.—Although the typical *G. arboreum* is at present cultivated as a field crop but very little, if at all, several races from its numerous varieties are of importance for field planting. Bengal, Jari, Deshi, Dacca, and the Garo Hill cotton may be mentioned.

Wild Species.—In Mexico, Lower California, Arizona, Central America, Brazil, Africa, Australia, on islands in the Pacific Ocean, and perhaps in other parts of the tropics are wild species of cotton. Most of these that have been described are named in the following list. Some few of the number are cultivated to a limited extent but are not of much economic importance.

Wild Species	Range
G. Davidsonii, Kellogg	Lower California
G. Klotschianum, Anderson	Galápagos Islands
G. Darwinii, Watt	Galápagos Islands
G. tomentosum, Nuttall	Hawaiian Islands
G. Harknessii, Brandegee	Lower California
G. Stocksii, M. Mast	India
G. obtusifolium, Roxburgh	India and East Indies
G. mustelinum, Miers	Brazil and Colombia
G. punctatum, Schumacher and Thonning	United States, Mexico, and West Indies
G. palmerii, Watt	Mexico
G. fruticulosum, Todaro	Mexico
G. Schottii, Watt	Yucatán
G. lanceolatum, Todaro	Mexico
G. microcarpum, Todaro	Mexico, South America, and Africa
G. taitense, Parlatore	Polynesia
G. Kirkii, M. Mast	Africa
G. mexicanum Todaro	Mexico
G. Hopi, Lewton	Arizona
G. Armourianum, Kearney	Lower California

References

1. Watt, Sir George: "Wild and Cultivated Cotton Plants of the World," Longmans, Green & Co., London, 1907.
2. Small, John K.: "Flora of the South Eastern United States," published by the author, New York, 1903.
3. Gammie, G. A.: "The Indian Cottons," *Mem. Dept. Agr. India*, vol. 2, No. 2, 1907.
4. Leake, H. M. and Ram Pershad: "Observations on Certain Extra Indian Asiatic Cottons," *Mem. Dept. Agr. India*, vol. 5, No. 5, 1912.
5. Evans, Walter H.: "Botany of Cotton," *U. S. Dept. Agr. Bull.* 33, 1896.
6. Lewton, Frederick L.: "Notes on the Taxonomy of the American and Mexican Upland Cottons," *Jour. Wash. Acad. Sci.*, vol. 5, No. 4, 1925.
7. Fletcher, F.: "Botany and Origin of American Upland Cotton," *Cairo Sci. Jour.*, vol. 3, 1909.

CHAPTER III

CULTIVATED VARIETIES OF COTTON

The term "variety" is used in this chapter in the sense in which the cotton grower uses it and not in the botanical sense. Varieties that the botanist knows are divisions of or offshoots from species. They differ from the regular species by one or more, perhaps slight but definite and permanent, hereditary characters. The planter's varieties are not distinguishable by such definite or permanent marks, although the apparent differences may be much more striking, as, for instance, the difference between Triumph, a big-bolled, low, vigorous-growing, large-leaved, short-staple variety, and Express, a small-bolled, tall-growing variety, with small leaves and long staple. Divisions of agricultural varieties are known as "strains." The plants belonging to a strain differ from others of the variety by one or more more or less well-marked characteristics. The agricultural varieties are not entirely uniform and do not breed true to type in all cases. Genetically considered, they are of hybrid constitution; various reversions and combinations of ancestral characters appear. Consequently, many different forms arise. W. A. Cook, a cotton breeder at Newman, Miss., noticed this 40 years ago. He said:

> I can take any of the so-called distinct varieties of cotton and in a few years develop all the known varieties from it. In other words, they will develop themselves in the course of time. All that is necessary is to watch the field from year to year, and when a "sport" is noticed, to save the seeds and plant them by themselves.

Probably Cook's idea of a "sport" was a little different from ours, but it may be understood that he meant that many new forms appear in cotton fields. This is known to be a fact. Dr. R. Y. Winters, of the North Carolina Experiment Station, has recently verified some of Cook's conclusions by actual experiments. He made selections from a commercial variety, King, for a series of years. From this one variety he obtained both long-staple and short-staple strains, and other strains differing in size and shape of boll, lint percentage, habit of growth, etc. This unstable condition of cotton varieties has come about from promiscuous mixing and crossing. Mutations probably occur occasionally. The mutants and different varieties and strains brought in from other parts of the Cotton Belt or from foreign countries when planted near local varieties hybridize with them. These hybrids cross with other plants, the new ones with still others, and so on, the final result being a hybrid complex unless selections are made and careful breeding work is carried on. Comparatively, very little breeding work was done prior to 1900.

Number of Cotton Varieties.—No one can determine just how many varieties there are in any cotton-growing country. Every country—in fact, almost every community—has its local varieties. Most of these so-called varieties, however, are simply duplicates of a comparatively small number of well-defined varieties. This multiplicity of varietal names has come about in different ways. If a planter gets seed of a certain variety, and it does well, he will continue to grow it for some years. His neighbors, seeing the good cotton, will purchase some of the seed. When they speak of the cotton they are growing, they are apt to give it the name of the man from whom they secured the seed. Thus, after a few years, there is a new varietal name. Sometimes new names are given old standard varieties by unscrupulous seed dealers in order to sell seed to a better advantage. New variety names are also originated by breeders for their new productions. This practice is entirely proper if the new strain differs from the old by sufficient marks to make it worthy of a distinct name.

Standard Commercial Varieties.—The confusion of names and the lack of varietal standardization have been a handicap to progress in cotton improvement and have resulted in financial loss to growers. In an effort to alleviate matters, a committee was appointed by the Southern Agricultural Workers Association to work toward the standardization of varieties. This committee, which consisted of 20 cotton breeders and agronomists from the various cotton-growing states, made a survey of all the varieties grown in the country in 1930 and chose the following 31 as standard commercial varieties. No variety was chosen that did not represent a fairly distinct type and was of commercial importance in 1930.

Acala-5.*	Half & Half.
Acala-8.	Kasch.
New Boykin.	Lone Star.
Cleveland-5.	Mebane.
Cleveland-884.	Missdel.
Piedmont Cleveland.	Station Miller.
Wannamaker Cleveland.	Mexican Big Boll.
Cook 307-6.	Oklahoma Triumph-44.
Delfos.	Pima.
Delta & Pine Land-8.	Rowden.
Delta & Pine Land-10.	Arkansas Rowden-40.
Deltatype Webber.	Toole.
Dixie-Triumph.	Stoneville.
Dixie-14.	Trice.
Express-121.	Wilds.
Lightning Express.	

* A brief description of each of these varieties is given in the list of varietal descriptions on a following page.

Origin of True Varieties.—As has been mentioned, the cotton plants in a field vary considerably. New forms are frequently appearing, because of the segregation and recombination of characters in hybrid strains, because of mutations, and because of climatic adjustments. When a strain is grown in a district for a period of time, the plants best adapted to the prevailing conditions produce the most fruit and consequently have the largest progenies. There is a natural adjustment, or a survival of the fittest. When this natural adjustment is aided by the intelligence of the breeder, desirable and well-adapted varieties may be secured.

The life of a variety is of rather short duration. Tracy[1] mentions the fact that of the 58 varieties named in the Tenth Census Report for 1880, only six were common in cultivation in 1895. These were Boyd Prolific, Dickson, Herlong, Peeler, Petit Gulf, and Texas Stormproof. None of these is grown to any extent at present, and most of them are extinct.

Of the 118 varieties listed by Tracy[1] in 1895, only 2, King and Truitt, are in cultivation today. Tyler,[2] in 1907, listed nearly 400 varieties. Of this number probably not more than 25 varieties are in existence at present, and only a few of these are cultivated extensively.

Registration of Cotton Varieties.—In an effort to standardize cotton varieties and reduce the number of varietal names, the American Society of Agronomy working in cooperation with the Southern Agricultural Workers Association, recently established a system for the registration of new cotton varieties of merit. To be worthy of registration, a new variety must show an improvement over existing varieties in respect to one or more characters and represent a relatively distinct new type. It is hoped that within a few years all the worth-while varieties will be registered and that varieties not registered will be considered as "scrubs" and have less standing than the registered ones.

Botanical Species Represented in the United States.—It is a question as to how many botanical species are represented in the varieties growing in the United States. The majority of the upland varieties are usually considered as belonging to the species *Gossypium hirsutum* L., but some of the varieties show evidence of "blood" of other species. Sea Island, and probably Egyptian, belongs to the species *G. barbadense* L. Some authorities consider that the Egyptian varieties are forms of *G. peruvianum* Cav.

As has been mentioned before, most American varieties are the product of repeated hybridization, and consequently few, if any, pure botanical species to be found among the cultivated varieties. Certain characteristics of a species may be noticed in a variety, and a conjecture made that that species is one of its ancestors. Watt considers that several other species and botanical varieties belong in the ancestry of American Agricultural varieties, among which may be mentioned *G. arboreum neglectum* Watt; *G. herbaceum* L., G. *punctatum jamaium* Watt., *G. purpurascens* Poir., and *G. Schottii* Watt. It is very probable that other than *G. hirsutum* and *G. barbadense* have entered into the composition of American varieties, since cottons have been imported from every section of the cotton-growing world and grown alongside the established varieties. There has been abundant opportunity for hybridization to take place. It is doubtful, however, if some of the species mentioned by Watt, *G. herbaceum* L. for example, have entered into the make-up of American varieties, for the reason that they are Asiatic species which can be crossed with the American varieties only with great effort. No fertile hybrids have been produced. (See Chap. II for further discussion of species.)

Relative Value of Varieties.—The relative value of a variety varies greatly with the season, market conditions, and adaptation to the environmental conditions where grown. The weather and market conditions that are to prevail cannot be foretold; consequently, there is no way of selecting in advance the variety that will be best suited for any particular locality. But the environmental factors of a locality are more or less permanent and fixed, and it is possible to make some varietal selections on the basis of average conditions. Boll weevils are worse in some places than in others. An early prolific variety is of most value in such regions. Some lands are less fertile than others; a hardy, vigorous-growing strain yields best on such lands. Rainfall and other factors also vary. It is, of course, most profitable to grow the variety best suited to meet all conditions. The safest and surest way of determining values for any plantation is to make tests on the place. Reliable information may be obtained if different varieties are grown side by side, under the same conditions, and comparisons are made.

Classes of Varieties.—Duggar[3] groups the upland varieties in eight classes: the Cluster, or Dickson, type; Semicluster, or Peerless, type; Rio Grande, or Peterkin, type; Kinglike, or King, type; Big-boll, or Truitt, type; Long-limb, or Petit Gulf, type; Intermediate, various types; and Long-staple upland, or Allen, type.

Duggar's system of classification is a good one in that an effort is made to group the various varieties according to their natural relationship, and various characters are considered. This system was very generally adopted and used for a number of years. With the passage of time, certain of the varietal types have been lost because some varieties have become extinct. Also some important new varieties which combine characters belonging to more than one of the old groups have arisen through hybridization or otherwise. These changes in varieties have made the Duggar system of classification less satisfactory than formerly.

The system proposed below is more or less artificial in that it is based on but few plant characters, principally boll size and staple length, but it is believed that it will be helpful in classing or grouping the varieties now grown.

Group 1. King type—early, small-boll, short-staple group.

Group 2. Dixie type—medium-late, small-boll, short-staple group.

Group 3. Cook type—round-boll, short-staple group.
Group 4. Triumph type—big-boll, medium staple group.
Group 5. Delfos type—small-boll, long-staple group.
Group 6. Webber type—big-boll, long-staple group.
Group 7. Various types—mixed or intermediate group.

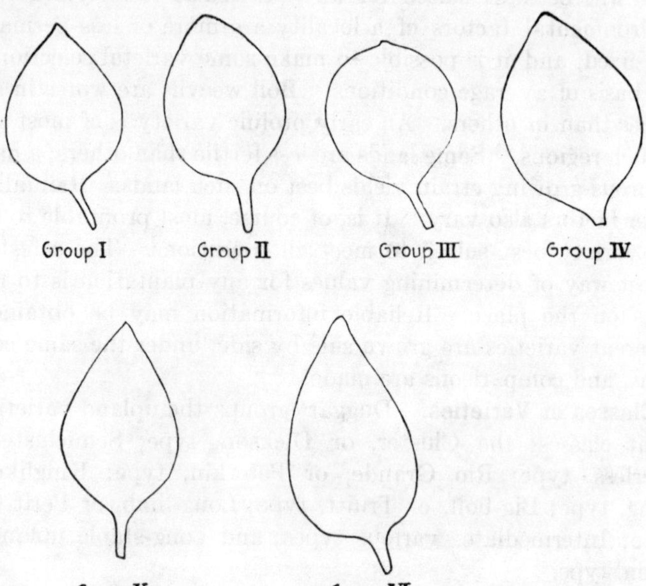

FIG. 7.—An outline sketch of a typical boll from each of the groups of upland varieties.

Early, Small-boll, Short-staple Group.*—The varieties in this group are characterized by their earliness, numerous small bolls, and short staple. Correlated with the earliness are several other characters which make the division fairly well defined. The plants are slender, mostly rather low, usually with one to three slender spreading basal vegetative branches; leaves small to medium size, softly hairy, with lobes narrower and deeper than those of the big-boll cottons; bolls small (see Fig. 7) three- to

* In this classification, staple from the shortest to 1 inch in length is considered short staple; $1\frac{1}{32}$ to $1\frac{3}{32}$ inch, medium-length staple; and $1\frac{1}{8}$ inch and longer, long staple. Bolls requiring fewer than 70 to make a pound of seed cotton are considered big bolls; bolls requiring 70 or more, small bolls.

five- but mostly four-locked; lint short and of good strength; seeds small to medium size, fuzzy, and greenish to brownish gray

King and Oklahoma Triumph-44 are good illustrations of the group. There are also selections from either King or its ancestor Sugar Loaf in cultivation (see Fig. 8).

Medium-late, Small-boll, Short-staple Group.—The varieties within this group are medium late and only medium prolific,

FIG. 8.—Plant of the King variety illustrative of the early, small-boll short-staple group. (*Courtesy R. Y. Winters.*)

but they are healthy and of vigorous vegetative growth. The group contains important wilt-resistant varieties, of which Dixie may be taken as a type (see Fig. 9). The plants are rather tall and slender with one to several long, slender, ascending to erect, basal vegetative branches; fruiting branches are slender and usually long-jointed; leaves small to medium in size, rather hairy when young, with lobes narrower and deeper than those of the big-boll, medium staple group; bolls small to very small in size, with three, four, or five locks, the locks remaining rather compact and firm for some time after the boll opens; lint medium short, of good strength, wiry, and elastic; lint percentage usually high, 35 or above; seeds, small to very small, some nearly smooth and brownish black, but the majority covered with a short, sparse fuzz.

Round-boll, Short-staple Group.—The varieties in this group are characterized by a medium-sized, rounded boll, short staple, and high lint percentage. The plants are medium early, of medium size, vigorous in growth, and range from semicompact to somewhat spreading in habit. Leaves are medium to rather large. Bolls rounded, 60 to 70 to a pound, mostly four-locked;

Fig. 9.—Plant of the Dixie variety illustrative of the medium-late, small-boll short-staple group. (*Courtesy R. Y. Winters.*)

lint per cent, 35 to 42; staple length, ¾ to 1 inch; seeds medium in size with fuzz of various colors.

Cook and Wannamaker Cleveland are good illustrations of the round-boll, short-staple group. They are cultivated extensively at present.

Big-boll, Medium-staple Group.—The main characteristic of this group is the large size of the boll, which is measured by the weight of the seed cotton within the boll. When plants are grown under normal conditions, this ranges from about 40 bolls to the pound for the largest to about 70 for the smallest bolls. The plants of the group are stocky, usually vigorous; the basal limbs strong and heavy, rather short, and two to three in number if the plants are not crowded. Fruiting branches are strong, ranging from very short- and irregularly jointed or semiclustered

Fig. 10.—Plant of the Rowden variety illustrative of the big-boll, medium-staple group. (*After Tyler.*)

to very long-jointed; leaves, large; bolls with four to five locks; seeds, mostly large, fuzzy, of various colors; lint, medium length, mostly $1\frac{1}{32}$ inches, soft and of good strength; lint per cent, 31 to 38, usually 35 or above (see Fig. 10).

Certain varieties belonging to the big-boll, medium-staple group have been selected in Texas for storm resistance. These are sometimes designated as the "stormproof" group. They are very sturdy and vigorous in growth, and the bolls mostly hang downward, so that the spreading burs form a sort of roof over them to shed off falling water. The locks are held in the boll securely but, being rather firm and protruding well, are easily picked.

Triumph, Long Star, and Rowden are good illustrations of the big-boll, medium-staple group.

Small-boll, Long-staple Group.—This group is characterized by the length and character of the lint and the small bolls belonging to its members. Many varieties have a number of characters that are correlated more or less with the staple characters, but the group also includes some varied forms. The majority of the older varieties were late, but recent ones are early. Most of the varieties have plants rather tall and slender, with two or three slender, upright basal vegetative branches and long, slender

Fig. 11.—Plant of the Delfos variety illustrative of the small-boll, long-staple group. This plant was grown under more favorable conditions than the others shown and consequently has more bolls.

fruit branches with joints of various lengths. The foliage is somewhat sparse; leaves small to medium in size, with narrow lobes; peduncles often long and slender; bolls small to medium in size, with three, four, or occasionally five locks; the cotton of each holds together in a rather loose mass; lint, 1⅛ to 1⅜ inches in length, weak to moderately strong, very soft, fine, and clinging, the best types resembling the lint of Sea Island cotton; seeds small to medium in size, sometimes naked or partly naked but usually covered with a brownish-gray or gray fuzz. The habit of growth, the lint characters, and the yellow pollen of many varieties within this group indicate relationship to Sea Island cotton. It is known that some of the varieties contain Sea Island "blood," and very probably many of them do (see Fig. 11).

FIG. 12.—Plant of the Deltatype Webber variety which belongs to the big-boll, long-staple group. (*Courtesy George J. Wilds.*)

Big-boll, Long-staple Group.—This group comprises principally varieties that have been produced comparatively recently by combining characters in one variety that were formerly found in different varieties, as, for instance, big boll and long staple. The varieties in this group are medium early and only medium in respect to prolificness. The plants are rather compact, stocky, short-limbed, and, in general, resemble the big-boll, medium-staple varieties in vegetative characters. The bolls are mostly more or less long-pointed and before opening appear to be very large, but the amount of seed cotton contained is not in proportion to the apparent size of the boll. The bolls are four- and five-locked, and 60 to 70 are required to make a pound of seed cotton. Seeds are of medium size and covered with fuzz; percentage of lint, 30 to 32; lint length, $1\frac{3}{16}$ to $1\frac{3}{8}$ inches; lint soft and silky and usually of excellent character.

Deltatype Webber and Missdel, varieties cultivated extensively at present, are good illustrations of the big-boll, long-staple group (see Fig. 12).

Intermediate Group.—This group contains varieties that possess one or more important characters which are more or less intermediate between characters of plants of other groups, or they have some definite characters of two or more groups. They can be placed only in a heterogeneous class.

Leading Cotton Varieties.—The following lists in Table I were prepared by the agronomists in the cotton-growing states. Varieties are listed in order of popularity, in general, but in many states this varies in different parts of the state.

Description of Upland Varieties.—Brief descriptions and histories of a number of upland varieties are given below. This list is not exhaustive. An attempt has been made to include only the most prominent old-time varieties, some varieties from which prominent present-day varieties have sprung, and the leading varieties being grown at present.

Allen Long Staple (Small-boll, Long-staple Group).—Developed about 1898 by J. B. Allen, Port Gibson, Miss. At one time this variety was grown extensively throughout the Delta of Mississippi and in other long-staple districts, but it is now displaced by earlier long-staple varieties.

Plants are tall and pyramidal in shape; semicluster in habit of growth; with one to three vegetable limbs near the base of stalk and short, irregularly jointed fruiting branches; bolls medium to small; lint very long and silky; seeds medium to small, fuzzy, white; bolls per pound, 78; seeds per

TABLE I.—LEADING COTTON VARIETIES GROWN IN THE UNITED STATES

State	Agronomist	Varieties	Acreage	Remarks
Virginia			54,000[a]	No report received on varieties grown
North Carolina	P. H. Kime	Farm Relief	200,000	1936 acreage
		Cleveland 884-4	140,000	
		Mexican	90,000	
		Cleveland-5	45,000	
		Wann. Cleveland	40,000	
		Humco Cleveland	10,000	
		Clevewilt	35,000	
		Dixie Triumph	25,000	
		Wilds	20,000	
		Foster	25,000	
		Total acreage	974,000	
South Carolina	E. E. Hall	Farm Relief	300,000	1934 acreage
		Dixie Triumph	269,000	
		Cleveland-5	150,000	
		Clevewilt	125,000	
		Cleveland 884-4	100,000	
		Foster	60,000	
		Wilds	50,000	
		Dixie-14	30,000	
		Wann. Cleveland	30,000	
		Lightning Express 8	25,000	
		Total acreage	1,299,000	
Georgia	R. P. Bledsoe	Toole	150,000	1934 acreage
		Half & Half	115,000	
		Piedmont Cleveland	70,000	
		Stoneville 2	50,000	
		D & P L 4-8	15,000	
		Farm Relief	6,000	
		Clevewilt	4,000	
		Cook	2,000	
		Total acreage	2,164,000	
Florida	W. A. Carver	Toole		1936 acreage
		Half & Half		
		Cleveland		
		Cook 307-6		
		Rhyne's Cook		
		Sea Island	4,000	
		Total acreage	90,000	
Alabama	H. B. Tisdale	D & P L Strains	300,000	1934 acreage
		Half & Half	250,000	
		Rucker	250,000	
		Cook-307-6	200,000	
		Rhyne's Cook	200,000	
		Other Wilt Resist. Cook	200,000	
		Non Wilt Resist. Cook	125,000	
		Dixie Triumph	20,000	
		Total acreage	2,144,000	
Mississippi	J. F. O'Kelly	D & P L strains	932,750	1935 acreage
		Delfos strains	533,000	
		Missdel strains	399,750	
		Stoneville strains	106,600	
		Express strains	53,300	
		Wilds	25,000	
		Total acreage	2,961,000	
Louisiana	C. W. Davis	D & P L Strains	480,491	1936 acreage
		Half & Half	273,097	
		Stoneville strains	147,866	
		Delfos strains	134,741	
		Missdel strains	19,718	
		Qualla	16,934	
		Coker Cleveland strains	15,620	
		Dixie Triumph	13,603	
		Wann. Cleveland	10,791	
		Total acreage	1,303,005	

[a] Most of the figures on acreage are from estimates collected by J. O. Ware.[10]

TABLE I.—LEADING COTTON VARIETIES GROWN IN THE UNITED STATES.—(*Continued*)

State	Agronomist	Varieties	Acreage	Remarks
Arkansas	J. O. Ware	Rowden strains	897,000	1934 acreage
		Half & Half	216,000	
		Acala strains	209,000	
		Missdel strains	163,000	
		Delfos strains	127,000	
		Mebane Triumph	79,000	
		D & P L strains	70,000	
		Stoneville strains	63,000	
		Wilson Type	41,000	
		Russell Big Boll	36,000	
		Total acreage	2,196,000	
Tennessee	Newman Hancock	D & P L strains	275,000	1936 acreage
		Half & Half	175,000	
		Stoneville strains	70,000	
		Cleveland strains	40,000	
		Acala, Delfos, Rowden, Trice and others	50,000	
		Total acreage	818,000	
Missouri	B. M. King	Stoneville 4 A		1936 acreage
		Stoneville 5		
		Deltapine (D & P L 11)		
		Rowden-40		
		Half & Half		
		Total acreage	382,000	
Oklahoma	L. L. Ligon	Mebane Triumph strains	727,000	1934 acreage
		Oklahoma Triumph	436,000	
		Half & Half	378,000	
		Acala-8	349,000	
		Rowden-40	290,000	
		Total acreage	2,909,000	
Texas	D. T. Killough	Mebane Triumph strains	6,232,000	1934 acreage
		Acala strains	751,000	
		Lone Star strains	558,000	
		Rowden	400,000	
		Half & Half		
		Delfos		
		Total acreage	9,418,000	
New Mexico	G. N. Stroman	Acala-8	75,000	1934 acreage
		Pardue's Acala	5,000	
		Young's Acala		
		Watson's Acala		
		Roger's Acala		
		Total acreage	100,000	
Arizona	C. J. King	Acala-8	153,000	1936 acreage
		Ambassador (Stoneville 4)	15,000	
		Pima Egyptian	31,000	
		S × P Egyptian	9,000	
		Total acreage	208,000	
California	Geo. J. Harrison	Acala-8	358,000	1936 acreage
		Stoneville	5,000	
		Mebane & Lone Star	5,000	
		Pima Egyptian	100	
		Total acreage	368,000	

pound, 3,800; length of lint, 1⅜ to 1$\frac{9}{16}$ inches; strength of fibers, 4.3 grams; lint percentage, 29. Adapted from Tyler.[2]

Acala (Intermediate Group).—This variety was developed from imported seed obtained by G. N. Collins and C. B. Doyle in 1906, at Acala, in the state of Chiapas, Southern Mexico. The present strain was developed from a selection of 20 plants made by Dr. D. A. Saunders, in 1911, from the original field grown at Waco, Tex. Acala is grown most extensively in Oklahoma, Texas, western Arkansas, New Mexico, Arizona, and California.

Plants are of medium height, with strong, erect main stem, few vegetative limbs; fruiting branches short-jointed; lower branches long, upper branches short; leaves of medium size, dark green; bolls of medium size, ovate, with short, blunt points; burs often pendant, stormproof; bolls per pound, 50

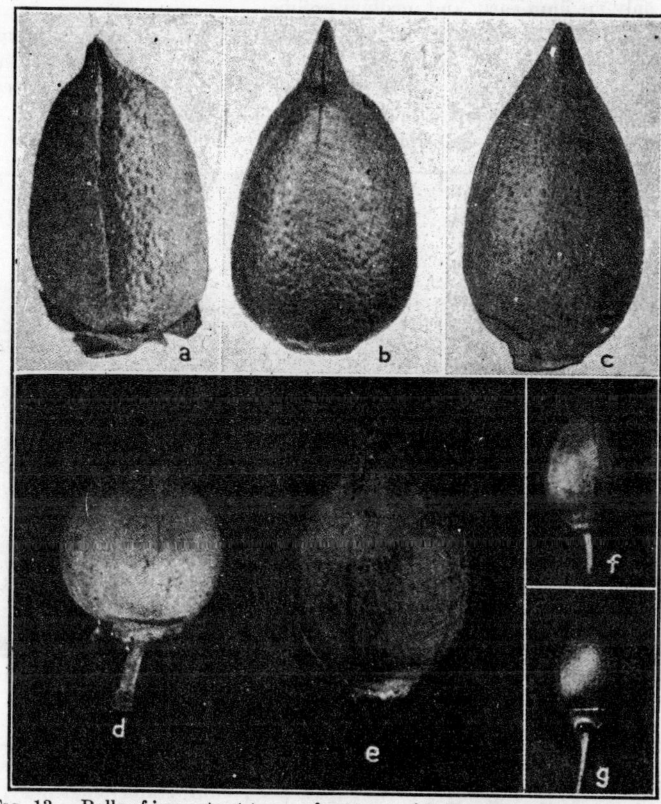

FIG. 13.—Bolls of important types of cotton. (*a*) Egyptian Mit Afifi. About natural size. (*b*) Egyptian Sakellaridis. About natural size. (*c*) American-Egyptian Pima. About natural size. (*d*) Upland short staple, Cook variety. Slightly larger than half natural size. (*e*) Upland big boll, Lone Star variety. Slightly larger than half natural size. (*f*) Upland long staple, Express variety. Somewhat less than half actual size. (*g*) American Sea Island. Somewhat less than half actual size.

to 55; length of lint, $1\frac{1}{16}$ to $1\frac{3}{16}$ inches; staple with good drag and extra strong; lint percentage, 32 to 35. After Oakley.[6]

Acala-5* (Intermediate Group).—Developed by C. N. Nunn from a plant selection made in a field of Acala cotton, at Okema, Okla., in 1914. The plants attain a height of 2 to 5 feet, depending on the supply of moisture

* Varieties so designated were chosen as standard commercial varieties by the Committee on Standardization of Varieties.

and soil fertility; they have a strong central axis and rather slender fruit branches with rather short internodes; leaves, medium-sized, slightly cupped; bolls, 65 to 75 per pound of seed cotton; staple length, 1 to 1$\frac{1}{16}$ inches; lint percentage, 33 to 37; seed, medium-sized, about 4,000 to a pound, with light-gray fuzz of moderate density; plants not disease-resistant and only medium early and medium prolific.

Acala-8* (Intermediate Group).—Developed by H. G. McKeever, of the U. S. Department of Agriculture, from a plant selection that was made in the same field of Acala cotton, at Okema, Okla., as the Acala-5 selection. Plants are somewhat larger and more spreading than Acala-5 plants, with larger leaves and bolls and longer staple.

Fig. 13*A*.—Field of upland cotton at blooming time. A scene in the Mississippi Delta.

Acala-8 plants attain a height of 3 to 5 feet under fair conditions; vegetative branches are few, fruiting branches are medium short-jointed; leaves, medium large, dark green; 60 to 70 bolls per pound; staple length, 1$\frac{1}{16}$ to 1$\frac{3}{16}$ inches; lint percentage, 35 to 38; seeds, light-gray color, medium-sized, about 3,500 to a pound; plants not disease-resistant; productive in regions where adapted.

Ambassador.† Registration No. 33. (Big-boll, Medium-staple Group).—This variety (formerly known as "Stoneville-4") was developed by C. A. Tate, of the Stoneville Pedigreed Seed Co., Stoneville, Miss. The original selection, made at Stoneville, in 1923, by H. B. Brown, came from Lone Star-65. Later selections purified the strain and increased its stability. The variety is now grown in Mississippi, southeast Missouri, northeast

* Varieties so designated were chosen as standard commercial varieties by the Committee on Standardization of Varieties.

† Varieties so designated have been registered by the American Society of Agronomy as new varieties of merit.

Arkansas, Texas, Arizona, and China. It is adapted to regions of moderate rainfall and medium soil fertility.

Under average conditions, plants are from 2 to 2½ feet tall, being somewhat lower and more stocky than most varieties. Branches are rather short-jointed and stout; stems, branches, and leaves hairy; vegetative branches few, leaves medium-sized, usually dark green and thick. Flowers are medium-sized, cream-colored, and anthers cream-colored; stigma projects beyond the anthers but little if at all; bracts rather large, cordate, and deeply toothed; bolls, rather large, 50 to 70 per pound under good conditions, rounded with a short, blunt point, mostly five-locked, stormproof; bolls pick well; staple length, 1 to 1 1/16 inches; seeds, medium-sized, brownish gray; lint percentage, 34 to 36; resistance to wilt and boll rot only fair; early and more prolific than most large-boll cottons.

Bohemian (Big-boll, Medium-staple Group).—Also known as Supak, Shupark, Shoepeck. Shuparch, etc. Originated over sixty years ago by a Bohemian settler named Supak, living in Travis, Austin County, Tex. Bohemian was a very popular cotton in Texas at one time but has been displaced by other varieties. It is the parent of Rowden.

Plants are large. basal limbs, two to three, often nearly prostrate; fruiting branches, numerous and long, somewhat drooping; joints, short and regular, making the plant very prolific; foliage, heavy; bolls, large, usually turned down by their weight, so that when the boll opens the lint is protected by a roof formed partly by the broad backs of the segments of the bur and partly by the large involucre; picks easily; lint, medium length; seeds, large, fuzzy, gray or brownish gray; bolls per pound, 55; seeds per pound, 3,240; average lint length, 15/16 inch; strength of fiber, 5.3 grams; lint percentage, 33 to 34. Adapted from Tyler.[2]

Boykin Stormproof (Big-boll, Medium-staple Group).—Originated by W. L. Boykin, Kaufman, Kaufman County, Tex. It is not grown extensively at present.

Plants are large and stocky; fruiting branches, long and long-jointed, late maturing; bolls, large, majority five-locked, holding cotton well during storms but easily picked, as locks cling together in one mass; lint, of good length; good percentage of lint; seed, large, fuzzy, brownish gray; bolls per pound, 50; seed per pound, 3,280; lint length 1 1/32 inches; strength of fiber, 5.2 grams; lint percentage, 34. Adapted from Tyler.[2]

*New Boykin** (Big-boll, Medium-staple Group).—This variety originated from a plant selection made in a field of Mebane by A. M. Ferguson, at Sherman, Tex., in 1913. New Boykin is medium early and yields better than the average under unfavorable growing conditions.

Plants are medium-sized to large and somewhat rangy and open; vegetative branches, 2 to 5, fruiting branches, 8 to 15; leaves, medium in size; bolls per pound, 75 to 80; staple length, 7/8 to 31/32 inches; lint percentage, 37 to 40; seeds, gray to white, medium-sized; disease resistance fair to good; rather productive, comparatively, in regions where adapted.

Bottoms (Intermediate Group).—This variety was developed by A. T. Bottoms, of Athens, Ala., about 1915, from a Texas variety. It has dwarfish

* Varieties so designated were chosen as standard commercial varieties by the Committee on Standardization of Varieties.

plants of a semicluster type, is extra early and fast in fruiting, and has bolls of medium size. The staple length is ⅞ to 1 inch, irregular. The variety is not grown extensively at present.

Brannon (Intermediate Group).—Brannon was developed from a selection made by G. W. Brannon, East Feliciana Parish, Louisiana, and further improved by N. B. Riddle, Riddle, La., a son-in-law of the originator, and by G. Brannon, Lindsay, La. It is intermediate between the upland long-staple and the Peterkin groups. It is not grown extensively now, if at all.

Plants are tall and slender; basal limbs, one to three or more; fruiting branches, long, rather long-jointed; bolls, medium to large; seeds, of medium size, covered with a short, sparse, brownish-gray fuzz, or nearly naked; bolls per pound, 66; seed per pound, 4,000; lint length, 1 inch; lint percentage, 32 to 37. Adapted from Tyler.[2]

Cleveland (Round-boll, Short-staple Group).—Known also as Cleveland Big Boll. This variety was developed by J. R. Cleveland, Stratton Miss., by 25 years of mass-selection work. The variety came into prominence about 1890. It is not grown extensively at present, having been supplanted largely by strains developed from it, as Wannamaker Cleveland and Piedmont Cleveland.

Plants are not uniform, being both semicluster and open in growth; joints of fruiting branches short, making the variety medium-early maturing; bolls, large, 50 per cent five-locked, not stormproof; lint, of medium length; seed, medium-large, fuzzy, light brownish gray; bolls per pound, about 60; seeds per pound 3,100; lint length, $1\frac{5}{16}$ inch; strength of single fibers, 5.5 grams; lint percentage, 35 to 37.

*Wannamaker Cleveland** (Round-boll, Short-staple Group).—This strain of Cleveland was developed by the plant-to-row method by W. W. Wannamaker and Sons, of Saint Matthews, S. C. The work was first begun in 1908. The variety came into prominence in 1916, the year the seed was placed on the market in a commercial way. It is earlier and more prolific, has smaller bolls, and is a lower growing and more spreading plant than the parent variety. It was formerly widely grown in hill-land districts east of the Mississippi River and to some extent in Arkansas and Louisiana.

Plants are of medium height with medium foliage, mostly with three or four basal vegetative limbs; medium early and fairly prolific; bolls, rounded with very blunt tip, 60 to 69 to a pound; lint percentage, 35 to 37; staple length, ⅞- to 1-inch; semiresistant to wilt.

Cleveland-54 (Round-boll, Short-staple Group).—This is a selection from Wannamaker Cleveland made by the Mississippi Experiment Station. It differs from the parent strain in having rather more compact plants and is slightly earlier and more prolific.

*Piedmont Cleveland** (Round-boll Short-staple Group).—This variety was originated by J. O. M. Smith and M. W. H. Collins from a plant selection made in a field of Cleveland Big Boll cotton, at Commerce, Ga., in 1914. The variety has been grown and distributed by the originator for more than 20 years and subjected to further selection and roguing. It has been earlier

* Varieties so designated were chosen as standard commercial varieties by the Committee on Standardization of Varieties.

and more prolific than most other short-staple Clevelands but has a lower lint percentage.

Plants are medium to large, rather compact; foliage is medium to heavy; bolls, 65 to 70 per pound; staple length, 7/8 to 15/16 inch; lint percentage, 34 to 36; seed, brownish gray, medium-sized; disease resistance, fair; earliness, medium; productiveness, good.

Cleveland-5* (Intermediate Group).—Cleveland-5 was developed by the cotton breeders of the Coker Pedigreed Seed Co., Hartsville, S. C., from a plant selection made in a field of Coker Cleveland cotton, near Hartsville, in 1921. It is probable that this original plant was a natural hybrid between Cleveland and some long-staple variety, since it had a longer staple than is usual for the Clevelands and also had some characters of another variety. Cleveland-5 has rather wide adaptability, a fair-sized boll, and staple of good character.

Plants are medium-sized, with two to four vegetative branches and medium-length fruit branches. Foliage is medium heavy; bolls, 65 to 70 per pound; staple length, 1 to 1 1/16 inches; lint percentage, 36 to 40; seed, brownish gray to light green, medium-sized, about 3,600 to a pound; disease resistance, earliness, and productiveness, medium.

Cleveland-884* (Intermediate Group).—This variety was developed by the Coker Pedigreed Seed Co., from a plant selected in a field of Coker Cleveland cotton, near Hartsville, S. C., in 1923. The variety is open in branching habit and has a small leaf for a Cleveland. It is rather early and productive and has fiber of good character.

When grown under average conditions, the plants attain a height of 2 1/2 to 3 feet. There are one to four vegetative branches and rather long fruit branches. Leaves are rather small; bolls, 65 to 70 per pound; staple length, 1 to 1 1/16 inches; lint percentage, 36 to 38; seed small, 3,900 to 4,200 per pound, gray to light brown; plants somewhat resistant to diseases in general but have only slight resistance to wilt.

Clevewilt (Intermediate Group).—Developed by Coker Pedigreed Seed Co., Hartsville, S. C., by series of selections from Cleveland-884, which was grown repeatedly on wilt-infested soil. It is highly resistant to wilt, productive; staple length, 1 1/32 to 1 1/16 inches; gin outturn, 37 to 38 per cent.

Cliett Superior (Big-boll, Medium-staple Group).—Introduced in 1924 by Cliett Cotton Breeding Farms, San Marcos, Tex. Grown from Mebane Triumph stocks. Considered to be best adapted to central and western Texas and western Oklahoma.

College No. 1 (Round-boll, Short-staple Group).—This is a selection from Sunbeam, a hybrid variety from a cross between Cook's Improved and Columbia. It is grown in Georgia.

Plants are of medium size, with no to one basal vegetative branch; fruiting branches long with medium-length joints, there being no indication of the cluster habit; bolls, large, oblong, 60 per pound; staple length, 15/16 inch; lint percentage, 38.

Columbia (Big-boll Long-staple Group).—A selection from Russell made by Dr. H. J. Webber while in charge of plant-breeding investigations,

*** Varieties so designated were chosen as standard commercial varieties by the Committee on Standardization of Varieties.**

U. S. Department of Agriculture. It was grown somewhat extensively for some years in the Delta sections of Mississippi and Arkansas and also in North Carolina and South Carolina, but has been displaced by Webber and its strains.

Plants are similar to Russell; bolls, large, long, ovate; lint, not uniform, not silky like Sunflower but stronger; seed, large, fuzzy, gray, a small proportion green; bolls per pound, 66½; seed per pound, 3,400; lint length, 1¼ inches; strength of single fibers, 5.6 grams; lint percentage, 31.7. Adapted from Oakley.[6]

Cook or Cook's Improved (Round-boll, Short-staple Group).—Originated by J. R. Cook at Ellaville, Ga., in 1895. This is a medium- to large-bolled variety, yielding a high percentage of lint, and is medium early. Selections made by the Alabama Experiment Station have displaced the parent strain of Cook to a large extent.

Bolls are quite round, 54 per cent five-locked; seed medium in size, fuzzy, greenish, or brownish gray; not stormproof and susceptible to boll rot; bolls per pound, 60 to 65; seed per pound, 4,000; lint length, ⅞ inch; strength of single fibers, 6.8 grams; lint percentage, 38½. Adapted from Tyler.[2]

Alabama Station Cook.—This is a strain of Cook developed by the Alabama Station. It differs from the parent variety in having smaller bolls and higher lint percentage and is more uniform.

Tri-Cook or Triumph Cook.—This variety was developed by M. R. Hall, James, Ala., in 1910, from a selection made in a field in which mixed seed of Cook and Triumph had been planted. The plants are not uniform; they resemble Cook largely. The strain is more resistant to wilt than Cook and has given more profitable yields on wilty soils.

Cook-1010.—This strain, originated by the Alabama Experiment Station, differs from the Alabama Station Cook in having smaller bolls, 66 to 70 per pound; higher lint percentage, 40 to 42, and shorter staple, ¾ to ⅞ inch. It is a good yielder.

Cook's Long-staple (Small-boll, Long-staple Group).—Developed by W. A. Cook, Newman, Miss. This variety is closely related to Allen. It was a popular variety in long-staple districts for several years but is not grown at present, having been superseded by earlier long-staple varieties.

Plants are tall and pyramidal in shape, with one to three limbs or often none; fruiting branches showing a tendency to semicluster but not so short and irregularly jointed as Allen; bolls, of medium size, pointed; lint, of good length, soft and silky; seeds, of medium size, fuzzy, gray; bolls per pound, 60; seed per pound, 3,650; lint length, 1 to 1¼ inches; strength of single fibers, 4.7 grams; lint percentage, 28.3. Adapted from Tyler.[2]

Cook-307-6* (Round-boll, Short-staple Group).—This is a wilt-resistant variety which was developed by E. F. Cauthen from a plant selection made in a field of nonresistant Cook cotton, near Auburn, Ala., in 1912. The variety yields well and is very hardy, easy to pick, and very wilt-resistant.

* Varieties so designated were chosen as standard commercial varieties by the Committee on Standardization of Varieties.

Plants are medium in size, with one to two basal vegetative branches and rather large fruit limbs; foliage is medium to light; bolls, medium-sized, 70 to 85 per pound; staple length, 7/8 to 1 inch; lint percentage, 35 to 38; seed, medium-sized, white to light green; earliness, medium, and productiveness, good.

Rhyne's Cook.—This is a strain of Cook-307-6 developed by Rhyne Bros., of Benton, Ala. It is very similar to the parent strain.

Culpepper (Intermediate Group).—Developed by L. E. Culpepper, Luthersville, Ga., by mixing seed of Wyche and Dickson. These dissimilar varieties have blended by hybridization into a fairly uniform intermediate cotton, still showing, however, a few plants approaching the parents in character of growth. The cross was made about 1890, and for some years the variety was rather common throughout the Cotton Belt. It is not grown very extensively now.

Plants are semicluster in habit of growth, with one to three basal vegetative limbs; fruiting branches, short and irregularly jointed; bolls, large, rounded; lint of medium length; percentage, good; seeds, large, fuzzy, greenish or brownish gray in color; bolls per pound, 50; seed per pound, 3,380; lint length, 29/32 inch; strength of single fibers, 6.7 grams; lint percentage, 35.1. Adapted from Tyler.[2]

Delfos (Small-boll, Long-staple Group).—This variety sprang from a single outstanding plant selected in a field of Foster-120 at the Mississippi Delta Experiment Station, in 1916, by H. B. Brown. It is grown extensively on alluvial lands in Mississippi, Louisiana, and Arkansas.

Delfos is very early and very prolific; it has low-spreading open plants, with comparatively slender main stem and branches; the fruiting branches are long and numerous; one to four basal vegetative branches are found on the plant if wide-spaced; leaves are small, having a pale-green color; bolls are narrowly ovate, short-pointed, four- and five-locked, open well, pick fairly well, and run 70 to 80 to the pound of seed cotton; lint percentage, 31 to 32; staple length, 1 1/8 to 1 3/16 inches.

Delta and Pine Land-8* (Intermediate Group).—This variety originated from a plant selection made by E. C. Ewing, of Scott, Miss., plant breeder for the Delta and Pine Land Co., in a field of Delta and Pine Land-4 cotton, at Scott, Miss., in 1921. (Delta and Pine Land-4 was produced by crossing Mebane and Polk, the latter a local variety.) Delta and Pine Land-8 is especially adapted to poor hill lands because the plants are vigorous in growth and have a high lint percentage and a fair staple length.

The plants are medium large, attaining a height of 3 to 6 feet under favorable conditions, compact, with rather short fruit branches; foliage is medium heavy; bolls, medium-sized, 70 to 80 per pound; staple length, 1 to 1 1/16 inches; lint percentage, 36 to 38; seed, medium-sized, gray; disease resistance, fair to good; earliness, fair; productiveness, fair.

Delta and Pine Land-10* (Intermediate Group).—This variety was produced by E. C. Ewing, at Scott, Miss., by crossing Express-122 and a noncom-

* Varieties so designated were chosen as standard commercial varieties by the Committee on Standardization of Varieties.

mercial hybrid containing Express and Mebane blood. It has proved to be productive over a large area of the central South and has a satisfactory staple length for a short cotton. The plants are medium-sized, being more spreading than Delta and Pine Land-8 plants but not so tall; leaves are medium to fairly large, distinctly light green; boll size, medium, 70 to 80 to a pound; staple length, 1 to 1$\frac{1}{16}$ inches; lint percentage, 33 to 36; seed, medium-sized, brownish gray; disease resistance, fairly good; medium early.

Deltapine.† Registration No. 32 (Intermediate Group).—This variety (formerly known as "D. & P. L.-11") originated from a plant selection made from a hybrid between Delta and Pine Land-10 and a noncommercial hybrid. It was developed at Scott, Miss., by E. C. Ewing.

The plants of the variety are of medium height and fairly open, stems and branches rather slender, mostly nearly glabrous and reddish in color; vegetative branches, not extensively developed; leaves, medium-sized, usually dark green, and only slightly hairy. Flowers, medium-sized, cream-colored; anthers, cream-colored; bolls, medium-sized, rather short-pointed, 70 to 80 per pound; staple, length, 1$\frac{1}{32}$ to 1$\frac{3}{32}$; lint percentage, 37 to 42; resistance to wilt and boll rots, only fair.

Dickson's Improved (Early Small-boll, Short-staple Group).—Developed about 1858 by David Dickson, of Oxford, Ga. This is an old variety not grown at present.

Plants are early-maturing, of the strict-cluster type, with one to three basal vegetative limbs; fruiting branches reduced to spurs by shortening of internodes, thus throwing the joints very close together; spurs, 2 to 6 inches long, usually longer in the middle of the stalk than at the bottom or at the top; leaves, very large; bolls, clustered, small, rounded shape; lint, of medium length, small, fuzzy, brownish gray; bolls per pound, 105; seeds per pound, 5,670; average lint length, $\frac{7}{8}$ inch; strength of single fibers, 5.1 grams; lint percentage, 29 to 32. Adapted from Tyler.[2]

Dillon (Medium-late, Small-boll, Short-staple Group).—Dillon is a wilt-resistant variety developed by W. A. Orton, of the U. S. Department of Agriculture, from selections from Jackson Limbless, at Dillon, S. C., in 1900. It is grown very little, if at all, at present.

Like all cluster varieties, Dillon is difficult to pick free from trash, but, owing to its resistance to wilt, its stormproof character, and its productiveness, it was popular on wilt-infected soils. Plants are tall, erect, with one or two large basal vegetative branches; fruiting limbs, reduced to short spurs, crowding the short-stemmed bolls into clusters; seeds, small, covered with close brownish-green fuzz; bolls per pound, 94; seeds per pound, 5,320; lint length, $\frac{7}{8}$ inch; lint percentage, 37. Adapted from Tyler.[2]

Dixie (Medium-late, Small-boll, Short-staple Group).—This variety was developed by W. A. Orton from a single plant selection at Troy, Ala., in 1902. The cotton from which the selection was made was grown on wilt-infected land. This variety was formerly considered one of the leading wilt-resistant varieties but is grown but little if any at present.

† Varieties so designated have been registered by the American Society of Agronomy as new varieties of merit.

Plants are vigorous, wilt-resistant, of medium height, pyramidal, of the Peterkin type, with two or more large basal limbs; fruiting limbs, long, slender, drooping; leaves, of medium size; bolls, of medium size, easy to pick but very stormproof; seeds, small, typically covered with a greenish-brown fuzz; bolls per pound, 75; seeds per pound, 4,100; lint length, about 7/8 inch; lint percentage, 34 to 35. After Oakley.[6]

Dixie-Afifi (Intermediate Group).—This is a hybrid between Dixie and Afifi, an Egyptian variety. It resembles Dixie except in staple length. Lint length, 1 3/16 inches; lint percentage, 30.

*Dixie-Triumph** (Medium-late, Small-boll, Short-staple Group).—Dixie-Triumph originated from a cross between Mebane Triumph and Dixie wilt-resistant cotton. The cross was made by W. W. Gilbert on Joe M. Johnson's farm, near Monetta, S. C., in 1908. Dixie Triumph resembles the Dixie parent in general plant form. It is a healthy, vigorous-growing variety and is very resistant to Fusarium cotton wilt (*F. vasinfectum*).

Plants were large and spreading, attaining a height of 3 to 6 feet; branches, long; leaves, medium-sized; bolls, medium-sized, 65 to 75 per pound; staple, length, 7/8 to 1 inch; lint percentage, 33 to 35; seed, medium small, tawny gray; medium late but earlier than Dixie. Productive if boll weevils are controlled.

Dixie-14* (Medium-late, Small-boll, Short-staple Group).—This variety was developed by S. P. Coker of Hartsville, S. C., from a plant selection made in a field of U. S. Department of Agriculture Dixie cotton in 1920. The special merits of the variety are its excellent wilt resistance and good production.

Under average conditions, Dixie-14 plants attain a height of 3 feet and are somewhat stocky; basal branches are small; fruiting branches are long, producing a spreading type of plant; leaves are medium to heavy; bolls, medium-sized, 65 to 70 per pound; staple length, 31/32 to 1 1/32 inches; lint percentage, 34 to 38; seed, medium-sized, greenish gray; disease resistance good; earliness, medium; productiveness, good.

Durango (Small-boll, Long-staple Group).—Developed from seed obtained by F. L. Lewton from the state of Durango, Mexico. The variety was grown in Texas and California rather extensively for a few years but has now been replaced by other more productive varieties.

Express (Small-boll, Long-staple Group).—This variety was developed from a single plant selected by Dr. D. N. Shoemaker, in a mixed field of so-called "Bohemian" cotton growing near Paris, Tex., in 1904. Although early and prolific, it was not popular in Texas, since it did not have large bolls, was not stormproof, and was lacking in drought-resistant qualities and high lint percentage. In 1911, E. C. Ewing introduced this variety into the Mississippi Delta in his varietal work at the Mississippi Delta Experiment Station. It did well in the Delta and soon spread over this whole section and over much of the alluvial land in neighboring states. Since 1920, Delfos has been displacing it.

* Varieties so designated were chosen as standard commercial varieties by the Committee on Standardization of Varieties.

Plants are very distinct in appearance, growing somewhat rank on rich soil, tall, pyramidal, open and somewhat spreading, with none to one basal vegetative branch; stem and branches, long with long internodes; fruiting limbs somewhat short-jointed; leaves, small to medium, light green in color; bolls, rather small and long though not so pointed as those of most long-staple cottons; seeds, small, covered with grayish-green fuzz; bolls per pound, 73 to 78; seeds per pound, 3,900; lint length, 1 3/16 inches; lint percentage, 28 to 29.

Express-350 (Small-boll Long-staple Group).—This strain is from a selection made by Ewing in 1913 and differs from the parent plant in that plants are shorter and more compact, more prolific, and more uniform.

Express-432 (Small-boll, Long-staple Group).—This strain is from a selection made by E. C. Ewing in 1914 and developed by the writer. It differs from Express in having larger and more spreading plants, a higher lint percentage (31 to 33), and a shorter staple (1 1/8 inches) and in being more disease-resistant.

*Lightning Express** (Small-boll, Long-staple Group).—This variety came from a plant selection made by the Coker Seed Co., in a field of Express-350 cotton grown near Hartsville, S. C., in 1922. The special merits of the variety were earliness, prolificness, and good length of staple. It is not grown extensively at present.

Plants are of medium size with few vegetative branches; fruit branches, of medium length; foliage, rather light; boll size, small, 75 to 85 per pound; staple length, 1 1/8 to 1 3/16 inches; lint percentage, 32 to 34; seed, small, gray; disease resistance, good; very early and prolific.

Express-121* (Small-boll, Long-staple Group).—This variety originated from a single plant selection made by W. E. Ayres in a field of Express-432 cotton, near Stoneville, Miss., in 1921.

Plants are medium tall and open but not so spreading as the parent strain; foliage, medium light; boll size, small, 75 to 80 per pound; staple length, 1 1/8 to 1 3/16 inches; lint percentage, 32 to 33; seed, medium-small, gray; disease resistance, good; earliness and productiveness, good.

Farm Relief (Intermediate Group).—Developed by the Coker Seed Co., of Hartsville, S. C., from a cross of Lightning Express and Cleveland made in 1921. It is early and a rapid fruiter; plants are open, foliage rather light; bolls, medium-large, 60 to 70 per pound; lint percentage, 35 to 38; staple length, 1 1/32 to 1 1/16 inches; quality, good.

Floradora (Small-boll, Long-staple Group).—This variety, which was probably Allen Long Staple to begin with, was taken from the Mississippi Delta section to Barnwell, S. C., by a cotton buyer named Coffin. For several years it was grown by Mrs. W. G. Bilmore Sims, of Barnwell, and sold under the name of Sims Long Staple. L. A. Stoney, of Allendale, Barnwell County, recognized its value and introduced it under the new name of Floradora. This cotton was grown rather widely a decade or so ago but more recently has been displaced by earlier long-staple varieties.

* Varieties so designated were chosen as standard commercial varieties by the Committee on Standardization of Varieties.

Bolls per pound, 80; seeds per pound, 3,900; lint length, 1³⁄₃₂ inches; strength of fibers, 4.5 grams; lint percentage, 30.5. Adapted from Tyler.[2]

Foster (Big-boll, Long-staple Group).—This variety was originated by D. A. Saunders in 1904 by crossing the Sunflower and Triumph varieties. The crossing and the subsequent selection work was done on a plantation belonging to a man named Foster, after whom the cotton was named.

Plants are rather small, stocky, and compact; bolls, large, ovate, sharp-pointed, 50 to the pound; foliage, dense; lint, strong and fine in quality, 1⅛ to 1⅜ inches in length; lint percentage, 34; seed medium to small with grayish-brown tuft.

Foster-120 (Big-boll, Long-staple Group).—This was a selection from Foster made by E. C. Ewing, in 1911, at the Mississippi Delta Experiment Station. The strain was grown rather extensively throughout the Delta of Mississippi and Arkansas for a number of years but has been discarded in favor of Delfos; it was earlier and had smaller plants than the parent strain.

Plants were rather low with medium-slender stems and rather light foliage, early and prolific; bolls, rather large, 56 to 62 a pound, pyriform, and long taper-pointed; lint percentage, 30 to 32; staple length, 1³⁄₁₆ to 1¼ inches; rather susceptible to wilt and anthracnose; well adapted to rich lands that are free from wilt.

Carolina-Foster.—This name has been applied by the Humphrey-Coker Seed Co., of Hartsville, S. C., to certain strains of Delfos that they have grown in South Carolina.

Coker Foster.—The Coker Pedigreed Seed Co., of Hartsville, S. C., has used this name for certain strains of cotton that they are breeding. These strains have the same ancestry as Delfos and are very similar to it.

Griffin (Big-boll, Long-staple Group).—This is a large-bolled variety of cotton originated by John Griffin, near Greenville, Miss., by crossing a big-boll variety with Sea Island and practicing selection for several years. The work was begun in 1867. The strain is grown very little at present.

Plants are large and vigorous with one to three basal vegetative limbs and medium-jointed fruiting branches; bolls, large; lint, long and silky but often weak; seeds, of medium size, fuzzy, gray; bolls per pound, 62; seeds per pound, 4,000; average lint length, 1⅜ inches; strength of fibers, 5 grams; lint percentage, 29.7. Adapted from Tyler.[2]

*Half and Half** (Round-boll, Short-staple Group).—Half and Half was originated by H. H. Summerour, of Duluth, Ga., about 1904. By individual plant selection and progeny-row testing, a strain was developed which has, according to figures obtained by Mr. Summerour, ginned 56.97 per cent lint. Cook was the parent strain. Cotton buyers object to Half and Half cotton on account of its short lint, but it has a very high lint percentage, which makes it popular with many growers. It is grown extensively in hill-land regions east of the Mississippi River and to some extent west of it.

* Varieties so designated were chosen as standard commercial varieties by the Committee on Standardization of Varieties.

Plants are medium-early, rather compact, leaves of medium size; bolls rounded, about 75 to a pound; lint length, ⅝ to ⅞ inch; lint percentage, 40 to 44.

Hartsville (Big-boll, Long-staple Group).—This variety was originated by D. R. Coker, of Hartsville, S. C., from a selection made from Jones's Big Boll. It was formerly grown to some extent in North Carolina, South Carolina, and the Delta regions of Arkansas and Mississippi. It is not prolific.

Fruiting limbs, rather long and long-jointed; the leaves and bolls are large; bolls per pound, 62; lint length, 1¼ to 1$^{5}/_{16}$ inches; lint percentage, 31.2 to 32. After Harper.[7]

Hawkins Improved (Early Small-boll, Short-staple Group).—This variety is said to have been developed by W. B. Hawkins, Nona, Ga., from a mixture of New Era, Peerless, Dickson, Herlong, and some other varieties. About 1907, it was grown over the Cotton Belt rather generally but is now grown but little.

Plants are fairly early in maturity, tall and pyramidal in shape, with one to three basal vegetative limbs; fruiting branches, numerous, short, and irregularly jointed; bolls, clustered to some extent, rather small to medium in size; lint, rather short; lint percentage, good; leaves, medium; seeds, small, fuzzy, light brownish gray; bolls per pound, 70; seeds per pound, 4,600; average lint length, $^{29}/_{32}$ inch; strength of single fibers, 5.3 grams; but percentage, 36.4. Adapted from Tyler.[2]

Jackson or African Limbless (Early Small-boll, Short-staple Group).—This variety was introduced in 1894 by T. W. Jackson, of Atlanta, Ga. It was similar to Dickson and Welborn's Pet but grew taller, and the leaves were somewhat larger. Like other cluster cottons, Jackson was very prolific on rich soils, where long-limbed varieties were too "weedy" in growth. It is not grown at present.

Plants are tall and slender, with one to three basal vegetative limbs; fruiting branches reduced to spurs from 1 to 6 inches long; leaves, very large; bolls, crowded together on the shortened branches, four- to five-locked, rounded in shape; lint, of medium length; seeds, medium in size, fuzzy, brownish gray; cotton very hard to pick free from trash; bolls per pound, 98; seeds per pound, 4,530; average lint length, ⅞ inch; strength of single fibers, 5.2 grams; lint percentage, 34.5. Adapted from Tyler.[2]

*Kasch** (Big-boll, Medium-staple Group).—Selected by Ed Kasch, of San Marcos, Tex., in 1912, from Mebane Triumph. The variety is noted for its large bolls, relatively high percentage of lint, good staple, and good production, comparatively, under Texas conditions.

Plants are medium to large and rather stocky; normally have 2 to 4 vegetative branches and 8 to 15 fruiting branches; leaves, medium to large; bolls large, 45 to 60 per pound; staple length $^{15}/_{16}$ to 1 inch; lint percentage, 38 to 40; seed, medium-sized, gray to white; disease resistance, earliness, and productiveness, medium.

* Varieties so designated were chosen as standard commercial varieties by the Committee on Standardization of Varieties.

Kekchi (Big-boll, Long-staple Group).—This variety was imported from Guatemala by the U. S. Department of Agriculture with the idea that it had boll-weevil resistance. It possesses some merit as a variety, but it has been found to be but little if any more resistant to weevils than many other varieties. It has not been grown extensively.

Keno (Small-boll, Long-staple Group).—This is a so-called "quarter" cotton which was originated many years ago by a negro, Mand Adkin, who was then living at Omega, La. Adkin developed it by a 3 years' selection of the best plants in a 50-acre field. Atkins, Mand Adkin, Eureka, Colthorp, and Dalkeith are other names for this variety. It is grown very little, if at all, at the present time.

Plants are tall and slender, pyramidal, open in growth; basal vegetative limbs, none to three, coming 6 to 8 inches from the ground; fruiting branches, long and slender and fairly short-jointed; bolls, rather small and pointed; lint, soft, fine, and silky; seeds, rather small, fuzzy, gray in color, a small percentage smooth and black; bolls per pound, 92; seeds per pound, 4,220; average lint length, $1\frac{5}{32}$ inches; strength of single fibers, 5.5 grams; lint percentage, 28.3. Adapted from Tyler.[2]

King (Early Small-boll, Short-staple Group).—In 1890, T. J. King, formerly of Louisburg, N. C., later of Richmond, Va., discovered a very prolific stalk of cotton in his field of Sugar Loaf, and from this stalk he developed what he called King, or King's Improved. Some years afterward he sent seed of the Sugar Loaf and of his new improvement to several experiment stations, and the reports convinced him that his new strain was practically identical with the parent variety. The variety is sold under the names King's Improved, King's Early, Mascot, Greer, Spotted Bloom, Ninety Day, and Little Texas. Simpkins and Simpkins' Ideal are very similiar if not identical. Adapted from Tyler[2] and others.

Lankart (Big-boll, Medium-staple Group).—Lankart was produced from a plant selection made in a field of Lone Star by C. S. Lankart, of Waco, Tex., in 1911. Plants are of the Mebane type; bolls, very large and storm resistant, 40 to 50 per pound; gin outturn 38 to 40 per cent; staple length, $\frac{31}{32}$ to $1\frac{1}{32}$ inches.

Lewis Prize (Intermediate Group).—This variety was developed by W. B. F. Lewis, Lewistown, La. It is not grown extensively now.

Plants are lacking in uniformity, some closely semiclustered, other more open and long-branched; bolls, of medium size, rounded; seeds, fuzzy brown; lint, of medium length and high in lint percentage; bolls per pound, 81; seeds per pound, 4,880; average lint length, $1\frac{5}{16}$ inch; strength of single fibers, 6.7 grams; lint percentage, 38.3. Adapted from Tyler.[2]

*Lone Star** (Big-boll, Medium-staple Group).—This variety was developed by D. A. Saunders from a single superior plant found in a field of Jackson cotton in the Colorado river bottom, near Smithville, Tex., in 1905. It is grown at present in Oklahoma and Texas.

* Varieties so designated were chosen as standard commercial varieties by the Committee on Standardization of Varieties.

Plants are of medium height with one to four basal vegetative limbs; main stem, short-jointed; fruiting branches, numerous, long, medium short-jointed; leaves, medium to large, very dark green; bolls, very large, round, or broadly ovate, with very short, blunt points; bolls per pound, 35 to 45; fiber, very strong; lint length, 1 to 1⅛ inches; lint percentage, 38 to 40.

The foregoing description is from *Cotton Seed Distribution Circular* issued by the Bureau of Plant Industry, U. S. Department of Agriculture, in 1919. Descriptions by others give the variety a boll size of 48 to 58 bolls per pound and a lint percentage of 35 to 38.

Meade (Big-boll, Long-staple Group).—This variety originated from a single plant selection made in 1912 by Rowland M. Meade, of the U. S. Department of Agriculture, from a field of Blackseed, or Black Rattler, at Clarksville, Tex.

Plants are erect, of average height, with regular internodes of medium length on both the main stalk and the vegetative branches; internodes of the fruiting branches, rather long, with little tendency to take the shortened cluster form; leaves, of medium size and rather thin texture; bolls, of medium size, with a thin bur, opening readily under humid conditions; seeds, large, about 3,000 to the pound, nearly naked after the lint has been removed, brownish black, slightly tufted at the end; lint length, 1½ to 1$^{11}/_{16}$ inches, uniform, with good luster, scarcely distinguishable from Sea Island lint when properly ginned; lint percentage, 26.

*Mebane** (Mebane Triumph). (Big-boll, Medium-staple Group).—Developed by A. D. Mebane, of Lockhart, Tex., from Boykin Stormproof, in 1889. The special merits of Mebane cotton are its large, stormproof boll, high lint percentage, and excellent staple. It is grown extensively west of the Mississippi River, especially in Oklahoma and Texas.

Plants of the variety are medium-sized, sturdy in growth, with two to four vegetative branches; leaves are medium large to large; bolls, 50 to 65 per pound; staple length, $^{15}/_{16}$ to 1$^{1}/_{16}$ inches; lint percentage, 37 to 40; seed, medium-sized, gray to white; disease resistance, fair; earliness, medium; productiveness, good under Texas conditions but rather poor, comparatively, in more humid regions.

*Mexican Big Boll** (Big-boll, Medium-staple Group).—The strain of cotton now recognized as Mexican Big Boll was obtained from an individual plant selection made by the cotton breeders of the North Carolina Experiment Station in a field of Hope's Mexican Big Boll, near Rocky Mount, N. C. in 1917. The variety has been reselected eight or ten times since that date, and increase plots have been grown in isolation.

The plants are medium-large, hairy, rather open, and have fairly heavy foliage. Bolls are large, 60 to 65 per pound; staple length, 1 to 1$^{1}/_{16}$ inches; lint percentage, 34 to 37; seeds, medium to large, gray; disease resistance, fair; earliness, medium; productiveness, good.

*Missdel** (Big-boll, Long-staple Group).—This variety originated from a plant selection made by H. B. Brown, of the Mississippi Experiment Sta-

* Varieties so designated were chosen as standard commercial varieties by the Committee on Standardization of Varieties.

tion, in a field of Foster-120 grown near Stoneville, Miss., in 1916. It was selected in the same field as Delfos and was known as Delfos-631 for some years. It was named Missdel in 1926, because it represents a distinct varietal type. The strain when first grown was lacking in uniformity, but it has been improved and made more uniform through further selection by the cotton breeders of the Mississippi Delta Experiment Station. Missdel has been characterized by good production, excellent staple, good boll size, and storm resistance.

Plants are medium-sized, attaining a height of 3 to 5 feet, with a rather prominent main stem and numerous medium-length fruit branches; leaves, medium-sized; bolls, medium large, 60 to 75 per pound; staple length, 1 5/32 to 1 3/16 inches; seed, medium large, white to gray; plants are susceptible to wilt but somewhat more resistant than Delfos; early and productive.

*Station Miller** (Big-boll, Medium-staple Group).—This variety originated from a plant selection made by J. F. O'Kelly, in 1926, in a field of Miller cotton, grown near State College, Miss. (The Miller cotton was very similar to Rowden if not identical.) The main merits of the variety consist in its prolificness for a large-boll cotton and in its wilt resistance. It is earlier than the parent variety, more prolific, and has a higher lint percentage.

The plants are medium-sized, with one to three vegetative branches; foliage is medium to heavy; bolls, medium sized, 60 to 65 per pound; staple length, 1 inch; lint percentage, 32 to 36; seed, medium to large, white to gray; disease resistance, good, especially wilt resistance; earliness, medium; produces well under favorable conditions.

Mortgage Lifter (Big-boll, Medium-staple Group).—This is a trade name for the variety Wyche (see description under that variety). Adapted from Tyler.[2]

Nankeen.—This is an old cotton almost extinct except in a few places where home weaving is carried on. It is hardly worthy of being called a variety since it was probably obtained by preserving the seeds of the yellow reversions or possible mutations, which sometimes occur in cotton of several different varieties. Except in color of lint, Nankeen resembles the common cotton grown in the same region. Adapted from Tyler.[2]

Okra Leaf (Early, Small-boll, Short-staple Group).—Okra-leaf plants have appeared in some 25 different varieties of cotton. Some of these were probably the product of crossing with okra-leaf varieties, but others were doubtless the result of mutations, for the plants have been known to appear in pure cotton strains grown continuously on a farm far removed from any other okra-leaf cotton. The Texas Experiment Station has produced an okra-leaf form from a broad-leaf variety by the use of X rays. There is record of an okra-leaf strain being grown as early as 1837.

The characters of an okra-leaf strain are, in the main, like those of the parent strain from which it sprang, but the leaves are of similar type in all cases; they are deeply dissected with slender linear to lanceolate lobes, thus to some extent resembling the leaves of an okra plant.

* Varieties so designated were chosen as standard commercial varieties by the Committee on Standardization of Varieties.

Parker (Early Small-boll, Short-staple Group).—This was a Mississippi variety grown by John M. Parker, at Maxime, Miss., and distributed to some extent. It later became mixed with Black Rattler and other varieties. It is probably now extinct.

Plants are rather tall and slender, with one to three long basal limbs and slender, fairly short-jointed fruiting limbs; leaves, of medium size; bolls, small, three-, four-, and five-locked; seeds, medium in size, fuzzy, gray, or greenish gray; bolls per pound, 96; seeds per pound, 3,800; average lint length, 1 inch; strength of single fibers, 5.2 grams; lint percentage, 30. Adapted from Tyler.[2]

Peeler (Small-boll, Long-staple Group).—This is an old variety said to have originated in Warren County, Mississippi, in 1864, and formerly grown extensively but now nearly extinct.

Plants are of medium height, two to three basal limbs, sometimes none, coming out 5 to 6 inches above the ground, making the plants somewhat long-shanked; fruiting branches, slender; joints, of medium length; bolls, small, three-, four-, and five-locked; seeds, medium in size, covered by sparse fuzz; lint, long, fine, and silky; bolls per pound, 121; seeds per pound, 3,850; average lint length, 1⅜ inches; strength of single fibers, 4.1 grams; lint percentage, 26.5. Adapted from Tyler.[2]

Peerless (Medium-late, Small-boll, Short-staple Group).—The origin of this variety is unknown. It was a popular and standard variety for several years but was allowed to deteriorate and is not grown now.

Plants are 3 to 4 feet high, pyramidal in shape; one to three basal limbs; fruiting branches, short and irregularly jointed; bolls, small to medium in size; seeds, rather small, fuzzy, greenish or brownish gray; bolls per pound; 69; seed per pound, 4,550; average lint length, ⅞ inch; strength of single fibers, 5 grams; lint percentage, 31. Adapted from Tyler.[2]

Peterkin (Medium-late, Small-boll, Short-staple Group).—This variety was developed by J. A. Peterkin, Fort Motte, Orangeburg County, S. C., about 1870. The original strain is grown but little at present, but Dixie Triumph and some other similar varieties are cultivated.

Plants are slender in growth, one to three basal limbs; fruiting branches, long, slender, with almost no tendency toward the semicluster habit; plants, somewhat late in maturity; bolls, medium to small, 70 per cent five-locked, opening widely, but cotton retained fairly well during storms; seeds, small, covered with a short, brownish-gray fuzz, a small percentage smooth and black; bolls per pound, 82½; seeds per pound, 5,300; average lint length, ⅞ inch; average strength of fibers, 5.8 grams; lint percentage, 39.6.

Texas Oak and Texas Wood are synonyms of Peterkin. Adapted from Tyler.[2]

Petit Gulf (Intermediate Group).—This variety was developed in 1840 by Col. H. W. Vick, of Vicksburg, Miss. By 1846, it had become very popular, and large quantities of seed were shipped from Petit Gulf, a small shipping point on the Mississippi River below Vicksburg.

The plant is late in maturing. It is large and straggling, with three or more vegetative limbs; fruiting branches, slender and long-jointed; leaves, medium in size, bolls, rather small; seeds, of medium size, mostly fuzzy, brownish gray; bolls per pound, 70 to 80; seeds per pound, 4,200; lint length, ⅞ to 1⅛ inches; lint percentage, 30 to 32. Adapted from Tyler.[2]

Pulnott (Intermediate Group).—This variety was originated by William Pulnott, formerly of High Shoals, Oconee County, Ga. At one time it was grown extensively in Georgia but has been recently largely superseded by Cook.

Plants are stocky and compact in growth, with one to three basal limbs; fruiting branches, rather short and irregularly jointed, leaves of medium size; bolls large, 66 per cent five-locked; seeds, medium in size, fuzzy, brownish or greenish gray; bolls per pound, 58; seeds per pound, 3,810; average lint length, 1 5/16 inch; strength of single fibers, 5.9 grams; lint percentage, 35.1. Adapted from Tyler.[2]

Qualla (Big-boll, Medium-staple Group).—This is a strain of Mebane developed by H. Conrads of San Marcos, Tex., about 1922. Staple length, 15/16 to 1 inch; lint percentage, 38 to 40; bolls, 60 to 70 per pound. Widely distributed in Texas.

*Rowden** (Big-boll Medium-staple Group).—This variety was developed from Bohemian cotton by Rowden Brothers, at Wills Point, Van Zandt County, Texas, about 1890. It is medium early in maturity. It is still a popular variety in parts of Texas and Oklahoma.

Plants are vigorous and stocky in growth; one to three basal limbs, fruiting branches vary in length from 2 feet at the base to 6 inches on upper part of plant; joints, regular and of medium length; bolls, large and hang downward when ripe, giving the cotton some protection from the weather; seeds, large, fuzzy, grayish white; bolls per pound, 50 to 60; seeds per pound. 3,360; average lint length, 15/16 inch; strength of single fibers, 6.3 grams; lint percentage, 35.4. Adapted from Tyler.[2]

Arkansas Rowden-40* (Big-boll, Medium-staple Group).—This variety originated from a single plant selection made by J. O. Ware, Fayetteville Ark., in a field of Rowden Brothers' Rowden, grown near Scott, Ark., in 1921. The variety was later turned over to R. L. Dortch for multiplication and seed sale. It has much vigor and yields better than many other varieties when grown under poor conditions.

Plants are rather large, with few basal limbs; leaves, medium-sized; bolls, 55 to 70 per pound; staple length, 1 to 1 1/16 inches; lint percentage, 32 to 36; seed, medium-sized, 14 to 16 grams per 100, whitish gray; disease resistance, fair to good; earliness, medium; productiveness, good.

Rucker (Early, Small-boll, Short-staple Group).—Developed about 1912 by Rucker Bros., of Alpharetta, Ga. It is thought to be a cross between King and Cook. The variety was formerly grown rather widely in Georgia and Alabama

Rublee (Early Small-boll, Short-staple Group).—This cotton was developed by C. A. Rublee, Seagoville, Tex. It is not grown extensively at present.

Plants are semicluster in habit of growth, imperfectly defoliate, many retaining their leaves and putting on squares late in the fall; bolls, medium

* Varieties so designated were chosen as standard commercial varieties by the Committee on Standardization of Varieties.

to small in size; seeds, medium in size, fuzzy, light greenish or brownish gray; bolls per pound, 70; seeds per pound, 3,600; average lint length, $\frac{7}{8}$ inch; strength of single fibers, 6.3 grams; lint percentage, 33. Adapted from Tyler.[2]

Russell (Big-boll, Medium-staple Group).—Known also as Big Boll Green Seed and Ozier Big Boll. Originated in 1895 by J. T. Russell from a single stalk found in a field of mixed Truitt at Alexander City, Ala. This variety was widely grown throughout the Cotton Belt in 1907 but more recently has been supplanted by other varieties.

Plants are large-growing, vigorous, with one to three basal limbs; fruiting branches, long; joints, of medium length; leaves, large; bolls, large, four- to five-locked; fairly stormproof; burs, very thick; seeds, large, covered by a dark-green fuzz; bolls per pound, 56; seeds per pound, 3,100; lint length, $\frac{31}{32}$ inch; strength of single fibers, 5.5 grams; lint percentage, 30.9. Adapted from Tyler.[2]

Salsbury (Small-boll, Long-staple Group).—This variety was developed by E. C. Ewing, at Scott, Miss., in 1916, from a cross of Wannamaker Cleveland and Express-122. It is grown but very little at present.

Plants are erect, of vigorous growth; foliage, intermediate between its parents Wannamaker Cleveland and Express-350; somewhat resistant to wilt; of rather rank growth on rich land; fruiting limbs, long and long-jointed; bolls per pound, 75 to 80; lint percentage, 31 to 31.5; lint length, $1\frac{3}{32}$ to $1\frac{1}{8}$ inches; strength of fiber good.

Southern Hope (Small-boll, Long-staple Group).—This is an old variety developed by Col. F. Robiew, of Louisiana, but it is not grown now.

Plants are tall, slender in growth, with one to three basal limbs; fruiting limbs, quite long and slender; leaves, medium to small; bolls, medium in size, opening well; seeds, fuzzy, light greenish gray; bolls per pound 73; seeds per pound, 4,160; average lint length, $1\frac{7}{32}$ inches; lint percentage, 31.2. Adapted from Tyler.[2]

Spruiell's Green Seed (Early Small-boll, Short-staple Group).—The variety was developed by A. M. Spruiell, Brompton, Ala., and is not grown extensively now.

Plants are not uniform, some as early as King, others later in maturity; bolls, medium in size; seeds, small; bolls per pound, 72; seeds per pound, 4,220; average lint length, $\frac{15}{16}$ inch; lint percentage, 32.8. Adapted from Tyler.[2]

*Stoneville** (Intermediate Group).—Stoneville originated from an individual plant selection made by H. B. Brown in a field of Lone Star-65, grown by the Stoneville Pedigreed Seed Co., near Stoneville, Miss., in 1923. The main features of merit of the variety are its earliness, prolificness, and uniformity of staple of fair length.

Plants of the variety are medium-sized, somewhat spreading, with two or more vegetative branches if not closely spaced; foliage is rather light; bolls, medium-sized, 70 to 80 per pound; staple length, 1 to $1\frac{1}{16}$ inches; lint per-

* Varieties so designated were chosen as standard commercial varieties by the Committee on Standardization of Varieties.

centage, 33 to 36; seeds, medium-small, gray; disease resistance, poor; very early and productive.

Sugar Loaf (Early Small-boll, Short-staple Group).—This is an old variety which has been grown in North Carolina for many years. Selections from Sugar Loaf, known commercially as King, or King's Improved, and Simpkins are grown to some extent at present in the southern and eastern sections of the Cotton Belt. The variety and its derivatives are characterized by blooms with red or purple spots at the base of the petals.

Plants are slender, one to three basal limbs; fruiting branches, slender and short-jointed but with little or no tendency to semicluster; leaves, medium to small in size; bolls, small, three-, four-, and five-locked, the majority, four-locked; seeds, small, covered with a short fuzz, brownish gray in color; bolls per pound, 93; seeds per pound, 5,600; average lint length, $^{29}/_{32}$ inch; lint percentage, 35. Adapted from Tyler.[2]

Sunflower (Small-boll, Long-staple Group).—Sunflower was developed by Marx Schaefer of Yazoo City, Miss., by means of mass selection. The original seed were bought at an oil mill. It is barely distinguishable from pure Floradora and some of the forms of Allen. Sunflower was a standard variety for a number of years, but since boll weevils have become widely distributed its use has been largely discontinued.

Plants are tall and pyramidal, with slight tendency toward the semicluster type; one to three basal limbs; fruiting branches, slender and irregularly jointed; leaves, medium in size; bolls, small, five-locked; lint, very fine, long, and silky; seeds, medium in size, fuzzy, gray, or light greenish gray; bolls per pound 90; seeds per pound, 4,320, lint length, 1⅜ inches; strength of single fibers, 4.9 grams; lint percentage, 25. Adapted from Tyler.[2]

Super Seven (Intermediate Group).—Developed by the Coker Pedigreed Seed Co., of Hartsville, S. C., in 1925, from a chance hybrid of Webber-49 and Dixie. It is wilt resistant, early, and prolific; bolls, medium small; staple length, $1\frac{1}{16}$ to 1⅛ inches. Not extensively grown at present.

Sunshine (Intermediate Group).—Sunshine is a strain of Rowden introduced by the J. W. Davidson Co., of McKinney, Tex., about 1918. It is earlier than the parent variety; staple length, $^{15}/_{16}$ to $1\frac{1}{32}$ inches; lint percentage, 34 to 37; bolls, 65 to 70 per pound.

Texas Bur (Big-boll, Medium-staple Group).—C. E. Smith, Locust Grove, Ga., introduced this cotton, which was probably a strain of the old Texas Stormproof. It is grown to a very limited extent at present.

Plants are stocky in growth, usually two basal limbs, rather heavy; fruiting branches with joints of medium length, leaves large; bolls, large, four- and five-locked; seeds, rather large, fuzzy, gray or brownish gray; bolls per pound, 67; seeds per pound, 3,680; average lint length, $^{29}/_{32}$ inch; strength of single fibers, 7 grams; lint percentage, 37.1. Adapted from Tyler.[2]

*Toole** (Medium-late, Small-boll, Short-staple Group).—This is a standard strain of Peterkin developed by W. W. Toole, Augusta, Ga. It is wilt-resistant and is grown in Georgia and other districts where cotton wilt is

* Varieties so designated were chosen as standard commercial varieties by the Committee on Standardization of Varieties.

troublesome. Plants are similar to Peterkin but with a slight tendency to semicluster.

Bolls are larger than those of Peterkin, 50 per cent five-locked; seeds, small, fuzzy, light brownish gray; bolls per pound, 73; seeds per pound, 5,110; average lint length, 1⁵⁄₁₆ inch; strength of single fibers, 6.7 grams; lint percentage, 37.5. Adapted from Tyler.[2]

Covington-Toole (Medium-late, Small-boll, Short-staple Group).—This strain is a selection from Toole, made by W. F. Covington, Headland, Ala. It shows a decided resistance to wilt and is planted rather extensively on wilty soils in the eastern states.

*Trice** (Early Small-boll, Short-staple Group).—This variety was developed by Prof. S. M. Bain from a single stalk found on the farm of Luke Trice, near Henderson, Chester County, Tenn., in 1907. It is considered one of the earliest maturing varieties grown today. It was formerly grown extensively in northern parts of the Cotton Belt, especially in Tennessee, Kentucky, and Missouri.

Plants are 2 to 5 feet high, frequently with few basal limbs, very prolific; fruiting branches, numerous, medium in length, and short-jointed; leaves, of medium size, light green; bolls, medium to large, ovate, often angular, three- to five-locked; seeds, medium to small, covered with whitish or brownish tuft; bolls per pound, 70 to 80; seeds per pound, 3,600; character of fiber, fair; staple length, ⅞ to 1 inch; lint percentage, 31 to 33. Adapted from Oakley.[6]

Mississippi Station Trice.—This strain differs from the parent in having slightly more compact plants, smaller bolls, 75 to 85 per pound, and longer staple, 1 to 1³⁄₃₂ inches. This strain is very early, a good yielder, and adapted to rich lands that are free of wilt. It was selected by the Mississippi Experiment Station.

Tidewater.—This name is applied to certain Acala selections made by cotton breeders of the U. S. Department of Agriculture, on James Island, S. C., about 1927. The variety has not been distributed extensively.

Triumph-406.—(Big-boll, Medium-staple Group).—Developed by A. M. Ferguson, of Howe, Tex., from Mebane Triumph, about 1908. Plants are rather early, small, low-branching; staple length, ⅞ to 1 inch; lint percentage, 36 to 39; bolls, medium to large, storm resistant. Grown principally in Texas.

Wacona (Big-boll, Medium-staple Group).—This variety was developed from Lankart by C. S. Lankart, at Waco, Tex., the first selection being made in 1921. Plants resemble Lankart but are somewhat earlier and more prolific; bolls, about 60 per pound; lint percentage, 33 to 35; staple length, 1¹⁄₃₂ to 1¹⁄₁₆ inches or longer.

Washington.† Registration No. 34. (Intermediate Group).—This variety (formerly known as Delfos-719) was bred by C. A. Tate, of the Stone-

* Varieties so designated were chosen as standard commercial varieties by the Committee on Standardization of Varieties.

† Varieties so designated have been registered by the American Society of Agronomy as new varieties of merit.

ville Pedigreed Seed Co., Stoneville, Miss., the original selection being made in 1927, at Stoneville, from Delfos 6102–324. Grown in Mississippi, Tennessee, Arkansas, and Texas.

On soil of average fertility, plants attain a height of 2 to 3 feet, being lower and more stocky than other strains of Delfos. Branches are of medium length and rather short-jointed; vegetative branches, few; leaves, slightly larger than the average for Delfos; bolls, 60 to 75 per pound, thick and short taper-pointed; staple length, 1$\frac{1}{16}$ to 1$\frac{3}{32}$ inches; lint percentage, 33 to 36; seeds, medium-sized, brownish gray; susceptible to wilt.

Webber (Big-boll, Long-staple Group).—Originated by D. R. Coker, of Hartsville, S. C., from a selection from Columbia. It is not grown at present.

Plants were vigorous, growing somewhat rank in rich soils; fruiting limbs, of medium length, long-jointed and drooping; leaves, very large and rather dark green in color; bolls, medium large; bolls per pound, 60; strength of fiber, good; lint length, 1$\frac{3}{16}$ to 1$\frac{1}{4}$ inches; lint percentage, 30 to 31.5. After Harper.[7]

Webber-49 (Big-boll, Long-staple Group).—This is a strain of Webber developed by the Pedigreed Seed Company at Hartsville, S. C. It differs from the parent strain in being lower and more stocky in growth, in being earlier and more prolific, and in having somewhat smaller bolls, 64 to 68 per pound; character of fiber, good; length of staple, 1$\frac{1}{4}$ to 1$\frac{5}{16}$ inches.

*Deltatype Webber** (Big-boll, Long-staple Group).—This variety was developed from a plant selected in a field of Webber 82, neart Hartsville, S. C., by the cotton breeders of the Coker Pedigreed Seed Co. Plants are vigorous and have large bolls for a long-staple cotton, and the staple has excellent character.

Plants are erect, rather stocky, 3 to 4 feet tall; fruit branches, short and numerous; foliage, rather heavy; bolls, 60 to 65 per pound; staple length, 1$\frac{3}{16}$ to 1$\frac{3}{8}$ inches; lint percentage, 31 to 33; seed, 3,750 per pound, light gray to brown; susceptible to wilt; medium late, but productiveness good for a variety of its staple length.

Welborn's Pet (Early Small-boll, Short-staple Group).—Originated by Jeff Welborn, of New Boston, Tex., in 1881. It is not grown at present.

Plants are tall, one to three basal limbs; fruiting branches, reduced to short spurs; leaves, large; bolls, rounded, four- to five-locked; seeds, medium in size, fuzzy, or brownish gray, a few nearly smooth, dark brown; bolls per pound, 70; seeds per pounds, 3,860; lint length, $\frac{7}{8}$ inch; lint percentage, 33.4. Adapted from Tyler.[2]

*Wilds** (Big-boll, Long-staple Group).—Wilds was produced by crossing Lightning Express and Deltatype Webber. The cross was made in 1919 at Hartsville, S. C., by the Coker Pedigreed Seed Co. The special merits of the variety are its productiveness and excellent long staple.

Plants are medium-sized and somewhat stocky, like the Deltatype parent; vegetative branches, one to four; fruit branches, rather short and numerous;

* Varieties so designated were chosen as standard commercial varieties by the Committee on Standardization of Varieties.

foliage, heavy; bolls, 60 to 75 per pound; staple length, 1$\frac{3}{16}$ to 1$\frac{3}{8}$ inches; lint percentage, 31 to 34; seed, large, gray to light brown; not wilt-resistant; earliness, medium; productiveness, good for an extra-staple variety.

Wyche or Jones Improved' (Big-boll Medium-staple Group).—This is one of the oldest big-boll cottons grown in America. It is said to have been introduced from Algiers by the Wyche brothers about 1857. After the death of Mr. Wyche, the variety was propagated, and seed were distributed by J. F. Jones; hence it is sometimes called Jones Improved. Truitt, Russell, Columbia, and Webber trace their ancestry to Wyche.

Plants are vigorous and prolific; wide-spreading branches near the base, upper branches usually short; bolls, large, ovate, blunt-pointed, five-locked, open well; seeds, medium-large, weighing 0.13 to 0.14 gram, covered with a grayish fuzz and well covered with lint; lint strong, 1 to 1$\frac{1}{8}$ inches; lint percentage, 31 to 32; season, medium late. Adapted from Webber.[3]

Egyptian Varieties in America.—Several Egyptian varieties have been grown experimentally in America, but none has proved satisfactory except the selections from Afifi listed below.

Yuma.—This is a variety of Egyptian cotton developed by the U. S. Department of Agriculture, in Arizona, in 1908. It arose as a sport or mutant from Afifi. It was grown in Arizona rather extensively for a time but has been displaced more recently by Pima. It differs from Afifi variety in having a longer length of lint, averaging about 1$\frac{1}{2}$ inches; it has a pale pinkish-buff color as opposed to the brown color of the Afifi cotton; also, the bolls are larger than the bolls of the Afifi variety. Adapted from U. S. Department of Agriculture *Bulletin* 38, 1913.

*Pima.**—This variety originated from a single stalk selection made in a field of Yuma cotton by U. S. Department of Agriculture cotton breeders, at Sacaton, Ariz., in 1910. It has finer, lighter colored, and longer staple than that of Yuma; staple length, 1$\frac{9}{16}$ inches. The entire acreage of American-Egyptian cotton in Salt River Valley, Arizona, was changed from Yuma to Pima in 1918. In 1920, over 92,000 bales was produced. Since that date, upland varieties have taken the place of some of the Egyptian. In 1935, there were but 17,600 bales of Egyptian.

S × P – 30.—This is the designation applied to a new hybrid variety produced by the U. S. Department of Agriculture cotton breeders by crossing Pima and Sakel (Sakellaridis), an Egyptian variety grown extensively in Egypt. The new variety seems to be superior to Pima both in quality of staple and in production and is replacing it.

Sea Island Varieties.—Varieties of Sea Island cotton are not so permanent nor so well known as upland varieties. To keep the strains up to a high standard it was necessary to do much breeding work. In many instances, the plantation selected and

* Varieties so designated were chosen as standard commercial varieties by the Committee on Standardization of Varieties.

continued to breed each year its own particular strain. In case seed were sold, which was not common, the variety designation given them was only the name of the breeder or grower. Inland growers usually bought seed each year and frequently from different parties. The plantation breeder grew from a new selection

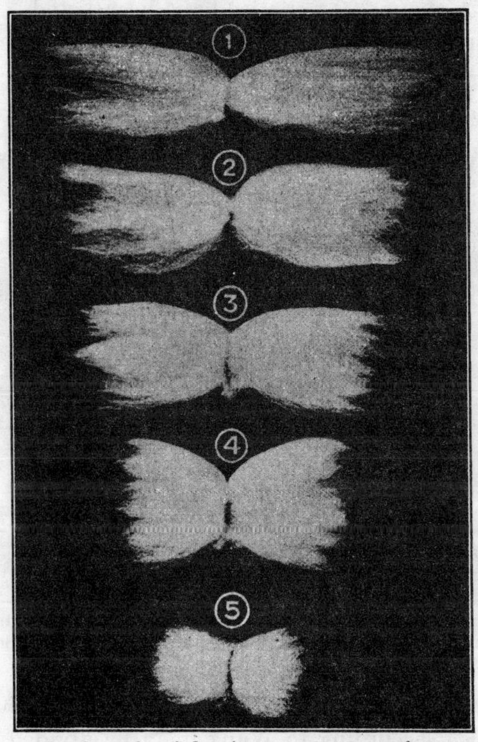

Fig. 14.—Length of staple of five important types of cotton, lint combed. (1) Sea Island; (2) Egyptian; (3) upland long staple; (4) upland short staple; (5) Asiatic type. (*After Meloy.*)

each year and frequently changed the type to some extent. Hinson, Seabrook, Rivers, and La Roche are names of prominent breeders mentioned by Orton.[4] These names have been used in the sense of variety names.

Foreign Varieties.—Each cotton-producing country has its own varieties; some countries have many. The following lists contain the more important varieties, with some brief data.

Varieties Grown in India.—The Report of the Indian Central Cotton Committee for 1936 lists some 40 varieties and gives brief data on each. Those listed in Table II are the ones grown most extensively.

Varieties Grown in China.—There are many native varieties of cotton grown in China and several imported ones. S. C. Wang, professor of Cotton Industries in the University of Nantung, estimates that about 200 varieties have been listed. He thinks that the majority of these are synonyms and that there are but

FIG. 15.—Various cotton varieties in breeding plot. (*a*) Upland, okra-leaf; (*b*) upland, Nankeen; (*c*) upland hybrid, Triumph × Express; (*d*) Sea Island; (*e*) Pima Egyptian.

a comparatively small number of true or distinct varieties. The varieties grown, according to Wang, may be divided into three groups: (1) the American group, which includes Trice, Stoneville, Lone Star, Cleveland Big Boll, Bank Account, King, Acala, and Columbia. Sea Island and Egyptians are grown experimentally but are of little economic value; (2) the Indian group, which includes the so-called "Chicken-foot" cotton of China. *Gossypium arboreum* Linn. var. *neglecta* Watt. is predominant, but var. *rosea* and var. *assamica* are also probably represented. These cottons were introduced long ago and have become thoroughly acclimated. The staple, in quality and length, is similar to that of American short-staple uplands, the length being about ⅞ inch. (3) The native group includes numerous varieties belonging to the species *G. nanking* Meyen. Most of these take the name of the locality in which they are grown. Nantungchow and Gar Gar are two varieties mentioned by Wang. The native varieties have a very coarse, short staple, averaging about 5/8 inch in length.

Other American varieties have been tried experimentally the last few years, and some have been introduced commercially.

TABLE II.—INDIAN VARIETIES

Varieties		Production, bales	Staple length, inches
Bengals	United Provinces	190,000	3/8 to 5/8
	Sind (Desi)	250,000	3/8 to 5/8
	Punjab (Desi)	725,000	About 5/8
	Mollisoni		
Broach	Broach	275,000	5/8 to 6/8
	Punjab-American 4 F	550,000	3/4 to 7/8
	Wagad	200,000	3/4 to 7/8
	Lalia		3/4 to 5/8
Oomras	Berar	600,000	1/2 to 6/8
	Central Provinces No. 1	200,000	5/8 to 6/8
	Central Provinces No. 2	150,000	5/8 to 6/8
	Central India	250,000	5/8 to 6/8
	Khandesh	350,000	4/8 to 5/8
	Mathia	250,000	5/8
Southerns	Westerns	200,000	5/8 to 1 3/16
	Western Farm (Hagari)		7/8 to 1 5/16
	Kumptas	160,000	7/8
	Jayawant		About 1
	Cambodia	210,000	7/8 to 1
	Tinnevellys	90,000	6/8 to 7/8
Total		5,505,000	

Ambassador (Stoneville-4) is the most prominent. Million Dollar, a recent selection from a native cotton, possesses some excellent qualities and is being multiplied and distributed extensively.

Egyptian Varieties.—Present-day Egyptian cottons are usually divided according to area of production into two general groups as "uppers" and "lowers" (according to whether grown in Upper or Lower Nile Valley). The Egyptian Ministry of Agriculture, in its official reports and estimates, mentions three groups:

long-staple, medium-staple, and short-staple. The long-staple group includes cottons with a staple length 1⅜ inches and longer; the medium-staple group, 1¼ to 1$\frac{5}{16}$; and the short-staple group, 1$\frac{1}{16}$ to 1$\frac{3}{16}$. It will be noticed that the length of the cotton in these groups is different from that of corresponding groups of American cotton. The short-staple group contains cottons nearly as long as the American long-staples.

EGYPTIAN VARIETIES AND AREA GROWN IN 1935[9]

Long staple		Medium staple		Short staple	
Varieties	Feddans*	Varieties	Feddans	Varieties	Feddans
Sakel	297,000	Nahda	5,000	Ashmouni } Zagora }	938,000
Maarad	82,000	Fouadi	32,000		
Giza-7	270,000	Giza-3	10,000	Others	5,000
Sakha-4	28,000				
Casouli	3,000				

* A feddan equals 1.038 acres.

African Varieties.—Cotton growing began in South Africa about 1850. A number of upland varieties have been tried—Allen, King, Cleveland, Toole, and Sunflower being among the number. Under the encouragement of the British Cotton Growing Association, some cotton is grown in Sudan, Uganda, Nyasaland, Nigeria, and other parts of Africa. Egyptian, Sea Island, Brazilian, and American upland varieties have been tried. Derivitives of some of these are now being grown, the most prominent probably being the variety designated as "U4." This variety, which is of the upland type, was selected from Uganda, a mixed variety of uncertain origin, at Bamberton, Transvaal, in 1925. It is early, prolific, jassid-resistant, has a staple 1⅛ inches in length, and has shown wide adaptation to various African districts.

Peruvian Varieties.*—At present, the principal cotton variety grown in Peru is Tanguis. It was estimated that 92 per cent of the 1933 crop was of this variety. Tanguis was developed in 1918 by Fermin Tanguis, who was seeking a wilt-resistant

* Gerard S. Klinge, of the Department of Agriculture of Peru, kindly furnished information on Peruvian varieties.

variety. Its ancestry is uncertain. Some authorities think that it came from a natural cross of some imported variety with a native species. Pima, an American-Egyptian variety, ranked second in acreage in 1933. Other Egyptian varieties, Sakel and Afifi, and also the Peruvian varieties Full-rough and Semi-rough and Egipto, or Suave, are grown to a lesser extent. According to Klinge,* the Full-rough variety belongs to the species *Gossypium peruvianum* Cav. and is a native of Piura, a province in northern Peru, while Semi-rough belongs to the species *G. microcarpum* Todaro, being a native of the province of Ica in southern Peru. The Egipto, or Suave, is probably derived from an American upland variety.

According to data furnished by Acting Commercial Dunn of Lima, Peru, Tanguis has a staple length of 1⁵⁄₁₆ to 1½ inches; Full-rough, about 1¼ inches; Semi-rough, 1³⁄₁₆ inches; and Egipto, or Suave, 1¹⁄₁₆ to 1⅛ inches. Some other authorities give the staple length of Tanguis as 1³⁄₁₆ inches.

Varieties in Argentina.†—There are no native cotton varieties in Argentina, but several American upland varieties have been imported and are being grown. Acala, Durango, Express, Foster, and other varieties were imported a decade or more ago. These were allowed to become mixed and have been acclimated to some degree. This mixture is designated as "Chaco," which is the name of the district in which they are grown most extensively.

More recently, Deltapine (D. & P. L.-11), Farm Relief, Acala, and Stoneville, newer American varieties, have been imported and are being introduced.

Brazilian Varieties.—Brazil has many varieties, both native and foreign. Long- and short-staple varieties have been grown together promiscuously and allowed to mix. Consequently, many of the varieties are badly hybridized and are lacking in uniformity. The following data on the leading cotton varieties in Brazil were kindly furnished by O. F. Lamartine, of Natal, Rio Grande do Norte, Brazil.

As was mentioned in Chap. I, several of the better American upland varieties have been tried experimentally in Brazil, and

*Gerard S. Klinge, of the Department of Agriculture of Peru, kindly furnished information Peruvian varieties.

† Information in regard to varieties grown in Argentina was given by Jorge R. Lorenzo, who is in charge of cotton experiment stations in that country.

Principal Varieties of Cotton Grown in Brazil

Variety	Type	Length staple, milli-meters	Lint, per cent	Notes
Mocó[a] or Seridó	*G. peruvianum*, tree type	35–45	30–32	Variety limited to the Serido section of northeast Brazil and adjacent valleys. Belongs to tree-cotton group; perennial, 5 to 8 feet high; bolls small; flowers lemon-yellow color. Plants very resistant to drought. Known to yield 450 to 480 pounds lint cotton per acre.
Rigueza or Verde	*G. peruvianum*, tree type	32–35	26–28	Native variety with fuzzy green seeds; lint very silky and strong. Requires a richer soil than Mocó. Grown in the northeastern states and along the S. Francisco River Valley.
Quebradinho	*G. vitifolium*, tree type	25–35	28	Very similar to Mocó. Known by different names in different states. Mostly grown in the central states (Minas), and in the extreme northern section (Maranhão, etc.).
Rim or Kidney cotton	*G. braziliense*, tree type	30	18–20	Lint very strong and sought by the natives for home spinning. Used to a limited extent only. Seeds small, adhering in a kidney-shaped mass. Grown in S. Francisco Valley in the states of Minas and Bahia.
Russell Big Boll Sunbeam Up Right Paulo Sousa Carioba Nova Paulista and others	Herbaceous type, imported American uplands	23–30	33–34	Grown in Central Brazil in the states of São Paulo and Minas principally; also in the northeastern section along the seacoast; are reported to mature there in 90 days or less. In São Paulo on rich soils yields as high as 800 pounds lint cotton per acre have been reported. The varieties Nova Paulista, Paulo Sousa, and Carioba are hybrids, coming from crosses made in Sao Paulo. Carioba is a cross between Webber and Russell Big Boll.

[a] In the Seridó district there is grown also a variety or strain called Mecó-zinho (Little Mocó), which resembles the true Mocó but has very small black seed and a fiber considered stronger.

within the last few years some of the better adapted ones have been introduced commercially in the southern states.

References

1. TRACY, S. M.: "Cultivated Varieties of Cotton," in "The Cotton Plant," *U. S. Dept. Agr. Bull.* 33, 1896.
2. TYLER, FREDERICK J.: "Varieties of American Upland Cotton," *U. S. Dept. Agr. Bur. Plant Ind. Bull.* 163, 1910.
3. DUGGAR, J. F.: "Descriptions and Classifications of Varieties of American Upland Cotton," *Ala. Agr. Exp. Sta. Bull.* 140, 1907.
4. ORTON, W. A.: "Sea Island Cotton: Its Culture, Improvement and Diseases," *U. S. Dept. Agr. Farmers' Bull.* 302, 1907.
5. CHILDS, R. R.: "College No. 1 Cotton," *Ga. State Col. of Agr. Bull.* 226, 1921.
6. OAKLEY, R. A.: "Distribution of Cotton Seed in 1915," *U. S. Dept. Agr. Bur. Plant Ind. Cir.* 1163, 1915.
7. HARPER, J. N.: "Experiments with Varieties of Cotton," *S. C. Agr. Exp. Sta. Bull.* 162, 1912.
8. WEBBER, H. J.: "Distribution of Cotton Seed in 1903," *U S. Dept. Agr. Bur. Plant Ind. Seed Distribution Leaflet*, 1903.
9. TODD, JOHN A.: "Cotton Statistics," *Empire Cotton Growing Review*, vol. 12, No. 4, 1935.
10. WARE, J. O.: "Extent of Improved Varieties of Cotton in the United States." Paper presented at meeting of Southern Agricultural Workers Association, Nashville, Tenn., 1937.

CHAPTER IV

COTTON PLANTS: THEIR FORM, STRUCTURE, AND GENERAL CHARACTERISTICS

The study of plant form is basic. It is only through, or by means of, a study of form that differences and resemblances may be recognized, and classification made. The physiologist observes form and structure as an expression of physiological laws and effects. The plant breeder, in endeavoring to produce new and more desirable varieties, studies closely differences in form appearing among the plants with which he is working and bases his selections on striking points of dissimilarity observed.

The cotton plants growing in a single field show considerable variation, but a much greater range of form is met when a survey is made of plants in the various countries in which cotton is found growing. Within the tropics, the natural home of cotton, the plant is commonly a perennial, and some species attain a height of 15 tc 20 feet, being known as "tree cottons." In the American Cotton Belt, the plant has been forced to assume the annual habit on account of the frosts and freezes of winter. Here it is a soft-stemmed shrub, much branched, and varying in height from 2 to 6 feet.

Framework of a Mature Cotton Plant.—A mature plant has a prominent erect main stem which has a growing point, or plumule, at the apex. The growth of the main stem is therefore monopodial, but the growth of secondary branches and branches of higher orders is either monopodial or sympodial. The length of the stem is determined mainly by soil and water conditions, but there is, of course, a difference due to species or variety. Certain varieties, such as Trice, King, and Delfos, are somewhat determinate in their stem growth—there is a tendency for the stem to stop growing in length comparatively early in the season.

The general shape or outline of plants ranges from columnar to rounded. The shape is determined by length of branches. Short branches throughout the whole length of stem produce the columnar form; long branches at the base with successively shorter branches toward the top produce the conical or sugar-loaf form; and long branches high on the stem, the rounded or apple-tree form.

Large leaves arise from the stem in alternate order and are arranged about the stem in regular spirals. The common arrangement, or "phyllotaxy," as it is called, for American cottons is

Fig. 16.—A well-fruited cotton plant with unopened bolls. (*a*) Main stem; (*b*) basal vegetative branches; (*c*) fruit branches.

three-eighths of a turn about the stem between successive leaves: or, expressed in another way, from one leaf to the next directly above it there are eight leaves, and the spiral passes around the stem three times. This is designated as 3/8 phyllotaxy. The course of the spiral is either to the right or to the left. Other types of phyllotaxy found among cottons are 1/3, 2/5, and 5/13. Old World cottons commonly have the 1/3 arrangement. The leaves on the branches have the same order of phyllotaxy as those on the main stem, but the spiral may run in the opposite direction on a part of the branches.

Each leaf has two buds, or at least the rudiments of buds, in its axil—the angle between the leaf stalk and the stem to which it is attached. One of these buds occupies a central position and is a "true" axillary bud. The other has a position to one side of the axillary bud and is known as a "lateral," or "extra-axillary," bud. It may be found on either side of the central

FIG. 17.—A cotton plant showing framework, open bolls, and both vegetative and fruiting branches arising from the same node on the main stem. (*After Duggar.*)

bud. The axillary buds produce only vegetative branches, that is, branches that bear flowers and fruit only after again branching (Fig. 16*B*). Most of the lateral buds produce fruit branches, which bear flowers and fruit directly (Fig. 16*C*). Under certain conditions, the lateral buds may make vegetative branches. Usually not more than one to four of the axillary buds on the lower part of the plant grow into branches. The others remain dormant. Both of the buds in the leaf axil may make branches (Fig. 17). Rainy weather and rich land may cause a higher percentage of vegetative branches to develop and thus produce extra vegetative growth at the expense of the fruit branches. Cook and Meade[2] report that vegetative branches are much more numerous on plants grown from newly imported seed than on acclimatized strains. While an excessive development of vegetative branches is not desirable, two or three are not objectionable.

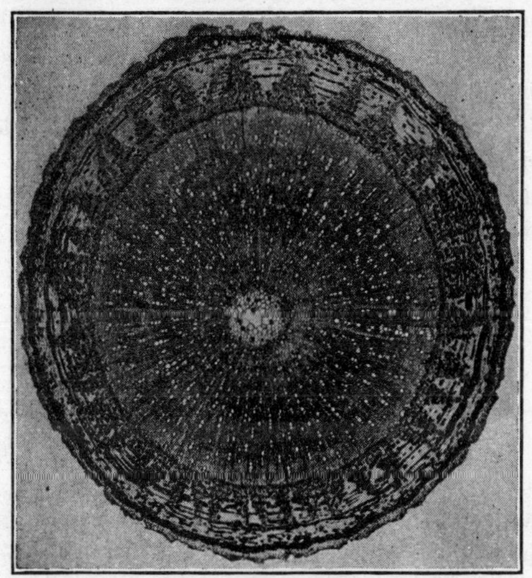

Fig. 18.—Enlarged cross section of stem of mature cotton plant. (*After Stanford and Viehoever.*)

Some of the most prolific varieties have from two to four vegetative branches to the plant.

The height and the spread of plants are determined largely by the length of internodes in stem and branches. It is generally thought that plants with short internodes are earlier than ones with long internodes, but the variety Express appears to be an exception to this rule. It is one of the earliest upland varieties; yet it is rather long-jointed.

Arrangement of Leaves and Flowers on Fruiting Branches.—The leaves of fruiting branches lack the spiral arrangement that is found on the main-stem and vegetative branches. The leaves appear to be somewhat two-ranked. The flower buds do not come from the leaf axils but appear to arise independently. It has been suggested by Cook and Meade[2] that fruit branches appear to be sympodial in their growth, that each flower terminates a joint, and that further growth is from a lateral bud. Leake,[3] in discussing the morphology of types of Indian cottons, says that the flowers are not concentrated in any well-defined inflorescence but are developed on the sympodial branches from terminal buds.

Structure of Stem.—Cotton stems have a moderately thick, tough bark in which bast fibers are prominent (Fig. 18). The bark has been used to a limited extent as a coarse fiber material, and it has been proposed as a possible source of fiber for cotton bagging for bales. The outer layer of the bark is more or less corky. It is of a yellowish-brown color on the older parts of stems and greenish to reddish on the younger.

The main portion of the stem consists of wood, or xylem, with prominent wood rays and water-carrying vessels, or ducts (Fig. 18). The wood is somewhat soft and rots quickly. Stalks turned under by a breaking plow rot and disappear within a few months. Cotton stalks may be macerated, and the material obtained used in making paper.

Leaves.—The leaves on various species of cotton vary greatly in size, shape, texture, and hairiness. Some are nearly smooth, or glabrous, whereas others, such as *Gossypium obtusifolium* Roxb., have decidedly hairy leaves. The leaf blade of most species is rather thin and papery, but some are thick and leathery. The size and the shape vary considerably on an individual plant; the range is much greater, however, on different species. Most varieties and species have leaves with five more or less well-defined lobes (see Fig. 1), but the actual shape ranges from a rounded entire leaf, *G. sturtii* of Australia, to the deeply cut leaves of the okra-leaved strain of the King variety of *G. hirsutum* (Fig. 15*a*). Many of the Asiatic varieties have deeply cut leaf blades, of which *G. arboreum* (Fig. 1*a*) is a good illustration. On the underside of cotton leaves are three to five prominent veins. Numerous branches from these ramify through the leaf. A small, shallow, cup-shaped nectary is usually found on the mid-part of the largest vein, and in some instances similar nectaries are found on two of the other larger veins.

Roots.—The main stem of a cotton plant is continuous with the main root, which is known as a "taproot." The transition takes place about the surface of the ground. The taproot has a proportionately thicker cortex, and the wood tends to arrange itself in a four-angle structure, with four more or less well-defined ridges showing on the outside. The branches that arise from the taproot are irregularly grouped in rows. The complete number of rows is four, but, because of twisting and irregularities, often not more than three can be recognized. Two of the rows of branches stand at right angles to the other two.

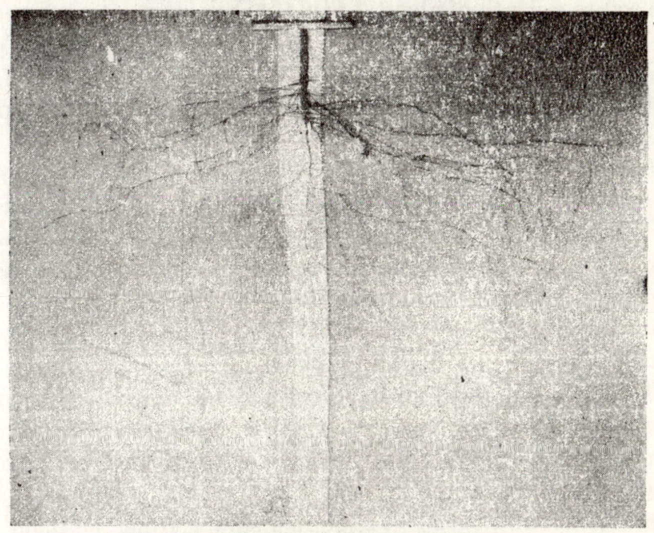

Fig. 19.—Root system of a cotton plant.

The taproot tapers rather rapidly. A foot beneath the surface of the ground it is no larger than the secondary branches. Two feet beneath the surface it has a diameter of about 1/16 inch. This slender root grows downward for a distance of several feet without diminishing in size appreciably. The depth reached probably depends, to some extent, on the texture of the soil and its moisture conditions. On Deer Creek silt-loam land, the writer, assisted by C. A. Tate, dug out the root system of a plant of Express cotton, an upland variety (Fig. 19). The end of the taproot died at a depth of 28 inches, but two laterals that arose just a short distance back of the tip pursued their course downward, one of them reaching a depth of 5 feet 11 inches. In their course, these roots passed through several layers of rather dense soil. These caused them to thicken somewhat and to pursue a slightly zigzag course, but they were not turned or stopped. Balls[4] says that the resistance offered cotton roots by soils of normal texture is negligible. In so far as the writer is aware, no one has dug out the root system of a cotton plant growing on tough buckshot or clay soil; so it is not known how well the roots can penetrate that type of soil.

Balls[4] dug out the root system of a plant belonging to one of the Egyptian varieties and found that the taproot penetrated the soil to a depth of about 7 feet 4 inches, where it was checked by soil saturated with water. This plant was grown in Egypt under irrigation. King[5] traced the taproot of a plant of the Pima Egyptian variety to a depth of 10 feet 8 inches. This plant was grown in Arizona under irrigation.

Root System of an Upland Variety of Cotton.—The root system (mentioned above) dug up by Mr. Tate and the writer on Sept. 15 belonged to a plant of the Express variety that was planted Apr. 7. There was no rain from June 1 until the time of the digging. The land was a silty loam of medium fertility. The plant was treated with nitrate of soda at the rate of about 300 pounds per acre and was carefully cultivated.

There were no roots in the upper 5 inches of soil. This condition was probably influenced by the dry weather. Other years, during rainy periods, the writer has seen numerous cotton roots appear above the surface of the ground. Beginning at a depth of about 5 inches, 21 laterals sprang from the main root within a space of about 10 inches. These laterals, several of which were large, radiated outwardly and slightly downward. Some of them were cut off in digging a trench about the plant 4 feet from the main root. Many branches sprang from the large laterals, and some of them grew downward. The main laterals did not appear to turn downward except in a few instances, and it is possible that in these cases the root running downward was a branch at the tip of the root rather than the main root itself. For a space of a few inches below the origin of the large laterals there were no large branches. Then below this point within a space of about 10 inches several rather slender laterals were given off. Below a depth of 28 inches there were very few branches arising from the few roots that reached that distance from the surface. Branches given off by laterals branched and rebranched until there was a network of fine threadlike roots running throughout and ramifying every inch of soil in a block more than 8 feet in diameter and 2½ feet in thickness. Below this block, or in the lower masses of soil reached by the roots, but very few branch roots were found, and it is doubtful if these functioned extensively. Balls[4] found that, in the Egyptian plants which he studied, the slender roots that grew down to the lower depths of soil there branched freely where moisture was found. They doubtless functioned normally.

Root System of Young Plants.—The root system of young plants develops rapidly and is surprisingly extensive. Balls[4] found that a seedling plant 7 days old had a total root length of 20 centimeters; one 14 days old, 2 meters; 3 weeks old, 4½ meters.

The following table from Balls[4] gives the depth length of roots of Egyptian cotton plants dug up at successive dates during a season. These plants were growing in a field.

	Length of Root, Centimeters
Sown Mar. 28	0
Measured Apr. 2	6
Measured Apr. 4	14
Measured May 15	55
Measured July 1	140
Measured Sept. 1	220

Flowers.—About six or eight flower buds are produced on each fruit limb, one arising at each joint. Unusually large plants may have many more. The buds appear at first as small, pyramidal-shaped green structures and are known as squares (Figs. 16, and 20*a*). Each square consists of three triangular-shaped leaflets known as "bracteoles," or "bracts," and the flower bud proper. The bracts completely enclose and protect the tender, growing flower parts. Flowers open about 21 days after the squares are first recognizable. The first flowers to open are low on the plant and near the main stem; the next are farther out on the branches or higher on the plant, the succession being both centrifugal and acropetal.

A flower bud that is to open the following day may be recognized by its enlarged but well-twisted corolla (Fig. 20*b*). Flowers used in hybridizing work are opened and emasculated when in this stage. Figure 20*g* shows the large, showy, open flowers.

They are somewhat bell-shaped and have a spread of about 2 inches. When it first opens, the corolla, which is the most conspicuous part of the flower, is creamy white to a deep yellow-

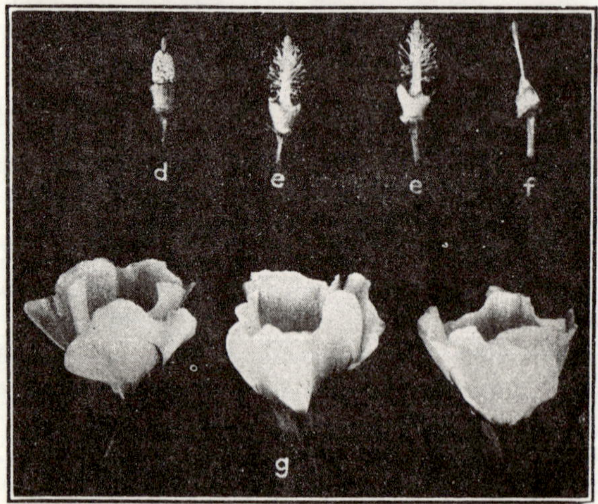

Fig. 20.—Flowers, flower parts and developing bolls. (*a*) Squares; (*b*) flower bud just before flower opens, bracts removed; (*c*) flower bud, bracts not removed; (*d*) flower with corolla removed showing stamens and pistil; (*e*) like *d* except that anthers have shed pollen; (*f*) pistil after other flower parts have been removed; (*g*) open flowers; (*h*) to (*l*) stages in development of bolls.

ish-cream color in American upland varieties, but toward the close of the first day it turns to a pinkish color. The second day it is an attractive red color; the third day it withers and usually falls, taking with it stamens and the upper part of the pistil. Flowers of some foreign species are deep yellow when they open; others are a brilliant red.

The parts of a flower as they succeed each other from the outside inward will now be studied (Fig. 21). First is the whorl of three green bracteoles, or bracts, which have been mentioned. Inside this is an inconspicuous green ring, the calyx, with five more or less well-defined irregular lobes, the sepals. Inside the calyx is the corolla, which consists of a whorl of five petals united at the base. Inside the corolla and attached to it at the

Fig. 21.—Diagram showing relation of parts in a cotton flower. (*After Cook.*)

base is a whorl of numerous stamens. The stamen filaments are united below to form a tube which surrounds the pistil (Fig. 20*d*). The anthers, on the outer end of the free part of the stamens, appear to be in 10 more or less well-defined rows. The stamen tube shows 10 ridges. The pistil consists of a small conical ovary (Fig. 20*f*), a slender threadlike style, and a stigma. The stigma appears to be simply the outer end of the style slightly enlarged, grooved or lobed, and covered with a stigmatic surface, which is receptive to pollen grains. There are as many lobes, or branches, at the tip of the stigma as carpels in the ovary or locks in the resulting boll. The pistil in upland varieties is formed from four or five carpels; in Egyptian and Sea Island, from three, as a rule.

A cotton flower also contains nectaries, which will be discussed later. Some species have about the base of the calyx some rudimentary structures the morphological significance of which is not well understood. Further details of floral parts will be given in Chap. VI.

Cotton Bolls.—The ovary enlarges, ripens, and makes a pod, or capsule, which is known as a "cotton boll." Different stages in the development of boll from ovary to maturity are shown in Figs. 20*h* to *l*. A young boll enlarges rapidly and grows to full size in a comparatively short time, but a considerable period further elapses before the boll is ripe, or ready to open. The seeds and lint within require time to develop and mature. The length of time between bloom and open boll ranges from about 45 to 65 days, varying to some extent with the variety but more with climatic conditions or with the time of season when the boll is developing. Less time is required during the warm weather of summer than during the cooler temperature of fall.

The size and the shape of cotton bolls differ greatly with different varieties. The variation in different species is even greater. Forty to fifty bolls of the largest-bolled varieties, such as Triumph and Lone Star, will make a pound of seed cotton, whereas from 80 to 110 bolls of the smaller-bolled sorts may be required. Most of the long-staple varieties have small bolls. The bolls are somewhat larger and there are more per plant on plants grown on rich land where there is plenty of moisture. On average land there are not more than 8 or 10 per plant if the cotton is closely spaced. Bolls with five locks yield slightly more seed cotton per boll than those with four locks.

When the boll is ripe, the capsule cracks or splits along the lines where carpels meet, and the cotton within expands greatly, pushing out beyond the walls of the capsule, or "burs," as they are called, in a white fluffy mass (see Fig. 22*a*). It is desirable that the locks cling together slightly, as in the boll shown in Fig. 22*a*. This facilitates picking, for bolls of this type can be picked at one grasp. The lower end of each lock is wedged in between two adjacent parts of the bur firmly enough to support it. The most satisfactory boll has the locks wedged tightly enough to keep them from falling to the ground, but they are not so securely fastened that they may not be pulled out entire. Each lock (Fig. 22*c*) maintains its identity or holds together because of the interlocking of its numerous fibers, which are attached to and surround the several seeds contained. The number of seeds in a lock varies somewhat but the normal number is nine. Figure 22*d* shows seeds with lint attached. The lint fibers have been combed out straight. Figure 22*e* shows lint fibers after removal from seeds. Each fiber is the outgrowth of a single epidermal

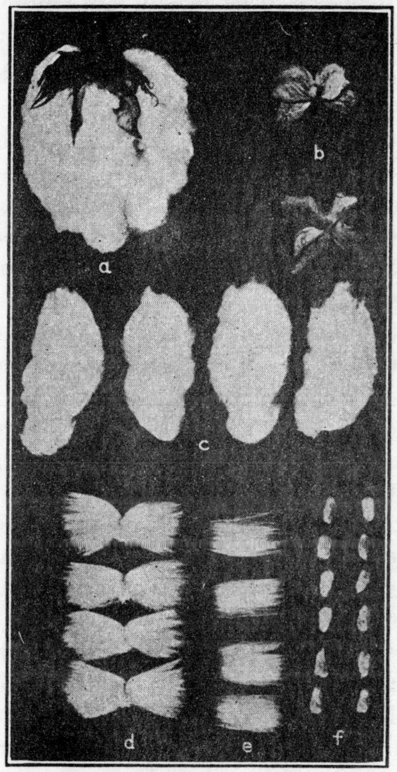

FIG. 22.—Open cotton boll and boll parts. (*a*) Boll; (*b*) empty burs; (*c*) locks; (*d*) seeds with lint attached; (*e*) lint; (*f*) seeds.

cell of the seed coat (for more complete discussion of cotton fibers, see Chap. VII).

The fibers shown in Fig. 22*e* are 1⅛ inches in length and are considered long-staple. Excepting Meade, the staple of long-staple American upland varieties ranges in length from 1⅛ to 1⅜ inches. The greater quantity of this kind of cotton is 1⅛ to 1³⁄₁₆ inches. Meade is an upland variety with a staple similar to that of Sea Island. The staple of both of these varieties averages about an 1½ inches in length, but some strains have considerably longer staple. Short-staple upland varieties have staples ranging in length from about ¾ to 1$\frac{1}{16}$ inches.

Storm Resistance.—The open bolls of some varieties of cotton hold the cotton well during storms or periods of rainy, windy weather. They are said to be "stormproof." These varieties have moderately firm locks which cling together well; they have a bur that is stiff, and on opening it assumes a cuplike form which will support and hold together the locks; the bolls tend to hang down, so that the bur serves as a roof over the locks, and rain water drains from the locks thoroughly; the lower tip of each lock is held rather securely between adjacent parts of the burs.

The bolls of some varieties do not open wide enough for the cotton to project. The burs of others are so twisted that the cotton is held very securely. In still other cases, the cotton is "tight in the boll," the lock being grasped so firmly at the base of the boll that it cannot be pulled from the boll except in pieces. All these types are stormproof in a sense, but the last three are not satisfactory on account of their bad picking qualities.

Certain varieties are weak in storm resistance, because the bur is thin and its divisions or fingers bend back so far after the boll opens that the locks are not supported or held securely. The lint of these varieties is usually soft, silky, and somewhat sparse on the seeds. The locks are not firm but stretch out and dangle in the wind.

Cotton Seeds.—Cotton seeds are irregularly pyriform in shape and approximately ⅜ inch in length (Fig. 22*f*). The size varies considerably with different species and varieties, the range being from about ¼ inch in length for the smaller to nearly ½ inch for some of the larger seeded sorts. Duggar[1] considers seed of upland cotton large if 100 seeds weigh 13 grams or more; medium, if 100 seeds weigh 10 to 13 grams; and small, if 100 seeds weigh less than 10 grams. A bushel of average-sized seed contains about 120,000 seeds.

The legal weight of a bushel of cotton seed varies in different states. For fuzzy-seeded upland cottons the range is from 30 to 33⅓ pounds. Much of the seed sold at present is sold by the 100 pounds, or in 100-pound sacks, and no mention is made of bushels. For smooth- or "slick-"seeded cottons, like Sea Island, 44 to 46 pounds is required to make a bushel.

Fuzz and Linters.—In most upland varieties, after the lint is removed there remains about the seed a woolly covering of short fibers, which is known as "fuzz" (see Fig. 22*f*). These are also epidermal hairs, produced by the outgrowth of epidermal cells. They differ from the lint fibers in being much shorter, under ¼ inch as a rule, and in being more or less colored. Lint is usually white in color, but the fuzz, or "linters," as that portion of the fuzz removed by seed-delinting machines is called, is frequently a greenish, brownish, or tawny color. Sea Island, Meade, and a few other upland varieties have no fuzz except a small tuft at the pointed end of the seed. Occasionally, a degenerate plant is found without lint or fuzz on its seeds.

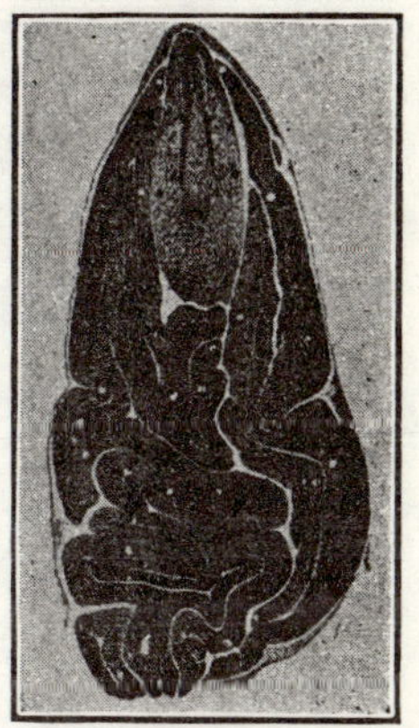

Fig. 23.—Enlarged longitudinal section of a cotton seed embryo. (*After Stanford and Viehoever.*)

Seed Structure.—Surrounding the seed proper is a stiff, tough, dark-brown covering, which is designated as the "seed coat." This, if examined closely especially during its development, shows two distinct layers which come from the integuments of the ovule, and one or two membranous or papery layers which have come from endosperm and nucellus of the ovule. Examined microscopically, there appears on the outside of the coat a well-defined epidermal layer among the cells which are the lint and linter hair cells. Under this are layers of pigment-bearing cells more or less compressed and crushed. Next comes a single layer of very thick-walled cells, which terminate the outer layer of the seed coat. Next toward the inside of the seed is a single layer of greatly elongated cells, the palisade cells. They have thickened walls and make up about half of the entire thickness of the seed coat. To the inside of the palisade cells are the remains of several layers of collapsed thin-walled cells. These in the mature seed coat are pigment-bearing. They mark the inner limit of the seed coat proper. Between the seed coat and the embryo are fragments of the nucellus and endosperm of the ovule. They appear as a very thin, brownish, papery layer.

The embryo, shown in longitudinal section in Fig. 23, makes up almost the entire kernel of the seed. It consists of two seed leaves, or cotyledons, and a small sprout, or caulicle. The cotyledons are large, comparatively thin, and much folded and twisted about the short cylindrical caulicle. The cotyledons are well filled with rich food, which is used by the young plant after germination. The embryo is nearly homogeneous in structure except for numerous oil glands which appear as dark dots throughout the cotyledons and the upper part of the caulicle. The lower portion that makes the radicle, or root, contains no glands.

Young Plants.—As the seed germinates, the caulicle of the embryo enlarges and lengthens rapidly. The lower end develops

Fig. 24.—Young cotton plants two or three days after germination. Cotyledons are very prominent.

the radicle, or primary root, which pushes its way out through the micropyle of the seed and rapidly plows its way downward in the soil. The portion of the caulicle between the radicle and the point of attachment of the cotyledons develops a primary stem, or hypocotyl. This stem enlarges and lengthens rapidly, becomes arched, and the top of the arch pushes its way upward and out of the ground. A day or so later, if conditions are satisfactory, the arched hypocotyl straightens out. This lifts the cotyledons and growing tip of the stem above the surface of the ground. Figure 24 shows the appearance of plants a few days old. The bud forming the stem tip between the cotyledons rapidly enlarges and unfolds, forming the stem proper, branches, and leaves.

Nectaries.—Cotton plants have two types of glands, external and internal. The external glands are known as "nectaries."

There are four sets of these, one inside the flower and three extrafloral. The number and the development of the extrafloral vary considerably in different species. The floral nectary forms a ring about $\frac{1}{16}$ inch wide on the inside of the calyx near its base. The secretory part of the nectary consists of numerous pluriseptate glandular hairs, or papillae, closely aggregated (Fig. 25). These papillae, according to Stanford and Viehoever,[6] are developed from modified epidermal cells. The floral and other nectaries excrete a sweetish liquid, which attracts bees, butterflies, and various other insects.

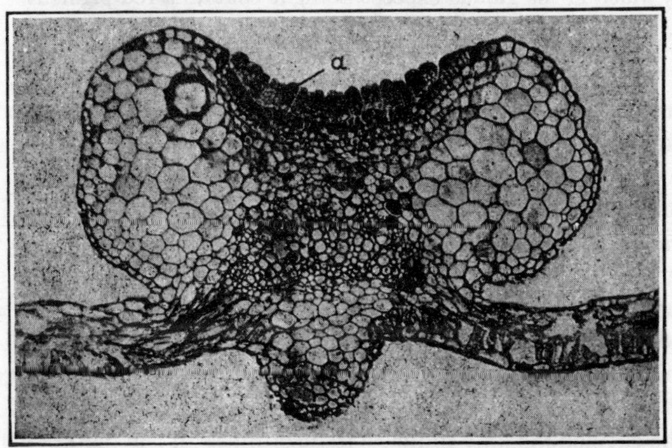

Fig. 25.—Section of cotton leaf and leaf nectary. (a) Glandular cells of nectary.

A second set of nectaries is found on the outside of the calyx near its base. This set consists, as a rule, of three irregularly triangular-shaped nectaries.

A third set consisting of three nectaries is situated on the pedicel of the flower, one being near the point of attachment of each bract. These nectaries are prominent. A part of the extrafloral nectaries may be lacking on early flowers.

On the large veins on the underside of foliage leaves is the fourth set of nectaries. There is usually one on the mid-part of the largest vein, and very frequently two others are in similar positions on the two other large veins of the leaf. Figure 25 shows a section through a leaf nectary.

The quality of the nectar from different nectaries may differ to some extent. It has been observed that bees visiting extrafloral nectaries do not go to the ones within the flowers.

Internal Glands.—Internal glands are found in all species of cotton plants and are thickly distributed over the entire plant, except the root, which does not have many. They appear as small, dark specks (Fig. 23) and are known as "black glands," "oil glands," "resin glands," etc. Each gland, according to Stanford and Viehoever,[6]

> . . . consists of an oblate or spheroidal central sac 100 to 300 microns in diameter, filled with a more or less homogeneous yellow or brownish secretion, surrounded by an envelope of one or more layers of flattened cells, which in the glands exposed to light contain a red pigment.

The glands exposed to light have anthocyanin in the border cells, and the secretion within the gland contains resins, ethereal oil, quercetin, and perhaps tannins. Glands that are not exposed to light have no anthocyanin in their border cells, but they contain a toxic, or poisonous, compound known as "gossypol."

It seems very probable that the function of the nectaries is to attract insects to the plants, but the function of the internal glands is not so evident. It has been conjectured that the substances in these glands, or at least most of them, are excretions useless to the plant, since they tend to remain in the glands indefinitely without change. Stanford and Viehoever[6] suggest that the changes in the gossypol which occur with the unfolding of the cotyledons may indicate its usefulness in the metabolism of the young seedling and that the pronounced toxicity of gossypol in the seed may be a protective adaptation against animal attack.

References

1. Duggar, J. F.: "Southern Field Crops," The Macmillan Company, New York, 1920.
2. Cook, O. F. and Meade, R. M.: "Arrangement of Parts in the Cotton Plant," *U. S. Dept. Agr. Bur. Plant Ind. Bull.* 222, 1911.
3. Leake, H. Martin: "Studies in the Experimental Breeding of Indian Cottons; Part 2. On Buds and Branching," *Jour. and Proc. Asiatic Soc. Bengal,* new series, vol. 5, No. 1, 1909.
4. Balls, W. L.: "The Cotton Plant in Egypt," Macmillan & Company, Ltd., London, 1919.
5. King, C. J.: "Water Stress Behavior of Pima Cotton in Arizona," *U. S. Dept. Agr. Bur. Plant Ind. Bull.* 1018.
6. Stanford, Ernest E. and Viehoever, Arno: "Chemistry and Histology of the Glands of the Cotton Plant, with Notes on the Occurrence of Similar Glands in Related Plants," *Jour. Agr. Research,* vol. 13, No. 8, 1918.

CHAPTER V

PHYSIOLOGY OF THE COTTON PLANT

Only a limited amount of research work has been done on the physiology of the cotton plant; consequently, there is a scarcity of data. The most extensive work has been by W. L. Balls in Egypt. He used Egyptian varieties mainly in his experiments, and the cotton was grown under irrigation.

Water Relations.—All living plants must be permeated by water. It functions in the solution and transportation of products of metabolism and the transportation of mineral nutrients from the soil to the plant, and its elements, hydrogen and oxygen, are made use of in nutrition.

Water Absorption.—Except for the enlarged portion of the taproot of cotton plants, the roots are very slender and inconspicuous; so much so, in fact, that it was not realized until comparatively recently that cotton is really a deep-rooted plant. When plants are pulled up, the smaller laterals and taproot are broken off so near the point where the root joins the stem that they appear to have been shallow-rooted. Balls,[2] King,[5] and some others excavated about cotton plants and painstakingly picked the soil away from the roots of a number of plants, so as to determine their extent and course. It has been found that the taproot, where conditions are favorable, sometimes penetrates to a depth of 10 feet and that the laterals, which are numerous, run outward for a distance, governed probably by the closeness of other plants, and thence downward almost as far as the taproot. The roots thus course very thoroughly through a large volume of soil. The larger the volume of soil available the larger the root development and the greater the plant's power for absorbing water and nutrient minerals. Other things being equal, this, of course, results in larger plants.

Although the taproot and the main laterals are for the most part small in diameter, they function actively. Each contains a few large vessels through which water moves freely, a quart or more being carried during a period of one day.

While the plant is small, it can absorb water only from the upper layer of soil; but as the root system develops, it draws its supply from the store deeper in the ground. The upper laterals die after a time, and new ones are grown farther down in the soil. During the dry weather of August and September, most of the water is drawn from sources several feet deep, unless there is a hardpan, or impermeable layer, near the surface.

The amount of soil water available is an important factor in the growth of cotton plants. A scarcity immediately checks growth, but if the scarcity is not too prolonged, a watering will revive the plants and quicken them into activity anew. An excess of water, as water-saturated soil, asphyxiates the roots and stops the growth of the whole plant. The rise in the water table or water level in the soil has done a great deal of damage to the cotton crop in Egypt. In 1909, it was estimated that the loss was equal to one-third of the crop.

Within certain limits there is probably a close relation between water supply and the length and strength of cotton fibers; Balls[1] is of the opinion that a single irrigation may have considerable effect.

Transpiration.—The amount of water evaporated from or given off by a field of large cotton plants in the course of a day is enormous. Balls[1] has estimated the loss for Egyptian cotton grown in Egypt to be 50 tons per acre per day, or 3 pints per plant. Ridgeway, in Alabama, and Lloyd,[4] in Arizona, obtained results showing somewhat smaller amounts lost. The water loss from the Egyptian plants was greater than the total amount supplied the land in irrigation. The difference was made up by ground water.

This heavy loss of water is due to the excessively large amount of evaporating surface exposed by the leaves in a field of cotton. When the stomata are open, the transpiring surface exceeds many times the area of the field.

One function of this transpiration current of water is to carry mineral nutrients from the soil to the various cells of the plant, where they are used in building up foods. Another function, one of much importance to cotton plants growing in hot, dry regions, is the cooling effect produced by the evaporation of water from the leaves and other parts. As has been shown in another place, this cooling effect reduces the temperature of cotton leaves several degrees. Lloyd[4] and Balls[2] have both shown that excessive loss of water causes cotton plants to stop growth temporarily, even if it is not sufficient to cause water stress or wilting.

It is a difficult matter to determine accurately the amount of water actually lost by cotton plants through transpiration when they are growing under normal conditions. Various methods have been devised, as cutting off a stem and immersing the cut end in a volume of water so arranged that the amount taken up can be measured or growing plants in pots and weighing the pots and plants frequently to determine losses. These methods are objectionable in that the plants are not in normal condition when the study is made or are not grown under normal conditions. A plant in a pot does not have a chance to make normal root growth, and one plant by itself is subject to stronger air currents and a drier atmosphere than a plant in a field. Probably the most accurate way is to make a series of successive weighings of a leaf of known area soon after it is removed from the plant. The weighing should be done among the plants in the field where the humidity of the atmosphere is normal. Ridgeway, in Alabama, reported by Lloyd,[4] measured a loss of 5.25 grams per 100 square centimeters per hour. The rate will, of course, vary greatly with environmental conditions. Lloyd observed that in Arizona under severe environmental conditions, there was a checking of transpiration about 9 a.m., but that in Alabama, under less severe conditions, it was later —11 a.m. to 1 p.m.

Relation of Water and Dry Matter Produced.—Photosynthesis and the elaboration of organic compounds begin in the leaves of cotton plants soon after sunrise, if light and temperature conditions are favorable, and continue until the roots have so reduced the supply of water available in their immediate vicinity that there begins to be an insufficiency in the plant. This causes the stomata to begin to close, and soon after there is a stoppage of photosynthetic activities in the leaf. The time at which this occurs varies considerably, depending on the wetness of the soil and the humidity of the atmosphere, but it is normally between 9 a.m. and noon. The plants are gaining in weight of dry matter only when new organic materials are being built up.

The field water requirements of cotton plants may be indicated by dividing the total weight of water lost by a given area during the season by the dry weight of the plants produced on the same area. In experiments conducted by King[5] with Egyptian cotton in Arizona, from 4,013 to 5,426 pounds of dry matter per acre were produced, the amount varying with different plots. The plot that received the most water made the most dry matter, but the percentage of dry matter for amount of water lost was smaller than that of a plot that received less water. Plot 1 used 4,855,184 pounds of water per acre and made 5,426 pounds of dry matter, the field-water requirement being 894.8. Plot 3 used 3,425,296 pounds of water per acre and made 4,013 pounds of dry matter, its field-water requirement figuring 853.6.

Water Supply and Normal Development.—A slight increase in the supply of moisture available to the plants may result in excess vegetative growth. This can be observed when an attempt is made to grow cotton on moist, rich bottom land. Increasing the supply of water in the soil further may hinder growth or even kill the plants. On the other hand, as is shown elsewhere in this chapter, the reduction in the supply of moisture even slightly below normal may cause a temporary stoppage in growth. A further restriction in the supply of water may result in the complete cessation of growth and the death of the plants. King[8] has recently shown that the crazy-top disorder of cotton plants in the Southwest is largely due to lack of a regular or sufficient water supply. Hubbard[9] has observed that cotton plants grown on light sandy soil at the U. S. Field Station at Shafter, Calif., may wilt and die suddenly because of taproot constriction. The soil becomes so dry and hard in places that the roots cannot enlarge. Irrigation water softens the soil and revives the plants so that they put out new roots and soon resume normal growth.

Effect of Water Supply on Flowering and Fruiting.—It was shown on a preceding page that there is a close relation between the growth of cotton plants and their water supply. All conditions being favorable, an increase in water, up to a certain amount, was followed by an increase in growth. This increase in growth results in taller plants, more and longer fruit branches, more nodes, more flower buds, and consequently more flowers and fruit. Balls[2] has shown not only that the flowering follows the growth but that if both are plotted the two curves coincide closely. Irregularities in flowering coincide with similar irregularities in growth that took place about 29 days previously. King[5] in his experiments with the water-stress behavior of Pima cotton found that the plot to which he applied the most water produced the greatest number of flowers per plant.

Pollination and Fertilization.—It is a commonly accepted belief among cotton growers that, if a rain occurs during the day a cotton bloom opens, the form or boll will be shed. This opinion has some basis in fact, but the damage done by rains is not nearly so great as is often supposed. If the pollen is wetted by a rain early in the day, or before pollination has taken place, it will not perform its function, and fertilization will not result. The stigma is receptive only during the first day. But the pollen is never wetted in all the flowers; some are pollinated early and some late; consequently, the rain damage, as a rule, is not great, rarely being more than 25 per cent. In some experiments made by the writer at State College, Mississippi, 13 per cent more bolls were shed where flowers were rained on in the forenoon than were shed from flowers opening the following day. Ewing[6] found at Holly Springs, in the northern part of Mississippi, that 15.6 per cent more bolls were shed where there was a morning rain than on the preceding and following days on which there was no rain. The shedding following afternoon rains was 6.7 per cent greater than on the preceding and subsequent days. Lloyd,[4] as a result of his observations of the effect of rain on open flowers, says that, at most, not more than 15 per cent of the shedding can be referred to the effect of rain. More depends on the time of the rain than on the amount.

The corolla, or bloom, of flowers that open during rainy weather often remains attached to the flower receptacle and clings about the young boll. If it decays in this position, many think that the boll will fall off, but this is probably a mistaken notion.

Germination.—Cotton seeds will not begin to germinate until they have absorbed a considerable amount of water—a volume equal in weight to half the weight of the seed or more. This absorption takes place rather slowly at first, because the seed is protected with a thin waxy covering which is somewhat impervious to water. The seeds will not germinate satisfactorily if kept on moist filter paper, such as is used in germinating most seeds. They must be thoroughly wetted and placed between the folds of wet cotton-flannel cloth to receive sufficient water. Germination in the soil is often slow, because of the scarcity of available water in the surface layer of soil. If the soil is moist, sufficient water will be absorbed in a few days. The abundance of water furnished by a warm shower makes optimum conditions for rapid germination.

The waxy covering of the seeds serves a useful purpose in that it hinders germination of seed in the field before cotton is picked. The seed cotton in the open bolls is rained on frequently, but seeds do not sprout badly unless the rains are prolonged and are so frequent that the locks of cotton are kept wet for several days at a time. Green seeds, or seeds from bolls that have opened within the past 60 to 90 days, will not germinate well. This appears to be another natural means for the protection of seed in the field.

Growth of Plants at Various Stages.—King,[5] by irrigating plots of Pima Egyptian cotton in Arizona at different rates and measuring the height of the growing plants each week, was able to show something of the relation between water content of the soil and stem growth which may be taken as an index of the growth of the whole plant. Plot 1 was irrigated at such frequencies that complete exhaustion of the available moisture in the upper 2 feet of soil did not occur during the period of plant development. Plot 2 was irrigated shortly before the depletion of all available moisture in the upper 4 feet of soil; and Plots 3 and 4 were given water when the available moisture in the upper 5 feet was near exhaustion. Plot 4 differed from Plot 3 in that it received an application of cottonseed meal and 16 per cent acid phosphate in the drill at the rate of 500 pounds per acre. The proportion of each constituent in the fertilizer is not given. Figure 26 shows the percentage of soil moisture at different periods and the corresponding growth responses in King's experiment.

Plots 1 and 2 were irrigated on June 10, and for this reason the plants on these plots made more rapid growth throughout the rest of June than did the plants on Plots 3 and 4, which were not irrigated until June 26. Throughout July and August, the plants in Plots 1 and 2 continued to grow more rapidly and appeared to be more luxuriant. It was observed during periods in the summer, when temperatures were high and transpiration was at its greatest, that plants on Plot 1 wilted more than the plants on any other plot.

It is a matter of common observation that plants grown during a season of moderate rainfall are more luxuriant than ones grown during a dry year. Texas cotton plants are considerably

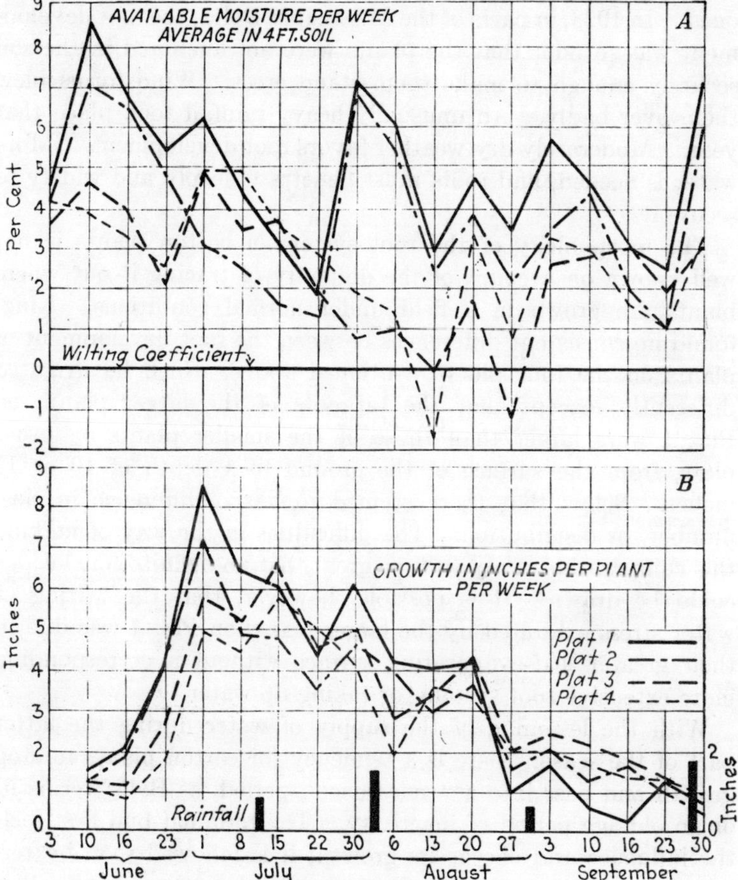

FIG. 26.—Average percentage of available moisture per week in the upper 4 feet of soil in four plots and the average increase in growth of the main stem of the plants per week. (*After King.*)

smaller than the ones grown in the Mississippi Delta. Difference in water supply is the main, but not the only, cause of the difference.

There is a close relationship between the root development of cotton plants and the moisture content of the soil. If the soil is saturated, root development is hindered; the roots fail to function properly, and the plant soon sickens. An excess of moisture in the soil part of the time, such as occurs during rainy seasons, tends to make the root system less extensive. Sufficient water can be secured by a smaller number of roots or by shorter ones. In 1923, in parts of the Mississippi Delta, the root development was so poor that the plants were not anchored in the soil securely enough to make them stand erect. Windstorms blew them over badly. An unusually heavy rainfall took place that year. Moderately dry weather favors root development. More water is needed, and roots must penetrate deeply and widely to secure it.

The morphology of the root system of cotton plants is not well known on account of the difficulty of tracing it out where plants are grown in a field under normal conditions. King[5] found no consistent differences between the root development of plants on the four plots, mentioned above, which he irrigated differently, except that the taproots of the larger plants on Plot 1 were larger than those of the smaller plants on other plots, from the surface of the ground to a depth of 10 or 12 inches. Below that there seemed to be no difference in size, number, or distribution. The difficulties in the way of making this comparative study were so great that no definite conclusions could be drawn. It is possible, however, that the wilting or water stress exhibited by the larger plants on Plot 1 was due to their greater leaf-evaporating surface without a corresponding more extensive root system for taking up water.

With the lessening of the supply of water during the latter part of the season there is a tendency for cotton plants to stop growth and pass into a "senescence" period, as Balls[2] terms it, or an old-age period of inactivity. The terminal bud first feels the influence and checks its growth; it is followed soon by terminal buds of branches and a little later by other parts. This stoppage of growth appears to be due to some internal influence which resembles the self-poisoning caused by excess heat. With the coming of fall rains in "rain countries" or with increase of water supply, as in Egypt when the Nile rises, there is often a rejuvenescence; the dormant buds awaken and grow, making new leaves, stem parts, and flowers.

Balls[2] found that the lower roots of Egyptian cotton plants were killed by the water if the water table rose high enough to cover them. But, following the recession or sinking of the water, new branches were sent out from the living roots higher in the soil. These new roots grew downward into the soil mass occupied by the old roots that had died.

Vernalization, or Iarovization, of Cotton Seed.—Vernalization is a method of seed treatment originated by certain Russian investigators. It is claimed that the treatment makes crops earlier and increases production. Ivanishin,[10] one of the Russian advocates of the method, describes it as follows:

> The first stage in vernalization is to moisten the seed to 80% (57 litres of water to moisten 100 Kg. cotton seed). The moistened seed is left in a heap, whereby the temperature in the heap rises to 23 to 35°C., under which conditions the seed is kept for 15 to 20 days; every means, including drying, cooling, etc., is used to prevent the seeds from germinating and to avoid the growth of molds. Not more than 1% to 3% of the seeds should be allowed to germinate. After this treatment the seeds are sown in the ordinary way. The results of experiments for the three years 1932, 1933, and 1934, show that after vernalization all the main developmental stages are accelerated and the production of lint is considerably increased.

Tashlanov, another Russian investigator, reports:

> Seedlings from vernalized cotton seed . . . emerged several days before controls, and certain Egyptian and American varieties respectively bloomed from 3 to 6 and from 1 to 3 days earlier and ripened from 5 to 11 and from 2 to 4 days earlier than controls. Substantial yield increases were noted in the first pickings, depending on variety. Usually no differences were noted in boll size, fiber length, and percentage.

Vernalization has been tried in an experimental way on certain crops in America, but, in so far as the author is aware, nothing of economic importance has been discovered.

Temperature Relations.—Cotton is by nature a tropical or warm-climate plant, but most of the commercial crop is grown in cooler regions. There the environmental conditions are not quite normal; consequently, special nurture must be given the plant to make it grow satisfactorily.

Germination Temperatures.—Cotton seed in germinators kept at a temperature of 60°F. will germinate slowly, provided moisture conditions are right. At 70 to 85°F. they germinate and grow almost normally. A complete test can be made in 5 to 6 days. If the germinator is kept at a temperature of 95 to 97°F.. the initial growth will be very rapid, and it may be possible to get some indication of germination percentage within 2 or 3 days. But these high temperatures inhibit growth after a comparatively short time, because of, according to Balls,[1] an accumulation of waste materials within the plant cells. Under the high temperatures, the metabolism within the cells of the plant causes these elements to accumulate faster than they can be thrown off. Their increase results in a self-poisoning which soon inhibits growth.

In the fields at planting time, soil temperatures are not likely to be high enough to cause trouble. If high temperatures do occur, they are only in the surface layer of soil and are of short duration.

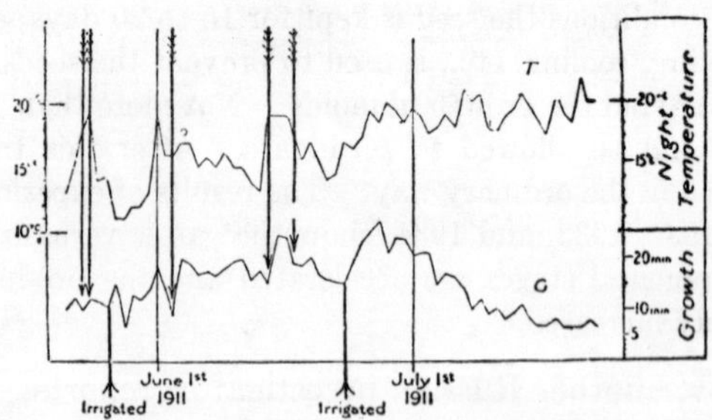

FIG. 27.—Graph showing relation of stem growth and night temperatures. The large arrows mark points where the growth was retarded by extreme heat the previous day. (*After Balls.*)

But it is a very common occurrence for temperatures to be too low for days at a time for satisfactory germination or for the growth of the young plants.

Temperature and Growth.—The temperature at which the cotton plants studied by Balls[2] made the best continuous growth was about 90°F.; that agrees with general observations made elsewhere. A temperature above 90° accelerates growth for a short time, as was mentioned above, but it apparently fosters the accumulation of materials within the growing cells that act as a poison and retard their activity. A temperature above 98°F. for a short time checks the growth of the main stem, and the effect lasts 24 hours or more. During the young and actively growing period of the plant, the rate of growth of the branches, leaves, and flowers follows closely that of the main stem. Later, under normal conditions, senescence strikes first the terminal bud or growing part of the main stem and checks its activity. The temperature and the growth of roots are controlled largely by the soil temperatures. These, for the deeper soils, are very uniform year after year and rarely become high enough to injure the roots. Often, early in the season, the surface soil is cold enough to retard growth.

In common with many other plants, the cotton plant makes most growth during the night or when the sun is not shining. Researches by Balls[1] showed a close relation between night temperatures and the length of growth of the main stem (see Fig. 27). Excepting the few drops in the growth curve due to the effect of extreme temperatures the day before, the two curves coincide almost perfectly until the first of July, a period of more than a month. After this time, certain inhibiting factors disturb the close relation between the two. Balls[2] figured a correlation of 0.7843 ± 0.0459 between length of growth of main stem and minimum night temperatures, if the daily periods where the maximum temperature had risen above 36°C. were excluded. With these periods included, the correlation was 0.5236 ± 0.0755.

Camp and Walker,[12] of Florida, also studied the effect of soil temperature on the germination of seed and the growth of young plants. They attempted to determine definitely the maximum and minimum temperatures of germination. They found that at temperatures above 35°C. there was a falling off in the rate and usually in the percentage of germination. There was some germination at 39°C. but none at 40°. At temperatures below 20° there was a falling off in the percentage of germination. The determination of the exact minimum tem-

Effect of Soil Temperature on Cottonseed Germination and Seedling Growth

Temperature		Time for 80 per cent germination, hr.	Total germination, per cent	Length of tops after 14 days, mm.	Relative rate of growth, per cent	Length of taproot after 14 days, mm.
°C.	°F.					
33	91.4	72	90	155	98	150
30	86.0	88	88	160	100	140
27	80.6	96	88	150	94	157
24	75.2	120	86	125	80	163
21	69.8	192	84	75	47	115
18	64.4	360 (60%)	68	50	32	85

perature for germination was found to be difficult because of fungous organisms which attacked the seed. Germination has been recorded at temperatures of 14.5 and 15°C., but none below 14°. The optimum temperature for germination, as determined by Camp and Walker, was about 34.5°C.

The table on p. 111 given by Arndt[11] shows the effect of a certain range of temperature on seed germination and on the growth of seedling plants for a period of 14 days.

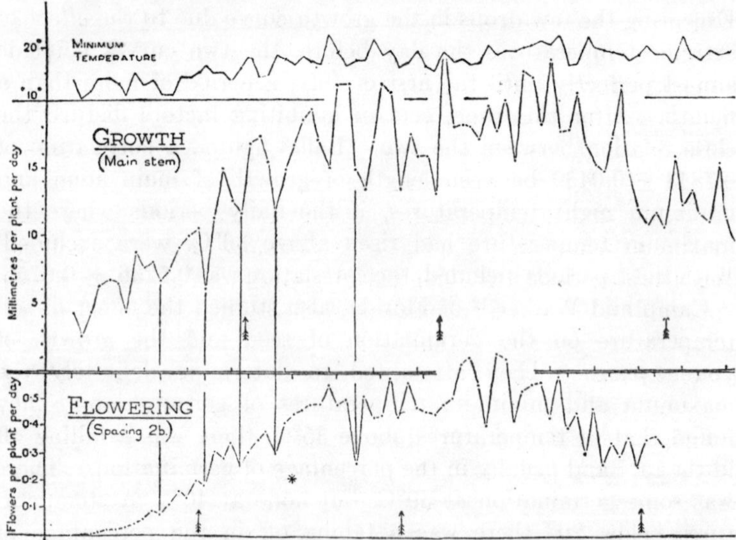

Fig. 27A.—Graph showing relation of growth of main stem and flower production. (*After Balls.*)

Temperature and Flowering.—It is a matter of common observation that cotton plants fluctuate widely in the number of flowers produced from day to day. Several environmental factors have a bearing on this phenomenon, and it is a difficult matter to determine the relative value of each, since the effect follows the cause by a period of several days or even weeks. Young squares or flower buds on Mississippi upland varieties may be recognized 20 to 23 days before the flower opens. Their primordia existed some days in advance of this. Probably they are more susceptible to the effect of environmental influences when very young than at any other time. Balls[3] found that there is a close relation between the number of flowers produced on any given day and the growth of the main stem of the plant on a corresponding day 29 days previous (see Fig. 27*A*). Temperature influences the growth of the stem immediately, as has just been shown. Consequently, flower production seems to be influenced by daily temperatures about 29 days before.

Temperature and Yields.—King and Loomis[13] report that the yields of Acala cotton at the Sacaton Field Station have been relatively low for several seasons that were characterized by high daily maximum temperatures during the fruiting period. The yields of Acala cotton were on the average 40 per cent more in the 3 years 1921, 1923, and 1925, when the mean maximum summer temperatures were less than 100°F., than during 1922, 1924, and 1926, when the mean daily average was above 100°. During the same period, no relation was found between summer temperatures and yields of Pima Egyptian cotton. The greater ability of the Pima variety to withstand high temperatures finds at least a partial explanation in the fact that the temperatures of healthy, functional leaves of the Pima cotton were about 5° cooler where exposed to the sun than the leaves of the Acala variety.

Tissue Temperatures.—The temperature within the tissues of young leaves and the younger parts of stems is usually below the shade temperature of the surrounding locality, the difference being in some instances as much as 10°F. The temperature of old leaves, on the other hand, is usually higher than the local shade temperatures, ranging, according to Balls,[2] from 5.4°F. below to 18°F. above. The rapid evaporation of water from the younger tissues tends to keep their temperature lower, but as they become older this regulatory power is largely lost. Water shortage in the soil will also cause a loss of this power. Evaporation from the leaves will be reduced if there is a scarcity of water about the roots. The closing of the stomata of a leaf will cause a partial or total loss of the thermoregulatory power in young leaves. This, according to Balls,[2] may result in a temperature increase ranging from 0 to 9°F.

Light Relations.—The cotton plant is generally considered a sun-loving plant and thrives best in the sunny South of the United States and in other countries that have abundant sunshine.

Sunlight and Growth.—Although it may seem strange, since cotton plants require abundant sunshine for satisfactory growth, recent research has shown that direct sunlight inhibits temporarily the growth of the plant. Photosynthesis and other physiological functions continue, of course, in the presence of sunlight and cannot go on without it. Lloyd,[4] experimenting with cotton plants in Alabama, found that there was not only a cessation of growth during the hours of sunshine but a shrinking of stem and leaves. Leaves accurately measured showed 10 per cent less area. King and Loomis,[13] experimenting with Acala cotton in Arizona, observed that the greater part of the growth was made at night. Their results are graphically shown in Fig. 28.

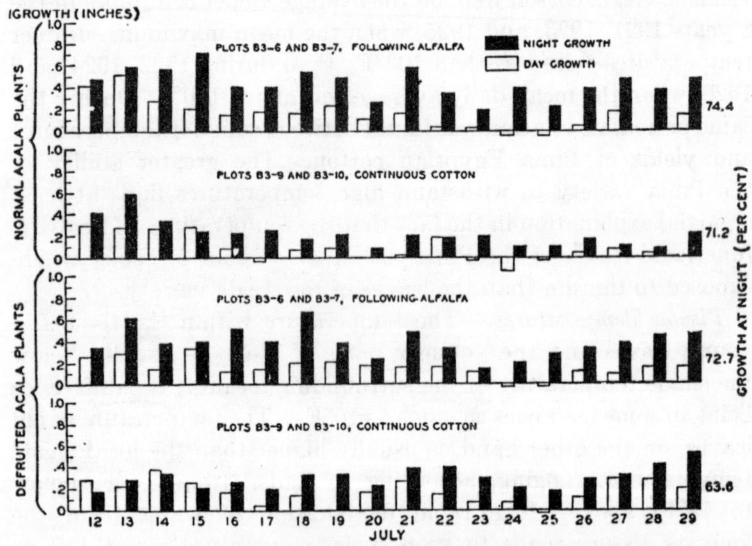

FIG. 28.—Average day and night growth of 10 normal and 10 defruited Acala cotton plants grown on two plots following alfalfa and two plots continuously in cotton for several years. (*After King and Loomis.*)

Balls,[2] working with Egyptian cotton, secured similar results. He proved that the growth inhibition apparently due to sunlight was directly due to a water shortage. He placed glass bell jars over plants in sunlight and found that their growth continued. The bell jar kept a humid atmosphere about the plant. He observed further that, if one-fourth of the lower leaves of the plant were removed, the lengthening of the main stem continued even if the plant was in sunlight. The leaf pruning so reduced transpiration that the root absorption was sufficient to supply all water losses and counteract the desiccating influence of direct sunlight. Root pruning had effect directly opposite to the leaf pruning.

Influence of the Period of Illumination.—Only a limited amount of work has been done on the problem of the effect of light periodicity on cotton. Konstantinov[15] studied the effect of varying the length of day on *Gossypium peruvianum* and *G. herbaceum*, two foreign species. He secured some acceleration in rate of flowering from the use of a 10-hour day. The reduced length of day had the greatest effect on the later maturing forms. The very early-maturing forms reacted hardly at all to the reduced light.

Berkley[16] studied the effect of different lengths of day with variations in temperature on vegetative growth and reproduction in cotton. He concluded that "cotton plants may produce fruits more readily with a specified length of day, thus being classed as a medium-day plant, but one is led to believe from the . . . data that temperature differences may be substituted for day lengths in certain combinations."

Photosynthesis.—The work of photosynthesis is normally most active during the first part of the forenoon and continues until the stomata close. This closure naturally prevents the chloroplasts within the leaf cells from receiving an adequate supply of carbon dioxide. Balls[2] found that from sunrise to 9 a.m. leaves increased in dry weight as a result of photosynthesis in accordance with the prevailing temperature; after that the increase was governed largely by stomatal action. The more the stomata were open during the day the greater the increase.

In reference to the amount of dry matter produced, Balls[2] says:

> The increase in dry weight in hourly series of observations, without correction for loss by translocation, has given record values at the highest point of the curve, namely 25($\pm$4) milligrams per square decimeter per hour. This rate is reached in June, between 8 a.m. and 11 a.m., when the stomata are at maximum dilatation, and the temperature is above 30°C.
>
> It is improbable that the intensity of illumination should ever be a limiting factor in the process except before sunrise.
>
> No definite signs of growth limitation by deficient photosynthesis have yet been discovered.

Cotton leaves have numerous stomata on both upper and lower surfaces. Balls[2] has estimated that for leaves of Egyptian cotton the number in the upper epidermis is 44 to 97 per square millimeter and 116 to 176 per square millimeter for the lower epidermis. The stem and hypocotyl contain about 20 per square millimeter, and the cotyledons contain about 200 on the upper surface and 275 on the lower surface. These figures show that the average number of stomatal pores in a square inch of leaf surface (including both sides) is more than 100,000. This immense number of pores exposes much cell surface, and consequently evaporation through the pores, when open, is considerable. A fully developed stoma is about 0.04 millimeter in length.

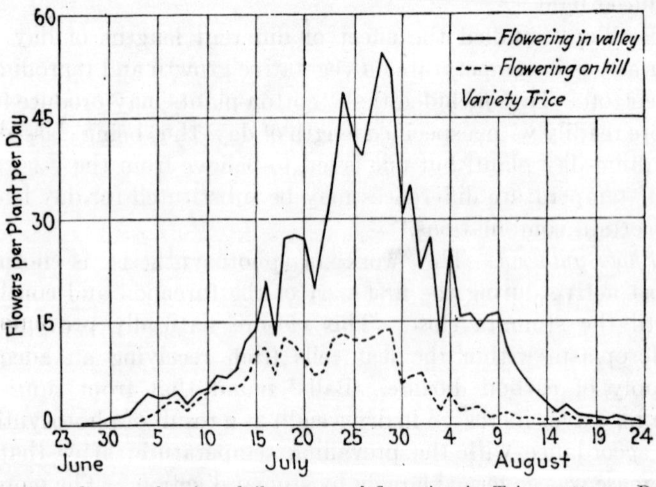

FIG. 29.—Curves showing daily range of flowering in Trice cotton. Represented as flowers per plant day, when grown under two different soil conditions. (*After Ewing.*)

Soil Relations.—Since considerable data are given on the influence of various fertilizers and soils on the growth of cotton plants in the special chapters dealing with those subjects, it will not be necessary to enter into a lengthy discussion of the subject at this point. It is well known, however, that the cotton plants will not make proper growth unless the soil is supplied with the common mineral nutrients in available form and in sufficient quantity; that there must be a free movement of soil water; that the soil must contain enough organic matter to give it good water-holding capacity and permit good aeration.

Varietal Differences in Absorption of Soil Solutions.—Harris[14] and his associates have found that there is considerable difference in different types of cotton in respect to their absorption of soil solutions. Egyptian cotton shows a preferential for chlorides, and upland cotton for sulphates. This power of selective absorption may help explain why Pima cotton shows a superiority over upland varieties in establishing seedling stands under the rather saline soil conditions of the Gila River Valley of southern Arizona. Sea Island cotton seemed to show a behavior similar to that of Pima.

Flowering and Fruiting.—Soil fertility is also an important factor in flower production and in fruiting of cotton plants. Observations made by Ewing[6] (see Fig. 29) show clearly the difference in effect that two soils within a few rods of each other have on the flowering of cotton plants. Trice cotton was planted on fertile valley soil and on infertile soil on top of a hill near by. On the rich soil the plants began to bloom first, bloomed longer, and produced three and a half times as many flowers in the aggregate as the plants on the hill. A high percentage of the flowers on the rich-land plants made bolls. These factors account for striking differences in fruiting and yields.

Cotton plants on sandy or sandy-loam land generally fruit earlier and more rapidly than those on clay lands. Well-drained lands also produce earlier plants. The practice of planting cotton on ridges or beds finds some justification in this fact.

The character of soil appears to have an effect on the length of boll period, or length of time elapsing between the opening of the flower and the resulting boll. Ewing[6] found that for plants growing on a poor sandy hill near State College, Mississippi, the period was 44.93 ± 0.18 days, while the period for plants growing on comparatively rich valley land near by was 48.38 ± 0.21 days, or a difference of 3.45 days. The plants on the high sandy land were more subject to drought. This probably caused the bolls to dry out more rapidly and open sooner. The shedding of forms and small bolls is influenced considerably by the nature of the soil, especially in its water relations. This subject is discussed at some length under the general topic of shedding.

Ecological Relations. *Plants.*—Various fungi live on or within the tissues of cotton plants. They enter the roots from the soil, as does the wilt-producing fungus *Fusarium vasinfectum* Atk.; or they enter leaves from the air or from drops of dew or rain, as does the angular leaf-spot organism *Bacterium malvacearum;*

or they attack and penetrate bolls, even completely ruining full-size bolls, as does anthracnose or the pink boll rot, caused by *Colletotrichum gossypii* Edg. In all instances, the advantage of the relationship is on the side of the fungus. It acts as a parasite, penetrating the cotton plant and drawing its food from the organic material that the cotton plant has built up. The cotton plant gets nothing in return. Under unfavorable conditions for growth, as during cold, rainy weather, the cotton plant is very susceptible to attacks of these fungi. Serious cotton diseases may result from them.

Cotton plants will not grow wild in much of the American Cotton Belt, as this is a region in which they are out of their natural habitat, and they cannot successfully compete with the weeds, grasses, and other plants native to the flora unless aided by man. Some planters apparently think that they can, if the amount of grass and weeds left growing in their fields be considered a basis of judgment. Clean culture and satisfactory environmental conditions are necessary for good production.

The cotton plants in a field where there is what is known as a "good stand," or a sufficient number of plants per acre to make a good yield, are in a continuous struggle with each other after the first few weeks of their existence. The roots of one plant interfere with the growth of the roots of its neighbors, and there is a struggle for the available soil water and the mineral plant nutrients that it holds in solution. A plant that has an abundance of room or that is not interfered with by its fellows produces well as an individual, but the acre yield will be greater if there are many plants—so many that the individual is crowded and its yield reduced considerably. In spacing his plants the planter must strike the mean that will give him the greatest production. This mean varies somewhat with the season, soil, insect pests, etc., but it is fairly uniform within certain limits. The problem of cotton spacing has claimed much attention from experiment stations and at present is fairly well understood (see Chap. XIII).

Insects.—Cotton plants have numerous nectaries and glands which are apparently for the purpose of attracting insects, since they excrete a sweetish liquid which is sought eagerly by many different kinds of insects. Insects entering the flowers to gather the nectar from the nectaries at the base of the calyx become dusted with pollen from the anthers of the flower. Some of this pollen may be deposited on the stigma of the same flower, but a part will be carried to the stigma of some other. These insect visitors are thus effective in securing either cross- or self-pollination. Insects also visit freely the glands on the outside of the flower bracts and on the underside of the leaves, but it is not clear what advantage to the plant this is.

Boll weevils, the pink bollworms, cotton leaf caterpillars, army worms, and several other harmful insects visit cotton plants. They destroy or injure the roots, leaves, flower bud, and bolls; no part of the plant is secure against their attacks.

Boll weevils have probably been studied more than any other cotton insect on account of the great economic importance of their depredations. Certain adjustments have been observed by means of which the plants overcome the harmful influence of the weevil to some extent. For instance, the square or flower bud may be shed so soon after the egg has been deposited in it that the square will dry up and the young weevil perish before coming to maturity; or, after the egg is deposited in the square or young boll, the plant cells in the tissue about the egg may multiply or proliferate so freely that the egg is crushed before hatching. Cotton plants produce many forms—more than twice as many as can grow into bolls. With this surplus the weevil can destroy a good many without doing any real damage to the plants. If weevils are so numerous that all or nearly all of the forms are destroyed, the plant quickens into renewed growth and square production which becomes extensive if there is a continued heavy infestation of weevils.

Life History Periods.—The time required for the cotton plant to pass through various stages in its life history and the length of time needed for the development of different parts vary greatly with the variety, the temperature, and other environmental conditions. Data on the growth and development of a variety under one set of conditions may be of some interest and value, however. Hammond, in *Bulletin* 33 of the U. S. Department of Agriculture, gives data on the development of plants in South Carolina, near Augusta, Ga. The season was cool and wet. Seeds were planted Mar. 29; the first plant appeared in 14 days; enough were up in 18 days to make a stand; none came up after 30 days; the third leaf made its appearance in 8 days after the plant came up; the fourth leaf appeared the next day. The planting date in this experiment was early for the latitude. If planted later under proper moisture conditions, the periods are shorter. Under weather conditions that ordinarily prevail at planting time, most plants are up in a week or 10 days after planting, and in some cases, under extremely favorable conditions they may be up in 4 days. If the soil is dry at planting time, the seed may be in the ground for several weeks before germinating.

Plants from later plantings made by Hammond showed much more rapid growth. The average for plants coming up at different times during April was 40 days from appearance of plant to appearance of square; this varied from 34 to 45 days; for plants coming up during May, the average was 29 days. Squares appearing in May, on the average, bloomed in 25 days; those appearing in June bloomed in 24 days; August squares bloomed in 21 to 27 days, averaging 25 days. June blooms made open bolls in 45 to 56 days, or 52 days on the average; July blooms, in 64 to 71 days, averaging 65 days.

For the northern part of Mississippi, Ewing[6] gives growth periods as follows: flowers from 7 to 10 weeks after planting; squares or flower buds 3 to 4 weeks before the flowers open; period of development of bolls, 40 to 70 days. The fruiting period, or the time during which most of the flowers open and the fruit is set, varies considerably, ranging from 40 to 80 days. The length of the period depends on the soil fertility largely, but other environmental factors also have a bearing. The length of the boll period also varies considerably, temperature and variety probably being the chief factors influencing it. According to Ewing's observations, flowers that opened from July 18 to 31 experienced an average daily temperature of 80°F. and produced open bolls in 48 days. Flowers opening Aug. 26 to Sept. 1, inclusive, experienced an average daily temperature of 70.8°F. and required 68.5 days to produce open bolls.

Some Factors Affecting Earliness in Cotton.—Ludwig[17] studied the developmental periods of squares and bolls of five varieties of upland cotton, grown at Clemson College, South Carolina, and certain factors that affected their development. He noted that the square period, that is, the time from the first appearance of the rudimentary flower bud to the open flower, was 22 to 26 days. There did not seem to be any difference in different varieties in the square period, but there was a difference of about 4 days in the two seasons when studies were made.

The boll period for the five varieties studied—the time from open flowers to open boll—ranged from 54 to 65 days. The varietal difference was about 4 days; and the seasonal difference, 4 to 11 days. Apparently the boll period had some effect on earliness.

Neither the amount of nitrate of soda applied as fertilizer nor the time of application had a perceptible effect on the square period or the boll period.

Variations in the spacing of the plants had no appreciable effect on either of these two periods. The earliness of the crop from close-spaced plants is due to other factors than the shortening of the boll period.

The usual cultivation period and a lengthened period produced an increase in square periods and boll periods over no cultivation. This may have been influenced by the increased number of bolls that matured late in the season.

Stripping forms from the plants had no perceptible effect on the square period. There was an apparent lengthening of the boll period as the date of stripping was delayed, but this effect may have been due to the time of the season. Stripped plants grew taller and produced more late blooms and bolls, matured a greater percentage of the late-set bolls, and remained green longer in the autumn.

Cleveland and Webber 49 bolls, which required 45 days or more to open, reached their full dimensions in about 18 days. The boll wall increased in thickness for a time and then became thinner. The resistance of the boll wall to weevil puncture increased from the time of blooming until near maturity.

Position of Bolls on Plant.—Studies made by Loomis[22] and others tend to show that the position of the square or boll on the plant has much to do with its rate of development and the quality of cotton produced. Loomis observed that among squares of the same date those on the outer nodes of fruiting branches required a longer period to develop into flowers than those on the inner nodes, closer to the main axis of the plant. The maturation period of bolls of the same flowering date was also longer on the outer than on the inner nodes of fruiting branches. The boll period on node 2 was apparently lengthened by the presence of a boll on node 1. When node 1 bore no boll, the boll on the second node developed in as short a time as a boll on the first node would have required. The percentage of square and boll shedding increased toward the outer end of the fruit branches, and there was more shedding on fruit limbs on vegetative branches than on fruit limbs attached to the main stem.

Venkataramanan,[18] of India, observed that there was a deterioration in the bolls as the season advanced. Seeds were lighter, lint lighter and shorter, and the maturation period of bolls longer. Spinning tests also showed that the earlier picked cotton was better and more suitable for a higher standard of warp counts.

Effect of Defloration on Yields.—King and Loomis[13] have found that for normal boll development it is necessary that there be translocated to each boll on a plant approximately 0.17 gram of organic material per day. The leaf area of a plant is an important factor in the determination of the daily quantity of carbohydrates manufactured. An early setting of bolls has been observed to decrease markedly plant growth and leaf expansion. On account of this fact, earliness in cotton may, under some conditions, result in lower yields. To test this matter experimentally, King and Loomis[13] removed all floral buds and bolls from alternate plants in a row of 200 Acala plants for the first three weeks in the flowering period, or until July 23. The deflorated plants made more vegetative growth immediately after the forms were removed and later did more blooming than the control plants. They yielded 30.6 pounds of seed cotton when picked, while the control plants yielded only 24.2 pounds.

Effect of Removing Bracts of Flowers and Bolls.—Kearney[19] found that cutting off the involucres of Pima cotton flowers on the day of anthesis caused very significant reduction in the external dimensions of the boll, in the weight of the dry contents of the boll, in the weight of individual seeds, and in the abundance of lint. The length of the period from anthesis to the opening of the bolls was shortened by 3 to 4½ days.

The reduction in the weight of individual seeds was virtually the same whether the involucres were removed from only 4 flowers on each plant or from all the flowers produced during a period of 6 weeks, an average of 48 flowers per plant. This indicated that the involucre contributes to the nutrition of the individual boll and but little to the nutrition of the plant in general.

It occurs to the writer that the lack of development of the bolls, where the involucral bracts were removed, may have been due in considerable part to lack of boll protection against sunlight and desiccation.

Effect of Radiation.—A number of experiments have been conducted in an effort to produce mutations in cotton by the use of X rays. These experiments and the general effect of radiation on cotton plants will be discussed in Chap. IX.

Shedding.—By shedding is meant the abscission, or dropping, of forms on the cotton plant. These consist of squares or flower buds, flowers, and bolls or fruits in various stages of development. Shedding is discussed as a separate subject because of its importance as a physiological problem, the interest that it has aroused, the investigational study given it, and its economic importance.

Mechanism of Abscission.—The following account of the anatomical structures involved in abscission of cotton forms is quoted by Lloyd[4] from Balls:

> The abscission layer across the flower stalk of cotton is very simple. No elaborate corky tissue is formed. A plate of cells one cell thick stretching across the stalk in a definite position, which is marked by a faint external groove, starts to divide from the periphery inwards. Each cell divides once, the two daughter cells separate instead of remaining united, and the continuity of the tissues is thus interrupted very quickly. The sole connection left is the vascular tissue of the wood, which readily breaks and the dead flower drops to the ground. The separation of the leaf is accomplished in the same way.
>
> In consequence of this mechanism, the effect follows the stimulus very quickly . . . (Balls, 1911, p. 227). . . . The dividing wall between the daughter cells splits immediately along its middle lamella. The daughter cells which are left on the face of the scar, after the stalk has broken away, bulge outward and form a simple callus (Balls, 1912, p. 69).

Shedding Due to Injuries.—Injury to squares, bolls, stem, or roots of a cotton plant may cause a shedding of forms. These injuries may be due to boll weevils or other insects, bacteria, or fungi, or they may be in the form of mechanical injuries that remove leaves or destroy roots.

Squares make response to injury, according to observations made in Alabama, by Lloyd,[4] throughout a period from 36 hours to 10 days. The majority fall the second day, but the high rates continue until the fifth day. Open flowers are rarely shed. Anthesis seems to inhibit abscission temporarily. Bolls may be shed in response to injury in 24 hours to 6 days. Small bolls are shed sooner than large ones, the maximum shedding being in 48 hours. Large ones are shed in from 3 to 6 days. Shedding of squares due to boll-weevil injury takes place irregularly, ranging from 1 to 20 days, with the majority falling in about 8 days. This is probably governed partly by the location of the weevil larva within the flower bud and the part eaten first.

Ewing[6] found that the length of time between flowering and shedding of bolls varied considerably with the variety and with the time of season. The mean persistence period for the variety Simpkins was 6.97 days; that of Columbia was 8.84 days. The period of persistence was longest when shedding first began and shortest at the close of the season.

Normal Shedding.—Much shedding of forms is due to causes other than diseases and mechanical injuries such as were mentioned above. These are of normal occurrence and are commonly considered as being caused by physiological troubles. This normal shedding is slight early in the season while plants are young but gradually increases during the season. For a time after blooming begins, flowering is considerably greater than shedding, but later the bolls shed per day may equal or even outnumber the flowers produced. Even where there are no insects to bother, more than half the flowers that open fail to make bolls that mature. The percentage of bolls set varies considerably with the variety. In counts made by the writer at State College, Mississippi, this ranged from 29 to 51 per cent for different varieties in the same test. Varieties like Trice, which produce a very large number of flowers, shed a higher percentage of bolls than varieties that have a smaller number. At Holly Springs, Miss., in 1914, Ewing[6] tagged 73,752 flowers. Only 40.31 per cent of these produced bolls. The weather and soil conditions there were very favorable, and a large crop was made—more than 500 pounds of lint per acre. Balls observed 40 per cent shedding in Egypt, while Harland, mentioned by Ewing,[6] observed 80 to 90 per cent in St. Vincent, in the West Indies.

Most of the normal shedding affects young bolls only. Squares are shed as a result of insect injury or fungous disease but rarely otherwise. Flowers are rarely shed.

Cause of Shedding.—As has been shown previously, rain during the period that cotton flowers are open interferes with effective pollination; this causes a higher percentage of the young bolls to fall on account of lack of fertilization. Hodson, mentioned by Lloyd,[4] showed that pollen grains burst when placed in water. Consequently, wetting pollen will ruin it. A rain during the morning after the anthers have opened but before effective pollination has taken place may do considerable harm.

It is probable that soil-water conditions have more bearing on shedding than any other factor. Ewing[6] and Lloyd[4] both observed that there was a gradual increase throughout the season. This, Lloyd considered, was probably caused by recession of water from plant roots in the deeper layers of soil. There are also minor fluctuations which are caused by daytime rains' interfering with pollination and by rains' causing changes in the moisture content of superficial layers of soil.

During periods of prolonged drought, when the water supply in the soil is scanty, shedding becomes excessive. Cutting a part of the roots of the plant so that the water supply is lessened causes an increase in shedding, but shading the plants after cutting the roots prevents a part of the loss. Irrigated plants shed more immediately before watering than just after. Increased shedding follows quickly if transpiration is in excess of absorption. The reduction of the water content of the plant results in the closure of the stomata; this interferes with the metabolic activities of the cells of the plant, and an increase in internal temperature occurs. Increased shedding is directly caused by these changes.

It has often been observed by planters that there is excessive shedding of forms during seasons of heavy rainfall. The excess water in the soil prevents the plant roots from functioning properly, and abnormal conditions prevail within the plant. Balls,[2] from his studies of the cotton plant in Egypt, showed that an excessive supply of water in the soil, such as resulted from an early rise of the water table in the soil along the Nile River, may cause root asphyxiation, interfere with absorption, and thus cause shedding.

According to studies made by Ewing[6] and Harland,[7] the productiveness of soils does not greatly affect the percentage of shedding. Plants growing on good soil produce many more forms than ones on poor soil, but, on the other hand, they shed many more forms. Lloyd[4] offers the suggestion that nitrate of soda may be influential in lessening the loss of forms. Manures and fertilizers make plants much larger and more vigorous and increase their capacity for production.

Hawkins *et al.*[20] recently made extensive studies of the physiological factors affecting cotton shedding in Arizona. They found that the amount of available soil moisture, through its influence on food conditions within the plants, is a major factor in regulating the fruiting behavior of cotton plants. When there was a scarcity of plant food within the plant available for the development of young bolls, shedding increased.

From the middle of the fruiting season until near its close there is a heavy drain on the energy and food materials and within the cotton plant to support the growing and developing bolls. When the plant has all the load it can carry, new forms must be thrown off. This probably accounts for at least a considerable portion of the increased shedding during the latter part of the season. Observations by Ewing[6] give credence to this theory. He found that the removal of half the flowers that opened reduced the rate of natural shedding to 8.7 per cent. The normal loss of the check plants was 60 per cent. The flower pruning was started July 27, a time when shedding was becoming active.

Kearney and Peebles[21] concluded from their studies of hybrid cottons that shedding is partly controlled by genetic factors. This is probably somewhat indirect. Strains of cotton with much vigor and low fruiting propensities will inherit these qualities and do less shedding.

References

1. Balls, W. L.: "The Development and Properties of Raw Cotton," A. & C. Black, Ltd., London, 1915.
2. Balls, W. L.: "The Cotton Plant in Egypt," Macmillan & Company, Ltd., London, 1919.
3. Balls, W. L.: "The Influence of Natural Environmental Factors upon the Yield of Egyptian Cotton," *Phil. Trans. Roy. Soc. London*, ser. B, vol. 208, pp. 157–223, 1916.
4. Lloyd, F. E.: "Environmental Changes and Their Effect upon Boll Shedding in Cotton," *Ann. N. Y. Acad. Sci.*, vol. 39, pp. 1–131, 1921.

5. King, C. J.: "Water-stress Behavior of Pima Cotton in Arizona," *U. S. Dept. Agr. Bull. Bur. Plant. Ind.* 1018, 1922.

6. Ewing, E. C.: "A Study of Certain Environmental Factors and Varietal Differences Influencing the Fruiting of Cotton," *Miss. Agr. Exp. Sta. Tech. Bull.* 8, 1918.

7. Harland, S. C.: "Manurial Experiments with Sea Island Cotton in St. Vincent with Some Notes on Factors Affecting the Yield," *West Indian Bull.*, vol. 16, No. 3, 1917.

8. King, C. J. and Loomis, H. F.: "Factors Influencing the Severity of the Crazy-top Disorder of Cotton," *U. S. Dept. Agr. Bull.* 1484, 1927.

9. Hubbard, J. W.: "Root Constriction of Cotton Plants in the San Joaquin Valley of California," *Jour. Agr. Research*, vol. 44, No. 1, 1932.

10. Ivanishin, I. A.: "Vernalization of Cotton Seed in New Cotton Regions," published at Tashkent, U.S.S.R., October, 1934.

11. Arndt, C. H.: "Effect of Soil Temperature on Cotton Seed Germination and Seedling Growth," *45th Ann. Rept. S. C. Agr. Exp. Sta.*, 1932.

12. Camp, A. F. and Walker, M. N.: "Soil Temperature Studies with Cotton," *Fla. Agr. Exp. Sta. Bull.* 189, 1927.

13. King, C. J. and Loomis, H. F.: "Agricultural Investigations at the U. S. Field Station, Sacaton, Ariz., 1925–1930," *U. S. Dept. Agr. Cir.* 206, 1932.

14. Harris, J. Arthur, Harrison, George J. and Lockwood, Edna K.: "A Criterion of the Differentiation of Varieties or of Experimental Areas with Respect to Their Capacity to Produce Seedling Stands of Cotton," *Jour. Agr. Research*, vol. 38, No. 11, 1929.

15. Konstantinov, N.: "The Influence of Period of Illumination on the Development of Cotton," Science Research Institute. Reviewed in *Empire Cotton Grow. Rev.* 13: 1, 1936.

16. Berkley, Earl E.: "Studies of the Effects of Different Lengths of Day, with Variations in Temperature, on Vegetative Growth and Reproduction in Cotton," *Ann. Mo. Bot. Gard.*, vol. 18, No. 4, 1931.

17. Ludwig, C. A.: "Some Factors concerning Earliness in Cotton," *Jour. Agr. Research*, vol. 43, No. 7, 1931.

18. Venkataramanan, S. N.: "The Characters of the Cotton Boll in Relation to Its Flowering Period and Position on the Plant," *Agr. Jour. India*, vol. 25, Part 3, 1930.

19. Kearney, T. H.: "Development of the Cotton Boll as Affected by Removal of the Involucre," *Jour. Agr. Research*, vol. 38, No. 7, 1929.

20. Hawkins, R. S., Matlock, R. L. and Hobart, Charles: "Physiological Factors Affecting the Fruiting of Cotton with Special Reference to Boll Shedding," *Univ. Ariz. Tech. Bull.* 46, 1933.

21. Kearney, T. H. and Peebles, R. H.: "Inheritance of Rate of Shedding in a Cotton Hybrid," *Jour. Agr. Research*, vol. 34, pp. 921–926, 1927.

22. Loomis, H. F.: "Development of Flowers and Bolls of Pima and Acalà Cotton in Relation to Branching," *U. S. Dept. Agr. Dept. Bull.* 1365, 1927.

CHAPTER VI

REPRODUCTION IN THE COTTON PLANT

FLOWER DEVELOPMENT, SPERMATOGENESIS, FERTILIZATION, AND THE DEVELOPMENT OF COTTON BOLLS

Double fertilization in flowering plants was discovered only a few years ago, comparatively. Since that time, oögenesis and fecundation have been studied in a number of species, but only a very small percentage of the whole number of species has been investigated. Oögenesis and fecundation in Egyptian cotton (*Gossypium barbadense* L.) was studied by Balls.[1] More recently, Gore[6] investigated the development of the female gametophyte and embryo in Sea Island (*G. barbadense* L.), Pima Egyptian (*G. barbadense* L.), and two varieties of upland cotton (*G. hirsutum* L.), Delfos 6102 and Mebane; and Reeves and Beasley[7] studied the development of the embryo in another variety of upland known as "Startex." Spermatogenesis and the development of cotton fibers have been investigated by a number of workers. From the studies made it appears that there are no marked peculiarities in the reproductive process of cotton.

Development of Flowers.—The rate of growth of cotton plants and of development of flowers varies considerably with the weather conditions prevailing in different seasons and in different parts of the same season. In the central part of the Cotton Belt, cotton planted the middle of April shows young flower buds or squares in 35 to 45 days. Cotton planted later in the season develops more rapidly. As has been shown in Chap. IV, flowers are borne on special branches, the fruit branches. Morphologically considered, the first flower bud terminates the fruit branch. Further extension is made by growth from a bud in the axil of the leaf just below. This is terminated by the next flower; and so on. From the appearance of mature fruit branches with bolls and flowers along the side, one is inclined to think they are simply lateral outgrowths.

The first step in flower formation consists in the outgrowth of a rounded lobe of embryonic cells at the tip of the stem. Next, from the sides of this first lobe, just a short distance back of the apex, three other lobes arise. These give rise to the three bracts which grow up around the developing flower parts. A second set of lobes just within the first forms the calyx. Similar lobes within the calyx make corolla and staminal column. According to Balls's[1] observation, the two latter are developed nearly simultaneously. Anthers arise as small lobes on the upper part of the staminal column. The pistil, the last part of the flower to develop, is made from three to five lobes which grow up in the central part of the flower (see Fig. 29*A*). These lobes grow together along the inner edge, each forming a hollow cone, which together form the compound ovary containing three to five cells. The tip portions grow into a beak, which lengthens into style and stigma.

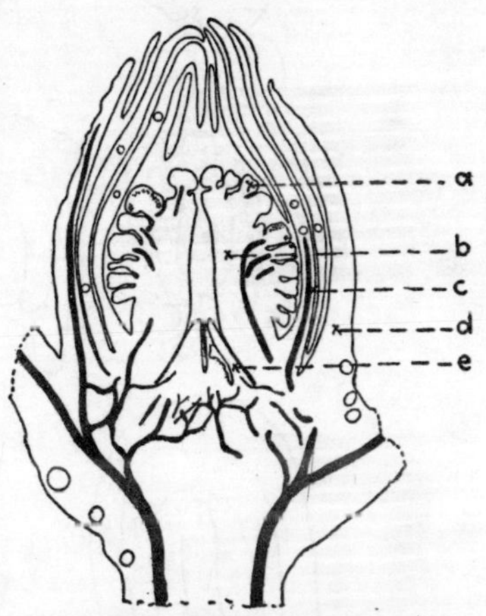

Fig. 29*A*.—Nearly median longitudinal section of flower bud of Mebane upland cotton when flower parts are in process of development. (*a*) Stamen; (*b*) staminal column; (*c*) petal; (*d*) sepal; (*e*) carpel. (*After Gore.*)

The inner edge of each carpel, or the part at the central axis of the boll, forms a ridge or placenta, from which ovules arise. The ovule appears at first as a mere bulge or lobe of embryonic cells on the side of the placenta. This lobe grows outward rather rapidly and forms the nucellus of the ovule, but before it has attained much size another outgrowth starts in the form of a ring about the base of the nucellus. This ring grows up around the nucellus as a protecting envelope and is known as an "integument." Before the first, or outer, integument has surrounded the nucellus, another similar integument appears inside and just above the first and is known as the "inner integument." The two integuments completely surround the nucellus except for a small opening left at its apex and known as the "micropyle" (Fig. 30*b*, *d*, and *i*). While the ovule is enlarging and developing its integuments, it apparently grows more on one side than on the other, especially at the base, thus producing a curved, or anatropus, ovule.

Oögenesis.—As observed by Gore,[6] the archesporial cell in cotton originates from a subepidermal cell of the nucellus and

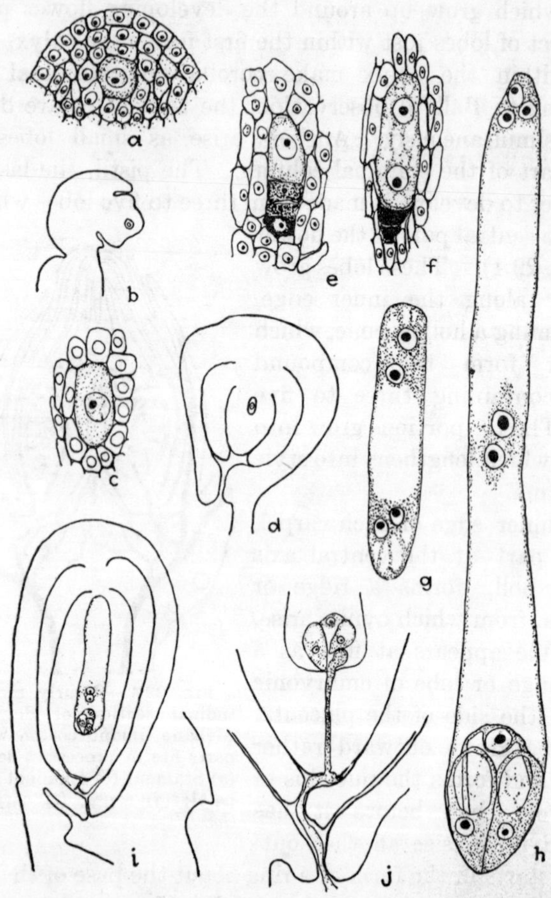

FIG. 30.—Early stages in the development of ovule and female gametophyte of cotton. (*a*) Primordia of ovule with sporogenous cell; (*b*) longitudinal section of young ovule showing young integuments and sporogenous cell; (*c*) megaspore mother cell just before reduction division; (*d*) young ovule from which drawing *C* was taken; (*e*) tetrad of megaspores, the three micropylar ones disintegrating; (*f*) two-nucleate female gametophyte; (*g*) four-nucleate female gametophyte; (*h*) completely organized female gametophyte showing the two synergids, egg nucleus, polar nuclei, and the antipodals about one week before flowering; (*i*) longitudinal section of ovule taken on the day of flowering; (*j*) portion of ovule showing growth of pollen tube through integuments and nucellus. (*After Gore.*)

arises before the differentiation of the integuments. From its division a primary parietal and a primary sporogenous cell are formed (Fig. 30*a*). This sporogenous cell enlarges and becomes the megaspore mother cell. The primary parietal and other cells about it multiply, thus causing the megaspore mother cell to lie deeper in the developing ovule (Fig. 30*d*). The megaspore mother cell next divides twice, the first of these divisions being the meiotic, or reduction, division in which the number of chromosomes is reduced one-half. As a result of these divisions, four megaspores are produced (Fig. 30*e*). The three next to the micropyle soon disintegrate, and the other one enlarges greatly to form the female gametophyte (Fig. 30*f*, *g*, and *h*). The nucleus of this enlarged embryo-sac cell divides three times, producing eight daughter nuclei. Four of these pass to either end of the sac. Then one from each group moves toward the center of the sac, where they meet and later fuse. The three nuclei left in the end of the embryo sac farthest from the nucellus disintegrate and function no further. The nuclei at the other end of the enbryo sac are formed into cells by the aggregation of cytoplasm about them and a limiting cell wall (Fig. 30*i* and *j*). One of these cells, usually the central one, enlarges more than the others and becomes the egg cell, or female gamete. The other two perform no important function, in so far as is known. In this condition, the embryo sac is fully developed and is ready to receive the male gametes or, in other words, is ready for fertilization. The development of the male gametes will next be traced.

Spermatogenesis of Cotton.—A number of investigators have studied spermatogenesis in cotton. Cannon,[2] about 1903, studied spermatogenesis in a first-generation hybrid between Sea Island and upland cotton. As it was a hybrid and the material used was grown in a greenhouse, some doubts have arisen as to whether or not the phenomena observed were of normal occurrence. Recent studies by Denham[3] have verified Cannon's work with few exceptions. Balls,[1] in 1905, published a paper in which was included a description of spermatogenesis in Egyptian cotton. His description differed from Cannon's in several particulars, especially in regard to the appearance of chromosomes and their number. Balls counted 20 chromosomes as the haploid, or reduced, number, whereas Cannon counted 28. Denham counted but 26 chromosomes in the Sea Island cotton that he studied. He considers that the peculiar forms observed by

Balls were due to faulty fixation and doubts the correctness of the count made by Balls. Beal,[8] Longley,[9] Webber,[10] and others have verified the counts made by Denham,[4] and it is now very well established that 26 is the haploid or reduced number of chromosomes in the cultivated American upland and in Egyptian and Sea Island varieties; and that 13 is the haploid number of the Asiatic cottons and the wild American species, *Gossypium klotzschianum* Anderss., *G. davidsonii* Kellogg, *G. harknessii* Brandeg., and *G. armourianum* Kearney.

In the following discussion, the views of Denham[3] are followed largely. About the time the young bracts meet at the top of the small flower bud or square, the developing anthers may be distinguished; and a week or 10 days before the flower opens, the microspores or young pollen grains are to be found within the anther. (The opening of the flower takes place about 3 weeks after the bud or young square can be distinguished.) The pollen grains do not reach mature form until 2 or 3 days before the flower opens.

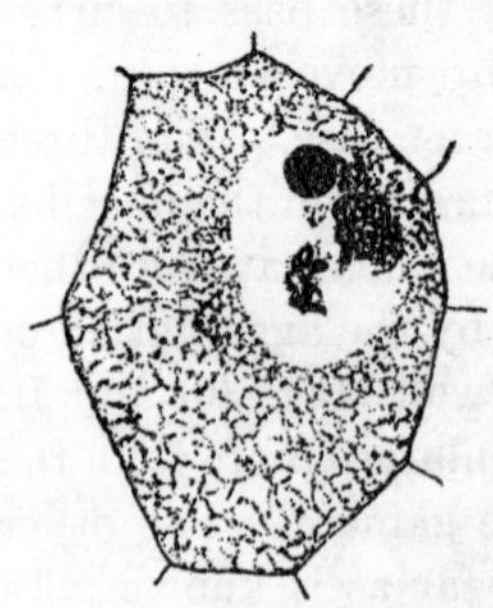

Fig. 31.—Mature pollen mother cell. (*After Cannon.*)

Portions of the subepidermal layer of cells in the young anther differentiate to form the archesporium. This archesporial layer of cells divides to form three distinct layers, the middle layer of which forms, after a series of divisions, the pollen mother cells; the other two layers form the tapetum, or nurse cells. The pollen mother cells enlarge rapidly and become filled with dense, granular cytoplasm (Fig. 31). The first indication of the beginning of the reduction division of the pollen mother cell is to be seen in the contraction of the cytoplasm of the cell. This contraction results in a rounding off of the cell and its separation from the surrounding tissue (Fig. 32). The reduction division of the cell nucleus reduces the number of chromosomes in the cell to half the number found in ordinary vegetative cells. This is the essential step in the development of distinct male sex cells, the male gametes. These, as will be seen later, are to be found in the pollen tube of the germinating pollen grain. As the nucleus of the pollen mother cell starts to divide, it enlarges, the network of fine achromatic threads becomes more prominent, and the small lumps of chromatin on the threads enlarge. A little later, these seem to break up into smaller granules which tend to locate on the network at points where threads intersect. In some plants, about this stage of nuclear division there is formed from the network mentioned a slender, much elongated thread, parts of which appear to become paired with other portions and to fuse. This is considered by some authorities to be the combining of elements derived from each parent, elements that have been carried as separate entities in the plant cells through their numerous generations from the fertilized egg to the pollen mother cell. Neither this continuous thread nor its pairing has been clearly distinguished in cotton.

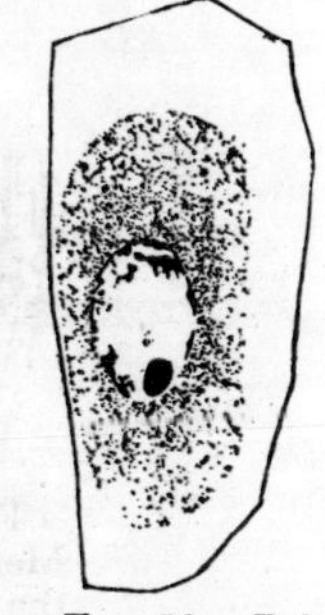

Fig. 32.—Pollen mother cell with cytoplasm withdrawing from the cell wall, an early stage of cell division. (*After Cannon.*)

In the next stage of nuclear division in cotton, the whole body of chromatin and thread network of the nucleus becomes massed at one side of the nucleus. This is known as "synapsis." This mass is so dense that nothing can be determined as to what is taking place within, but it is generally thought that in this knot there is a conjugation of parental elements and that this stage in the division is of much genetic significance. This stage of the nuclear division is of longer duration than any other. Passing from the synapsis stage there is a loosening of the synaptic knot, and from it there comes a continuous thick thread, or spireme, which spreads itself in a more or less looped fashion throughout the nuclear cavity. Following this, there is normally a contraction or shortening of the spireme; this is soon followed by its segmentation or breaking into 26 pieces, the double or bivalent chromosomes. These are in the form of irregular loops or rings. Denham considers that conjugation has taken place telosynaptically and that two parental chromosomes are arranged end to end in the bivalent chromosome.

The bivalent chromosomes are next moved to a plane across the central part of the cell; a bipolar spindle of achromatic fibers is formed, some of the fibers of which are attached to the chromosomes. The fibers of the spindle appear to shorten and pull apart the halves of the bivalent chromosomes (Fig. 33). The halves are drawn to the poles of the spindle (Fig. 34) and there disintegrate as daughter nuclei are formed. This marks the end of the heterotypic phase of the reduction division. The second, or homotypic, follows soon after.

In the homotypic division, a spireme is formed which segments into 26 univalent chromosomes. These are moved to the equatorial plate of the cell, as mentioned; a bipolar spindle is formed, and the chromosomes split, one-half of each going to each spindle pole. Here the chromosomes disintegrate, forming

FIG. 33.—Metaphase of the reduction division. (*After Cannon.*)

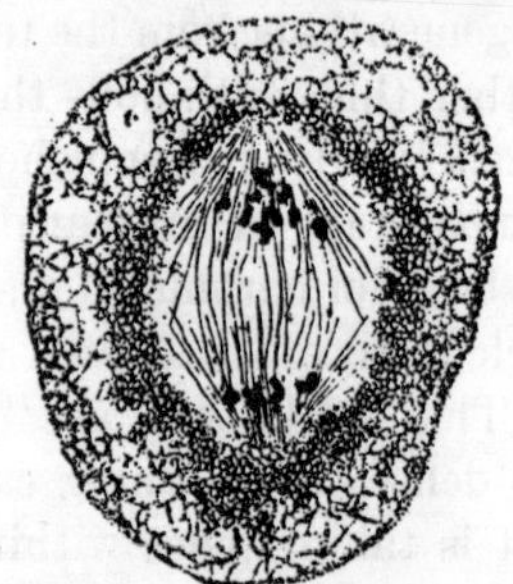

FIG. 34.—Anaphase of the reduction division. (*After Cannon.*)

small granules, which are scattered throughout the two daughter nuclei formed about the poles of the spindle.

The two divisions, heterotypic and homotypic, take place in a comparatively short time. Balls has estimated that not more than 12 hours elapse between synapsis and the tetrad of nuclei.

In cotton, no cell plate separating the cells organized about the nuclei of the tetrad is formed. The cells of the tetrad are separated by furrows, or constrictions, which appear on the periphery of the cytoplasm, midway between the nuclei, and deepen until they meet in the center. After the formation of the tetrad of cells, their cell walls thicken rapidly and form the elaborately figured spiny coat seen in the mature pollen grain (Fig. 35). Two or three days before the flower opens, according to Balls,[1] the tetrad cell divides to form a large tube cell and a small generative cell. The generative cell later divides to form two male gametes, or sperm cells.

Pollination.—As has been shown, pollen grains are produced in abundance by the numerous anthers of cotton flowers. A mature grain is spherical and spiny, has irregularly sculptured walls, and is barely visible to the naked eye (Fig. 35).

Quantities of pollen are deposited on the stigmas of cotton flowers (Fig. 36). Most of it, as is shown in Chap. X, comes from anthers in the same flower. It is deposited either mechanically or by insects visiting the flowers.

Fertilization.—Pollen grains germinate soon after they land on the surface of the stigma. A tube is sent out from the grain; it enters the stigma, grows down through the style, enters the ovary, and keeps growing until it has entered an ovule. It squeezes through the micropyle of an ovule and then plows its way through the tissue of the nucellus until the embryo sac is reached. At this point, the end of the tube bursts, and the two germ nuclei, or male gametes, are emptied into the sac. One of the nuclei moves to the egg cell, enters it, and fuses with the nucleus of the egg cell, thus bringing about fertilization. The other male gamete fuses with one of the two polar nuclei that have moved to the center of the embryo sac. The two polar nuclei now fuse, and there is produced a triple-fusion nucleus, which is known as the "endosperm nucleus." This divides to produce the endosperm of the ovule.

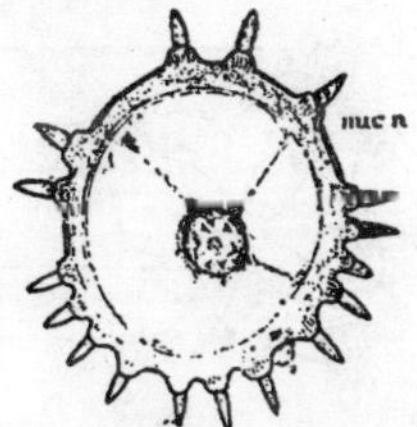

FIG. 35.—Cross section of mature pollen grain showing spines on surface of grain. (*After Balls.*)

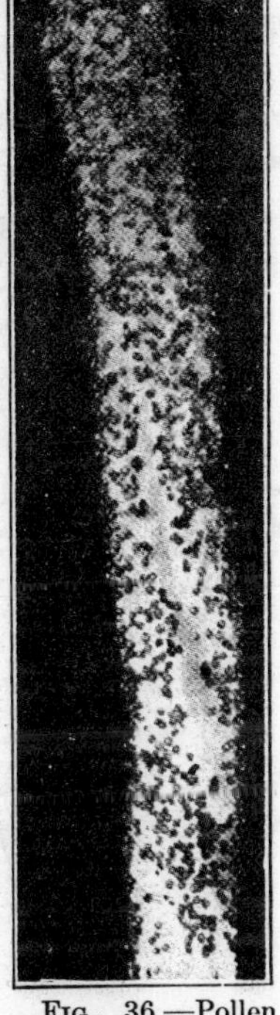

FIG. 36.—Pollen grains on stigma of a cotton flower. This was taken from an open pollinated flower. (*After Kearney.*)

The growth of the pollen tube and fertilization take place in remarkably short time, when it is considered that the pollen tube must grow 2 inches or more. Balls[1] and Gore[6] have observed that fertilization has taken place within 30 hours after the flower opened.

Endosperm.—As interpreted by Balls,[1] the triple nucleus, resulting from the fusion of two polar nuclei in the embryo sac

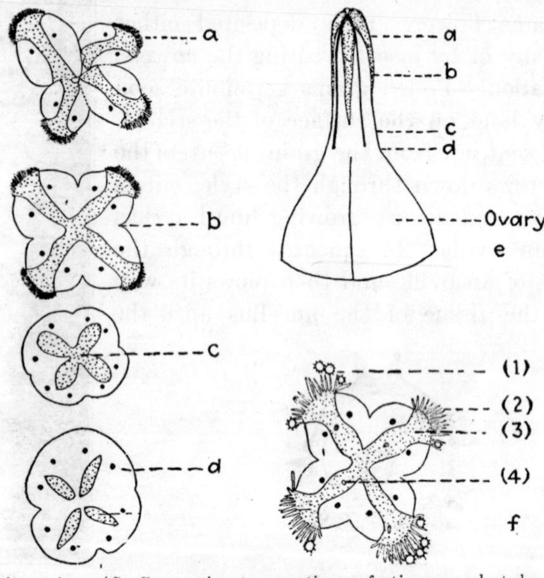

FIG. 36*A*.—(*a*) to (*d*), Successive transections of stigma and style of Mebane upland cotton; (*e*) habit sketch of half-grown pistil showing where transections were made, stigmatic region stippled; (*f*) transection of upper part of style of upland cotton made on day of blooming; (1) pollen grains; (2) stigmatic surface; (3) conducting region; (4) transection of pollen tube. (*After Gore.*)

and the second male nucleus, divides on the afternoon of the day following pollination. The nucleoli of the parent nuclei retain their individuality for a few divisions, showing two large nucleoli and one small one which came from the male gamete. The chromosome behavior is very irregular. Nuclei multiply rapidly and arrange themselves in a single layer in the protoplasm lining the embryo sac. The central part of the embryo sac is occupied at this time by a large vacuole filled with cell sap. Soon cell walls form between the free nuclei, beginning at the end of the sac next to the micropyle. Nuclear division and cell-wall formation continue until all the lower end of the embryo sac is filled with thin-walled cells, and the young embryo is surrounded. As the embryo enlarges, it appears to digest the surrounding endosperm cells and to occupy their space. In the ripe seed, the endosperm and the nucellus have so nearly disappeared that only a vestige remains—a thin, papery membrane.

The Embryo.—According to Balls's[1] observations, the egg cell of Egyptian cotton divides about 60 hours after fertilization is effected. The first division is at right angles to the axis of the embryo sac. The next division is at right angles to the first. Further divisions come in rapid succession. By the time the young embryo is a week old it contains hundreds of cells and has assumed a somewhat cordate form, which indicates the beginning of cotyledons. The cotyledons grow out into broad, thin sheets, become much folded, and fill almost all the space within the seed coats (see Fig. 23).

The cells of the cotyledons of the full-size embryo contain a quantity of oil and other concentrated foods. From kernels of cotton seeds, which are principally embryo cotton plants, are obtained cottonseed oil, cottonseed meal, and other important products.

Polyembryonic seeds have been observed by different workers Harland[11] recently reported finding 20 seeds containing two embryos in several thousand Sea Island seed that he examined. One of the embryos was usually much smaller than the other. Sixteen pairs of mature plants were raised from 20 seeds mentioned. Fourteen of the pairs were found to consist of one haploid and one diploid plant. Both of the plants in the other two pairs had the diploid number of chromosomes. These haploid plants produced no viable pollen but were somewhat fertile as a female parent, producing a few seed when pollinated with normal Sea Island pollen and when pollinated with some of the 13-chromosome species, such as *G. arboreum* and *G. armourianum*.

Development of Cotton Seeds.—When the egg cell in an ovule is fertilized by the male gamete, the ovule becomes a young seed. It is then less than 1/16 inch in length but grows rapidly, attaining full size, or a length of about 2/5 inch, in about 20 days. The bulk of the mature seed consists of embryo. The integuments of the ovule become the seed coats. They enlarge and thicken as the seed grows, differentiating various parts. (For details of form of embryo and structure of seed coat, see discussion of seed structure in Chap. IV.)

The epidermis of the outer integument has cells with a thick basal wall and thinner outer and side walls. The outer wall is covered with a thin cuticle. Certain of the epidermal cells bulge outward; the protoplasm and nucleus of the cell move toward the outgrowth and later enter it (Fig. 37*a*, *b*, *c*, and *d*). The outgrowth of the cell continues to grow until a length of an inch or

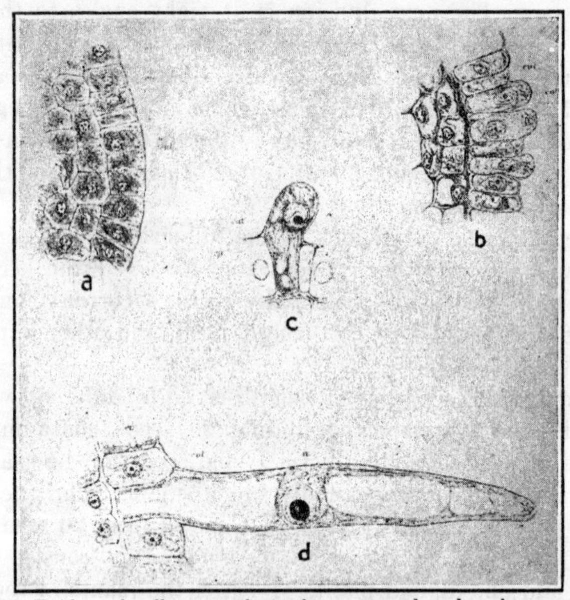

FIG. 37.—Epidermal cells on surface of young seed and early stages in the development of young fibers: (*a*) Undifferentiated epidermal cells (row along the right edge); (*b*), (*c*), and (*d*) early stages in fiber development. (*After Balls.*)

more is attained in some instances. The longer outgrowths make lint hairs, and the shorter ones make fuzz. The development of lint and fuzz will be discussed in more detail in Chap. VII.

There are commonly nine ovules in an ovary cell—the space occupied by a single lock of cotton. Under good conditions, these nine ovules make nine seeds, the usual number per lock, but it frequently happens that one or more fail to develop. The failure may be due to lack of nutrition or to lack of fertilization. They frequently make rudimentary seeds, which have a hardened seed coat and fuzz but contain no embryo or kernel. Such rudimentary seeds form an impurity in the lint cotton and are known as "motes." Since bolls of upland cotton contain four or five locks, they contain approximately 36 or 45 seeds each.

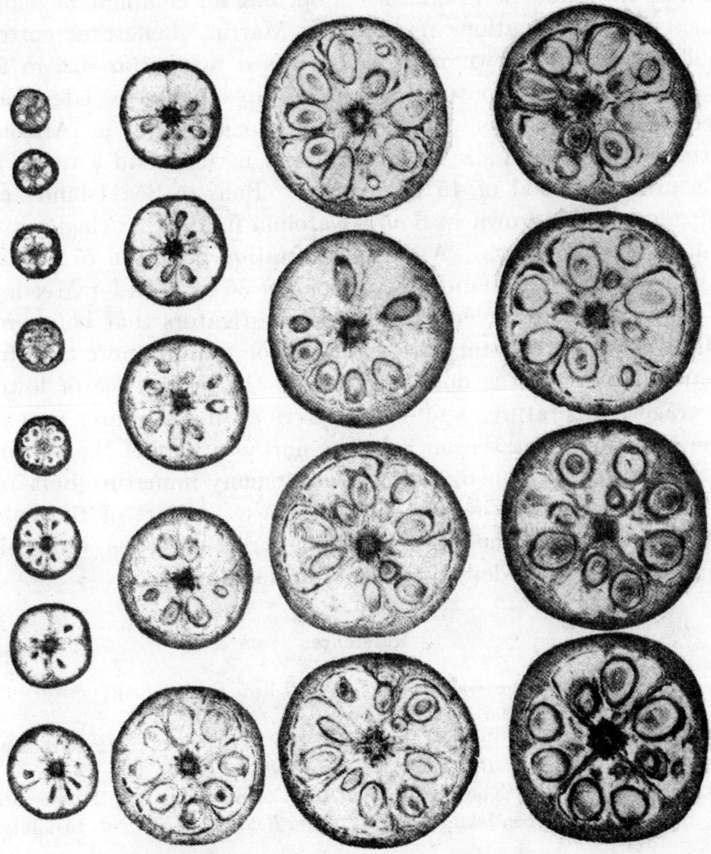

FIG. 37A.—Median cross sections of bolls of *Gossypium hirsutum* L. showing daily increase in size from the date of flowering to the twenty-first day of development. (*After Farr.*)

Development of Bolls.—As has been seen, the cotton boll is the enlarged and developed ovary of a cotton flower. The length of time required for the development and ripening of a boll varies somewhat with the variety of cotton and the time of the season at which blooming occurs. The maturation period ranges normally from about 45 to 65 days. During the first few days following the opening of the flower, the young boll enlarges rapidly (see Fig. 37*A*). It attains full size in about 20 days, but at this time the seeds and lint are very immature. They require almost the whole period until boll opening for complete development. In observations reported by Martin,[5] Lone Star cotton bolls on plants grown in Texas attained maximum size in 20 days; early bolls matured in 42.57 days, whereas late bolls required 44.55 days. Pima cotton bolls grown in Arizona attained maximum size in 25 days and in 1921 had a range in maturation period of 45 to 80 days. Bolls of Sea Island and Meade cottons grown in South Carolina in 1922 reached maximum size in 21 days. A mean maturation period of 57.6 days was noted for Sea Island, and a mean of 56.14 days for Meade.

It has been observed by several investigators that bolls produced during the latter part of the season require more time for maturation. This is doubtless due to the prevalence of lower average temperatures and shorter days during the later part of the cotton-growing season. In the northern part of the Cotton Belt, it usually happens that there are many immature bolls on cotton plants when a killing frost comes. A part of the bolls that have not had the full maturation period will open, but their lint is more or less defective, weak, or discolored.

References

1. Balls, W. L.: "The Sexuality of Cotton," *Year Book*, Khedivial Agricultural Society, 1905.
2. Cannon, W. A.: "Studies in Plant Hybrids: The Spermatogenesis of Hybrid Cotton," *Bull. Torrey Botan. Club*, vol. 30, p. 133, 1903.
3. Denham, J. H.: "The Cytology of the Cotton Plant. 1. Microspore Formation in Sea Island Cotton," *Ann. Botany*, vol. 38, No. 151, July, 1924.
4. Denham, J. H.: "The Cytology of the Cotton Plant. II. Chromosome Numbers of Old and New World Cottons," *Ann. Botany* vol. 38, No. 151, July, 1924.
5. Martin, R. D., Ballard, W. W. and Simpson, D. M.: "Growth of Fruiting Parts in Cotton Plants," *Jour. Agr. Research*, vol. 25, No. 4. July, 1923.
6. Gore, U. R.: "Development of the Female Gametophyte and Embryo in Cotton," *Amer. Jour. Botany*, vol. 19, pp. 795–807, 1932.
7. Reeves, R. G. and Beasley, J. O.: "Development of the Cotton Embryo," *Jour. Agr. Research*, vol. 51, No. 10, 1935.

8. BEAL, J. M.: "A Study of the Heterotypic Prophases in the Microsporogenesis of Cotton," *Cellule*, vol. 38, pp. 247–268, 1928.
9. LONGLEY, A. E.: "Chromosomes in Gossypium and Related Genera," *Jour. Agr. Research*, vol. 46, No. 3, 1933.
10. WEBBER, J. M.: "Chromosome Number and Meiotic Behavior in Gossypium," *Jour. Agr. Research*, vol. 49, No. 3, 1934.
11. HARLAND, S. C.: "Haploids in Polyembryonic Seeds of Sea Island Cotton," *Jour. Hered.*, vol. 27, No. 6, 1936.

CHAPTER VII

COTTON FIBERS

In a sense, cotton fibers are the units from which cotton fabrics are made and around which the whole cotton industry centers. A study of the structure and nature of cotton fibers is therefore fundamental.

Development of Cotton Fibers.—Early stages in the development of cotton fibers were traced in Chap. VI. It was shown that each fiber is produced by the outgrowth of a single epidermal cell. Until the cell bulges out on the outer side, it appears the same as all other epidermal cells. Certain of the epidermal cells produce outgrowths that continue lengthening until the long lint hair, or fiber, is made; other cells produce outgrowths which are similar to the fiber outgrowths at first but which stop growing before attaining much length. These short epidermal hairs are known as "fuzz." Certain varieties have a low percentage of lint hairs in comparison with the number of fuzz hairs. Others have a relatively high percentage of the long hairs. The percentage for any particular variety is about the same from year to year, although the environmental conditions vary.

Balls,[1] in his studies of the growth of lint hairs in Egyptian cotton, found that the lint cells attained full length during the first half of the maturation period of the boll—about 25 days. The cell wall remained very thin during this time, the nucleus persisted, and the diameter of the cell was about the same as in the mature hair. During the second half of the maturation period of the boll, a period of about 25 days, the fiber cell wall was thickened by deposits of cellulose formed on the inside of the primary wall. This appeared to be put down in layers, some fibers showing as many as 25 concentric layers. These layers were thought by Balls to represent daily growth rings similar to annual rings in trees. He found that the number of layers in a hair wall coincided almost exactly with the number of days cell-wall thickening had been in progress. It was thought that a thin layer was deposited during the night or early morning but that the noonday sun stopped the process in a way similar to the way it stopped growth of stems and branches (see Chap. V).

Since Balls's work on the development of the cotton fiber was published, fiber studies have been made by Farr,[13] Hawkins and Serviss,[14] Armstrong and Bennett,[15] Gulati,[16] Kerr,[25] Anderson and Moore,[26] and others. Their work has been on American, Egyptian, and Indian varieties and has, in general, confirmed Balls's conclusions. Kerr,[25] however, is of the opinion that the greater mass of the fiber layer is formed during the day. He found evidence to indicate that there was a relation between the warmer temperature prevailing during the day and the mass of fiber wall formed. If a warm night intervened between two successive days, the layers then formed were not so distinct or sharply defined. This was evidence that the heavier wall was laid down during the day rather than at night, but it did not prove that it was the heat rather than the light that was operative.

Anderson and Moore[26] grew cotton plants under constant light and temperature and found that no growth rings were formed in the wall of the cotton fibers, the whole fiber wall being a homogeneous mass. When the light periods were alternated with 12-hour periods of darkness, the growth rings in the fiber wall appeared, although the temperature of the plants had been held the same throughout the 24 hours. This indicated that there was a relation between the light the plant received and the formation of the growth rings.

The earlier workers thought that the fibers all, or nearly all, started growing in length the day the bloom opened. More recent studies have indicated, however, that new fibers continue to arise from the epidermal cells on the surface of the ovule for a period of 10 days or more after the flower opens. Gulati[16] reports observing epidermal cells in the process of division 10 days after the flower opened. Farr[13] saw a large number of dividing cells on the twelfth day and estimated that there was a thirty-two-fold increase in the surface of the young seed during the first 20 days. This increase in surface area furnishes room for the new fibers that arise later during the period of seed development. Lang, in some unpublished work, finds evidence that, in case of the fuzzy-seeded cottons studied, the true fibers nearly all arise the first day or two but that the fuzz fibers arise about 6 days later.

Armstrong and Bennett[15] observed that fibers had practically all reached full length in 25 days, although some were still lengthening after 30 days. Hawkins and Serviss[14] state that the elongation of fibers in Acala cotton was complete in 21 to 24 days after flowering, and in Pima Egyptian in 27 to 30 days.

Kerr[25] was able to tell just when the secondary wall started to form by staining the young fibers with chlorozinc iodide, which colors the primary wall a faint blue and the secondary wall a deep violet. The secondary thickening started from the 17th to the 20th day and continued for 30 to 50 days, the length of time depending upon the climate or part of the season in which the boll was developing. Fibers developed late in the season made slower growth and contained many more growth rings.

The minute structure of the wall thickenings of the cotton fiber is difficult to determine. The concentric layers mentioned above contain more or less distinct small fibrils arranged in irregular and reversing spirals. These fibrils, in turn, according to Farr,[18] are composed of cellulose structural units, or micellae, held together by a pectin matrix.

The growth rings of the cotton fiber may be rendered visible by treating the fibers for a short period with CS^2 and NaOH or with cuprammonium hydroxide (Schweizer's reagent), which causes the cell wall to swell several times its original thickness. The layers may be seen with a microscope that magnifies 250 diameters, but much depends on the light used.

The fiber cells are cylindrical until about the time the boll begins to open, and they contain living protoplasm until this time. With the opening of the boll there is further loss of water, and the fibers become dry. As water is lost, the cylindrical cells collapse, assuming a flattened, ribbon-like form, which shows a few to many twists (see Figs. 38 and 38*A*). The twisting of the fibers as the boll opens causes the lock of seed cotton to expand greatly.

It has been shown that the course of the twists in the fiber bears a definite relation to the course of the delicate spiral fibrils that are found in the secondary wall. When the fiber dries, the twists or convolutions follow the angle of the spiral, and reversals occur whenever the direction of the spiral is changed.

Form and Structure of a Mature Fiber.—When magnified, a mature cotton fiber appears as a flattened, more or less twisted tube, the length ranging from 1,000 to 3,000 times the diameter. The form of a cotton fiber is sometimes compared to that of an empty fire hose; the comparison is apt except that the fiber is usually considerably more twisted (Fig. 38*A*). The edges of the fiber appear somewhat thickened, but this is due to the folding of the wall. The diameter is slightly less toward the end that was attached to the seed, and the other end narrows to a point in the last fourth of the fiber length. Throughout most of its length the fiber is fairly uniform in diameter.

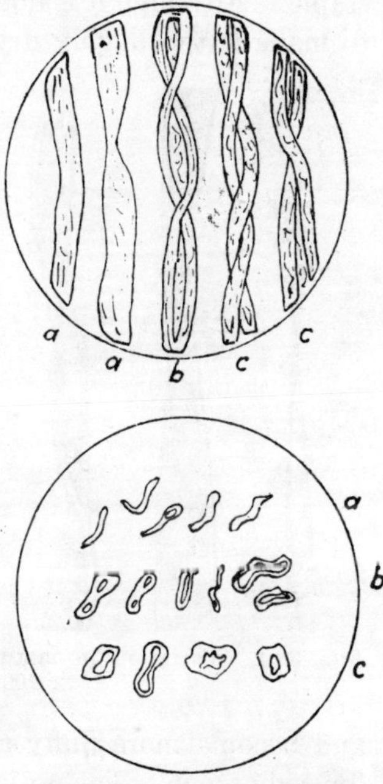

Fig. 38.—Cotton fibers in longitudinal view and cross section: (*a*) unripe fibers; (*b*) half-ripe fibers; (*c*) ripe or mature fibers. (*After Beal.*)

A sample of lint cotton, on the basis of maturity or development, shows various sorts of fibers. These are sometimes classed as (1) unripe, (2) half ripe, and (3) ripe. The unripe fibers have a very thin wall and lack twist (Fig. 38*a*). Such fibers are known technically as "dead cotton." They are of but little value. Being weak and brittle, they break up in the process of manufacture, increase the waste, and weaken the yarn. The ripe fibers are the normal type with thickened walls and good twist. In a good cotton, most of the fibers are of this type. The twists are sometimes in one direction and sometimes in the other. They occur at irregular distances. As a rule, the smaller the diameter of the fiber the larger the number of twists it has. Sea Island, a type with very fine fibers, has much-twisted fibers. The twists cause the fibers to cling to each other and thus improve their spinning qualities. The half-ripe fibers are intermediate in character and value. If cotton bolls open prematurely because of drought, defoliation of plants by insects, or any other reason, the fibers are likely to be unripe or half ripe. The fibers produced by diseased plants or diseased bolls are likely to be faulty. Certain bolls on every plant fail to develop properly on account of their location and consequently have poor fibers. This is especially true of the first bolls to set on a plant. Some fibers on every cotton seed fail to develop properly and therefore produce faulty lint. Exposing seed cotton to sun and wind for a few hours after picking is thought by some to improve the staple. Although it is doubtful if this helps the staple more than to make it thoroughly dry, it does make the cotton look better

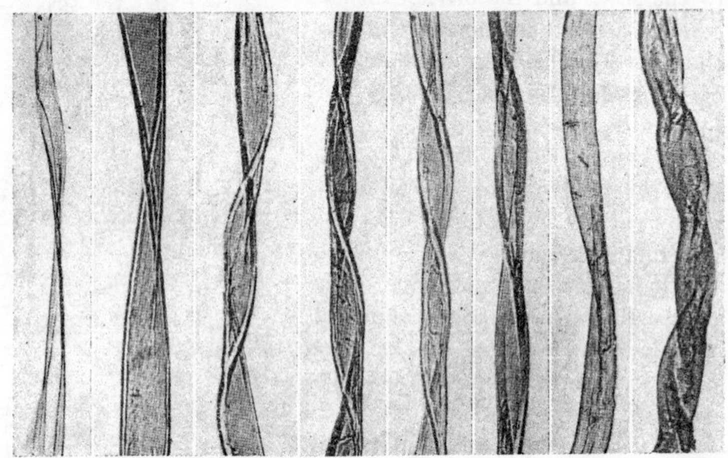

FIG. 38*A*.—Variation in maturity of fibers from American upland cotton. × 200. (*After Richardson.*)

and become more fluffy and probably improves the grade to a limited extent. The notion is prevalent, also, that the character of cotton staple is improved by storing the seed cotton some weeks before ginning. It is said that the fibers become a rich creamy yellow by taking up oil from the seed and that their tensile strength is increased. The North Carolina Experiment Station[17] reports that seed cotton stored from early October until late in February did not change in weight, ash, or moisture content of fibers and that the storage did not influence the grade and staple length. There was a slight increase in strength of fibers, but this increase was very slight when the lint was made into yarn. The change observed did not warrant the risk and expense of storing seed cotton on the farm to improve the quality of the fibers.

Levine,[3] in his study of the structure of cotton fibers, identified five different parts in a mature fiber:

1. The integument, or outer layer. This is commonly known as the "cuticle," or "waxy layer." Levine found it to contain a mixture of cutinous, pectinous, gummy, fatty, and other unidentified bodies.

2. An outer cellulose layer, which is largely the original cell wall.

3. The layer of secondary deposits. This is nearly pure cellulose. Balls,[1] as was mentioned above, observed numerous concentric layers in this portion of the fiber wall.

4. The walls of the lumen, a spiral structure surrounding the central cavity of the fiber and more dense than any other part of the fiber.

5. The substance in the lumen, structureless and of a nitrogenous nature.

Fuzz.—After the lint fibers are removed from cotton seeds, there remains on the seed, in most varieties, a velvety covering of short cotton hairs known as fuzz. The fuzz hairs, as has been mentioned, are produced by the outgrowth of epidermal cells in a way very similar to the growth of lint hairs. They are much shorter than lint hairs, ranging from ⅛ to ½ inch or more in length. The longer hairs found in linters are mostly regular lint hairs that the gin failed to remove at the first ginning. The fuzz fibers, as a rule, have a greater diameter than the lint hairs, but, as the cell wall is thin, they resemble immature or undeveloped fibers. They are light and weak in body.

Effect of the Environment of Cotton Plants on Staple.—For cotton plants to produce fiber of normal length and strength, they require rather favorable environmental conditions continuously. If the conditions are not favorable, the lint will not grow to full length. Good culture makes conditions better, especially by regulating the supply of moisture. Cook[5] expresses the belief that the Delta regions of Mississippi and Louisiana are well adapted to the growth of staple cottons because there is a more uniform supply of moisture than in many other regions. Balls[6] found, by measuring the lint length of the cotton in certain dated bolls of Egyptian cotton, that there was a definite relation between periods of irrigation and lint length. The bolls that had lint lengthening at the time the water supply was most abundant made the longest lint. A difference of nearly 2.5

millimeters was noted. The effect of the irrigation showed on the third or fourth day after watering and continued until about the tenth day. Kearney[7] reports that complaints are sometimes made by the consumers of Pima cotton that there are bales of mixed staple. This is a very uniform variety of cotton when grown under uniform conditions, but it was found that in fields where there were sandy, alkali, or more or less barren spots the lint produced by plants grown on those spots was decidedly inferior. This explained the finding of mixed staple in certain bales, since seed cotton from all parts of the field was picked and put together.

In 1923, the writer obtained from the Staple Cotton Cooperative Association Classers of Greenwood, Miss., the classing data in regard to staple cotton grown on four large plantations in the Mississippi Delta that year. Seventy-seven per cent of the bales were classed as full 1⅛ inches. This was a rainy season, rainfall for the year being 81 inches at the nearest weather observation station. In 1924, which was an abnormally dry year with a rainfall of about 41 inches, the summer being exceptionally dry, similar data were collected. The variety of cotton, the plantations, and classers were the same both years. In 1924, 70 per cent of the bales were classed as straight 1⅛ inches. There was a difference of 1/32 inch, due very probably to the difference in water supply. The cotton was better cultivated in 1924 than in 1923.

Sturkie[19] found that soil type, temperature, or humidity did not affect length of lint, but that the amount of available moisture in the soil influenced it markedly, a low moisture content resulting in short lint. The critical period in elongation of lint was about 16 days from bloom. The length of lint could be shortened 3 millimeters (⅛ inch) by reducing the soil moisture to the critical point. Armstrong and Bennett[20] found that lint produced by small plants on unfertilized plots was of practically the same length as that from vigorous plants on rich soil. Reynolds and Killough[21] found no significant correlation between length of fiber and amounts of nitrogen, phosphoric acid, or potash or rates of application of fertilizer in central Texas. A number of others have obtained similar results.

Length of Cotton Fibers.—The length of cotton fibers is a hereditary character. It is influenced to a limited extent, as has been seen, however, by environmental conditions. The principal species and varieties of cotton vary much in length of staple. Table III from Cook[8] gives the staple length, color, relative yarn counts for which they are commonly used and the relative values of the principal types of cotton. Pictures of seeds of the foregoing types of cotton with combed lint attached are shown in Fig. 14.

TABLE III.—A SELECTED LIST OF COTTONS, SHOWING CHARACTER AND RELATIVE PRICES

Country	Variety	Average length, inches	Relative value	Counts up to	Color	Remarks
America	Sea Island (South Carolina)	2	230	300	Cream	Silky and regular
	Sea Island (Georgia and Florida)	1⅝	215	200	Cream	Silky and regular
Egypt	Sakellaridis	1½	173	150	Dark cream	Silky and soft
	Nubari	1⅜	160	100	Light brown	Silky and weak
America	Long-staple Upland	1¼	160	60	White	Soft and strong
	Short-staple Upland	1	100	40	White	Soft and strong
India	Tinnivelly	⅞	95	30	White	Best of Indians
	Surat, Broach, etc.	¾	91	30	Light brown	Harsh, strong
	Sind	⅝	71	10	Dull white	Poor
China		¾	88	20	Dull white	Rather harsh

The length of fibers within a single lot of cotton varies considerably, being greater for the cottons with extra staple length and less for well-bred varieties with which an effort has been made to make the staple length uniform. On seeds from strains that have not been selected for lint uniformity, very frequently the fibers on the pointed end of the seed are considerably shorter than those on the larger end, the difference being ¼ inch, or more in some cases. The range is often greater than the difference between varieties, average lengths being considered. There is frequently a difference of ⅟₁₆ inch or more in the staple length of different bolls. This probably depends to some extent on the weather conditions prevailing when the particular bolls were developing. In some observations made by the writer, the first formed bolls, which were low on the plant, had shorter staple than the average. The ones above these were longer; then there was shorter lint toward the top of the plant and toward the outer end of the branches.

Kearney[22] found that the fiber on the lower half of certain Pima Egyptain plants was 1.62 millimeters shorter than that on the upper half. Armstrong and Bennett[15] report decided difference in staple length between different bolls near together on the same plant and even on different seeds in the same lock. They noted also a distinct tendency for shorter lint to be produced near the top of the plant when grown in the field. Their general conclusion was that conditions seemed to be favorable for the production of long lint in bolls from all flowers during certain periods of time. Hancock, of Tennessee, in some unpublished work on staple length in different parts of the lock, finds a higher percentage of long fibers on seeds near the middle of the lock.

There is also, of course, considerable variation in staple length between different plants of a strain—the better bred a strain the less the variation.

The diameter of cotton fibers ranges from about $\frac{1}{1562}$ inch (16.3μ) for Sea Island to $\frac{1}{1185}$ inch (21.5μ) for the coarsest Indian varieties. Upland varieties are intermediate, being about $\frac{1}{1310}$ inch (19.4μ) in diameter. As a rule, the longer a fiber is the smaller its diameter.

Measuring the Length of Fibers.—The cotton classer estimates the length of cotton fibers after drawing a tuft of lint from the sample and "pulling" it for a few moments to straighten and even the fibers in the tuft, or he may place a rule over the tuft to measure it.

The cotton breeder ordinarily combs out the fibers while they are attached to the cotton seed and estimates or measures their length.

Comparatively recently several different fiber-sorting machines have been devised. With most of these the fibers are straightened and made to lie parallel by drawing them through sets of fine combs. In the process the fibers of different lengths are grouped. They may then be measured with a rule, and the bunches weighed to get the comparative amount of fibers of each length. The operation of the Suter-Webb sorter will be described so as to give the reader some conception of the process.

The first step in the process is to take a composite sample of the lint that is to be sorted. As the lint comes from the gin, the fibers are in a tangled mass. They may be straightened somewhat by drawing out and lapping over small pulls somewhat as a cotton classer performs the process of stapling cotton, but no fibers are thrown away. After the tuft is partially

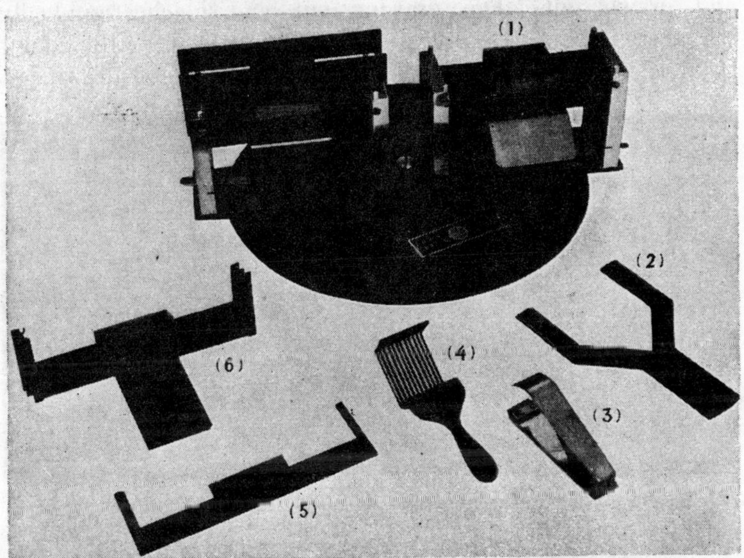

FIG. 38*B*.—Suter-Webb cotton fiber sorter and equipment. (1) Combs in place in sorter; (2) tool for lifting combs; (3) wide-tipped forceps for grasping end of fibers; (4) tool for pressing fibers into combs; (5) single comb; (6) combs in rack.

straightened, a definite amount is weighed out on a delicate scale, say 100 milligrams, and placed in the combs of the sorter [Fig. 38*B* (1)], the ends of the fibers being allowed to project about $\frac{1}{16}$ inch beyond the combs. The end of these projecting fibers is grasped with the wide-tipped forceps, and a small bunch of the fibers drawn out through the combs. This bunch of fibers is placed in the combs on the opposite side of the sorter. Other bunches are likewise drawn and transferred to the other combs. After no more fibers project far enough to be grasped, the outer comb is dropped. This allows the fiber length between the first and second combs to project. These projecting fibers are grasped, drawn out, and transferred as previously mentioned.

Other combs are dropped, and the process is continued until all the cotton in the first set of combs has been transferred. Next the sorter is rotated one-half turn, and the fibers are drawn from the second set of combs as from the first, except that this time the drawing is from the opposite end of the fibers. This procedure allows the longer fibers in the tuft to be grasped first and drawn out. The drawing out and placing gradually straightens the fibers and groups or sorts them. After about three transfers have been made, the fibers may be drawn out

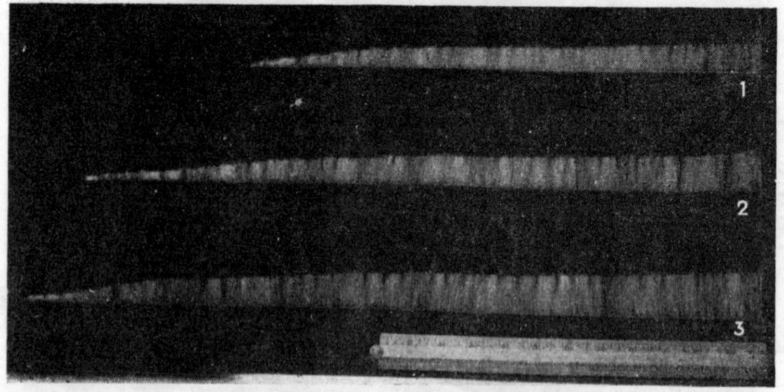

Fig. 38*C*.—Arrays of fibers from three different samples of cotton lint. (1) Half and Half; (2) Cleveland-5; (3) Delfos.

and placed on black cloth in an array, as in Fig. 38*C*, showing the lint of a short-staple variety (1), a medium-length staple variety (2), and a long-staple variety (3). The fibers in array 2 show less uniformity of length than those in the other two and is thus the poorest array.

Strength of Fibers.—Cotton graders estimate the strength of cotton fibers by breaking small tufts clasped between thumb and forefinger of each hand. That their judgment is not altogether reliable is shown by the fact that the results from spinning tests may not agree with their classing and that figures obtained by breaking individual fibers with a machine do not tally with their conclusions in many instances. Balls,[6] from actual tests made, concludes that graders' "strength" and "breaking strain" are utterly disconnected and have nothing to do with each other. He considers that what the grader determines is nearer "impact" resistance than actual strain that the fibers can withstand.

The actual breaking strain of individual fibers may be determined accurately by clasping the ends of a single fiber between the jaws of a fiber-testing machine (see Fig. 39) and applying the strain gradually.

More accurate strength tests may be made by what is known as the "Chandler bundle" method. In testing by this method, a bundle of definite size containing, say, 10,000 fibers, which have been made to lie parallel by combing, is tightly wrapped with

FIG. 39.—Fiber tester with a capacity of 0.1 to 50 grams, used in determining the strength of cotton fibers. (*After Dewey.*)

thread and broken by a machine able to exert a pull of more than 100 pounds between the jaws that clamp the bundle of fibers. Table IV from Dewey[9] gives the breaking strain in grams of fibers of several of the most important types of American cotton. The figures for the highest and the lowest are averages of 20 fibers.

The actual strength shown by fibers varies considerably, depending on humidity of the atmosphere, the part of the fiber exposed between the jaws of the machine, the part of the seed from which the fiber came, and the boll or plant from which the fiber came.

The strength of a fiber depends principally on the thickness of its cell wall. The strength is not necessarily proportional to the diameter, since some fibers with thin walls may be large in diameter but weak in strength. The strength of mature, well-developed fibers is, in general, in proportion to their diameter.

There is much variation in the strength of cotton fibers on account of changes in the environmental conditions under which the plants were grown. The different environmental factors mentioned on a preceding page as affecting length also affect

TABLE IV.—TENSILE STRENGTH OF COTTON FIBERS

Variety of cotton	Breaking strain, grams		
	Highest	Lowest	Average
American Upland:			
Big Boll Stormproof group	9.0	5.2	6.67
Big Boll group	11.6	4.6	6.60
Cluster group	7.7	5.1	6.00
Semicluster group	7.1	4.7	5.86
Peterkin group	6.0	5.0	5.70
Early group	6.9	5.2	5.63
Long-staple group	5.6	3.5	4.72
Sea Island	7.6	4.7	6.14
Egyptian from California and Arizona	8.0	5.6	6.65

the strength of fibers. Grimes[23] has shown that the exposure of cotton fabrics to sunlight for 375 hours caused a loss in breaking strength in all cases. The average loss ranged from 8 to 47 per cent in the warp and from 18 to 58 per cent in the filling. There is also some evidence that the exposure of seed cotton to the sun in piles in the field after picking may result in a loss in fiber strength.

Relation of Yarn Strength and Fiber Strength.—The strength of yarns spun from cotton fibers is not determined altogether by the strength of the individual fibers contained. When a yarn breaks, not many of the fibers are broken; they are simply pulled apart. Much then depends on the way the fibers twist about one another and cling together. Long fibers with numerous twists or convolutions make the strongest yarns, other things being equal.

Webb, of the U. S. Department of Agriculture, has recently shown that yarns spun from fibers with small diameter are stronger than yarns from thicker fibers, even if the two sets of fibers are of the same length. The gain in strength appears to be due to the fact that there is more fiber-surface tissue in the yarn from the fine fibers.

Fineness.—The lint of certain kinds of cotton, such as Sea Island, has a soft or silky feel and is considered fine lint. The lint of other kinds, such as the Indian wool, or Garo Hill, cotton (*G. arboreum*, var. *assamica*), has a very harsh feel and is considered coarse. A microscopical examination of the fibers of the two types of cotton shows that the Sea Island cotton has fibers of somewhat smaller diameter than the other, but the chief difference is in the much thicker cell wall of the Indian type, which gives it the character of coarseness.

Hurst,[24] giving results from a Russian investigator at Tashkent, says that the number of fibers per seed in five American varieties varied from 7.8 to 14.7 thousand and that the weight of 1,000 fibers varied from 4.4 to 7.6 milligrams. The seed weights varied from 97.0 to 167.2 milligrams. On strains of the Asiatic species *G. herbaceum*, the variation in number of fibers per seed was from 3.6 to 9.2 thousand.

Uniformity of Fibers and Spinning Qualities.—The machines in cotton mills are set for certain fiber lengths. If there are many fibers in the lint cotton which are considerably shorter than those for which machines are set, a considerable amount will go into the waste. The ideal fibers for spinning purposes are those that are of uniform length, diameter, cell-wall thickness, and twist. Such fibers may be spun economically; if their diameter and twist or convolutions are uniform, they will fit together closely, thus making a strong thread; if their cell walls are uniform, they will take dyes more uniformly.

Since many varying factors have a bearing on lint uniformity, it will never be possible to secure lint with a near approach to uniform fibers; they do not and cannot grow that way. By growing pure strains of cotton that have been especially selected for uniformity of staple, considerable improvement may be made over the present supply.

Color and Luster of Lint Cotton.—The lint of the wild cottons is, as a rule, a brownish or khaki color. Reversions or sports with lint of this color are occasionally found in nearly all varieties of cotton. Less frequently, plants are found which have a green lint similar in hue to that of the green fuzz on some seed. This lint is of a clear-green color when the boll first opens, but on exposure to light it fades to a tawny brown.

The color of the standard cultivated races of cotton ranges from pure white, as found in American uplands; cream, illustrated by Sea Island; dark cream, like Egyptian Nubari; to brown, as found in Egyptian Ashmouni.

Cotton that is exposed to the weather for a long time before it is picked tends to take on a dull-white or bluish color. This is objectionable, because such cotton does not bleach or dye well. Some cotton is spotted, yellow stained, or tinged, as a result of insect or frost injury.

The luster of lint is commonly thought to be due to variations in the reflection of light from different parts of the cuticularized walls of the exposed fibers. Balls[6] considers that luster is more than simply reflection of light. Luster is temporary and easily lost; yet the cuticle which is supposed to form the reflecting surface is very resistant. He believes that the fibers which are translucent refract a certain amount of light, and there is reflection also from concave surfaces within the fiber.

Lint Percentage.—The ginning outturn, or lint percentage, is determined by at least three factors: the number of fibers per unit area of seed surface, the size of the fibers, and the weight of the seed. The variation of different varieties with respect to the number of fibers per unit area of seed surface has not been investigated thoroughly, but it is known that some of the short-staple varieties with high lint percentage do have many more fibers per seed than do the varieties with longer staple and lower percentage. It is probable that, in general, the number of fibers varies directly with the lint percentage. The size of the individual fibers is a factor of relatively little importance in its effect on lint percentage. The long fibers have a small diameter, and the short ones are relatively thick; so these characters tend to balance each other. The size or weight of the seed is also another important factor in determining lint percentage. Varieties with the highest lint percentage have small seeds, whereas some varieties with low percentage have large seeds.

Growers are partial to a cotton that turns out well at the gin, and within certain limits buyers prefer staple from cotton that turns out well. Although it may be possible to reduce the vitality of seeds by selecting small-seeded strains for a series of years in an effort to raise the lint percentage, it has been the writer's experience that, everything else being equal, the small-seeded strains are preferable on account of their quicker germination, greater ease in getting a stand of plants, higher oil content,

TABLE V.—RANK OF VARIETIES IN LINT FREQUENCY, LINT INDEX, LINT PERCENTAGE, AND LENGTH OF LINT (*After Hodson*)

Variety	Rank in lint frequency	Rank in lint index	Rank in lint percentage	Rank in length of lint
Boykin	1	2	2	25
Cook	2	11	9	24
Triumph	3	1	3	16
Roundnose	4	5	6	18
Cleveland	5	12	7	19
Christopher	6	8	12	15
Texas Bur	7	6	5	11
Culpepper	8	9	8	12
Lone Star	9	3	4	8
Covington-Toole	10	13	10	22
Spruiell	11	7	16	9
Acala No. 5	12	4	1	6
Half and Half	13	16	13	20
King	14	15	14	13
Rublee	15	14	15	17
Rowden	16	10	11	7
No Chop	17	18	18	21
Simpkins	18	21	17	23
Trice	19	17	19	14
Dixie-Improved	20	23	20	10
Foster	21	20	21	5
Express	22	24	24	4
Columbia	23	19	22	2
Webber-49	24	22	23	3
Dix-Afifi	25	25	25	1

and better lint percentage. Rowden, a variety with very large seeds, has usually given poor stands.

Cotton grown on rich land or on land that has been given a liberal application of nitrate of soda has a lower lint percentage than cotton grown on less productive soils. The difference in this case is due to the richer lands producing heavier, plumper seeds. The better lands produce longer fibers and probably slightly more lint per seed, but this is more than overbalanced by the increase in the weight of seeds. In tests conducted by the writer, of 11 different varieties on four different types of soil, the plantings on the poorer soils gave higher lint percentages consistently. In 1923, the average difference was 1.7 per cent.

Considering that there is danger of causing the deterioration of varieties if selections are made on the basis of lint percentage, Cook[11] suggests that the weight of lint produced by 100 seeds, a lint index, be used instead. But it is doubtful if this is of any considerable value to cotton breeders, since it does not give any measure of the yield of lint per acre, which is really the principal feature.

Hodson[12] suggested the term "lint frequency," which he defined as "the weight in grams of the fiber of uniform length produced per square centimeter of seed surface." "Lint frequency," however, does not differ greatly from "lint index"; only in the former more stress is placed on seed-surface area. Table V from Hodson[12] shows the relative rank of certain varieties in lint frequency, lint index, lint percentage, and lint length. A study of this table gives some idea of the relation of these different characters.

Composition of Lint Cotton.—Johnson[27] gives the composition of lint cotton as follows: cellulose, 90 per cent; water, 7 to 8 per cent; wax and oil, 0.4 per cent; nitrogenous matter, 0.6 per cent; and mineral matter, 1.0 per cent. Ahmad and Sen report a wax content of 0.23 to 0.47 per cent on 11 Indian varieties and one foreign variety studied. The composition of cotton fibers is given in more detail in Tables XII and XIX.

References

1. Balls, W. L.: "The Existence of Daily Growth-rings in the Cell Wall of Cotton Hairs," *Proc. Royal Soc. of London*, ser. B, vol. 90, 1919.
2. Miller, T. S.: "Cotton Trade Guide and Students Manual," Steck, Austin, Tex., 1915.
3. Levine, B. S.: "The Structure of the Cotton Fiber," *Science*, new series, vol. 40, No. 1042, 1914.
4. Evans, W. H.: "Botany of Cotton," in "The Cotton Plant," *U. S. Dept. Agr. Bull.* 33, 1896.
5. Cook, O. F.: "Cotton Improvement under Weevil Conditions," *U. S. Dept. Agr. Farmers' Bull.* 501, 1912.
6. Balls, W. L.: "The Development and Properties of Raw Cotton," A. & C. Black, Ltd., London, 1915.

7. KEARNEY, T. H.: "The Uniformity of Pima Cotton," *U. S. Dept. Agr. Dept. Cir.* 247, 1922.
8. COOK, O. F.: "Commercial Types of Cotton," in "Atlas of American Agriculture," U. S. Deparment of Agriculture, 1918.
9. DEWEY, L. H.: "The Strength of Textile Plant Fibers," *U. S. Dept. Agr. Bur. Plant Ind. Cir.* 128, 1913.
10. BOWMAN, F. H.: "The Structure of the Cotton Fiber," Macmillan & Company, Ltd., London, 1908.
11. COOK, O. F.: "Danger in Judging Cotton Varieties by Lint Percentages," *U. S. Dept. Agr. Bur. Plant Ind. Cir.* 11, 1908.
12. HODSON, E. A.: "Correlation of Certain Characters in Cotton," *Ark. Agr. Exp. Sta. Bull.* 169, 1920.
13. FARR, WANDA K.: "Cotton Fibers. I. Origin and Early Stages of Elongation," *Contrib. Boyce Thompson Inst.*, vol. 3, No. 3, 1931.
14. HAWKINS, R. S. and SERVISS, GEORGE H.: "Development of Cotton Fibers in the Pima and Acala Varieties," *Jour. Agr. Research*, vol. 40, No. 11, 1930.
15. ARMSTRONG, G. M., and BENNETT, C. C.: "Some Factors Influencing the Variability in Length of Cotton Fibers on Individual Plants as Shown by the Sorter Method," *Jour. Agr. Research*, vol. 47, No. 7, 1933.
16. GULATI, A. N.: "A Note on the Differentiation of Hairs from the Epidermis of Cotton Seeds," *Indian Jour. Agr. Sci.*, vol. 4, Part 3, 1934.
17. North Carolina Agricultural Experiment Station: "Storage of Seed Cotton," *U. S. Dept. Agr. Rept. on Agr. Exp. Sta.*, 1927, p. 32.
18. FARR, WANDA K. and ECKERSON, SOPHIA H.: "Formation of Cellulose Membranes by Microscopic Particles of Uniform Size in Linear Arrangement," *Contrib. Boyce Thompson Inst.*, vol. 6, No. 2, 1934.
19. STURKIE, D. G.: "A Study of Lint and Seed Development in Cotton as Influenced by Environmental Factors," *Jour. Amer. Soc. Agron.*, vol. 26, pp. 1–24, 1934.
20. ARMSTRONG, G. M. and BENNETT, C. C.: "Effect of Soil Fertility, Boll Maturation Period, and Early and Late Production of Bolls on the Length of Cotton Fibers," *Jour. Agr. Research*, vol. 47, pp. 467–474, 1933.
21. REYNOLDS, E. B. and KILLOUGH, D. T.: "The Effect of Fertilizers and Rainfall on Length of Cotton Fiber," *Jour. Amer. Soc. Agron.*, vol. 25, pp. 756–764, 1933.
22. KEARNEY, T. H.: "Length of Fiber from Bolls at Different Heights," *Jour. Agr. Research*, vol. 28, No. 6, 1924.
23. GRIMES, MARY ANNA: "The Effect of Sunlight and Other Factors on the Strength and Color of Cotton Fabrics," *Tex. Agr. Exp. Sta. Bull.* 474, 1933.
24. HURST, C. C.: "Recent Work in Plant Breeding," *Empire Cotton Grow. Rev.*, vol. 13, No. 2, 1936.
25. KERR, THOMAS: "The Structure of the Growth Rings in the Secondary Wall of the Cotton Hair," *Protoplasma*, vol. 27, No. 2, 1937

26. Anderson, Donald B. and Moore, J. H.: "The Influence of Constant Light and Temperature upon the Structure of the Walls of Cotton Fibers and Collenchymatous Cells," *Amer. Jour. Botany*, vol. 24, No. 8, 1937.

27. Johnson, W. H.: "Cotton and Its Production," Macmillan & Company, Ltd., London, 1926.

CHAPTER VIII

VARIATION, HEREDITY, AND CORRELATION OF CHARACTERS IN COTTON PLANTS

A discussion of heredity and related subjects in a book of this type finds its justification in the following facts: The amount of cotton produced in a field or in any locality depends on the yielding power of the plants grown. This yielding power is influenced greatly by certain inherent traits that the plants possess. Good-yielding strains may be maintained by cotton breeders who choose plants that have these desired traits and propagate from them. They are helped greatly in making their choices by a knowledge of the laws of variation. They should be able to distinguish between mere fluctuations caused by difference in environmental conditions and really significant variations. They need to know what characters are hereditary and what laws govern their inheritance.

Variations.—No two cotton plants are exactly alike. The differences that the individuals of a variety of species exhibit are known as "variations." Variations are of different types, but they may be grouped in two general classes: continuous, or nonheritable; and discontinuous, or heritable. The continuous variations are mere fluctuations in plant characters caused by differences in the environmental conditions under which the plants grow. The discontinuous variations, on the other hand, are not due to external influences but to some change in the genetic constitution of the plants.

Wild species of cotton, like the wild species of most other plants, are relatively fixed and stable. The variations they show are almost all of the continuous type. Cultivated varieties of cotton are more variable, the nonheritable variations fluctuating through a wider range, and the heritable occurring more frequently. This is partly due to the fact that cultivated plants have better growing conditions. They are not crowded by grass, weeds, and other plants, as the wild plants are. They are plowed and hoed, which makes moisture and nutrient elements more available and more plentiful. Each plant has a better chance to show its individuality.

The heritable variations fall into two classes: mutations and new combinations of characters resulting from segregation in strains of hybrid ancestry. Mutations are marked variations in the characters of individuals which are caused by sudden changes in the hereditary constitution of the parent organism or by chromosome aberrations. They will be discussed further in a subsequent paragraph.

Heritable variations due to the segregation and recombination of factors are common in cotton. Varieties and even species having markedly different characters are frequently allowed to hybridize. The cross brings together various factors that belong to the two parents. These factors enter into different combinations, some of which produce characters such as are found in one parent; some, like ones found in the other parent; whereas there may be still other characters that are different from anything found in either parent. The number of apparently independent characters of cotton plants is large—probably in the hundreds—and the factors responsible for these characters are even greater in number. With the large number of factors, almost innumerable factor combinations are possible. This accounts for the number of various character combinations found in a field of cotton. None of the commercial varieties of cotton is genetically pure, in the sense of being a pure line. All are of more or less hybrid constitution, and it may be expected that segregation will occur frequently and new forms will appear often. This may be a help to the cotton breeder in that desirable new combinations or plant types may be found easily, but it is also a disadvantage in that the desirable new strains may not be very permanent.

Humbert and Mogford[18] studied closely the length of lint in all the bolls of a cotton plant and the variation in the progeny of this plant. They found variation in lint length to occur in different parts of the plant, in different parts of the same boll, and in the same lock. There was also similar variation in the progeny produced but no consistent correlation between the variations in length of lint on the parent plant and in the progeny. The mean lengths of the lint in parent and progeny were about the same. It was concluded that the variations observed were due to environmental rather than genetic influences.

Mendel's Laws.—The first investigator to formulate definite laws of heredity was Mendel, whose work, published in 1866, was brought to the attention of the scientific world in 1900. He discovered that genetic factors tend to *segregate* in the offspring of hybrid plants and that the segregation which occurs between the members of one pair of factors is *independent* of that in others. These two principles, with certain modifications and amplifications which have since been made, particularly a recognition of the occurrence of linkage and of the operation of multiple factors, have made it possible to understand and explain many of the complexities of inheritance and to put the problems of plant breeding on a scientific basis. Most of the important types of Mendelian inheritance have been found in cotton.

Heredity.—It is a matter of common observation that a plant inherits its characteristics from its parents or resembles them in most respects. The tendency is for the parent forms to be reproduced true to type, or very nearly so, and the stability of our agricultural varieties depends on this fact.

Mendelian Inheritance in Cotton.—The application of Mendel's law to inheritance in cotton has been studied by Kearney,[1] Balls,[3] Fletcher,[4] Tyson,[5] Harland,[6] Leake,[7] Leake and Prasad,[8] McLendon,[9] and others. These investigators secured similar results in general. If the character was controlled by a single factor, as are most of the color characters, simple Mendelian ratios were obtained. But with the majority of the traits considered, simple ratios were not found. Kearney[1] plotted frequency distributions for each of the 39 characters he studied in the second generation of a hybrid between Holdon, an upland variety, and Pima, an Egyptian variety. But three of his graphs are bimodal or multimodal; hence only the three characters, petal spot, anther color, and mid-lock furrow of the boll, showed direct evidence of segregation in Mendelian proportions.

Leaf Color.—Simple Mendelian ratios were obtained by Leake,[7] Ware,[19] McLendon,[9] and the author in the study of the inheritance of leaf-blade color where a red-leaf variety was crossed with one with green leaves. The F_1 plants had reddish-colored leaves, but the color was not so dense as in the red-leaf parent.

Leaf Shape.—Ware,[19] the author, and others have found that a cross of okra-leaf plants with normal broadleaf varieties gives an F_1 plant that is intermediate between parents in shape. The F_2 gives simple Mendelian monohybrid ratios.

Hutchinson[20] found much more complicated inheritance in his studies of the inheritance of leaf shape in certain Asiatic cottons. He secured evidence of a series of multiple factors governing leaf shape and of a linkage between the leaf-shape characters studied and brown lint.

Short Branch.—Kearney[29] crossed a strain of Pima Egyptian cotton, which had fruiting branches reduced to a single internode, with normal Pima plants. The F_1 had fruit branches of intermediate length. In the second generation, there was segregation into three classes: short branch, intermediate, and normal, or long, branch. The proportions of the several classes suggested a 1:2:1 ratio, indicating absence of dominance in the heterozygous condition.

Petal Color.—Results obtained from the study of the inheritance of petal color in cotton hybrids have varied considerably. Kearney[1] found his F_1 plants to have petal color intermediate between the light-colored petals of the Holdon parent and the yellowish petals of the Pima parent. His F_2 plants gave a unimodal frequency curve extending from one parental extreme to the other, with the majority of the individuals on the side next to the lighter colored flowers.

Tyson[5] reported complete dominance of yellow in the F_1 generation of a cross of white- and yellow-flowered Indian cottons. The F_2 generation showed 156 individuals classed as pale and 411 classed as white.

Balls[3] states that in crosses of upland and Egyptian cottons the F_1 plants always have intermediate petal color. Certain crosses gave a near approach to the 1:2:1 ratio in the F_2 generation, but further study of segregation in later generations indicated a more complex inheritance. From all the evidence he secured Balls decided that the petal color of crosses of upland and Egyptian cottons is controlled by not less than three pairs of factors.

Petal Spot.—Balls[3] reports that a cross of Egyptian cotton, which has a bright-red spot, or "full spot," on the claw of the petal, with upland, which has no spot, gives an intermediate F_1.

In the F_1, the spot is smaller and has a vaguer outline than in the Egyptian parent. Some of Balls's crosses gave an approximation of the 1:2:1 ratio in the F_2, as, for instance, 23 full spot, 42 intermediate, and 31 spotless; in other crosses, the numbers were much different, as, for instance, 2:40:115. Balls says:

> The only decision at which we can arrive with certainty is that the presence or absence of the petal spot in these Egypto-American crosses is not determined by a single pair of allelomorphs.

Leake and Prasad[8] reported complete dominance of petal-spot color in the F_1 in a cross of Chinese cotton without a petal spot and an Indian cotton with one.

McLendon[9] considered the petal spot to be intermediate in the F_1 of a cross between Sea Island and upland cottons; most of his F_2 progenies showed a ratio approaching 1:2:1.

Kearney[1] found the petal spot in the F_1 of the Holdon × Pima hybrids he studied to be dominant but reduced in size. The F_2 generation gave a very close approach to a 3:1 ratio.

Anther Color.—Hybrids of Egyptian and upland varieties made by Balls[3] showed the F_1 anther color to be intermediate, and a true 1:2:1 ratio was obtained in the F_2 generation. No exceptions were noted in later generations.

Kearney[1] secured only partial dominance for the yellow color in the F_1 of the Holdon × Pima cross he made. The F_2 gave some evidence of bimodal distribution.

Boll-lock Number—The average number of locks to the boll in the F_1 generation in the Holdon × Pima hybrid studied by Kearney[1] was exactly intermediate between the averages of the parents. The F_2 generation gave a unimodal distribution, which indicated that several factors were active in determining the boll-lock number. Balls[3] secured similar results when studying this character.

Boll Shape.—Boll shape is not due to a single factor but is determined by several factors. Kearney,[1] using ratio of diameter to length as a character index, found the mean of F_1 to be intermediate between the means of the parents (Fig. 40), though the F_2 frequency-distribution curve did not show clear segregation. This probably indicated a series of multiple factors. Balls[3] secured similar results in his studies of this character.

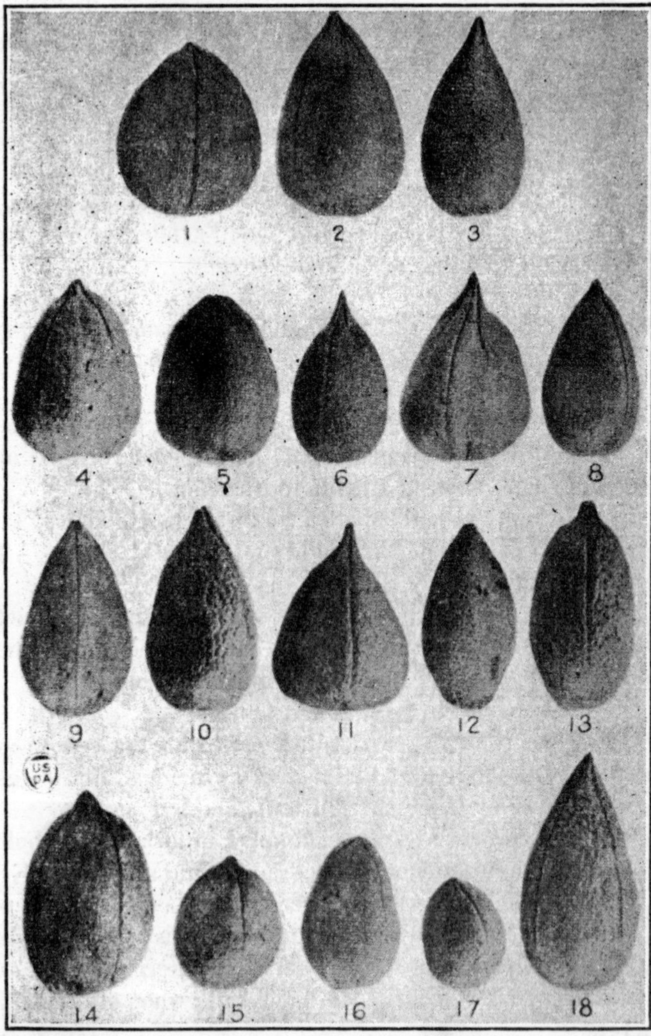

FIG. 40.—Bolls of the parental and hybrid cotton plants. (1) The Holdon parent, showing the relatively large maximum and apical diameters, almost complete absence of a beak, short mid-lock furrows, light color, and smooth surface, the oil glands being very inconspicuous. (2) The first-generation hybrid, intermediate in shape, relative diameter and character of the surface, but resembling the Pima parent in having a well-developed beak and no mid-lock furrow. (3) The Pima parent, showing the relatively small maximum and apical diameters, conspicuous beak, absence of a mid-lock furrow, dark color, and pitted surface. (4) to (18) Second-generation hybrid individuals, showing variations as follows: (4) The most uplandlike; (8) the most Pimalike; (4) and (5) very light and very dark colored; (5) and (6) beakless and strongly beaked, (9) and (10) very smooth and very rough surface; (11) to (16) extreme variations in shape; (17) and (18) very small and very large; (7) and (9) presence of mid-lock furrow in different degrees; (16) and (18) termination of the lock furrows at considerable distances from the base of the boll. (*After Kearney.*)

Fiber Length.—McLendon,[9] in crosses between Sea Island and Cook and between Sea Island and Hastings Big Boll (Cook and Hastings being upland short-staple varieties), secured intermediate fiber length in the first generation. The second generation segregated into about three-fourths long to one-fourth short.

Balls,[10] with crosses between Egyptian and upland varieties, obtained results similar to McLendon's.[9] In a cross of Egyptian Afifi with a short-staple upland variety, F_1 plants were obtained which had a staple length greater than the Afifi parent's. The writer secured a similar intensification in F_1 hybrids of a cross between Trice and Triumph, both upland short-staple varieties. The hybrid had staple 1/8 inch longer than either parent.

In the Holdon × Pima cross, Kearney[1] obtained F_1 plants with a mean fiber length nearly equal to the mean fiber length of the long-fibered Pima parent. His F_2 plants gave a unimodal curve which was nearly symmetrical.

Fiber Color.—First-generation hybrids of a cross made by the writer between Cleveland, a short-staple variety with white lint, and the so-called "Nankeen" cotton, a short-staple variety with brown lint, had uniform brownish, cream-colored lint, intermediate between the two parents in color. Selfed seed from the F_1 plants gave an F_2 generation which segregated in almost an exact 1:2:1 ratio. Ware[19] secured almost identical results from a like cross.

Balls[3] crossed brown-linted Egyptian varieties with uplands having white lint. The first crosses showed an intermediate cream-colored F_1 and an F_2 with 12 brown, 21 creamy, and 11 white. The browns and whites bred true up to the fifth generation, while the creamy-colored ones broke up. Later crosses made by Balls gave more complicated results.

Seed Fuzziness.—McLendon, from crosses of smooth- and fuzzy-seeded varieties, reports fuzziness dominant to smoothness in the F_1 generation. In the F_2 generation the ratios ranged from 1:1 to 18.6:1 for different crosses. This indicates that segregation is complex.

Kearney[1] found the fuzziness of the Holdon parent to be dominant over the smoothness of the Pima parent in the F_1 plants. The F_2 generation gave no evidence of definite segregation (Fig. 41).

Winters, in some unpublished work on the study of fuzz inheritance, finds indications of a single factor difference, with the fuzzless character dominant. He says in a personal letter, "There exists a fuzz pattern character which may be confused with

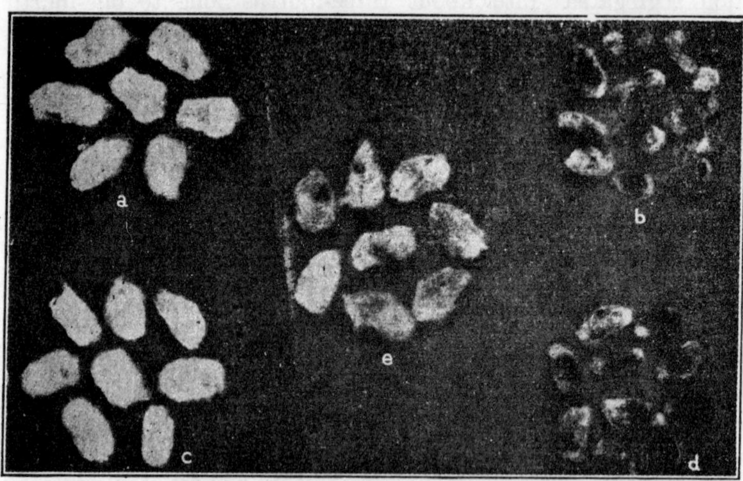

Fig. 41.—Seeds, after removal of the fiber, of the parental and hybrid cotton plants. (*a*) The Holdon parent, having the seed completely covered with long white fuzz; (*b*) the Pima parent, having the seeds partly covered with short greenish or brownish fuzz; (*c*) and (*d*) the fuzziest seeded and the smoothest seeded second-generation hybrid plants; (*e*) the first generation hybrid, having the seed completely covered with long green or white fuzz. (*After Kearney.*)

the fuzzless character. This behaves quite differently and may account for the varying results."

Balls,[3] in discussing the inheritance of seed fuzziness, says:

In crosses of any Egyptian with American upland, we meet with complications. The entire fuzz is dominant, and the F_2 has given such 15:1 ratios of "entire:slight," as 97:6, 180:11, etc.

Kearney[28] more recently made a rather thorough study of the inheritance of fuzziness in crosses between fuzzy and fuzzless strains of Pima Egyptian, between fuzzy and fuzzless strains of upland, and between upland and Pima, using the same strains in more than one combination. He found that crosses within the Pima variety showed the fuzzless character to be dominant in the F_1 and a near approach to a 3:1 ratio in the F_2. The same results were obtained when a smooth-seeded upland strain was crossed with a fuzzy-seeded strain but, when the fuzzy-seeded upland was crossed with the fuzzless-seeded Pima, the fuzzy character was dominant, and the F_2 showed approximately one smooth-seeded plant to three fuzzy. When the fuzzy-seeded Pima was crossed with the smooth-seeded upland strain, the smooth-seed character was dominant, and the F_2 generation con-

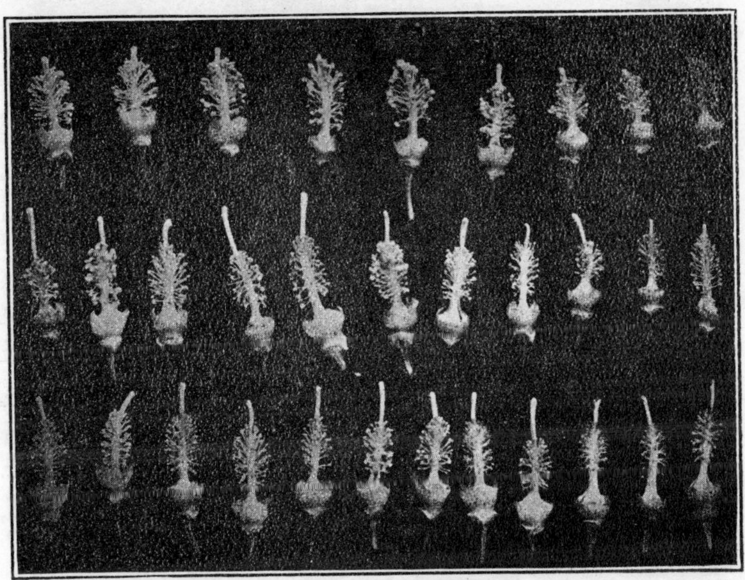

Fig. 42.—Cotton flowers showing stamens and pistil. The first flower in the upper left-hand corner of the figure is from upland cotton; the last one in the lower right-hand corner is from Sea Island; the rest are from F_2 hybrids between upland and Sea Island.

tained 12 smooth-seeded plants and 9 fuzzy. This work indicates that fuzz inheritance is mainly of the simple Mendelian type, but the influence of the factors for smoothness and fuzz varies in different combinations.

Inheritance of Other Characters.—The inheritance of a number of other characters of the cotton plant has been studied by Kearney,[1] Balls,[3,10] McLendon,[9] and others. These additional characters include leaf shape, color of fuzz on seeds, prevalence of glands on different parts of plants, pitting of boll surface, pistil length, stamen length, calyx dentation, etc. (Fig. 42). The results obtained did not differ materially from the ones outlined above, except that perhaps, in general, the segregation was not so definite There are many other characters that have not been carefully studied.

Inheritance of Quantitative Characters.—The inheritance of quantitative characters, such as size, weight, and height, is of much importance, economically considered, for values are largely determined by the quantities produced. The inheritance of quantitative traits is difficult to study because, as a rule, each character is under the control of several factors, and because the characters are subject to nonheritable fluctuations caused by differences in environmental conditions. These fluctuations may be wide and such as to render the placing of individuals in proper classes difficult. It was thought at first that quantitative characters did not follow Mendel's law in their inheritance, since for most characters there seemed to be a continuous graduation in the F_2 hybrids from one parental form to that of the other. After it was discovered that two or more factors are necessary, in some cases, for the production of certain simple characters, it was but a step to use this information in working out the inheritance of quantitative traits.

Comparatively little work has been done on an analysis of quantitative characters in cotton. Kearney,[1] however, shows clearly that they tend to resemble, in their inheritance, similar traits in other plants, the F_1 of a cross between two pure parents tending to be more or less intermediate between them and no more variable than either, but the F_2 showing a greatly increased variability, thus indicating the occurrence of segregation in a group of multiple factors. In a cross between the Holdon and Pima varieties, for example, Kearney finds that the mean fiber length of the former is 2.1 units, with a standard deviation of 0.5; of the latter, 7.4 units, with a standard deviation of 0.6; of the hybrid offspring (F_1), 6.6, with a standard deviation of 0.7; and of the second-hybrid generation (F_2), 5.0, with a standard deviation of 1.5.

Linkage of Characters.—It is the common belief among students of heredity that the physical basis of each hereditary factor is to be found in the chromosomes of the cell nucleus. The chromosomes are fairly permanent structures, each species or variety of plants having a regular number in its cells. Since the number of hereditary factors belonging to an individual, either plant or animal, is usually several times the number of its chromosomes, it seems evident that certain factors must occur together in the same chromosome and be transmitted together, or *linked*. That this is true has been well demonstrated by Morgan and his associates in their breeding experiments and cytological studies with the fruit fly *Drosophila*. This insect has only four pairs of chromosomes, but more than 300 factors have been identified, most of which have been proved to be linked.

In cotton, according to the studies of Denham and Beal (see Chap. VIII), there are 26 chromosomes. Since the genetic factors in cotton are doubtless far more numerous than this, there must be many cases of linkage between factors. But little attention has been given this subject by students of heredity in cotton. Thadani[17] has observed what he considers well-defined linkage between certain factors. A naked-seeded (A), sparse-linted (b) variety was crossed with a fuzzy-seeded (a), abundant-linted (B) variety. A naked-seeded, abundant-linted type was obtained in the F_1 generation. In the F_2 generation, the segregation was as follows: 103 AB:54 Ab:53aB:0ab, thus indicating that there is linkage between these two pairs of factors.

Hutchinson,[20] as was mentioned above, found linkage between certain genes for leaf shape and for brown lint. Harland[21] reports linkage between green lint and crinkled leaf and between red coloration of certain types of upland and the cluster habit of some upland cottons.

Mutations in Cotton.—As was mentioned on a previous page, mutations are heritable variations which are not the result of segregation and recombination of genetic factors. They are marked by the sudden appearance of new forms among the progeny of pure races. The mutant may differ from other individuals of the progeny in respect to but one trait, but often it has several rather striking character differences.

The cause of mutations is not well known. They may be due to alteration in the genetic factors of body or germ cells or to deviations in the number of chromosomes. The first class is probably the more important.

In common with most other cultivated plants, mutations occur in cotton, but it is a question as to how frequent they are. Because there is hybridism in the ancestry of most cotton varieties, it seems reasonable to believe that many of the new forms or types that appear in cotton fields are the result of segregation and new combinations of factors from parents that were hybridized in previous generations. Whether the new forms arise from mutations or from hybrid segregation is very difficult to determine, but it appears probable that mutations may be expected in cotton as frequently as in other cultivated crops.

A number of new forms or types have appeared and behaved in such a way that they fit the definition of mutants. Kearney[12] offers the following argument in favor of the theory that certain new strains or varieties of cotton have originated as mutants:

(1) The derivation of each from a single plant discovered in a field of very different cotton; (2) the distinctness of their botanical characters, especially in the recently developed Nubari and Sakellaridis (Egyptian varieties); and (3) their tendency to remain uniform.

In the cotton-breeding fields of the Mississippi Experiment Station, several distinct and unique plants have been found, from some of which new strains or varieties have been developed. One appeared in a field of Lone Star, the selected plant being designated as "Lone Star-132." This new plant differed from the plants about it in habit of growth, being rather tall and narrow; in shape of leaves; in size and shape of boll; in lint length, being about ⅛ inch longer than that the other plants; and in lint percentage. The other plants had a rather high lint percentage, running about 35, whereas the new variety had a low lint percentage, running about 28. Progenies grown from this plant were unusually uniform, considering the fact that the original plant and progenies were subject to open pollination. The new variety developed from this plant was uniform as long as it was grown (about 4 or 5 years) and yielded well, but it was discarded on account of some undesirable qualities. The original appears to have been a mutant. It resembled no known variety.

Another remarkable plant appeared in a field of Express cotton, from which was developed the strain Express-432. This plant also differed from the other plants of the field by marked characteristics and gave rise to a well-defined, uniform new strain which is grown commercially. It has been grown 10 years and is still essentially uniform. It possesses certain characters widely different from the parent variety and different from others known. Numerous selections made from the 432 strain have not proved to be an improvement over the original. Kearney[12] reported similar experience with selections made from Pima cotton, which is thought to have originated as a mutant.

In 1930, the writer found a peculiar plant with irregular rounded leaves growing in a field of Express-317 cotton—an upland variety with normal foliage. Seed of the round-leaf plant were saved, and progenies grown from 1931 to 1936. It

Fig. 42*A*.—Round-leaf plants are shown in the central part of the picture and normal broad-leaf plants on either side.

was found to breed true to the original type. The plants possessed an unusual habit of growth (Fig. 42*A*) and rather striking boll and lint characters. They did but very little blooming until late in the season, when they bloomed profusely. The plant found in 1930 was evidently a well-defined mutant.

A large number of the present varieties of upland cotton have been developed from certain unusual individual plants. Lone Star, Triumph, Webber, and Express may be mentioned as examples.

Each of the more important recent Egyptian varieties is regarded by Kearney[12] as arising from a mutant—"from an individual plant that showed an abrupt and definite change in the characters expressed." Four distinct mutants have appeared in the plots or fields of Egyptian cotton grown in Arizona. Two of these gave rise to Yuma and Pima, commercial varieties of considerable importance.

Cotton seed and flower buds have been treated with X rays by Horlacher and Killough[22] and by others in an effort to produce mutations artificially. Plants grown from treated seed have shown many abnormalities in form and growth. In some cases, the plants died after emerging from the ground. Most of the changes seemed to be in the nature of deficiencies. In some cases, the new forms reproduced themselves in a later generation and may be considered as mutations artificially produced. Horlacher and Killough obtained some mutations that may be classed as retrogressive and some that were apparently progressive. The change of a forked-leaf (okra-leaf) strain to a normal-leaf type and the change of a strain with virescent yellow leaves to a normal green-leaf type may be considered examples of progressive mutations. The investigators say:

> Seeds containing embryos known to be heterozygous for leaf shape were x-rayed and plants grown from these. The bolls from these plants were selfed and the individual boll progenies grown separately. Normal segregation for leaf shape gave 1 normal leaf, 2 intermediate leaf, 1 forked leaf. A mutation from forked to normal in a branch of the heterozygous plant (Nn to NN) would produce a boll from which the progeny would all have normal leaf shape. Likewise a mutation from normal to forked (Nn to nn) would produce a boll from which all the progeny would have forked leaves. A total of 525 individual boll progenies were grown. Among these a number of bolls produced only normal leaf progeny, and a few produced only forked-leaf progeny.

Bud Mutations.—Bud mutations are common in many species of plants, but not many have been found in cotton. Hutchinson[23] reports that a plant of *Gossypium arboreum*, var. Burma, a laciniated-leaved strain, grown at the Cotton Research Station in Trinidad, gave rise to a broadleaf bud mutant.

Singh, of Cawnpore, India, in a personal letter to the writer mentioned what appears to be a bud mutation. A hybrid plant in his genetic plots had one branch with green leaves, yellow flowers, and spotless petals. Other branches of the plant had red leaves, red flowers, and spotted petals, which were normal for the crossed strain.

Apparent bud mutations have appeared on several plants in the progenies of the round-leaf strain mentioned above. These appear in the form of a branch bearing apparently normal leaves

FIG. 42*B*.—Round-leaf plant with a branch bearing apparently normal broad leaves.

and flowers (Fig. 42*B*). Seed from selfed bolls on these branches have, however, produced only round-leaf plants.

Correlation of Characters in Cotton.—There are several characters of cotton plants that are generally considered to be positively correlated or connected in some way in their inheritance. Among the most common ones are:

1. Large bolls and large seed.
2. Long staple and small bolls (some of the long-staple varieties of recent origin, like Webber, for instance, have apparently

large bolls, but the seed cotton within does not weigh in proportion to the boll length and diameter).

3. High lint percentage and short lint.
4. Large seeds and low lint percentage.
5. Small seeds and high oil content of seeds.
6. Length of stem axis and branches and date of maturity.

On the other hand, there are several characters that appear to be antagonistic, or negatively correlated. These are listed by Duggar[14] as follows:

1. Extreme earliness is opposed to large bolls.
2. Extreme earliness is usually not associated with the highest yields of lint, except when the fruiting season is shortened by early frost or by the presence of the boll weevil.
3. Great length of lint excludes the probability of a high percentage of lint.

Although statements have often been made concerning the correlation or antagonistic relation of several of the aforementioned characters, there has been a scarcity of reliable data on the subject. Characters were said to be correlated, but no statistical data were given to prove it. Within the last few years, valuable data on the subject have been published by Kearney,[1] Dunlavy,[2] and Hodson.[15] Their papers give statistical data on the correlation of a large number of the characters of cotton plants, but only a limited amount of this information is of direct agricultural value.

Plant breeders must depend largely on the physical characters exhibited by plants as a guide in making selections or in isolating new strains. But it is often impossible to tell by observation alone what character combinations will be best or will result in the most productive progenies. A certain amount of aid may be gained from correlation statistics, if available. On this subject, Babcock and Clausen[16] say:

> Biometricians consider the correlation coefficient the most powerful tool the agricultural investigator can have, since it is a most excellent measure and is applicable to an immense range of variables. Remembering that this constant is an index of the mutual relation that exists between the variations of any two characters, we realize that, if it is high, it indicates that they are in some way closely related, and, if it amounts to unity, it shows that one is the cause of the other or else both are the result of the same causes.

From Hodson's[15] work it seems that the degree of correlation between certain characters of the cotton plant varies considerably with the variety used and with climatic or other conditions prevailing during different years.

TABLE VI.—CORRELATION BETWEEN CHARACTERS IN 167 INDIVIDUAL COTTON-PLANT SELECTIONS, TOGETHER WITH MEAN, STANDARD DEVIATION, AND COEFFICIENT OF VARIABILITY (*After Dunlavy*)

Pairs of characters	Correlation	Number of plants	Mean	Standard deviation	Coefficient of variability
Staple	−0.445 ±0.0419	167	32.2	1.31 ±0.0484	4.08 ±0.1498
Lint, per cent			39.39	1.10 ±0.0405	2.79 ±0.1028
Weight, seed	0.426 ±.0428	167	12.98	0.813 ±0.0300	6.25 ±0.2300
Staple			32.2	1.31 ±0.0484	4.08 ±0.1498
Weight, seed	−0.529 ±0.0376	167	12.98	0.813 ±0.0300	6.25 ±0.2300
Lint, per cent			39.39	1.10 ±0.0405	2.79 ±0.1028
Lint, index	0.153 ±0.0512	167	8.43	0.45 ±0.0166	5.34 ±0.1965
Staple			32.2	1.31 ±0.0484	4.08 ±0.1498
Lint, index	0.203 ±0.0501	167	8.43	0.45 ±0.0166	5.34 ±0.1965
Lint, per cent			39.39	1.1 ±0.0405	2.79 ±0.1028
Lint, index	0.704 ±0.0208	167	8.43	0.45 ±0.0166	5.34 ±0.1965
Weight, seed			12.98	0.813 ±0.0300	6.25 ±0.2300
Boll size	0.214 ±0.0568	127	7.94	0.6 ±0.0254	7.55 ±0.32
Staple			32.3	1.41 ±0.0595	4.35 ±0.1840
Boll size	−0.394 ±0.0506	127	7.94	0.6 ±0.0254	7.55 ±0.3200
Lint, per cent			39.3	1.06 ±0.0448	2.69 ±0.1138
Boll size	0.664 ±0.0338	127	7.94	0.6 ±0.0254	7.55 ±0.3200
Weight, seed			12.96	0.8 ±0.0339	6.17 ±0.2610
Boll size	0.480 ±0.0461	127	7.94	0.6 ±0.0254	7.55 ±0.3200
Lint, index			8.39	0.415 ±0.0175	4.95 ±0.2094
Seed, per lock	−0.161 ±0.0586	126	8.71	0.332 ±0.0141	3.81 ±0.1619
Staple			32.3	1.22 ±0.0517	3.78 ±0.1606
Seed, per lock	0.099 ±0.0596	126	8.71	0.332 ±0.0141	3.81 ±0.1619
Lint, per cent			39.5	1.12 ±0.0476	2.83 ±0.1204
Seed, per lock	−0.149 ±0.0587	126	8.71	0.332 ±0.0141	3.81 ±0.1619
Weight, seed			12.9	0.804 ±0.0341	6.24 ±0.2652
Seed, per lock	−0.082 ±0.0596	126	8.71	0.332 ±0.0141	3.81 ±0.1619
Lint, index			8.39	0.456 ±0.0193	5.43 ±0.2308
Seed, per lock	0.257 ±0.0749	71	8.63	0.299 ±0.0169	3.46 ±0.1962
Boll size			7.76	0.473 ±0.0268	6.09 ±0.3453
Per cent five-lock bolls	−0.109 ±0.0555	144	41.8	15.85 ±0.6288	37.89 ±1.500
Staple			32.3	1.37 ±0.0544	4.24 ±0.1691
Per cent five-lock bolls	0.036 ±0.0561	144	41.8	15.85 ±0.6288	37.89 ±1.500
Lint, per cent			39.3	1.09 ±0.0435	2.77 ±0.1095
Per cent five-lock bolls	0.114 ±0.0555	144	41.8	15.85 ±0.6288	37.89 ±1.500
Weight, seed			12.94	0.789 ±0.0313	0.610 ±0.2420
Per cent five-lock bolls	0.0784 ±0.0558	144	41.8	15.85 ±0.6288	37.89 ±1.500
Lint, index			8.39	0.436 ±0.0174	5.20 ±0.2075
Per cent five-lock bolls	0.533 ±0.0577	70	30.7	16.28 ±0.9296	53.03 ±3.028
Boll size			7.75	0.513 ±0.0293	6.62 ±0.3780
Per cent five-lock bolls	−0.045 ±0.0720	72	31.5	16.49 ±0.9267	52.35 ±2.94
Seed, per lock			8.64	0.304 ±0.0171	3.52 ±0.1978

Table VI, after Dunlavy,[2] gives the correlation existing between a number of different characters in a Texas cotton. The variety is not named, but it is taken to be one of the Texas big-boll medium-staple cottons, since they are the principal cottons grown in that state. The data were taken from choice plants selected for progeny-row work. The number used ranged from 70 to 167. The fact that only select plants were used limits the application to such plants.

Most of the correlations Dunlavy[2] gives are significant, and a number of them are of interest and of direct value. Among the most interesting, agriculturally considered, are: staple length and lint percentage, a rather high negative correlation being shown; weight of seed and length of staple, a rather high positive correlation (the lighter seeds here probably tended to be defective or undeveloped and had shorter staple); weight of seed and lint percentage, a high negative correlation, as was to be expected; lint index and weight of seed, high positive correlation; boll size and staple length, medium-high positive correlation (the better developed bolls within the variety had better staple length); boll size and lint percentage, rather high negative correlation (the larger bolls had larger seed, hence lower lint percentage); boll size and weight of seed, high positive correlation; boll size and lint index, high positive correlation (the larger seed in the large bolls had more lint on individual seeds); per cent five-lock bolls and boll size, high positive correlation.

Table VII, adapted from Kearney,[1] shows significant correlations for a large number of characters. These data were collected from 182 second-generation hybrid plants of the Holdon × Pima cross. Although this list of correlations is a long one, none is between the same pairs of characters studied by Dunlavy. Comparison is, consequently, not possible.

The correlations in Kearney's list most interesting to cotton growers are boll length and fiber length, which gave a small positive correlation, and boll diameter and lint index, which gave a somewhat higher positive correlation.

Hodson[15] gives data on the correlation of characters of several varieties of cotton grown in Arkansas, namely, Trice, Foster, Express, Triumph, and all the varieties grown in the variety tests two different years. Data were collected on 5 different years. This work is interesting in that it shows much difference in correlation of the same pairs of characters when found in different varieties or in the same variety in different years. In most instances, the correlation coefficient was too small to be significant. Larger numbers might have given more consistent results for different varieties and perhaps for different years.

TABLE VII.—COEFFICIENTS OF CORRELATION IN THE SECOND GENERATION OF THE HOLDON-PIMA HYBRID WHICH ARE REGARDED AS SIGNIFICANT

Character pair	Correlation coefficient	Character pair	Correlation coefficient
Axis length and—		Corolla length and—	
Internode length	+0.831	Petal color	+0.214
Internode number	+0.715	Petal spot	+0.244
Fruiting-branch length	+0.620	Stamen length	+0.242
Fruiting-branch internode	+0.226	Pistil length	+0.470
		Petal color and	
Pedicel length	+0.218	Fiber length	+0.277
Fiber color	+0.178		
		Stamen length and—	
Internode length and—		Anther color	−0.254
Internode number	+0.300	Pistil length	+0.197
Fruiting-branch length	+0.596	Boll diameter	+0.180
Fruiting-branch internode	+0.307	Pistil length and—	
Leaf length	+0.190	Lock number	−0.182
Pedicel length	+0.259		
		Lock number and—	
Internode number and—		Boll diameter	+0.202
Fruiting-branch length	+0.388		
		Boll length and—	
Fruiting-branch length and—		Boll diameter	+0.289
		Fiber length	+0.175
Fruiting-branch internode	+0.362	Boll diameter and—	
Pedicel length	+0.183	Lint index	+0.214
Boll diameter	+0.182		
		Fiber length and—	
Leaf length and—		Fiber color	−0.230
Boll diameter	+0.205		
Pedicel length and—			
Bract length	+0.208		
Bract number of teeth	+0.187		
Bract length and—			
Boll length	+0.186		
Boll diameter	+0.187		

Among the most interesting of the significant correlations obtained by Hodson were the following:

1. Number of base limbs and number of bolls per plant. This gave a positive correlation for every variety every year studied.

In 1917, when Foster cotton was used, the correlation coefficient was 0.4348 ± 0.0938.

2. Number of days from planting to open boll and height of plant.

This gave a positive correlation coefficient of 0.4012 ± 0.0817.

3. Lint percentage and weight of seed. The correlation was negative every year except one, the figures being too low that year for them to be significant; in 1917, the correlation coefficient was −0.4007 ± 0.0818. Dunlavy[2] secured a negative correlation of −0.529 ± 0.0376 for these same characters.

4. Length of lint and lint percentage. The correlations were all negative and mostly low. In 1918, the coefficient was −0.2445 ± 0.0680 for the short-staple varieties and −0.6146 ± 0.0409 for both short- and long-staple varieties considered as one population. The latter figures are similar to ones obtained by Dunlavy[2] for the same characters, −0.445 ± 0.0419.

Kearney,[24] in a paper more recent than the one mentioned above, gives further data on the correlations of boll, seed, and fiber characters of cottons that he has studied, and he gives also a résumé of the more important work of other investigators. Killough and Hafner[25] report negative correlations between yields of lint and length of lint and also between length and percentage of lint. They obtained positive correlation between lint yield and percentage, but the correlations were not high. Griffee, Ligon, and Brannon[26] studied correlations in upland cotton grown at Stillwater, Okla., and some interesting relations were observed. Yield of seed cotton, when considered in relation to other characters mentioned below, showed correlation coefficients as follows: length of stem internode, −0.573 ± 0.104; area of the largest leaf, −0.379 ± 0.133; number of vegetative branches, −0.458 ± 0.122; lint length, −0.465 ± 0.121; lint percentage, 0.522 ± 0.113; yield of lint, 0.864 ± 0.039.

Metazenia in Cotton.—By metazenia is meant the influence of pollen on tissues of the mother plant. Harrison[27] pollinated flowers on certain Pima Egyptian plants with pollen from Hopi cotton and flowers on other Pima plants with Pima pollen. The

Pima variety has staple about 1⅝ inches in length, while the Hopi is about ⅞ inch, and there is a difference of about 18 days in the boll period of the two varieties. The boll period of flowers on Pima plants that had been pollinated with Hopi pollen averaged 1.1 days shorter than the flowers pollinated with Pima pollen, and there was a significant reduction in length of lint of a full 1/16 inch as a result of the pollination with Hopi pollen. Other characters studied showed similar effects.

References

1. Kearney, Thomas H.: "Segregation and Correlation of Characters in an Upland-Egyptian Cotton Hybrid," *U. S. Dept. Agr. Bull.* 1164, 1923.
2. Dunlavy, Henry: "Correlation of Characters in Texas Cotton," *Jour. Amer. Soc. Agr.*, vol. 15, No. 11, 1923.
3. Balls, W. L.: "The Cotton Plant in Egypt," Macmillan & Company, Ltd., London, 1919.
4. Fletcher, Frederick: "Mendelian Heredity in Cotton," *Jour. Agr. Sci.*, vol. 2, 1907.
5. Tyson, P. F.: "Some Experiments in the Hybridizing of Indian Cottons," *Mem. Dept. Agr. India*, vol. 2, No. 6, 1908.
6. Harland, S. C.: "A Study of Inheritance in the Cotton Hybrid, Sea Island and Native St. Croix," *St. Croix Agr. Exp. Sta. Rept.*, 1913–1914.
7. Leake, H. M.: "Studies in Indian Cotton," *Jour. Genetics*, vol. 1, p. 205, 1911.
8. Leake, H. M. and Prasad, Ram: "Studies in Indian Cottons," *Mem. Dept. Agr. India*, vol. 6, p. 115, 1915.
9. McLendon, C. A.: "Mendelian Inheritance in Cotton Hybrids," *Ga. Agr. Exp. Sta. Bull.* 99, 1912.
10. Balls, W. L.: "Studies of Egyptian Cotton," *Year Book*, Khedivial Agricultural Society, 1909.
11. East, E. M.: "Studies on Size Inheritance in Nicotiana," *Genetics*, vol. 1, p. 164, 1916.
12. Kearney, T. H.: "Mutation in Egyptian Cotton," *Jour. Agr. Research*, vol. 2, No. 4, 1914.
13. Hays, H. K. and Garber, R. J.: "Breeding Crop Plants," McGraw-Hill Book Company, Inc., New York, 1921.
14. Duggar, J. F.: "Southern Field Crops," The Macmillan Company, New York, 1911.
15. Hodson, E. A.: "Correlation of Certain Characters in Cotton," *Ark. Agr. Exp. Sta. Bull.* 169, 1920.
16. Babcock, E. B. and Clausen, R. E.: "Genetics in Relation to Agriculture," McGraw-Hill Book Company, Inc., New York, 1918.
17. Thadani, K. I.: "Linkage Relations in the Cotton Plant," *Agr. Jour. India*, vol. 18, pp. 572–579, 1923; reviewed in *Bot. Abs.*, vol. 13, No. 5, p. 523.

18. Humbert, E. P. and Mogford, J. S.: "Variation in Certain Lint Characters in a Cotton Plant and in Its Progeny," *Tex. Agr. Exp. Sta. Bull.* 349, 1927.
19. Ware, J. O.: "Genetic Relations of Nankeen Lint to Plant Color and Leaf Shape in Upland Cotton," *Ark. Agr. Exp. Sta. Bull.* 300, 1934.
20. Hutchinson, J. B.: "The Inheritance of Leaf-shape in Asiatic Gossypium," *Jour. Genetics*, vol. 28, pp. 437–513, 1934.
21. Harland, S. C.: "Two Cases of Linkage in New World Cottons," *Trop. Agr.*, vol. 11, p. 12, 1934.
22. Horlacher, W. R. and Killough, D. T.: "Progressive Mutations Induced in *Gossypium hirsutum* by Radiations," *Amer. Naturalist*, vol. 67, p. 713, 1933.
23. Hutchinson, J. B.: "Bud Mutation in Cotton," *Trop. Agric.*, vol. 6, p. 275, 1929.
24. Kearney, T. H.: "Correlations of Seed, Fiber, and Boll Characters in Cotton," *Jour. Agr. Research*, vol. 33, No. 8, 1926.
25. Killough, D. T. and Hafner, V. E.: "Varieties of Cotton for the Gulf Coastal Plains of Texas," *Tex. Agr. Exp. Sta. Bull.* 354, 1927.
26. Griffee, Fred, Ligon, L. L. and Brannon, L. H.: "Biometrical Analysis of Upland Cotton Grown at Stillwater, Okla," *Okla. Agr. Exp. Sta. Bull.* 187, 1929.
27. Harrison, George J.: "Metazenia in Cotton," *Jour. Agr. Research*, vol. 42, No. 9, 1931.
28. Kearney, Thomas H., and Harrison, George J.: "Inheritance of Smooth Seeds in Cotton," *Jour. Agr. Research*, vol. 35, No. 3, 1927.
29. Kearney, Thomas H.: "Short Brarch, Another Character of Cotton Showing Monohybrid Inheritance," *Jour. Agr. Research*, vol. 41, No. 5, 1930.

CHAPTER IX

COTTON BREEDING

All efforts put forth to better or improve hereditary qualities of cotton plants may be considered cotton breeding. The cotton plant is susceptible of being changed or molded to fit a variety of conditions or to produce a varied product. The breeding or production of desirable varieties that yield well and have an excellent quality of fiber is a very important part of crop production.

History of Cotton Breeding.—It is impossible to say who was the first cotton breeder or where the work first started, as more or less work on cotton improvement has been carried on since early times. The Danish colonist Rohr, mentioned by Watt,[1] probably did some technical cotton-breeding work on the island of St. Croix about 1790. In his book, which was published at that time, he speaks of methods he used in the selection and hybridization of cottons. The introduction of Jumel's plant in 1820 (see Chap. I) may be considered the real beginning of cotton improvement work in Egypt. From that date until the close of the century, new or improved varieties were introduced from time to time, but distinct advances as a result of scientific breeding work were not made until after 1904.

Sea Island cotton was introduced into the mainland of North America about 1786. It is said that the plant at that time was perennial, larger than it is at present, and less productive. The growers early learned that by choosing the best plants, or the ones that best suited their needs, and propagating from them, they could grow more profitable crops. Consequently, they selected seed from the most compact, most productive plants, with the longest and finest staple, and continued the selection from year to year. As a result of this work, the plant was changed to an annual, made more productive, and the fiber was improved in length and character. But breeding work must be continued to hold the variety up to the high standard of excellence that it has attained, that of being the finest cotton in the world.

Upland cottons were introduced into Virginia and other southern colonies soon after their founding. New varieties originated from time to time, and some good improvement work was done by a number of men interested in cottons, among whom J. B. Allen, of Port Gibson, Miss.; W. A. Cook, Newman, Miss.; John Griffin, Greenville, Miss.; T. J. King, Louisburg, N. C.; J. A. Peterkin, Fort Motte, S. C.; and J. R. Cleveland, Stratton, Miss., may be mentioned.

The year 1898 marks the beginning of scientific cotton breeding in the United States. That year Dr. Herbert J. Webber started cotton breeding for the U. S. Department of Agriculture. Methods in use by the growers of Sea Island cotton were studied, improved, and adapted to the breeding of upland cottons.

Other breeding methods, which included hybridization and acclimatization, were developed. The majority of the most valued varieties grown at present were either produced by Dr. Webber or have been originated by some of his associates in the U. S. Department of Agriculture or by his students.

Prior to 1898, some cotton-breeding work was done at several of the experiment stations, but most of the work was not extensive, and but little of permanent value came from the efforts. Mell and Newman, of the Alabama Station, reported by Ware,[16] did some cotton hybridization work in 1886; and Speth, of the Georgia Station, crossed some cottons about 1889. Some selection work was started by Newman at the South Carolina Station and by Bain at the Tennessee Station about 1900. A limited amount of breeding work was started by the Texas Station also about 1900.

Between 1900 and 1910, scientific cotton breeding was started by the Pedigreed Seed Company of Hartsville, S. C., and by Clemson College in South Carolina. The Mississippi Agricultural Experiment Station started breeding work in 1910, and since that time experiment stations in several other states have started similar work. Several commercial seed-breeding plants have also been established. Some of these do careful breeding work and put out a supply of well-bred seed. These seed companies are especially helpful to the cotton growers, because they produce and make available quantities of pure, high-class seed. The U. S.

Department of Agriculture and the state experiment stations originate good varieties and grow a small amount of good seed, but they lack facilities for producing the latter in quantity.

The Object of Cotton Breeding.—As was pointed out in Chap. VIII, many of the varieties of cotton grown at the present time are of hybrid origin, or constitution, and are consequently subject to much variation. Many of the varying forms produced are inferior, and they cause a general deterioration of the whole population of plants. Breeding work is necessary to hold varieties up to their present standard or condition, to say nothing of improvements. There is room for definite improvement in the character or qualities of all cottons. They may be changed so as to be better adapted to the environmental conditions where they are grown; they may be made earlier; the yield may be increased; the lint percentage improved; the lint improved in spinning qualities and length; the lint improved in uniformity on the various plants of the variety, on different parts of an individual plant, or even on different parts of an individual seed; the form of plants may be improved; bolls may be made larger, and other picking qualities improved; storm resistance may be increased without interfering with the picking qualities; many varieties need to be made more disease resistant, especially more resistant to wilt and anthracnose. It is not to be expected that any one variety can be improved along all lines or made to approach perfection in every respect. There are apparently some antagonistic qualities, as were pointed out by Duggar.[2] Plants that have unusually large bolls cannot be extremely early, varieties with long staple will not have high lint percentage; extreme earliness is opposed to high lint percentage; a high lint percentage is seldom found in varieties that have large seeds. To the list of opposed qualities mentioned by Duggar may be added wilt resistance versus extreme earliness and prolificness. Some of the semiresistant varieties are early or medium early, but none of the regular resistant varieties is early.

Cotton-breeding Methods.—Three different types of work are carried on by persons seeking to make improvement in cotton plants, or, in other words, to do cotton breeding. These are selection, which consists in searching out and propagating from certain outstanding plants; hybridization, the essential part of which consists in crossing different races with the idea of combining in one plant desired qualities found in different plants; and acclimatization, which consists in transferring strains or varieties from one region to another and growing them in the new habitat long enough for them to become adjusted to it or acclimated. Each of these methods will be discussed separately and in some detail. (This discussion of cotton breeding applies particularly to American upland cotton.)

Cotton Breeding by Selection.—The selection method of cotton breeding is used more than any other method, and probably better results have been secured with it than with any other. The other methods, hybridization and acclimatization, also include selection work and would not be successful without it.

Mass Selection.—Mass selection is the simplest type of selection and is one of considerable value to the cotton grower who wishes to do some improvement work on his cotton but does not have the opportunity to do more complicated breeding work.

The cotton grower who wishes to do mass-selection work goes into his cotton field about the first of September, or after there are some open cotton bolls on nearly every stalk, and tags a large number of what he considers good stalks. He must have in mind a definite type of plant, size of boll, length of staple, etc., and choose only plants that make a near approach to his ideal. Furthermore, he must hold to the same ideal year after year to make progress. After the bolls on the tagged stalks are all open, they are picked, the cotton from all being put together. This seed cotton must be so ginned that there will be no mixing with other seed. This can be done easily by feeding the cotton by hand into one gin of a large ginnery and catching the seed on the floor in front of the gin. These seed should be planted the following year in a seed patch isolated from other cotton. The land should be good so that the quantity of seed produced will be large, and the plants should be widely spaced so that each plant will have a chance to show its individuality. New selections should be made in this patch for planting the next year's seed patch. The rest of the seed grown can be used in planting the general crop.

By continuing the plan outlined for a series of years, a fairly uniform strain of cotton may be produced. Sunflower, at one time one of our leading long-staple cottons, was produced in this way by Mark Schaeffer, of Yazoo City, Miss. He started with mixed seed obtained from an oil mill. Cleveland Big Boll is another important variety developed by mass selection. While this plan will give results eventually, it is slower than other methods. One of its disadvantages is to be found in the fact that a part of the fine-looking plants chosen are first-generation hybrids which will not breed true the next generation. Other good-looking plants fail to transmit their good qualities, and their seed becomes mixed with that of the better plants.

Individual Plant Selection.—The selection of individual plants followed by plant-to-row testing is the plan usually followed by plant breeders. In looking for outstanding plants to select, the breeder starts work as soon as the cotton has begun to open freely and walks over large fields, row by row, of the variety with which he has decided to work. Chances of success are better if the land is rich and the plants are well fruited and spaced far enough apart for each one to develop well. The breeder surveys two to four rows as he walks along. Any plants that, at first glance, appear promising are examined in some detail; their characteristics are noted, lint combed out so that its length may be judged, and the lint percentage roughly estimated. A breeder who has had some practice can usually estimate to within 2 per cent of the correct lint percentage. Only a few plants are found each day that measure up to the requirements in mind. These plants are carefully tagged, and it is the writer's practice to write on the tag a brief description of the plant chosen. Later when the bolls are all open, the plants are picked separately, the seed cotton of each being put in a paper bag, along with the descriptive tag that was on the stalk.

The outstanding plants chosen in making selections fall into three classes: (1) Exceptionally well-formed and well-fruited typical specimens of the variety with which the breeder is working. If the variety is relatively pure or well bred, this class will include most of the selections made. (2) Plants that contain good character combinations which are segregates of characters from hybrid ancestors. (3) Mutants, which are striking new forms that have appeared, and which have certain characteristics not possessed by the parental strain.

Plant Characters Observed When Making Selections.—As a cotton breeder walks along a row of cotton looking for superior plants, certain characters of the plants hang out as signboards to

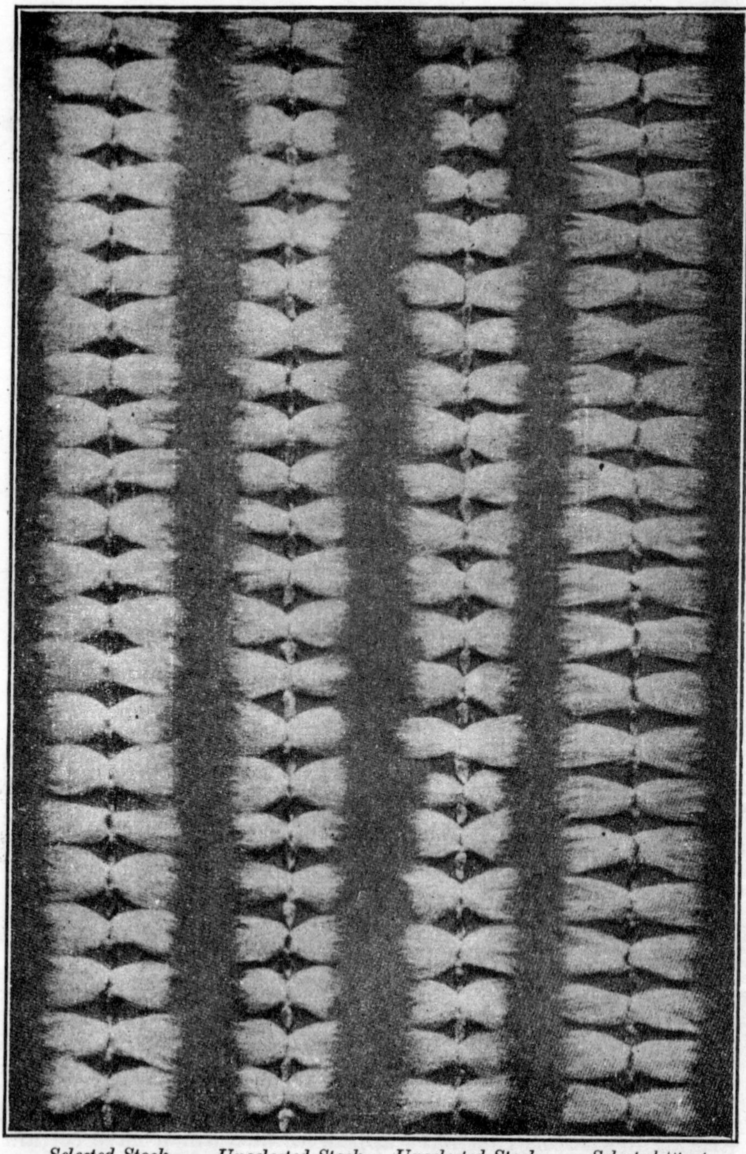

FIG. 43.—Samples of lint, showing the effect of selection on the uniformity of Meade cotton. The two center rows show combed lint from consecutive plants in the original unselected stock. The two outside rows show combed lint from consecutive plants in selected stock. Note the uniformity of fiber and seed in the selected stock as compared with the irregularity of both fiber and seed in the unselected stock. (*After Meloy and Doyle.*)

attract attention. Of these the most conspicuous is prolificness, or amount of fruit on the plant, and it is probably the most important character to consider in making a choice. Other desirable qualities that the breeder of upland cotton looks for are compactness of plants, low branches, and short internodes (plants with extremely short internodes in branches, such as are found in the cluster varieties, are not desirable); light foliage, which is especially desirable in humid regions or in regions of heavy boll-weevil infestation; fair-sized bolls (small-bolled cot-

Fig. 44.—A small roller gin for ginning cotton from individual plants. (*After Childs.*)

tons are more prolific and may yield more but are harder to pick); freedom from disease; satisfactory lint length; satisfactory lint percentage; and good picking qualities. With some varieties and with certain species of cotton the branching habits are of considerable importance. Leake[3] has shown that certain races of Indian cottons bear cotton blooms only on sympodial branches. If the secondary branches that the plant produces are sympodial, blooms will be formed early; but if they are monopodial, flowers will not be formed until the tertiary branches appear, which are sympodial. This makes the flowering much later. Varieties with this type of branching are not suited to the more northern portions of the cotton belt in India. Among the upland varieties, plants with more than three or four vegetative branches are to be avoided. Figure 11 shows a good type of upland cotton plant.

Testing Selections.—After the selected plants are picked, as was explained above, they are taken to a laboratory or to some other suitable place, where the cotton lint on some of the seeds is combed out, as shown in Fig. 43, and measured. Some breeders place the seed with the combed lint attached in one of the more simple fiber sorters, such as the Pressley or the McNamara, and sort out roughly and weigh fibers of different lengths. This process gives an idea of the proportion of different fiber lengths on the seed.

The seed cotton from each plant is ginned by itself on a small gin—either a 10-saw saw gin or a small roller gin made for the purpose (see Fig. 44)—and the seed saved. Most breeders weigh the seed cotton and seed from each plant and then compute lint percentage. All the notes and data on each plant are recorded on pedigree cards or sheets and preserved.

The following season the seed from the selections is planted in a plant-to-row test. If seed is plentiful, two short rows may be planted. If these are planted in different parts of the breeding block or cut, it adds to the value of the test. If a strain does well in two places, the chances are greater that it has merit. The seeds must be planted by hand, preferably in hills about 2 feet apart and in rows 100 to 200 feet in length. To insure a good stand, not fewer than 10 seed should be planted in a hill. At thinning time, the plants should be thinned to one in a hill. It is desirable to plant seed from plants with similar characters in adjoining rows.

During the growing season the different strains in the progeny rows are observed from time to time, and note made of marked peculiarities. In the fall after all the bolls have opened, the rows are studied in detail, and notes made on each. Each row is picked separately, and the cotton weighed. On the basis of this yield weight and the other data gathered, a few of the best strains are saved, usually 5 to 10 per cent. The others are discarded or saved for planting in the general crop. If the poorest of the rows are discarded, and the seed from the others massed, it makes a good quality of seed to plant, considerably better than seed from simply mass-selected plants, because the poorer strains have been eliminated from this.

The second year after the original selection is made, a portion of the seed from each of the choice progeny-row strains that were saved is planted in another test, which is sometimes called a "new-strains test." This test is handled in much the same way as was the progeny-row test discussed above, except that, as more seed is available, the rows may be planted with a planter. Besides being more rapid than hand planting, the planter places the seed more uniformly if the land is in good condition. A planter with a hill-dropper attachment and a roller wheel behind to regulate depth is to be preferred. Since more seed are available it is possible to have more replication of rows. It is the writer's practice to have four series in the test, which means that there are four rows of each strain. These are in different parts of the test plots and afford a more reliable trial than would a single row.

Results from the new-strains test will show clearly any strains that have special merit. If any such have appeared (in most instances, they have not), the progeny-row remnant seed, that is, the rest of the pure seed saved from the progeny row, is planted in an isolated patch the following year, and a start made to increase the strain. If seed are available, it is a good plan to put the few strains that were outstanding in the new-strains test in a variety test the following year, where they may be compared with standard commercial varieties. If they are not an improvement over the varieties already in use, they should be discarded. Seed from the new-strains test may be planted in the variety test. Plenty of seed of the new strains is available for this test, and more replications may be used. Randomized blocks, or the Latin-square method of testing, add to the reliability of the test.

Increase Fields.—As a result of the three tests—progeny-row, new-strains, and variety—the merits of new strains are determined rather conclusively. Any strains that have ranked comparatively well for 3 years must be worthy of multiplication and should be increased as rapidly as possible. The remnant of pure seed saved from the progeny row should be planted on good land in a field isolated from other varieties of cotton as far as is convenient. A barrier of at least 25 rows of corn or of some other crop should be used. The seed should be made to plant as great an acreage as possible. The increase plots or fields should be inspected row by row at least twice each season for the first two seasons, and all off-type plants rogued or pulled out. In so far as it is practicable to do so, all fields from which seed is to be saved for planting purposes should be rogued (see Fig. 45).

Practical Results from Selection Work.—If the origin of the leading cotton varieties being grown at the present time is studied, it is found that most of them have arisen from plant selections, the product of simple selection work or selection work

Fig. 45.—Roguing Meade cotton. (*After Meloy and Doyle.*)

combined with hybridization or acclimatization. Lone Star, Mebane's Triumph, Delfos, Dixie, Half and Half, Webber, Pima, and others were produced by straight selection work. Foster, Salsbury, Delta and Pine Land-8 and -10, Deltapine, Cleveland-5, Wilds, Stoneville, and some others came from selection combined with hybridization. Acala, Durango, and Express are the result of acclimatization combined with selection work. (See Chap. III for further details concerning the origin of varieties.)

Through or by means of selection, various characters of cotton plants may be made much more uniform. This is especially true of the important character of lint length. The effect of selection on this character is well shown by Meloy and Doyle[4] in work with Meade cotton (see Fig. 43).

Pima, an American-Egyptian variety, arose from a single remarkable plant which was probably a mutant. It gave a progeny of comparatively uniform plants, and all its descendants have been more uniform than is usually the case among cotton progenies, yet Kearney[5] has shown that the strain was susceptible of change when subjected to selection and roguing for

Fig. 45A.—Breeding plots and trial grounds of the Stoneville Pedigreed Seed Company, Stoneville, Miss.

7 years. The fiber was increased in length 3.8 millimeters, or more than ⅛ inch; the amount of lint on 100 seeds was raised approximately 1 grain; and several other slight heritable variations were observed. A new strain arose within one of the progenies which lacked the petal spot found on most Pima plants and had a higher percentage of four-lock bolls. The selection work on Pima cotton during the 7 years was apparently not sufficient to reduce the strain to gametic purity. It was consequently still susceptible to change by selection.

Cotton Breeding by Hybridization.—The primary object in hybridization, or crossbreeding, is to combine in one individual characters or qualities at present found only in different individuals. Following the crossing, there is segregation and recombination of characters in many combinations. The breeder searches among the F_2, or second-generation, plants for the combination desired and, when he has found it, selects and propagates from the plant showing the characters wanted. It is usually necessary to select and reselect within the strain two or more times to secure satisfactory uniformity and stability within the new hybrid strain. Some authorities have expressed the belief that mutations are more common in strains of hybrid ancestry. Although this may be true, it is difficult to determine whether the new forms are due to mutations or to segregation and new combinations of characters.

Although selection has been the most effective method in cotton breeding, it is probable that hybridization assists indirectly in many instances. It is impossible to say how many of the outstanding plants that are selected owe their good qualities to the segregation of characters in hybrids produced by natural crossing that took place years before. As has been shown by Johannsen and others, selection within a pure line has practically no effect. If cotton strains are line bred, or inbred until they approach genetic purity, selection probably will have but little effect. Crossbreeding is thus a necessary prerequisite, even if several years removed, to most effective selection work. The fact that a character may be transferred from one variety to another by crossing makes this type of breeding promising, even if it is usually very difficult to get just the combination of characters wanted in a fixed or stable race of plants.

Limits of Hybridization.—All American upland varieties belonging to the species *Gossypium hirsutum* will cross with one another, and they will also cross with Sea Island, Egyptian, and the cultivated South American varieties, producing fertile progeny in all cases. The American upland varieties can be crossed with the more prominent Asiatic varieties belonging to the species *G. herbaceum* and *G. arboreum* only by great effort, and persisting fertile hybrids are very rare. Harland[17] reports a cross between *G. arboreum* and *G. hirsutum*. The first-generation hybrid was very nearly sterile, but by a series of backcrosses this sterility was modified to complete fertility. Other crosses between New and Old World cottons have been reported by Skovsted,[18] Sadao,[19] and one or two others.

As was shown in Chap. VI, the cultivated New World cottons have 26 chromosomes, whereas the Old World species have but 13. This accounts for their lack of compatability in crossing.

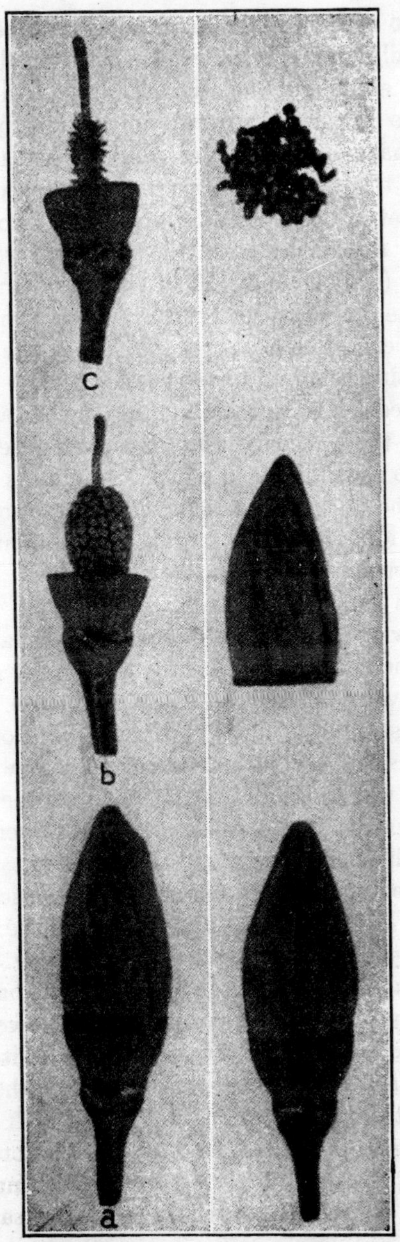

Fig. 46.—Flower buds of Pima cotton. The successive stages of the process of emasculation during the evening preceding anthesis: Intact buds (*a*); bud with corolla removed, showing the tightly closed anthers (*b*); anthers removed, flower ready for pollination (*c*). (*After Kearney.*)

Manipulation of Flowers in Artificial Crossbreeding.—Cotton flowers are easily cross-pollinated, because the flower parts are large, pollen is abundant, the stigma is large and receptive, and the plants are normally susceptible to crossbreeding. If a cross is to be made on a certain day, during the afternoon of the day previous the breeder chooses for emasculation certain flower buds that are scheduled to open the following morning. These may be recognized from the enlarged, yet twisted and folded, corolla (see Fig. 46*a*). The corolla is removed with the small blade of a knife or a scalpel, parts shown in Fig. 46*b* being left. The anthers, or male part of this flower, are next removed, either by scraping them off or by peeling off the filaments with anthers attached. The anthers contain pollen and should not be broken lest the pollen be scattered over other flower parts. It is not safe to use a flower in which the anthers have begun to open. After the anthers are removed, the flower presents the appearance of Fig. 46*c* and is said to be "emasculated." Only the pistil, or female part of the flower, is left. The stigma is the club-shaped portion at the top.

A small paper bag is next tied over the emasculated flower bud and left until the next day. To provide a supply of pure pollen, bags are tied over similar flower buds on plants of the variety that is to be crossed on the first. By nine o'clock of the following day, if the sun is shining, the emasculated flower will be ready to be crossed. A whole flower is removed from the plant that is to furnish the pollen. The anthers in this flower will be open by this time, showing a mass of yellow pollen. The brush of anthers is rubbed against the stigma of the emasculated flower. If the work is well done, pollen can be seen on the stigma. The paper bag is again tied over the crossed flower to keep away insects and left at least 2 or 3 days. It will do no harm to leave it on until the resulting boll opens. This may serve to keep boll weevils from the boll. After the cross is made, it is, of course, necessary to attach a tag to the flower, so that the boll may be distinguished later.

If the crossing work is carefully done and is done fairly early in the blooming season, the period when shedding is lowest, a high percentage of the crosses will be successful, say 75 per cent or more.

Viability of Pollen.—Work by Kearney[13] showed that pollen of both upland and Egyptian varieties (Durango and Pima) was most viable between the hours of 10 a.m. and 2 p.m. By 6 p.m., only 3 or 4 per cent of the grains seemed to be alive. Pollen from flowers enclosed in paper bags all day was still nearly normal at 6 p.m. A fair percentage of the pollen grains was still viable at 6 a.m. the next morning. The writer has found the pollen of various upland varieties to be very satisfactory for use in the afternoon of the day the flower opened, provided the flower had been bagged.

Fertility of Bagged Flowers.—In breeding work, it is often desirable to have seed from many self-pollinated plants. To self-pollinate by hand is burdensome if there are many plants. It has been found that, if flower buds are enclosed in small paper bags the day before opening, they will be self-pollinated, and a fair percentage will set bolls. Kearney[13] reports more than 90 per cent as compared with normal open-pollinated flowers, but the number of seeds in a boll was considerably lower in the selfed bolls. In work carried on by the writer, not more than 50 per cent of the bagged flowers made bolls, but some of the failures were due to natural shedding.

Some cotton breeders fasten paper clips over the tip of the flower bud to prevent its opening and thus secure self-pollination. Others wrap wire or rubber bands around the bud. Any one of these methods may be satisfactory if it prevents the flower bud from opening.

Natural Crossing in Cotton.—Different varieties of cotton are very frequently planted in close proximity, as, for instance, in adjacent fields. In nearly every locality there are bees, butterflies, and various other insects which visit the cotton flowers and may carry pollen from one field to the flowers of plants in another, thus bringing about cross-pollination and hybridization. This, as has been shown elsewhere, has been partly, if not largely, instrumental in causing our cotton varieties to contain many hybrids, to be impure, and to deteriorate freely.

In breeding work and in practical cotton growing, the question often arises as to what percentage of crossing takes place between two varieties of cotton when planted side by side in a field, in adjoining fields, or separated various short distances. Considerable experimental work to obtain data in answer to these questions has been carried on by Webber,[6] Balls,[7] Allard,[8] Shoemaker,[9] McLendon,[10] Brown,[11] Leake,[12] Kearney,[13] and others.

Webber,[6] from his experiments in South Carolina and in other southern states, estimated that there was from 5 to 10 per cent of crossing between different varieties of upland cotton when grown in adjacent rows. Allard[8] found that, when Keenan and Okra Leaf or Red Leaf cottons were planted in alternate rows in northern Georgia, 20 per cent of the bolls on the Keenan cotton were hybridized. The seed of these bolls produced a varied number of hybrid plants, ranging from one to as many as there were seeds.

Ricks and Brown,[11] in experiments carried on at State College, Mississippi, found that when green-leaf varieties were planted in rows alternating with rows of Willett's Red Leaf, 4.9 to 11.1 per cent of the ovules were crossed. The percentage varied slightly with the different green-leaf varieties. The percentage of crossing on Trice and Lone Star-132 was lower than on Triumph, Cleveland, and Lone Star-15. This was doubtless due to the fact that the first two varieties were slightly earlier than the red-leaf variety and did a larger percentage of their blooming early in the season. Individual boll progenies of Trice showed that 36 per cent of the bolls were hybridized to some extent; and 44 per cent of the Lone Star-15 bolls were hybridized. But few of the seeds in any one boll were hybridized; the number ranged from one to nine.

When the varieties mentioned were planted in alternate hills in alternate rows, the crossing was 18.5 per cent. In a plot where a green-leaf variety was separated from the red-leaf by a barrier of 10 rows of corn, two crossed plants were found in a progeny of 244 plants. This gives a moderately high percentage, considering the position of the two varieties. If a larger number of plants had been grown, the percentage might have been lower. The two crossed seed from which the two hybrid plants were grown were probably the result of a single insect visit to 1 of the 100 bolls gathered for seed.

In other natural-crossing experiments conducted by the writer at State College, Mississippi, in 1918 and 1919, Winesap cotton, a red-leaf variety slightly earlier than Willett's Red Leaf, was planted in various positions with respect to green-leaf varieties. Fifty or more stands of honeybees were kept within a half mile of the experimental plots. Many honeybees visited the flowers; 112, according to actual count, entered a single flower during the first day it was open. Counts of progenies from bolls on rows next to a red-leaf row showed 3.9 to 5.6 per cent crossing; second rows from the red-leaf row showed from 0.68 to 0.78 per cent; third row, 0.4 per cent; fourth row, 0.51 per cent; fifth row, 0.78 per cent; sixth row, 0.17 per cent; seventh row, 0.14 per cent; eighth row, 0.19 per cent; ninth row, 0.00 per cent; tenth row, 0.00 per cent; eleventh row, 0.00 per cent; twelfth row, 0.00 per cent.

In 1919, twenty-seven rows of Winesap cotton were planted at one end of a cut, and the rest of the cut was planted in Cleveland, a green-leaf variety. That year there was much more crossing, due probably to the larger number of plants grown. The figures ran as follows:

Bolls from the green-leaf row next to the Winesap rows produced 14.8 per cent hybrid plants; second row, 6.5 per cent; third row, 6.9 per cent; fourth row, 4.0 per cent; fifth row, 3.2 per cent; eighth row, 1.3 per cent; tenth row, 1.9 per cent; sixteenth row, 0.6 per cent; twentieth row, 0.5 per cent; twenty-fifth row, 0.4 per cent; thirtieth row, 0.17 per cent; thirty-fifth row, 0.12 per cent; fortieth row, 0.23 per cent; fiftieth row, 0.00 per cent; sixtieth row, 0.05 per cent; seventieth row, 0.24 per cent; eightieth row, 0.00 per cent; ninety-fifth row, 0.06 per cent; one hundred and fourteenth row, 0.06 per cent; one hundred and nineteenth row, last row in the cut, 0.00 per cent.

Four green-leaf plants grown at different points in the red-leaf plot had more than 50 per cent of their ovules crossed. The numbers were 57.1, 66.2, 81.4, and 61.7 per cent.

Data obtained by other investigators have not varied greatly from that given on the foregoing pages. McLendon[10] in Georgia observed 1.9 per cent crossing between adjoining rows; Shoemaker, 10.9 per cent in Texas; and Kottur, 6 per cent in India.

The data presented show that the amount of crossing that takes place between two varieties only a few yards apart is limited; yet there is some, and there may be some crossing between fields ¼ or ½ mile apart.* The only practicable way to prevent

* T. H. Kearney, in a personal letter to the writer, says that at Sacaton, Ariz., there is a limited amount of crossing each year between plants in fields of Pima and upland cotton ½ mile apart. In 1927 there were about four hybrid plants per acre.

this very limited amount of crossing is to grow only one variety in a community.

The percentage of crossed seed depends on: (1) The relative rate of blooming of the neighboring varieties. (2) The position of stigma and stamens in the flower. As Kearney[13] has shown, the flowers of some varieties are especially adapted to cross-pollination, and the flowers of others are better adapted to self-pollination. (3) The relative abundance of insects. Insects are much more abundant in some regions than in others, but difference in number does not appear to make a corresponding difference in cross-pollination. In natural crossing experiments made by the writer in the Mississippi Delta in 1918, where insects were very much less abundant than at State College, not more than two or three visiting a flower in the course of a day, practically as much crossing was obtained as at State College in similarly conducted experiments.

Characteristics of Hybrids.—First-generation hybrids are frequently larger, more vigorous, and more productive than their parents. Many of them are very interesting and very promising, but no great amount of importance can be attached to them, because they are not likely to reproduce their valued qualities.

In the second generation, as has been mentioned in Chap. VIII, there is a breaking up of types and a segregation of characters into many combinations, but by far the larger part of these are worthless from an economic point of view. The breeder selects a few plants in the plot that make the nearest approach to his ideal and grows their progenies in a plant-to-row planting the following year. Some of these will segregate further, but some few may breed true for certain desired character combinations. Further selections may be made in the third and later generations, but the chances are that nothing will ever be secured that is very satisfactory. Only one cross in many proves to be of high value.

Backcrosses.—If the first cross does not give satisfactory results, it may be worth while to cross the hybrid with one of the original parents, or "backcross it," as it is called. By backcrossing, the relative number of characters belonging to either one of the parents may be increased in the hybrid, and a more desirable form secured.

Results from Hybridization.—As may be seen from the preceding discussion, it is a very simple matter to crossbreed cottons, and it is comparatively easy to secure desirable types in the first generation; but it is a task to secure the wanted character combinations in types that breed true in later generations. Notwithstanding this difficulty, several good varieties have been produced in this way (see Chap. III). Still other good varieties have been produced by selection in varieties that were hybridized many years before or by the selection of mutants in strains of hybrid ancestry.

Although, as has been shown, uncontrolled hybridization is the cause of much deterioration in cotton, it is also probably true that maximum yields are due to hybridization in many instances. Certain varieties appear to have dropped in comparative yields after being subjected to close selection for 10 years—selection that made them more uniform and tended to reduce them to genetic purity. Two strains of Express grown by the writer which were self-pollinated and line-bred for 3 years, when crossed gave a progeny in which the plants were larger than their parents, produced more flowers, and yielded more cotton. Leake and Prasad, in India, found that certain cottons of the Asiatic type retained fewer bolls, comparatively, if inbred one or more generations but that later crossing tended to overcome this effect.

In some inbreeding work carried on in Louisiana by the writer with eight different varieties for a period of 8 years, most of the inbred lines have produced fewer blooms, had smaller bolls, and produced slightly less cotton than cross-pollinated strains of the same varieties.

Kearney,[13] in careful experiments performed with Pima cotton in Arizona, found that

> . . . inbreeding during seven generations brought about no reduction in the daily rate of flowering, in the percentage of bolls retained, in the size, weight, and seed content of the bolls, in the weight and viability of the seeds, or in the abundance of the fiber, as compared with those of the continuously open-pollinated stock. It is concluded that the deleterious factors had been eliminated in the ancestry of the Pima variety and that its present state of fertility is due to segregation rather than to heterosis.

Further experiments conducted by Kearney[13] showed that in Pima and Acala cotton a large proportion of the pollen deposited on the stigma of a flower comes from the anthers of the same flower, being either transferred by insects or deposited mechanically by the flower itself. It may be inferred from this that a large percentage of cotton ovules are naturally self-fertilized.

Results from Acclimatization Work.—Certain cotton varieties are especially adapted to a given set of conditions but not well adapted to many others. Todd[14] notes that many of the finer Egyptian varieties in Egypt are at their best only in particular limited districts.

By transferring varieties from one region to another and growing them in the second for a number of years or until

FIG. 47.—Numbers 1 and 2, American cotton from seed just imported. No. 3, degenerate American cotton from seed grown in China only three years. No. 4, Chinese cotton from selected seed. (*After Griffing.*)

they become adjusted to the new environment, certain new types may be found which are of unusual value to the second region.

When a cotton is transferred from one region to another with different environmental conditions, as from the American Cotton Belt to China, for instance, the first crop grown produces well and is satisfactory. If grown longer, it seems to be the tendency for the variety to "go to pieces," as it is termed. Many off-type or worthless plants appear, and deterioration is rapid. Griffing[15] found this to be true for Trice, Acala, Miller, and College No. 1 taken from the United States to China (see Fig. 47). If selections are made within the variety, and new strains built up in this way, the results will be satisfactory, and well-adapted cottons may be obtained. Trice handled in this way had given

Griffing[15] very satisfactory results in regions to which it is suited. In coastal regions it was found to be inferior to native cottons.

It appears that the removal of cottons to new environmental conditions not only stimulates the plants to extreme variation and the production of many inferior forms but also produces some exceptionally good types. Griffing noticed some of these in both the Acala and Trice varieties with which he was working in China. Kearney made similar observations on the Egyptian cotton with which he was working in Arizona. The Pima, Yuma, and one or two other less known varieties sprang from these variant forms found in fields in Arizona.

Results from Acclimatization Work in America.—During the 300 years that cotton has been grown in the American Cotton Belt, cottons have been imported from nearly every part of the world that grows cotton. Many of these importations were, of course, complete failures; others were probably crossed with American varieties and lost their individuality but left their impress; still others imported were found to be desirable and adaptable to certain regions and were preserved. Sea Island, Egyptian, Acala, and Durango belong to the latter class. Sea Island was imported more than a century ago, selected and acclimated to the conditions prevailing in Georgia and South Carolina, and made the best long-staple cotton in existence. Egyptian cotton was imported and grown experimentally by the U. S. Department of Agriculture in irrigated valleys of southern Arizona in 1902. As the climate of that region resembles that of Egypt, it was thought that a cotton adapted to Egypt might do well there. Several Egyptian varieties and strains were tried. The Afifi variety gave better results than any other; so the others were dropped. The method of selection and progeny-row testing was followed for several years without much gain apparently. In 1908, a superior and distinct new type appeared which indicated a real advance and gave rise to the variety Yuma. This was followed a little later by another still better new form which gave rise to the Pima variety. This proved so good that it was increased as rapidly as possible and introduced commercially. It soon largely displaced other Egyptian varieties in Arizona and California.

Acclimatization of American Varieties in Foreign Countries.—Since American varieties of cotton are superior in yield and lint characters to many of the native varieties of other countries, many importations of American varieties have been made.

About 1840, the East India Company, in order to foster the growing of American cotton in India, not only imported seed but employed 10 American cotton growers to go to India to introduce American methods of culture. This plan met with very little success, and the Americans soon returned home. The Dharwar American cotton grown about the town of Dharwar in the Bombay Presidency is from some of the American cotton imported at that time and marks one of the few successful attempts at introduction. Cambodia, or Tinnevelly American, probably the most valuable cotton grown in India, is also of American origin.

Failure has resulted in most cases when American cottons were taken to other countries, because the cotton was not handled properly. It was allowed to deteriorate or become mixed, and few if any attempts were made at selecting better adapted strains which appeared the first few years after introduction or during the usual period of deterioration. Within the past few years, certain American varieties have been successfully introduced into parts of China by Griffing[15] and by others, who used modern methods of cotton breeding. American cottons have been introduced into Chosen (Korea) by Japan and are apparently being grown successfully. Brazil, Argentina, and other South American countries are now introducing American uplands.

Harland,[20] in discussing the acclimatization of cottons in new areas, says:

> In adapting a cotton to a new habitat it acquires a set of genes that adapt it to the particular region. American upland has been adapted to the southern states of America. It is thus unlikely that its introduction into the tropics would be attended with much success. For the past 60 years or more, attempts have been made to introduce the upland cotton into various parts of tropical Africa. The history of such introductions is one, on the whole, of failure. There are some instances in which a form of upland has become a standard cotton in certain districts, as for instance U_4 in South Africa, Allen Long Staple in Nigeria, and Nyasa upland in Nyasaland. The commercial U_4 is markedly different from typical upland. . . . Modern uplands are apparently unable to acclimatize themselves. Why is this? Probably at the time the Punjab stock came to India, the upland cotton of the Southern States

had a good deal more genetic diversity than at present. The Punjab American is said to resemble closely the Petit Gulf cotton of New Orleans common 100 years ago.

Breeding Cotton Varieties for Wilt Resistance.—Distinct advances in one type of cotton breeding, that of breeding strains resistant to cotton wilt, a disease caused by *Fusarium vasinfectum* Atk., have been made comparatively recently by W. A. Orton and his associates in the Bureau of Plant Industry, U. S. Department of Agriculture, and by A. C. Lewis and others in Georgia and Alabama. New varieties developed may be grown profitably on land badly infected with wilt. In breeding for wilt resistance, the usual method of selection and plant-to-row testing is followed, the only difference being that all the work is done on wilt-infected soil.

Asexual Reproduction of Cotton Plants.—Beckett[21] and others have shown that widely different species of *Gossypium* may be successfully united by budding or by grafting. In Beckett's work, *Thurberia thespesioides* and plants of three other genera were successfully united with species of *Gossypium.* Of the budding trials, 38.1 per cent were successful, and 73.4 per cent of the grafts. Rea[22] reports that from 2,000 cuttings made from different parts of cotton plants—old wood and soft parts of stem, vegetative branches, and fruit branches—19.4 per cent of the cuttings calloused over and 6.2 per cent put out roots. The cuttings were kept in moist sand until calloused and then transferred to earth in pots.

Defloration as an Aid in Cotton Production.—Cotton growers often notice that when their plants become well loaded with fruit, vegetative growth is checked or even stopped almost altogether and that this stoppage of growth may occur when the plant is too small to bear many bolls. The stoppage works to an apparent disadvantage in such cases, but, on the other hand, if the plants have a tendency to become too rank or to make too much vegetative growth, as they do in humid regions where the soil is fertile, it may be a real advantage. Eaton,[23] at Sacaton, Ariz., carried on an experiment with both Acala upland and Pima Egyptian varieties to determine whether or not yields could be increased by delaying boll setting for a few weeks and whether or not the quality of lint would be better if fewer bolls were grown on a plant. All flowers were removed from certain plants during the first 25 days of flowering, and other plants grown with these were undisturbed. In the case of the Acala variety, the untreated plants produced 16.8 mature bolls per plant, whereas the ones deflorated early in the season produced 20.1. The Pima untreated plants produced 42.4 bolls on the average, and the treated 49.9. In another experiment conducted by Eaton, all bolls, flowers, and larger squares were removed from 100 plants 8 days after flowering started. These plants matured 24 per cent more cotton than another 100 plants alternately spaced with the treated plants. This treatment to increase yields may not be of practical use in cotton production, but it is of value to the cotton breeder if he has crossed or selfed bolls on the plants that he wishes to save. A much higher percentage of the late bolls is retained if the early forms are removed. Then, too, the destruction of all early forms on cotton plants by cotton flea hoppers may not be harmful.

Eaton[23] found further that where the number of bolls per plant was limited, by allowing but one boll to develop on a branch the boll weight was increased 12 per cent in the case of Acala and 43 per cent in the case of Pima. The fiber was slightly longer on the one-boll-per-branch plants of both varieties.

Judging Cotton Plants.—It is comparatively difficult to judge the relative merits of different varieties of cotton when individual or a few plants are offered in an exhibit. Certain plants grown under very favorable conditions may appear much better than others grown under slightly less favorable conditions but would appear at a decided disadvantage if the two were put in a field side by side. As has been shown by recent research, larger yields per acre may be obtained if plants are closely spaced in the row, and the most desirable variety is the one that will yield the most cotton when so planted; yet the individual plants of the variety, especially if grown under close spacing conditions, may make a poor showing in the exhibit room. The important considerations are quantity and quality of lint produced per acre. Certain general characteristics of cotton plants are considered desirable, however, and the degree of their expression may be judged in a cotton exhibit. Exhibits stimulate interest in growing better cotton and are educative to some extent.

The following score card, adapted from *Bulletin* 294 of the U. S. Department of Agriculture, may be used in judging cotton.

SCORE CARD FOR THE COTTON PLANT

	Perfect Score
Plant form	25 points
Size: medium to large, as influenced by soil, location, season, and variety	5
Form: symmetrical, spreading, conical to rounded, height and spread according to variety, soil, season, etc	5
Stalk: minimum amount of wood in proportion to fruit	5
Branches: springing from base, strong, vigorous, short-jointed	5
Head: well-branched and filled, fruited uniformly	5
Fruiting	25 points
Bolls: abundant, uniformly developed, plump, sound, firm, and well-rounded	5
Number of bolls: according to variety, soil, and season	5
Bolls per pound of seed cotton: big-boll varieties, 45; medium-sized bolls, 60; long-staple varieties, 70	5
Picking qualities (the perfect boll has large, stiff, protruding locks that cling together and are easily removed from the boll)	5
Storm resistance (the perfect type has drooping bolls, which have large, stiff burs that hold the locks well and act as a shed over the locks below, but spread apart, so that they do not interfere with picking)	5
Yield	29 points
Seed cotton per acre (standard 1 bale per acre)	12
Per cent lint (40 per cent standard for short-staple varieties and 32 per cent for long-staple)	12
Seeds: 45 per boll, plump, color according to the variety, and germination not less than 95 per cent	5
Lint	21 points
Strength: tensile strength good and uniform throughout	5
Length: standards, short-staple, 1 inch; benders, 1⅛ inches; long-staple, 1 3/16 inches or longer	5
Character of staple: fibers fine, soft, silky, and of good spinning quality	5
Uniformity: all fibers of equal length, strength, and fineness	5
Purity: fiber white and free from stain, dirt, and trash	1
Total points	100

Deterioration of Cotton Varieties.—As has been shown elsewhere, and as everyone who has grown cotton any length of time knows, cotton varieties will deteriorate in yield and in other qualities unless some effort is made to prevent it. As cotton is commonly grown and handled, there is crossing between different varieties in adjoining fields and much mixing of seed at the gin. Sanders and Cardon, in Texas, found that, if the roll of seed from the gin breast is not removed between bales, some seed from a given bale are held in the roll and mixed with the seed of the second bale following it. In trials made by Ballard and Doyle, in Texas, it was found that a certain amount of seed from a particular bale remained in the auger conveyor after the three following bales had been ginned. Some of this was mixed with the seed of the fourth bale. Much of the seed planted is from unselected stock and is inferior. If this is mixed with good seed, it soon lowers the value of the whole lot; or if there is crossing between adjacent fields, similar effects result.

Even if all field and gin mixing is prevented, some deterioration will take place in a variety of cotton in course of time. As has been shown previously, many variant forms appear in fields of cotton from year to year. Some few of these are valuable, but the majority are of an undesirable type. These faulty plants cross with others, making hybrid progenies and, after a time, numerous inferior plants. Unless some breeding work is done in the way of roguing out these inferior plants when they appear or in selecting the better type of plants to breed from, the variety will not maintain its present status.

Duggar[2] estimates that seed of a good variety of cotton will yield 20 per cent more cotton than mixed, mongrel seed. Figures from almost any experiment-station variety test show that this is a conservative estimate.

The deterioration of a variety may be greatly lessened by preventing mixing in the field and at the gin. Mass selection, if done properly and continued for a period of years, will prevent deterioration in a variety and may bring about some improvement. Plant-to-row work from individual plant selections may be used to isolate high-producing strains of good type, but propagating from individual plant selections is apt to change the type of the variety; and unless very careful testing is done before the new strain is increased, a new cotton may be had that is not so good as the old variety. The variety type may be held fairly satisfactorily if seed from a number of the progeny rows that are good representatives of the type are massed.

Inbreeding and the selection of pure-line strains will result in strains or varieties that are more stable and that deteriorate less, but it is probable that these will not yield quite so well as less homozygous strains.

Plantation Variety Tests.—A cotton grower should grow only well-bred cotton of a variety adapted to his land and locality. The surest way for him to determine just what he should grow is to conduct a variety test, one or more, on his land where several of the leading varieties are planted, cultivated, and grown under exactly the same conditions. The soil on which the test is made should be as uniform as is to be found on the plantation; the rows should be of equal length and width; the varieties should be thinned alike and picked and weighed separately; etc.

After a choice of variety is made, the question arises, How can a supply of choice seed of this variety be secured? If he has the time and inclination, the planter may produce it himself on the plantation. He can produce as good seed as can be bought, provided he, or someone employed by him, is willing to give this work the time and energy necessary to develop something worth while. Cotton breeding is, as mentioned by Duggar,[2] the work of a specialist, and no one is likely to succeed unless he is willing to specialize. He must, in addition to having mental traits suited to the work, have considerable energy and be willing and have the strength to endure considerable physical hardship. He must spend day after day walking up and down cotton rows with July or August sun beating down upon him.

It is usually much less trouble and more satisfactory for a cotton grower to secure at least a start of good seed of the variety desired from the U. S. Department of Agriculture, from an experiment station that is doing cotton breeding, or from a reputable commercial breeder.

One-variety Communities.—In certain sections of most of the states of the Cotton Belt, the cotton growers of communities have organized and agreed to grow but one variety of cotton. North Carolina, Arizona, and California were pioneers in this work, but at present Georgia, Mississippi, and Texas have the most one-variety communities. In 1935, Georgia had 108 such communities.

The community system has several good features. A better variety will be grown by at least a part of the growers. With but one variety in a community, cross-pollination in the fields and gin mixing of seed are very largely eliminated. The lint cotton may be marketed more advantageously, since there is a considerable quantity of uniform staple. This may be sold cooperatively.

On Apr. 13, 1937, President Roosevelt signed the bill passed by Congress which provides for the official classing of cotton from communities organized for the improvement of their product. The act further authorizes the Secretary of Agriculture

> . . . to collect, authenticate, publish, and distribute, by telegraph, radio, mail, or otherwise, timely information on the market supply, demand, location, condition, and market prices for cotton, and cause to be prepared regularly and distributed for posting at gins, post offices, or in public or conspicuous places in cotton growing communities, information on prices for the various grades and staple lengths of cotton.

This act, when put into force, should encourage the growing of better quality cotton and the formation of more one-variety communities, since it is necessary for a community to be organized to obtain this classing service. In many areas there has been but little encouragement to growers to improve the quality of their cotton, because about the same price was paid for various cottons in their market. If growers know the grade and staple of their cotton and the market price for the various grades and staples, they are in a much better position to demand and get a premium for quality cotton. If they are reasonably sure of getting better prices for better quality cotton, they will be stimulated to grow more of it.

Effect of Long-draft Spinning on Cotton Breeding.—Within recent years, the long-draft system of spinning has been adopted by many mills that make cotton yarns, and it is being introduced in others. At present, approximately one-third of the active cotton spindles in the United States are of the long-draft type. It is claimed that uniformity of staple length is of less importance where this type of spinning is used, and the claims are considered reasonable by some of the textile authorities. However, conclusive experimental tests to determine the exact difference to be secured have not been made.

If it is found that cotton with irregular staple length can be handled satisfactorily in the cotton mills, cotton breeders can pay more attention to yields and less to uniformity of fiber.

Registered, or Certified, Seed.—Many states have seed laws designed to prevent the sale within the state of seed that is impure, not true to varietal type, below standard in germination, or undesirable in other ways. Official boards charged with the enforcement of these laws have drawn up sets of rules covering the production; inspection; registration, or certification; and sale of seed. Rules vary in different states, but all have the same end in view—the prevention of fraud in the sale of seed.

Formerly the term "pedigreed seed" was used often in connection with seed offered for sale, but more recently that term has ceased to have much significance. Anyone could use the term in connection with seed offered for sale, and there was no good way of checking up on the pedigree involved. Unless the integrity of the breeder is known, pedigreed seed put out by a seed company does not mean much. Certified, or registered, seed bearing tags issued by a qualified official inspector, who has inspected the cotton in the field, inspected the seed in the bin, and tested the germination, is dependable.

Handling Planting Seed.—It is perhaps not necessary to say that, if cotton seed are to be saved for planting purposes, the cotton must not be picked while it is wet, nor must a large amount of green seed cotton, that is, cotton gathered during the first few weeks of picking, be packed in a cotton house. If more than one kind of cotton is grown on a plantation or a change of varieties is made, great caution must be observed to see that cotton houses, picking sacks, wagons, etc., are thoroughly cleaned. If more than one variety is ginned at a gin, it is not safe to let the seed pass through a screw conveyor. If it is not possible to have belt conveyors, the seed should be allowed to fall on the floor in front of the gin stands and be sacked there. Small lots of seed cotton from breeding plots, that is, lots of 100 to 500 pounds of seed cotton, may be fed into the breast of one gin stand by hand, and the seed caught in front of the stand. Smaller lots of seed should be ginned on a small gin such as was described elsewhere in this chapter.

It is better in most instances to sack seed at the gin in even-weight bags, say bags that will hold 100 pounds and that are made from 8-ounce burlap. Heavier or finer woven fabric may cause some seed to heat. While the seed are green, or during the first few weeks of ginning, it is best to stand the sacks on end over the floor of the warehouse. They should not be tied up or piled on top of one another. After 4 or 5 days, the sacks may be tied and stacked. They should be stacked in a dry, airy place and in such a way that air can circulate between them.

In the more humid parts of the Cotton Belt, cotton seed are often so damaged by rains before the cotton is picked that the germination percentage is low. In these areas, seed do not retain their vitality well even if the germination is good when they are stored. If the seed are normal, the vitality may hold up well for 2 years, but after that, there is a rapid deterioration. Simpson[24] says:

Storage experiments with Sea-Island and upland cotton seed under the humid conditions prevailing at James Island, South Carolina, showed that in ordinary storage cotton seed deteriorates rapidly after two years. A definite relation is indicated between the moisture content of the seed during storage and the rapidity of deterioration. Sea-Island seeds, with a moisture content reduced below 8 per cent, when stored in tin containers to prevent the rapid reabsorption of moisture, retained their germination percentage with only slight impairment for 4½ years. Upland cotton seed stored under various conditions and containing from 8.75 to 13.78 per cent moisture deteriorated rapidly when the moisture in the stored seed remained above 10 per cent. Dried seed stored to prevent reabsorption of moisture showed only slight deterioration after 2½ years. Seed containing 13.78 per cent moisture and stored to prevent drying were all dead nine months after the beginning of storage.

Delinting Planting Seed.—Some authorities recommend delinting cotton seed that are to be used for planting purposes. The delinting may be done by subjecting the seed to the fumes of certain gases, by placing them in concentrated sulphuric acid for a few minutes to remove the fuzz, or by passing them through delinting machines such as are used in cottonseed-oil mills. Seed that have been delinted plant somewhat more uniformly and germinate sooner if there is a scarcity of moisture in the soil; and if the delinting is done with acid, there is some measure of disease control, because all disease-producing organisms on the outside of the seed are destroyed by the acid.

But there are some disadvantages. The delinting makes the seed more expensive. Delinting with strong acid is somewhat dangerous to the men doing the work. If the delinting is done with machines, there is usually the possibility of getting seed mixed if different varieties are delinted on the same set of machines. Green seed are much more apt to heat if they have been delinted. If green seed that have been delinted are packed in a sack, the seed in the center of the sack may heat even if it is lying on the floor by itself. Delinted seed rot in the ground after planting more frequently than normal seed if cold, rainy weather follows immediately after planting.

If seed are ginned closely and then passed through a seed cleaner with screens and fans for removing light or immature seed and stray locks of cotton, trash etc., very satisfactory planting seed are secured without delinting.

Germination Testing.—It is usually advisable for a cotton grower to have his planting seed tested for germination before planting. This is especially advisable if the weather the fall before was rainy. The seed may have heated or have been injured in some way unknown to the planter. All the leading seed companies make a careful test of all the seed they sell.

Prime seed germinate 90 per cent or better. Seed that germinate below 85 per cent are considered below standard, but may be satisfactory for planting if a larger quantity per acre is used. Ordinarily, cotton seed will not germinate satisfactorily under 60 to 90 days after the bolls open. Many of the seed pass through a period of dormancy.

Cotton seed are more difficult to test than corn seed, oats, or most other agricultural seeds. They cannot be tested satisfactorily between moist blotting papers, in moist sawdust, in sand etc., kept at room temperature. The temperature must be fairly high and uniform; the seeds must be kept moist but not wet; and they must have air. In greenhouses or in other rooms where these conditions obtain, they may be tested in soil satisfactorily. Cotton seed may be tested in germinators heated by a gas jet or by an oil lamp, but with these it is difficult to hold the temperature between the proper limits. Where current is available, an electric germinator is preferable.

On account of the difficulties met in testing cotton seed, home testing is not very satisfactory or advisable. It is better to send samples of seed to a laboratory that is equipped for the work. The U. S. Department of Agriculture has two or more seed-testing laboratories where seeds will be tested free of charge. Most agricultural colleges are equipped for the work and also some county agents. There are, in addition, some private laboratories that will make tests for a moderate charge.

When a test is to be made, small samples of seed, say ½ pint, should be taken from various sacks (preferably with a sampling tube so that seeds may be taken from the central part of the sack) or from various parts of the pile of seed if not sacked, so that a representative sample may be secured. The samples should be numbered, and corresponding numbers placed on the sacks from which they were obtained.

Making the Test.—The first step in testing samples of seed is to count out a given number of seeds from each sample, say 100 or, better, 200, so that a duplicate test may be made. Each of these hundreds is next thoroughly wetted. This may be done by immersing them in water while they are held in the hand and squeezing and relaxing several times to drive air out of the fuzz. The seeds are next spread out in a single layer between the folds of moist outing cloth, cotton flannel, or moist absorbent paper toweling and placed on a wire tray of the germinator. After the seeds have been in the germinator 2 or 3 days, a count should be made, and all sprouted seeds removed. If allowed to remain in the germinator, they may lift the cloth from other seeds that have not yet germinated, and these will not have enough moisture. A second count may be made about the sixth day. Usually this is enough to end the test, but in some cases it is advisable to make a third count.

Water should be kept in the lower part of the germinator so that the air will be very humid. The cloths in the upper part of the germinator tend to become too wet and must be wrung out. The ones in the lower part have a tendency to become too dry and need sprinkling. The seeds should be well wetted at first but after that should be kept moist, not wet. The temperature may range from 20 to 30°C. (66 to 86°F.).

The rules for seed testing adopted by the Official Seed Analysts of North America advise the use of an alternation of temperatures in the germinator—30°C. for 6 to 8 hours, then 20°C. for the remainder of the day.

References

1. Watt, Sir George: "Wild and Cultivated Cotton Plants of the World," Longmans, Green & Co., New York, 1907.
2. Duggar, J. F.: "Southern Field Crops," The Macmillan Company, New York, 1911.
3. Leake, H. M.: "Studies in the Experimental Breeding of Indian Cottons, an Introductory Note. Part 2. Buds and Branching, *Jour. Proc. Asiatic Soc. Bengal*, new series, vol. 5, No. 1, 1909.

4. Meloy, G. S. and Doyle, C. B.: "Meade Cotton an Upland Long-staple Variety Replacing Sea Island," *U. S. Dept. Agr. Bull.* 1030, 1922.

5. Kearney, T. H.: "Heritable Variations in an Apparently Uniform Variety of Cotton," *Jour. Agr. Research*, vol. 21, No. 4, 1921.

6. Webber, H. J.: "Improvement of Cotton by Seed Selection," *Year Book U. S. Dept. Agr.*, 1902, p. 365.

7. Balls, W. L.: "Some Cytological Aspects of Cotton Breeding," *Amer. Breeders' Assoc. Rept.*, vol. 5, 1908.

8. Allard, H. A.: "Preliminary Observations concerning Natural Crossing in Cotton," *Breeders' Mag.*, vol. 1, p. 247, 1910.

9. Shoemaker, D. N.: "Notes on Vicinism in Cotton in 1908." *Amer. Breeders' Assoc. Rept.*, vol. 6, p. 252, 1909.

10. McLendon, C. A.: "Mendelian Inheritance in Cotton Hybrids," *Ga. Agr. Exp. Sta. Bull.* 99, 1912.

11. Ricks, J. R. and Brown, H. B.: "Cotton Experiments, 1916," *Miss. Agr. Exp. Sta. Bull.* 178, 1916.

12. Leake, H. M.: "Studies in the Experimental Breeding of the Indian Cottons, an Introductory Note," *Jour. Proc. Asiatic Soc. Bengal*, new series, vol. 4, p. 13, 1908.

13. Kearney, T. H.: "Self-fertilization and Cross-fertilization in Pima Cotton," *U. S. Dept. Agr. Bull.* 1134, 1923.

14. Todd, John A.: "The World's Cotton Crops," A. & C. Black, Ltd., London, 1923.

15. Griffing, J. B.: "Observation on the Behavior of Cotton Plants Especially during Acclimatization," Agriculture and Forestry Series, University of Nanking, China, vol. 1, No. 6, 1923.

16. Ware, J. O.: "Plant Breeding and the Cotton Industry," *U. S. Dept. Agr. Yearbook* 1936, p. 697.

17. Harland, S. C.: "Homologous Genes for Anthocyanin Pigmentation in New and Old World Cottons," *Jour. Genetics*, vol. 30, No. 3, 1935.

18. Skovsted, A.: "New Interspecific hybrids in the *Genus Gossypium*," *Jour. Genetics*, vol. 30, No. 3, 1935.

19. Sadao, Nakatomi: "Hybridization between Old World and New World Cotton Species and the Chromosome Behavior in the F_1," *Japan. Jour. Botany*, vol. 5, pp. 371–383, 1931.

20. Harland, S. C.: "Acclimatization of Cottons in New Areas," *Empire Cotton Grow. Rev.*, vol. 9, No. 4, 1932.

21. Beckett, R. E.: "Budding and Grafting Trials with Cotton and Related Plants," *U. S. Dept. Agr. Cir.* 267, 1933.

22. Rea, H. E.: "Asexual Reproduction of Cotton," *Jour. Hered.*, vol. 19, pp. 356–357, 1928.

23. Eaton, Frank M,: "Early Defloration as a Method of Increasing Cotton Yields, and the Relation of Fruitfulness to Fiber and Boll Characters," *Jour. Agr. Research*, vol. 42, No. 8, 1931.

24. Simpson, D. M.: "Relation of Moisture Content and Method of Storage to Deterioration of Stored Cotton Seed," *Jour. Agr. Research*, vol. 50, No. 5, 1935.

CHAPTER X

CHEMISTRY OF THE COTTON PLANT

Many chemical analyses have been made of commercial products coming from the cotton plant, but only a comparatively small number have been made of the plant itself or of its different parts.

Composition of Mature Plants.—The relative weight of the different parts of a cotton plant will vary greatly with the season, the richness of the soil, and the variety, and these differences will affect the chemical composition in so far as quantity of different compounds is concerned. Figures given are only averages from a number of determinations and are not accurate for specific cases.

McBryde[8] reports relative weights of different parts of mature, dry cotton plants as follows:

	Grams	Per cent
Roots	14.55	8.80
Stems	38.26	23.15
Leaves	33.48	20.25
Burs	23.49	14.21
Seed	38.07	23.03
Lint	17.45	10.56
Total	165.30	100

The leaf weight of the foregoing was taken before the plants had shed many leaves. The plants were grown in South Carolina. In other states farther west, the percentage of stalk, roots, and leaves, as compared with seed and lint, is considerably greater. Fraps,[7] of Texas, reports the percentage more than twice as great. In districts where boll weevils are prevalent, the vegetative parts of plants are large in proportion to the seed and lint.

Young cotton plants contain a high percentage of water. As they become older, the water content gradually becomes less until maturity is reached. Anderson and Ross[6] report the water content of fresh young plants 5 weeks old as being 84.7 per cent; of fresh mature plants, 63.7 per cent; of air-dried mature plants, 7.4 per cent. White[1] made analyses of the fertilizing constituents of cotton plants of one variety, Cleveland, for 4 successive years. The land on which the plants were grown was a moderately productive "good cotton soil." It was fertilized with a complete fertilizer containing 70.2 pounds of available phosphoric acid, 18 pounds of potash, and 20.8 pounds of nitrogen.

Table VIII, compiled from White's results, shows fertilizing constituents of the aboveground portion of mature plants. It will be noticed that there is close agreement between figures obtained in different years.

TABLE VIII.—FERTILIZING CONSTITUENTS IN ABOVEGROUND PORTION OF MATURE COTTON PLANTS
(Dry Matter)

Year	Per cent ash	Per cent nitrogen	Composition of ash				
			Per cent phosphorus	Per cent sulphur	Per cent potassium	Per cent calcium	Per cent magnesium
1910	6.15	1.56	3.77	1.70	11.80	13.21	5.25
1911	6.85	1.43	3.10	1.41	11.47	10.82	4.80
1912	7.10	1.50	3.62	1.61	9.41	10.04	4.10
1913	7.25	1.54	2.66	1.55	11.19	11.61	3.99
Average	6.83	1.50	3.28	1.56	10.96	11.42	4.53

Table IX, compiled from results published by different experiment stations, shows fertilizing constituents in entire mature cotton plants grown in different states. These results are figured on a different basis from that given by White in Table VIII and so are not exactly comparable.

Table X, from McBryde,[8] shows a proximate analysis of an entire mature plant grown in South Carolina.

From Tables VIII and IX some idea may be obtained of the draft that cotton plants make on the fertility of the soil. This draft is not in direct proportion to the size of the crop. It is greater, comparatively, for small yields and smaller for large ones. Fraps[7] showed that, if a large amount of potash is present in the soil, an excess may be taken up by plants.

TABLE IX.—FERTILIZING CONSTITUENTS OF MATURE COTTON PLANTS

Station	Per cent ash	Per cent nitrogen	Per cent phosphoric acid	Per cent potash	Per cent lime
Alabama	6.07	1.56	0.55	1.26	1.65
South Carolina	6.27	1.58	0.47	1.43	1.53
North Carolina Department of Agriculture	7.28	1.31	0.47	1.34	0.76
Texas	....	1.71	0.45	1.50	2.07

TABLE X.—PROXIMATE ANALYSIS OF MATURE COTTON PLANT

Per cent water	Per cent ash	Per cent protein	Per cent fiber	Per cent nitrogen-free extract	Per cent fats
7.36	5.81	9.13	30.94	42.84	3.92

Probably a better idea of the drain that cotton crops make on the soil can be obtained by considering the pounds of fertilizing constituents found in the lint and seed obtained from an acre of good cotton. In Table XI, compiled from McBryde,[8] the fertilizing constituents in corn, oat, and cotton crops are shown in form for comparison.

The cotton yield used in the preceding table, 954 pounds seed cotton per acre, is above the average for the Cotton Belt, but it is only a fair crop for well-cultivated cotton on good land. The lint percentage used is a little low for short-staple cotton but high for long-staple; therefore it averages well. Average crops for the South are about 30 bushels of oats per acre and 20 bushels of corn per acre. It will be observed that the fertility removed from the soil when only the lint is taken away is very little—only about 3 pounds. This drain is very much less than that made by oats and corn (see columns 2 and 3 in Table XI), when only the grain is removed. At present, however, cotton seed is removed also and is rarely returned in manures. Oat straw, as a rule, is likewise taken from the land, but corn stover is usually left. The figures, then, in columns 3, 4, and 5 are best for comparison. These indicate that the cotton crop takes more nutrients from

TABLE XI.—FERTILIZING CONSTITUENTS IN THE PRODUCTS OF AN ACRE OF CORN, OATS, AND COTTON
(Pounds per Acre)

	300 pounds lint cotton	30 bushels oats	20 bushels corn	300 pounds lint and 654 pounds seed	30 bushels oats and 1,515 pounds straw	20 bushels corn and 1,634 pounds stover
Nitrogen..............	0.72	17.50	16.50	20.80	26.50	29.80
Phosphoric acid (P_2O_5)	0.18	6.00	6.75	6.84	9.00	9.50
Potash (K_2O)..........	2.22	5.25	5.30	0.85	36.00	30.40
Soda (Na_2O)..........	0.08	0.15	0.18	0.20	1.65	1.92
Lime (CaO)...........	0.46	0.75	.....	1.68	5.25	7.93
Magnesia (MgO).......	0.41	2.25	2.50	3.67	5.25	8.25
Sulphurous acid (SO_3)..	0.26	0.45	0.18	1.10	2.70	
Insoluble matter.......	0.08	9.00	0.12	0.23	27.00	23.40

the soil than does the grain of a corn crop but considerably less than does the whole crop of corn or oats.

Composition of Different Parts of the Cotton Plant.—Since the different parts of cotton plants differ greatly in composition, and since some parts are used for one purpose and some for another, it is of value to know the definite composition of each part. Table XII, compiled from Anderson and Ross,[6] shows the complete analysis of different parts of a mature cotton plant.

It will be noticed that the fertilizing constituents of roots and stem are proportionally about the same. The leaves are conspicuously richer. Also, the bolls are comparatively rich in nutrient elements, but this is to be expected, since they contain the seed. The empty burs are moderately rich in potash and phosphoric acid but low in nitrogen. The lint is low in all fertilizing elements.

The roots, stems, and bolls are all high in fiber, but the leaves are comparatively low. Roots and stems average about the same in carbohydrates and proteins, but the roots contain considerably more fat. The leaves are rich in all the principal foods—proteins, fats, and carbohydrates—and are nutritious

TABLE XII.—COMPLETE ANALYSIS OF PARTS OF MATURE COTTON PLANTS
(Percentage of Constituents in Water-free Material)

	Nitrogen	Phosphoric acid	Potash	Lime	Magnesia	Oxide of iron	Soda	Sulphuric acid	Silica	Ash	Protein	Fiber	Fat	Carbohydrates
Roots	0.48	0.26	0.90	0.45	0.44	0.25	0.44	0.14	0.64	3.72	3.00	40.62	2.78	49.88
Stem	0.64	0.21	0.85	0.78	0.28	0.21	0.30	0.14	0.16	3.09	4.00	45.31	1.11	46.49
Leaves	2.25	0.48	1.09	5.28	0.94	0.43	0.66	1.05	1.70	12.55	14.06	8.71	8.49	56.19
Bolls	1.83	0.78	1.60	0.51	0.55	0.15	0.23	0.42	0.21	4.74	11.44	45.21	9.81	29.07
Seed	3.54	1.40	1.13	0.32	0.30	0.03	0.28	0.11	0.02	3.65	22.13	11.91	23.05	39.26
Lint	0.18	0.09	0.59	0.07	0.14	0.16	0.07	0.09	0.07	1.25	1.12	87.02	0.61	10.00

feed for stock. Distasteful organic compounds present in small quantities probably account for the fact that they are not relished more by animals. The seed, too, contain food elements in large quantities and, as is well known, serve as excellent feed for cattle and some other kinds of livestock. Cotton lint is often spoken of as being pure cellulose, but an examination of Table XII will show that it contains several other materials in considerable quantity. It is lower in fertilizer and food elements than all the other main parts of the plant, but it ranks highest in fiber.

McHargue[11] has also made a study of the mineral constituents of the different parts of the cotton plant, studying especially the occurrence of some of the rarer elements, such as iron, copper, and manganese, not reported by other analysts. Where the work is comparable, the results check rather closely with the results obtained by White,[1] Anderson and Ross,[6] and Fraps.[7] Table XII*A*, from McHargue,[11] gives the pounds per acre of elements contained in an average crop of air-dry cotton plants.

McHargue[11] found copper in all parts of the cotton plant that were examined, the kernels containing the highest percentage.

The largest amount of iron was found in the leaves, the most manganese in the hulls, and the most zinc in the kernels. The leaves contained 2.2 per cent sulphur, which was more than five times as much as was found in the stalks or kernels or in any other part of the plant.

The water content of cotton fibers, according to Bowman,[9] varies with the season, ranging from 1 to about 4 per cent in the new crop and less as the season advances. Above 2 per cent is

TABLE XIIA.—ELEMENTS CONTAINED IN AN AVERAGE CROP OF AIR-DRY COTTON PLANTS
(Pounds per Acre)

Parts of plant....	Leaves	Stalks	Kernels	Hulls	Fiber	Total
Pounds per acre..	800	1,000	300	300	300	2,700
Copper (Cu)......	0.010	0.008	0.015	0.004	......	0.037
Iron (Fe)........	0.448	0.103	0.042	0.067	0.057	0.727
Manganese (Mn).	0.040	0.009	0.004	0.037	......	0.090
Zinc (Zn)........	0.026	0.042	0.090	0.006	......	0.164
Calcium (Ca).....	30.600	12.910	0.522	0.370	0.040	44.442
Magnesium (Mg).	3.600	2.533	1.061	0.350	......	7.544
Phosphorus (P)...	2.112	1.573	5.040	0.157	0.084	8.966
Potassium (K)....	15.040	22.030	3.225	3.096	......	43.391
Sodium (Na).....	4.160	4.060	2.010	2.279	......	12.509
Sulphur (S)......	15.200	4.140	1.080	0.105	......	20.525
Nitrogen (N).....	22.000	11.750	21.120	1.680	......	46.550
Total..........	93.236	59.158	34.209	8.151	0.181	184.945

excessive even in a new crop, and when more than this amount is present it is probably due to a wet season and the cotton's being packed before drying or to the addition of water artificially. McBryde[8] found that if samples of apparently dry cotton were heated to 212°F., they lost from 5 to 7 per cent in weight but that, if they were put back in the room from which they had been taken, they gradually regained all the weight they had lost This was the so-called "water of hydration."

Composition of Plants at Different Stages of Growth.—The composition of cotton plants varies considerably at different stages of growth. White,[1] of Georgia, made analyses of plants in four different stages of development—first form, first bloom, first open boll, and mature plant—for 4 successive years. He used but one variety, Cleveland Big Boll, and grew plants on good cotton land near Athens, Ga. Before planting, 634 pounds of a mixed fertilizer, consisting of 468 pounds of superphosphate, 36 pounds of muriate of potash, and 130 pounds of nitrate of soda, were applied in the drill. First forms appeared 34 to 40 days after the seed sprouted; blooms, 24 to 32 days later; open bolls, 57 or 58 days after blooms; and the plants were mature in 90 to 100 days after open bolls appeared.

Tables XIII and XIV show results from analyses made by White in 1910. Figures for other years were very similar.

TABLE XIII.—WEIGHT AND COMPOSITION OF PLANTS AT DIFFERENT STAGES OF GROWTH

Average Weight of Plants
(Dry-matter weights; only aboveground parts used)

	Grams
First form	20.83
First bloom	48.65
First open boll	82.43
Mature plant	172.35

Nitrogen in Dry Matter

	Per Cent
First form	4.27
First bloom	3.60
First open boll	2.67
Mature plant	1.56

Ash in Dry Matter

	Per Cent
First form	13.35
First bloom	14.25
First open boll	11.38
Mature plant	6.15

TABLE XIV.—CHEMICAL COMPOSITION OF THE ASH

	First form, per cent	First bloom, per cent	First open boll, per cent	Mature plant, per cent
Phosphorus	5.35	4.28	4.03	3.77
Sulphur	2.86	2.16	1.70	1.70
Potassium	16.07	12.40	10.20	11.80
Calcium	20.00	15.05	12.24	13.21
Magnesium	7.86	5.40	5.20	5.25

From Tables XIII and XIV it will be observed that cotton plants make a large gain in weight as they pass from the first-form stage to maturity and that the percentage of important minerals decreases gradually. Table XV shows that the total mineral weight increases greatly as the plant increases in size.

TABLE XV.—QUANTITATIVE COMPOSITION OF A SINGLE PLANT

	First form, grams	First bloom, grams	First open boll, grams	Mature plant, grams
Dry matter	20.83	48.65	82.43	172.35
Ash	2.78	6.93	9.38	10.60
Nitrogen	0.89	1.75	2.20	2.69
Phosphorus	0.15	0.30	0.38	0.40
Sulphur	0.08	0.15	0.16	0.18
Potassium	0.45	0.87	0.96	1.25
Calcium	0.55	1.05	1.15	1.40
Magnesium	0.22	0.38	0.49	0.56

TABLE XVI.—PERCENTAGE OF MINERALS TAKEN UP DURING DIFFERENT LIFE PERIODS

	To first form	First form to first bloom	First bloom to first open boll	First open boll to mature plant
Nitrogen	34	32	18	16
Phosphorus	37	40	18	5
Sulphur	43	32	10	15
Potassium	35	28	13	14
Calcium	33	41	10	16
Magnesium	38	31	20	11

From the analyses made during the 4 years' study, White[1] calculated the percentage of mineral nutrients the cotton plant takes up during each of the four periods mentioned above.

From the foregoing and from other data, he estimates that, under conditions similar to the ones under which his plants were grown, a cotton plant takes up approximately one-third of its total mineral nutrients during the first period of 30 days; a second third during the second period of 30 days terminating with the formation of the first bloom; 85 to 90 per cent by the end of the third period, or by the opening of the first boll. This leaves only 10 to 15 per cent to be added during the maturing period of 90 to 100 days. Most of the total dry matter of the plant, however, is produced during the last period. According to White's figures, approximately one-eighth is produced the first period; one-eighth the second; one-fourth the third; and one-half the last, or maturing, period.

Relative Composition of Cotton Plants as Affected by Fertilizers.—It has been demonstrated often that an application of commercial fertilizers will increase cotton yields and produce larger plants. White,[3] in his experiments in Georgia, found that, on the average, a mature plant receiving a moderately heavy application of complete fertilizer contained 155.2 grams of dry matter; one receiving no phosphate, 135.5 grams; one receiving one-half the standard amount of phosphorus, 137.4 grams. He found also that a mature plant receiving the complete fertilizer contained 6.85 per cent ash (on dry-matter basis); one receiving no phosphorus, 6.10 per cent; one receiving one-half the quantity of phosphorus, 5.85 per cent. The last figures are not according to expectation.

The relative amount of nitrogen, potash, and phosphoric acid in plants does not seem to be influenced greatly by fertilizers applied to the soil; this is especially true if the land is moderately rich already. White,[3] as a result of his rather extensive experiments extending over 2 years, was unable to see any definite relation. He concludes:

> The obvious indication of these results, as compared with those yielded by the application of the standard fertilizer, is that substantial diminution in amount of either of the principal ingredients of the plant's food—phosphorus, calcium, potassium, or nitrogen—involves (1) substantial reduction in the weight (dry matter) of the plant in its entirety and in the several periods of growth, particularly after setting of the form; and (2) increase in time period from form to bloom, from bloom to open boll, and in the maturing of the plant. The proportion of ash to dry matter at any stage of growth or relative amounts of nitrogen and mineral ingredients do not appear to be seriously affected by the amount of the food supply. This would seem to indicate the important fact of the absence of a power in the plant to store food in any particular period of growth beyond the needs of the plant for the period.

Anderson,[6] in Alabama, secured results similar to White's. The nitrogen and phosphate fertilizers appeared to give no consistent increase of those elements in the plants, but the potash appeared to be an exception. Its use in fertilizers gave an increase in the plants regularly. Potash also gave an increase in White's experiments but was used in only two tests. Table XVII, from Anderson's work, shows the composition of plants grown on rich garden soil and on less fertile field soil with various fertilizers applied to each. (The amount of fertilizer used per acre is not given. The seed-cotton weights given in Table XVII are the yields from plots 10 feet square.) The effect of the potash fertilizer is more marked on the poorer soil.

TABLE XVII.—EFFECT OF FERTILIZERS ON THE COMPOSITION OF COTTON PLANTS

Fertilizers used	On field soil				On rich garden soil			
	Potash, per cent	Phosphoric acid, per cent	Nitrogen, per cent	Seed cotton, ounces	Potash, per cent	Phosphoric acid, per cent	Nitrogen, per cent	Seed cotton, ounces
None	1.256	0.788	1.883	9.29	2.538	0.758	2.352	130.83
Nitrate soda and kainit	2.123	0.345	1.969	30.00	2.026	0.741	2.436	120.00
Nitrate soda and phosphoric acid	1.051	0.537	1.883	23.21	1.494	0.688	2.064	96.25
Kainit and phosphoric acid	2.119	0.488	1.841	29.17	2.751	0.900	2.442	132.86
Nitrate soda, kainit, and phosphoric acid	2.562	0.557	1.833	37.50	3.054	0.696	2.339	145.34
None	.....	.....	.....	12.50	2.683	0.724	2.273	141.25

Results secured by White[3] (Table XVIII) indicate that the fertilizer used may have considerable effect on the storage of fat in the seed. Lack of phosphorus appears to have the most influence as a deterrent.

Effect of Soils on Composition of Cotton Plants.—It is well known that the character of the soil on which plants are grown determines largely the amount of growth that they make and their yields. The effect that various soils have on the relative amount of various chemical compounds in the plants in not so well known. Anderson[6] analyzed some plants grown on rich garden soil and some grown on field soil of fair fertility but low in organic matter. The garden-soil plants, as is shown by Table XVII, were much richer in potash, phosphoric acid, and nitrogen.

Fraps[7] analyzed plants from seven different regions in Texas. A wide variation in composition was obtained, but much of this was probably due to climatic differences. No definite relation to different soil types was established. White[3] analyzed plants grown in different parts of Georgia but found no wide divergence in composition.

Kearney[12] examined for salt content the fiber of Sea Island and upland cottons grown on irrigated land in Arizona in comparison with the fiber of the same varieties grown in South Carolina. The fiber from a very salty field contained slightly more salt than fiber from good soil.

TABLE XVIII.—EFFECT OF FERTILIZERS ON STORAGE OF FAT IN COTTON SEED

	Average amount of seed, grams	Average amount of kernel, grams	Per cent kernel	Per cent fat in kernel
1911				
Standard fertilizer	0.146	0.066	45.5	42.8
No phosphate	0.118	0.052	44.2	36.6
One-half standard phosphate	0.130	0.058	44.8	40.2
No potassium	0.120	0.059	45.0	42.0
One-half standard potassium	0.125	0.056	44.9	41.8
No nitrogen	0.115	0.050	43.8	39.8
One-half standard nitrogen	0.125	0.056	44.6	40.1
1912				
Standard fertilizer	0.155	0.071	45.8	38.2
No phosphate	0.131	0.059	45.0	31.5
One-half standard phosphate	0.145	0.065	44.6	36.5
No potassium	0.135	0.061	45.5	38.3
One-half standard potassium	0.140	0.062	44.2	38.0
No nitrogen	0.135	0.060	44.8	35.6
One-half standard nitrogen	0.142	0.064	45.1	37.5

Cotton grown on rich land has lower lint percentage than that grown on poor land, on account of the better developed seed. The larger, or more perfect, seed contain a slightly higher percentage of oil. Oil mills pay $5 more a ton for seed grown in the Mississippi Delta than for seed grown in the hill section of the state. However, the Delta-grown seed contain more oil mainly because the long-staple varieties are grown there. Most of the long-staple varieties have small seed comparatively rich in oil.

Difference in Composition of Different Varieties.—Cotton varieties differ considerably in yields of lint and seed and in lint percentage. This makes some difference in the organic composition of plants, but there seems to be but little difference in the relative percentage of the fertilizer constituents contained. White[3] analyzed varieties in a variety test in Georgia but found only slight differences in chemical elements. Rast, of Georgia, and Brown, of Mississippi, found that seed of different varieties differ greatly in oil content, the difference being as much as 10 gallons per ton, in some instances, for different varieties grown in the same field.

Other Organic Compounds in Cotton Plants.—Numerous organic compounds are found in small quantities in cotton plants. Among these may be mentioned gossypol, which, as determined by Withers and Carruth,[2] is the toxic substance in cottonseed meal and is present in the secretion of internal glands widely distributed over cotton plants; quercimeritrin and isoquercitrin, isolated from petals by Viehoever, Chernoff, and Jones[10]; quercimeritrin in leaves; an ethereal oil distilled in small quantities from seedling or squaring plants by Viehoever and others.[10] This oil has the property of attracting boll weevils, and it is possible that it is to them a guide to the plants. There is sugar in the nectaries of flowers and leaves, and other compounds are found in the plants in small quantities, but to discuss these does not fall within the province of this work.

References

1. White, H. C.: "The Feeding of Cotton," *Ga. Agr. Exp. Sta. Bull.* 108, 1914.
2. Withers, W. A. and Carruth, F. E.: "Gossypol, the Toxic Substance in Cotton Seed Meal," *Jour. Agr. Research*, vol. 5, No. 7, 1915.
3. White, H. C.: "The Feeding of Cotton II," *Ga. Agr. Exp. Sta. Bull.* 114, 1915.
4. Williams, C. B.: "Cotton Plant," *N. C. Dept. Agr. Bull.*, September, 1906.
5. Handy, R. B. *et al.*: "The Cotton Plant," *U. S. Dept. Agr. Bull.* 33, pp. 1–433, 1895.
6. Anderson, J. T., Ross, B. B. *et al.*: "The Cotton Plant Considered in Some of Its Chemical Relations," *Ala. Agr. Exp. Sta. Bull.* 107, 1899.
7. Fraps, G. S.: "The Chemical Composition of the Cotton Plant," *Tex. Agr. Exp. Sta. Bull.* 247, 1919.

8. McBryde, J. B.: "A Chemical Study of the Cotton Plant," *Bul'. Tenn. Agr. Exp. Sta.*, vol. 4, No. 5, 1891.
9. Bowman, F. H.: "The Structure of Cotton Fibre," Macmillan & Company, Ltd., London, 1908.
10. Viehoever, Arno, Chernoff, Lewis H. and Jones, Carl O.: "Chemistry of the Cotton Plant with Special References to Upland Cotton," *Jour. Agr. Research*, vol. 13, Nc. 7, 1918.
11. McHargue, J. S.: "Mineral Constituents of the Cotton Plant," *Jour. Amer. Soc. Agron.*, vol. 18, No. 12, 1926.
12. Kearney, Thomas H.: "The Salt Content of Cotton Fiber," *Jour. Agr. Research*, vol. 28, No. 3, 1924.

CHAPTER XI

FERTILIZERS, MANURES, AND ROTATIONS FOR COTTON

Cotton is grown on many fields year after year. As it is a clean-culture crop, there is not much chance for an accumulation of organic matter in the soil; and since the lands are usually bare during the winter, the period of heaviest rainfall, they are subject to much erosion and leaching. Bacterial action in the soil is active during most of the year. This soon destroys the organic matter that is added or that was present in the fresh soil. The lands of the Cotton Belt are, for the most part, low in productivity. Nearly all are low in humus and nitrogen; many have insufficient phosphorus; and some have but little potassium.

Commercial fertilizers are being used considerably on cotton lands, but often they are not applied judiciously. Expensive nitrogenous fertilizers are bought when legumes should be grown; complete fertilizers are frequently used where but one or two constituents are needed. Cotton planters must learn to handle their lands so as to increase the supply of nitrogen and organic matter in the soil; they must learn just what elements are needed and the most economical source from which to purchase; and they must learn how to prevent loss of nutrients already in the soil by leaching and erosion.

History of Cotton Fertilization.—Cotton growers in America have long known the value of stable manures, cotton seed, and clovers as fertilizers, but these were not used extensively until comparatively recently. Formerly, rich, new lands were cheap, and labor was cheap and abundant. It was less difficult to acquire and clear fresh lands than to fertilize or build up "worn-out" ones. As a consequence, lands were robbed of their plant nutrients until worn out; then new lands were taken up. The animal manures produced were of little consequence, because not many animals were kept—very few besides the mules and horses used in cultivating the cotton—and since the animals that were kept were allowed to graze in pastures a large part of the time, manures did not accumulate. Cotton seed was sometimes scattered broadcast over the land and then plowed under. Often it was piled in heaps and left exposed to the weather for several months to rot; then it was scattered.

According to White,[1] few fertilizers were used on cotton prior to 1860. The natural richness of the soil was depended on, and "the virgin soil skimmed." Peruvian guano was introduced into the United States about 1840 and first tried on cotton by David Dickson of Hancock County, Georgia, in 1846. This was probably the first use in the United States of a concentrated fertilizer on cotton. When first tried by Dickson, guano gave surprisingly good results. Other planters began using it with splendid results at first, but later results were not so satisfactory in many instances, on account of the poverty of the soil and the poor cultural practices followed.

Following Liebig's discovery of the value of chemical fertilizers as nutrients for plants, the opening of potash-salt beds in Germany, and the discovery of phosphate deposits in Tennessee, Florida, and South Carolina came the use of artificial manures on a commercial scale. Cotton growers began to use them extensively just after the Civil War. The war took their cheap labor from them. No new lands could be cleared, but there was a heavy demand for cotton at good prices. The worn-out lands would not produce well without some aid. Commercial fertilizers met the need exactly. They could be bought and applied without much trouble, and they gave good returns. Their use soon became widespread in the older states of the Cotton Belt. Unfortunately, many growers began to expect too much from the fertilizers. They allowed the soil to become low in organic matter from failure to use proper rotations and neglected seed-bed preparation and culture, expecting the fertilizers used to make a crop in spite of these handicaps. The failures naturally resulting caused many farmers to consider commercial fertilizers positive evils. They were frequently denounced in public meetings, and some efforts were even made to secure legislative measures to prohibit their sale. Later, better information was distributed, and the proper use of fertilizers learned. The use at present is confined principally to states east of the Mississippi River.

Soil Elements Removed by Cotton.—Soils in cotton-growing districts are usually poor, but this is not due to the drain that cotton plants make on the soil. An examination of Table XII and other tables in Chap. X and of Table XIX below should convince anyone that the draft is comparatively light.

TABLE XIX.—WEIGHT OF PLANTS AND PLANT NUTRIENTS IN PLANTS THAT PRODUCE A BALE OF COTTON

Part	Per cent of weight	Weight in pounds	Pounds nitrogen	Pounds phosphoric acid	Pounds potash
Roots	8.8	417	2.00	1.08	3.75
Stems	23.15	1,096	7.01	2.30	9.33
Leaves	20.25	959	21.58	4.60	10.45
Burs	14.21	673	5.52	3.23	20.79
Seed	23.03	1,090	38.58	15.26	12.32
Lint	10.56	500	0.90	0.45	2.95
Total	100.00	4,735	75.59	26.92	59.59

All of the cotton plant is returned to the soil except the seed and the lint. The fertilizer constituents in the seeds are returned in part in manures from animals fed cottonseed meal and in cottonseed fertilizers. The fertilizers in a 500-pound bale of lint weigh less than 5 pounds, and more than half of this is potash, a constituent that is plentiful in most soils.

Maintenance of Soil Productivity.—With proper handling, soils may retain their richness and continue to produce well indefinitely. Fields in parts of Europe that have been in cultivation hundreds of years are today producing better than they were a hundred years ago. Fields in China that have been cultivated for 4,000 years are still producing well.

Elements that are scarce in the soil naturally must be supplied in the form of manures or artificial fertilizers. Elements present in the soil may be rendered more available by the introduction of organic matter.

It is not profitable to crop poor land. If land is not productive, it should be made so as soon as possible. After it is built up, the land should be so managed that valuable elements are not lost in excessive amounts by leaching and erosion.

Response Cotton Makes to Fertilizers.—Cotton plants respond promptly and liberally to proper fertilizers, whether barn manures, legumes in the form of green manures, or chemical fertilizers. Five loads of stable manure per acre or a crop of cowpeas turned under may double yields on poor soils. Even a light application of commercial fertilizers has an effect and is profitable if the fertilizers are judiciously applied. The fact that it is so sensitive to fertilizers leads to the use of the cotton plant as a test plant in soil-fertility work. Fertilizers not only increase growth and fruitfulness but in some cases hasten maturity and shorten the entire life period.

Chemical Fertilizers.—The chief function of manures is to furnish plants with nitrogen, phosphoric acid, and potash. Natural, or barn, manures furnish these essential elements but not in the best proportion. If they are used in quantity sufficient to supply needed phosphoric acid, the nitrogen may be in excess—so much so that the cotton plants become very large but fail to fruit well. Barn manures are very valuable in that they increase the supply of vegetable matter in the soil and improve its mechanical condition. Constituents already present are made more available. Combined with sufficient phosphoric acid and potash (if needed in the particular soil), barn manure makes an ideal fertilizer. Unfortunately, the supply of barn manure is very limited in the cotton states, and chief dependence must be placed on artificial or chemical fertilizers. Many experiments have been carried on during the last 30 years to determine the value of the three important chemical fertilizers as manures for cotton. Much information of practical value has been obtained as to the amount of the different constituents to use on various soils and where or in what form the different constituents can be obtained most economically.

Phosphatic Fertilizers.—The phosphate fertilizers are in several different forms—superphosphate, treble superphosphate, metaphosphate, raw-rock phosphate, ground bones, basic slag, etc.

Raw-rock Phosphate.—This material, as the name implies, consists of the untreated phosphatic rock, which is dug from the ground. The greater part of this rock comes from the bones of millions of animals that lived ages ago. The rock is finely ground before being put on the market and is sometimes designated as "floats." At present, rock phosphate is obtained in commercial quantities from beds in South Carolina, Tennessee, and Florida. There are large undeveloped deposits in Idaho, Wyoming, and Montana. Rock phosphate, although containing about 30 per cent phosphoric acid, is objectionable as a fertilizer for cotton, as the phosphoric acid is in an insoluble form and is not readily available. When used on soil containing much organic matter or along with a crop of green manure turned under, it may give fair results. Most experiments show acid phosphate to be a better fertilizer for cotton. Being less soluble than acid phosphate, rock phosphate remains in the soil longer, and the residual effect is probably greater.

Superphosphate.—Acid phosphate is made by treating rock phosphate with an equal weight of sulphuric acid. The resulting composition contains 14 to 20 per cent phosphoric acid in soluble form. In this form it is readily available to plants and when used as a fertilizer for cotton gives quick results.

Basic Slag and Other Phosphates.—In addition to the sources of phosphoric acid discussed above, there are several others of lesser importance. Among the number may be mentioned basic slag, a by-product in the manufacture of steel from iron ores high in phosphorus. It is produced and used in Europe chiefly. When used as a fertilizer for cotton, it gives fair results, being better than rock phosphate but not so profitable as superphosphate. It contains 10 to 25 per cent phosphoric acid.

Guanos rich in phosphoric acid were formerly used extensively in fertilizing cotton and gave excellent results. Only an occasional shipment from islands in the Pacific or Atlantic oceans is received at present.

Raw bone, bone tankage, bone black, bone ash, dried ground fish, ammonium phosphate, and several manufactured phosphates are also sources of phosphoric acid, but these are not used extensively in fertilizing cotton.

Need for Phosphatic Fertilizers.—Practically all the cotton lands east of the Mississippi River, except the Yazoo-Mississippi Delta lands and other alluvial or bottom lands, are low in available phosphoric acid. West of the Mississippi, most of the sandy land, and much of the rolling clay land as well, is improved by an application of a phosphatic fertilizer. On account of the low content of phosphatic constituents in the mass of the soils of the cotton states, it will probably always be desirable to supply these elements artificially.

Amount of Phosphoric Acid to Apply.—The most profitable amount of phosphoric acid to use varies in different regions, depending on the natural condition of the soil, its previous handling, etc. Considerably more is used in eastern than in western states, with apparently more profitable results. In South Carolina, increasing the application up to 600 pounds of superphosphate may give increased yields, but frequently profitable returns are not secured if more than 300 pounds are used. In Texas, the maximum amount is about 300 pounds. The South Carolina Experiment Station[4] secured a gain of nearly 200 pounds of seed cotton per year for 13 years in a test in which 600 pounds of 8 per cent phosphoric acid was used alone. A fertilizer consisting of 600 pounds of 16 per cent superphosphate, 3 per cent ammonia, and 3 per cent muriate of potash gave an increase of 748 pounds of seed cotton over the average of the checks. Ames,[5] in the brown-loam section of north Mississippi, secured a gain of 399 pounds of seed cotton from the use of 200 pounds of 16 per cent superphosphate. Ferris,[6] on the sandy coastal-plain soils of South Mississippi, secured as a 6-year average a gain of 345 pounds of seed cotton from an application of 100 pounds of 16 per cent superphosphate. An average gain of 409 pounds resulted each year from using 200 pounds for 2 years. Fraps,[7] in 151 fertilizer experiments in Texas, secured a gain from the use of phosphoric acid in 74 per cent of his experiments. The average gain, for the experiments that gave gain, was 113 pounds of seed cotton from an application of 150 pounds of superphosphate.

The Louisiana Experiment Station, in some unpublished work with fertilizers on prairie soils at Crowley, La., secured an average gain of 321 pounds of seed cotton per acre from the use of 4 per cent phosphoric acid. This was applied in a 600-pound application of a 5-4-4 fertilizer for an 8-year period. An average gain of 224 pounds of seed cotton per acre for a 3-year period was obtained from the same fertilizer application of Vidrine, La., and an average gain of 213 pounds of seed cotton from the same application at Sunset, La. Increasing the phosphoric acid to 8 per cent gave an additional increase of 66 pounds, the average of the three sets of experiments being taken. Further increase

TABLE XX.—FERTILIZERS RECOMMENDED FOR COTTON

Agronomist and state	Region	Application
E. T. Batten, Virginia	Southeastern part of state	500 to 800 lb. per acre of a 4-8-6 fertilizer, all applied before planting. On very light sandy soils, 8 or 10% potash may be needed. On the poorer sandy soils, an additional application of soluble nitrogen at chopping-out time is advised
W. H. Rankin, North Carolina	Coastal plains	600 to 800 lb. of a 4-8-4 fertilizer and on most soils; 15 to 20 lb. of a soluble nitrogen as a side application
	Piedmont	600 lb. 4-10-4 fertilizer and 10 to 15 lb. of a soluble nitrogen as a side application. On the blackjack soils, a 600-lb. application of a 4-8-8 fertilizer and no side application of nitrogen
	Second bottom lands	600 to 800 lb. of a 4-8-4 fertilizer and 10 to 15 lb. of a soluble nitrogen as a side application
H. P. Cooper, South Carolina	....	200 to 600 lb. of a 4-8-4, 5-7-5, or 5-10-5 is recommended for cotton in this state. It is customary to use a side dressing of nitrogen at a rate of about 100 lb. per acre of such materials as sodium nitrate, ammonium sulphate and calnitro
R. P. Bledsoe, Georgia	....	An application of enough fertilizer to supply 32 lb. each of nitrogen, phosphoric acid, and potash. This may be applied in 500 or 600 lb. of a 6-6-6 or in 400 lb. of a 4-8-8 plus 100 lb. of nitrate of soda as a top dressing
J. D. Warner and W. E. Stokes, Florida	Northwestern part of state	Use 400 to 600 lb. of a 3-8-5 fertilizer before planting, and side dress with 18 to 27 lb. of readily available nitrogen at chopping out
J. W. Tidmore, Alabama	....	Use a neutral or basic 6-8-4 fertilizer on practically all of the soil types of the state. This grade of fertilizer or its equivalent is recommended up to 600 lb. per acre
Clarence Dorman, Mississippi	All Delta soils	24 to 40 lb. nitrogen
	Black prairie soils	300 to 400 lb. of a 6-8-0 fertilizer. Where cotton rusts, use a 6-8-4 fertilizer
	Upland valley soils	300 to 600 lb. of a 4-8-4, 6-8-4, or 4-8-8 fertilizer

TABLE XX.—FERTILIZERS RECOMMENDED FOR COTTON.—(*Continued*)

Agronomist and state	Region	Application
Clarence Dorman, Mississippi	Hill soils	300 to 600 lb. of a 4-8-4, 6-8-4, or a 4-8-8 fertilizer
	Gulf coastal soils	500 to 600 lb. of a 4-8-4 fertilizer
H. E. Hendricks, Tennessee	Highland rim soils and red soils of the Central Basin	500 lb. of a 5-10-5 fertilizer
	Red land of west Tennessee and yellow land of eastern third of Tennessee	300 lb. of a 5-10-4 fertilizer
	White land of west Tennessee	300 lb. of a 5-10-10 fertilizer
	Yellow land of the western two-thirds of west Tennessee	160 lb. of a 5-10-10 fertilizer
H. H. Krusekopf, Missouri	Alluvial soils in southeastern part of the state	On older sandy lands, 300 to 400 lb. of a 4-10-6 fertilizer
D. J. Burleson, Arkansas	Coastal-plain section in south Arkansas	200 to 400 lb. of a 4-10-4 fertilizer
	Hill section of north Arkansas	200 to 400 lb. of a 4-12-4 fertilizer
	Lighter soils, lowland section	200 to 400 lb. of a 4-10-4 fertilizer
	Heavy soils, lowland section	200 to 400 lb. of a 4-8-4 fertilizer
Horace J. Harper, Oklahoma		200 lb. of a 4-12-4 for average conditions. Because of limited rainfall, it is hazardous to apply fertilizers to crops that mature in the summer. They can be used more profitably on small grains and winter legumes and on sweet clover and alfalfa. On some soils of the state 150 lb. of superphosphate per acre is the most economical fertilizer
Franklin L. Davis, Louisiana	Mississippi and Red River alluvial lands	30 to 40 lb. nitrogen
	Prairie lands	400 lb. of a 4-8-6 fertilizer
	North Louisiana hill lands	300 lb. of a 8-8-8 fertilizer
	Benchland soils	400 lb. of a 7-6-6 fertilizer

TABLE XX.—FERTILIZERS RECOMMENDED FOR COTTON.—(*Continued*)

Agronomist and state	Region	Application
	Soils of the East Texas timber country	200 to 400 lb. of a 4-6-4 or 4-8-4 or a fertilizer furnishing similar ratios and amounts of nutrients. On the better and more level lands, applications up to 600 lb. per acre can be used with profit. On the lands with open permeable subsoils, a part of the nitrogen may be applied to advantage as a side dressing. On the Lufkin fine sandy loam, a 4-12-4 fertilizer is suggested for land that has not been fertilized before
E. B. Reynolds, Texas	Soils of the Gulf coast prairie	200 to 600 lb. of a 4-8-0 or 4-10-0; or 150 to 200 lb. of a 16-20-0 fertilizer
	Soils of the blackland prairie	Fertilizers are not in general use on these soils. None of the fertilizer experiments here has shown large profits from the use of fertilizers. 100 to 200 lb. of an 11-48-0 or 16-20-0 ammonium phosphate is suggested tentatively
	Soils of other parts of the state	Fertilizers are not recommended for parts of the state lying west of the blackland prairie and the Grand Prairie on account of the dry land conditions
J. C. Overpeck, New Mexico	....	Experiments have shown no returns from the use of superphosphate; the use of ammonium sulphate has not proved profitable; and potash fertilizers are not needed. The problem is chiefly one of organic matter. Manure and green manure crops give good returns
R. S. Hawkins, Arizona	....	But little fertilizer is used on cotton in Arizona. Since commercial fertilizers have not been used much, the amounts and analyses needed are not known definitely. We do know that the soils do not require potash. The following three fertilizers have given good results: 200 lb. of a 11-48-0 fertilizer; 350 lb. of a 4-8-0 (an organic fertilizer); 200 lb. of a 10-38-0 (acidulated tankage base)
B. A. Madson, California	....	Fertilizers are used on cotton but very little. A small amount of nitrogen is used in some sections. In the main, the soils upon which cotton is grown have not shown a need for fertilization

in the amount of phosphoric acid gave slight gains but not enough to pay for the additional fertilizer.

The margin of profit from the use of phosphate fertilizers is governed not only by the increase in yields but by the market price of the fertilizers and the value of cotton seed and lint. If 100 pounds of 16 per cent superphosphate gives a gain of 100 pounds of seed cotton, it is profitable and should be used. When used with nitrogenous or potassic fertilizers or with both, greater gains may be secured. Some idea in regard to whether or not it would be profitable to use a phosphate—or any other fertilizer in any particular region—may be obtained from Table XX, which gives recommendations made by agronomists in the several cotton-growing states.

Time and Method of Applying Phosphate Fertilizers.—Since all forms of phosphate fertilizers remain in the soil well, there is no object to be gained by applying them after the planting. Before planting, they can be put in more economically and in better condition. They may be distributed by hand in the furrow or with a fertilizer distributor and then bedded on; or they may be put with a distributor in beds already made. The fertilizer should be put down deep enough so that it will not come in direct contact with the seed to be planted later.

Nitrogenous Fertilizers.—Nitrogen is a necessary element in the growth of plants. In its pure form, a gas of the air, it is not available to any plants except bacteria and certain other microorganisms. It exists in nature in many different organic and in several inorganic compounds. In a nitrate, an inorganic compound, it is directly available for plants. Other compounds must be changed to this form before the nitrogen contained can be used.

Fertilizers That Contain Nitrogen.—The more important sources of nitrogen for fertilizing cotton are legumes, such as clovers, cowpeas, and soybeans; farm manures; nitrate of soda obtained from extensive beds in Chile, South America (the commercial products containing 15.5 to 16 per cent nitrogen); synthetic nitrate of soda (containing about 16 per cent nitrogen); synthetic urea (containing about 46 per cent nitrogen); sulphate of ammonia, obtained principally as a by-product in the manufacture of coke and illuminating gas from coal (containing about 20 per cent nitrogen); calcium nitrate (a manufactured product containing about 12 to 14 per cent nitrogen); calcium cyanamid (containing, in the commercial product, about 21 per cent nitrogen); dried blood from large slaughterhouses (containing 6 to 14 per cent nitrogen); tankage, dried animal wastes from large slaughtering establishments (containing from 4 to 12 per cent nitrogen); cottonseed meal, the residue left from cottonseed kernels after oil is extracted (containing 6 to 8 per cent nitrogen); cotton seed (containing about 3.5 per cent nitrogen).

Relative Value of Nitrogenous Fertilizers.—The value of any of the various nitrogenous fertilizers depends, of course, on the amount of nitrogen contained and its availability. Results obtained by different experimenters have varied to some extent,

TABLE XXI.—RELATIVE EFFECT OF NITROGEN FROM DIFFERENT SOURCES ON COTTON

(*Alabama Agriculture Experiment Station Results*, 1921)

Fertilizer per acre		Yield of seed cotton, pounds			Average per acre		Per cent of availability
Amount, pounds	Kind	1917	1918	1920	Yield of seed cotton, pounds	Gains from nitrogenous fertilizer in 1918–1920, pounds	
140	Nitrate of soda	823	1,312	800	981	306	100
351	Cotton seed meal	825	1,178	758	920	213	94
425	Peanut meal (with hulls)	779	1,161	678	873	165	89
754	Velvet bean meal (with hulls)	550	1,132	639	774	131	79
	No nitrogen	...	1,070	440			

but, in general, there is not a great amount of difference between fertilizers of the same type, a pound of nitrogen in one being about equivalent to a pound in another. Nitrate of soda, ammonium sulphate, and ammonium nitrate have ranked near together, but calcium cyanamid has given much poorer results in many experiments, as shown by fertilizer tests. The organic fertilizers, cottonseed meal, cotton seed, peanut meal, velvet-bean meal, etc., have ranked only fair as compared with nitrate of soda, but their residual effect is greater. Tables XXI and XXII from the Alabama[8] and Louisiana experiment stations throw light on the relative value of nitrogenous fertilizers.

Calcium cyanamid must be applied cautiously if used as a side dressing. If it reaches the leaves of young plants or the moist leaves of older ones, it burns the foliage.

Barn manures are valuable as a source of nitrogen. Results following their use are better than the quantity of nitrogen contained would indicate. Clovers, peas, vetches, and other legumes, where used in proper rotations, constitute the cheapest and best source of nitrogen.

Amount of Nitrogenous Fertilizers to Use.—The most satisfactory amount of nitrogenous fertilizer to use varies with the

TABLE XXII.—TEST OF COMPARATIVE EFFICIENCY OF NITROGEN FROM VARIOUS SOURCES—LINTONIA SILT LOAM
(*Louisiana Experiment Station, Baton Rouge,* 1927–1936)

Source of nitrogen in a 5-8-8 fertilizer applied at the rate of 600 pounds per acre	Yield of seed cotton, pounds per acre 1936	Increase in yield from nitrogen, pounds seed cotton	10-year average yield seed cotton, pounds	10-year average increase from nitrogen, pounds	Rank
No fertilizer	731	...	839		
No nitrogen[a]	875	...	911		
Nitrate of soda	1,567	692	1,355	444	2
Sulphate of ammonia	1,338	463	1,210	299	9
⅔ nitrate of soda, ⅓ sulphate of ammonia	1,401	526	1,294	383	4
Granular cyanamid	1,506	631	1,291[b]		
Calcium nitrate	1,560	685	1,278	367	5
Cyanamid	1,487	612	1,230	319	7
Urea	1,381	506	1,226	315	8
½ nitrate of soda, ½ cottonseed meal	1,487	612	1,309	398	3
Nitrate of soda-potash	1,716	841	1,360	449	1
16-20 Ammophos (B)	1,382	507	1,239	328	6

[a] Superphosphate was used as the source of phosphoric acid and muriate of potash as the source of potash.

[b] Three-year average yield: leuna saltpeter previous to 1934.

soil and with the season. Ordinarily, it is not advisable to use less than 16 pounds of nitrogen per acre. Increasing the amount to 40 pounds usually increases profits, but the use of more than that amount is uncertain in results. It may be conducive to somewhat larger yields, but the margin of profit is apt to be less. If too much is used, the plants may make too much vegetative growth and fruit poorly. Table XXIII from the Delta Branch

Experiment Station[13] gives valuable data on the use of various amounts of nitrate of soda under cotton grown on loam soil in the Mississippi Delta.

TABLE XXIII.—RATES OF APPLYING NITRATE OF SODA, 1921–1936

Rate	Seed cotton per acre, 16-year average, pounds	Seed-cotton increase, 16-year average, pounds	Increase, 16-year average, per cent
Check, no fertilizer	1,172		
47 pounds	1,306	135	11.48
94 pounds	1,391	219	18.72
141 pounds	1,509	338	28.82
Check, no fertilizer	1,172		
188 pounds	1,635	463	39.52
234 pounds	1,706	534	45.61
281 pounds	1,729	557	47.55
Check, no fertilizer	1,172		

Ewing,[9] on loam soil in the Mississippi Delta at Scott, Miss., secured a gain in yield of 13.6 per cent from the use of 100 pounds of nitrate of soda and a gain of 16.8 per cent from 100 pounds of nitrate of soda on buckshot soil. One hundred and fifty pounds of nitrate of soda gave a gain of 19.9 per cent on loam soil and 20.6 per cent on buckshot. These data were based on 4 years' work.

Time to Apply Nitrogenous Fertilizers.—Although results obtained from using nitrogen-bearing fertilizers at different stages in the growth of the plant vary considerably, owing to soil, and especially seasonal, differences, it is the general opinion of agronomists that nitrate of soda and easily soluble fertilizers should be applied fairly early, either a part before planting and the rest about the time when squares appear or in one application soon after the cotton is thinned and hoed out. If applied before planting, there is danger that some of the fertilizer may be lost by leaching; or, if there are frequent rains early in the season, much of the fertilizer may be used in stimulating grass and weeds to extra growth and the extra cotton plants that are to be taken out. One hundred and forty pounds of nitrate of soda applied just after chopping out gave Cauthen[10] 80 pounds more seed cotton per acre than the same amount applied before planting. Nitrate applied when the first squares appeared or when the first blooms showed gave lower yields by 40 or 50 pounds than when it was applied just after chopping out. These figures represent averages from tests running through 6 years. If the nitrate is applied late in the growth of the cotton plant, say after blooming has started, and a period of dry weather ensues, not much of the fertilizer will be absorbed, or, if it is absorbed, the effect may come too late to help in the production of fruit. Consequently, but little benefit is derived.

Cottonseed meal, cotton seed, and all other organic fertilizers should be applied before planting. It is necessary for them to decay and go through certain chemical changes before the nitrogen contained is available.

Potassic Fertilizers.—The use of fertilizers containing potash usually gives less profit with cotton than nitrogenous or phosphatic fertilizers, since most cotton soils are naturally well supplied with potash, and, as cotton stalks are commonly left on the land, the cotton crop removes but a small amount of potash from the soil.

The chief potassic fertilizers used on cotton are kainit (containing about 12 per cent potash), muriate of potash (containing approximately 50 per cent potash), and sulphate of potash (containing about 50 per cent potash). The bulk of these fertilizers comes from extensive salt beds in the region of the Harz Mountains in northern Germany.

There seems to be little difference in the fertilizing value of the different forms of potash mentioned. A pound of potash in one form has as much effect as a pound in any other. This being the case, it behooves the grower to use the form that will yield him a pound of potash for the least money.

Gains Due to the Use of Potassic Fertilizers.—In most areas of the Cotton Belt, the original supply of potash in the soil has been sufficient to meet the needs of cotton plants. This is especially true of areas west of the Mississippi River. As the land is cropped in cotton, the area requiring potash enlarges. Twenty-five years ago, not much potash fertilizer was used in Mississippi and Alabama, but now it is used rather generally. In parts of the Coastal Prairie of southern Louisiana, potash now gives greater returns than any other fertilizer ingredient. The urgent need of a potash fertilizer is indicated by the plants' dying with so-called "rust," or "potash hunger." This trouble appears most frequently on old, run-down soils.

Amount of Potash to Use.—From 50 to 100 pounds of muriate or sulphate of potash or 100 to 200 pounds of kainit is recommended for cotton in areas where potash is needed. The larger applications are best where cotton rusts badly. A 600-pound application of a complete fertilizer, such as a 5-8-8, will meet the needs for a heavy potash treatment.

Potash fertilizers should be applied before planting.

Lime.—Lime as a fertilizer for cotton has given no immediate returns in the majority of experiments. Duggar[2] made a number of tests in Alabama but secured no consistent gains. South Carolina reports no benefit. Lime used in a cotton, corn, oats, and cowpeas rotation at State College, Mississippi, gave no increase in yields. Similar results were obtained in a cotton, corn, and soybeans rotation in Louisiana. More recent work, however, has shown that, under some conditions, or when the lime is applied at a moderate rate to certain lands, there was some direct benefit of the cotton.

Lime corrects the acidity of soils and fosters the growth of certain legumes, such as melilotus and alfalfa, and may thus increase the cotton production indirectly if it follows a heavy growth of a legume.

Cotton is not affected adversely by slight acidity of the soil.

Barn Manures.—The composition and value of barn manures vary much, depending on the feeds the animals have consumed and the care that has been used in handling the manures. Average barn manure from cities contains per ton, according to Voorhees,[3] 8 to 10 pounds of phosphoric acid, 6 to 8 pounds of potash, and 8 to 10 pounds of nitrogen. Animals fed legume hay will produce manure rich in nitrogen. Manure collected under cover with but little bedding being used is considerably richer than the city manure mentioned above. Trotter, of the Mississippi Experiment Station, reporting the composition of mule manure collected under cover and relatively pure, gave available phosphoric acid, 1.1 per cent; nitrogen, 0.82 per cent; and potash, 0.80 per cent.

The secondary value of barn manure is of much importance to poor cotton soils low in organic matter. It improves their mechanical condition, increases their water-holding capacity, promotes bacterial action, and possibly supplies growth-promoting substances, such as auxins. The Mississippi Experiment Station and a number of other experimenters report larger gains from the use of barn manures than from the use of any other fertilizers.

Although there is abundant evidence that barn manure is a splendid fertilizer for cotton, its scarcity on the farms of the South renders its use a matter of little consequence. Possibly the mention of its valuable qualities will influence some farmer to keep more livestock and produce more manure.

Green Manures.—As has been mentioned frequently, most soils of the Cotton Belt could be improved greatly by the addition of a quantity of vegetable matter. Since the supply of barn manure is limited, the only practicable source of vegetable matter is green manures. Green manures fall into two groups: legumes and nonlegumes.

The legumes include true clovers, vetches, bur clover, cowpeas, soybeans, and velvet beans. Legumes are valuable in that they increase the supply of nitrogen in the soil, increase organic matter, and serve as a cover crop. The nonlegumes used for green manures are rye, barley, wheat, and oats. These will make quicker and greater growth during winter than the legumes and are better as a winter cover crop; but, considered as a whole, they are not so valuable as the legumes for use as green manures.

Duggar,[2] in Alabama, in 1899, secured a gain of 696 pounds of seed cotton from a green manure crop of cowpeas and a gain of 536 pounds from a crop of velvet beans. The peas and beans were picked, and the vines turned under. Velvet bean plots in which only the stubble was turned under gave a gain of 424 pounds. This is a good gain from the use of the stubble alone. Turning under the stubble is really more profitable than turning under the whole crop. Other southern experiment stations have secured similar results, but the margin of gain is usually not so large as is indicated in the experiments just mentioned.

The Louisiana Agricultural Experiment Station, in some unpublished work with cover crops on Lintonia silt-loam soil, secured a 3-year average increase of 625 pounds of seed cotton per acre after turning under vetch or Austrian winter peas and an increase of 457 pounds after *Melilotus indica*. An annual application of 36 pounds of mineral nitrogen gave an average increase of 456 pounds of seed cotton per acre. In addition to the effect of the legumes on the crops immediately following them, the residue from the *M. indica* increased the yields of cotton by 377 and 237 pounds of seed cotton per acre, respectively, the two following years. The average increase from the residue of the vetch or Austrian winter peas amounted to 234 pounds per acre the first year and 190 the second. There was little or no residual effect from the applications of mineral nitrogen.

Guanos.—The true guanos are deposits of dung from various species of seafowl, bats, etc. They have been used as fertilizers for centuries, but it was not until near the middle of the nineteenth century that their true value was recognized by the commercial world. The supply, however, was practically exhausted in a few years. Since the amount of guano accumulated each year is limited, the production at present is a matter of small consequence.

The main supply of guano came from the coast of Peru and adjacent islands. The best of this product contained 12 to 14 per cent nitrogen and 12 to 14 per cent phosphoric acid and was excellent fertilizer for cotton. It was first used on cotton about 1845. For a few years, surprisingly good results followed its use. At present, the term is used to a limited extent as a trade name for certain mixed chemical fertilizers, which may or may not contain true guano.

Composts for Cotton.—From about 1870 to 1880 or a little later, composts for cotton were widely advocated by Farish Furman, of Georgia, and by a few other writers. Their use soon became rather general. The composts were made by piling together in successive layers barn manure, cotton seed, acid phosphate, kainit, and sometimes earth and other fertilizer materials. The compost was moistened and allowed to ferment for a few weeks. Then it was thoroughly mixed and applied in the drill before the planting. Careful experiments made by the Alabama Experiment Station[2] and by other stations have failed to show positive gains from the use of the composted material as compared with that of equal amounts of the same materials in fresh condition. Composts for cotton are used very little at present by practical growers.

Cotton Fertilizers Recommended by Agronomists of the Southern States.—Recommendations for the more important soil belts in the various cotton-growing states are shown in Table XX. These recommendations are rather conservative and are based on averages from experimental results and various observations of results from practical farming operations. They will serve as a safe guide for planters to follow.

Best Mode of Applying Fertilizers.—For all moderate amounts, applying fertilizer in the drill will give better results than broadcasting applications. Williams,[11] of North Carolina, secured 156 pounds more seed cotton per acre from fertilizer applied in the drill than when it was broadcasted. This result was the average of three different tests, all of which gave better returns from the drill application.

Fertilizer placement tests indicate that best results are obtained from fertilizers placed near the seed but far enough away so that there is no hindrance to germination. The distance will depend to some extent on the texture of the soil, moisture content, etc. Ordinarily, the best placement is about 2 inches below the level of the seed and about 2 inches to one side. If the soil is sandy, the fertilizer should be placed farther away.

Dividing the application is of no benefit in the use of any except the easily soluble nitrogen fertilizers, and divided applications of these will not always give better returns than one large application early in the season. Dividing the application makes extra work and is often not practicable.

Effect of Fertilizers on Maturity.—All fertilizers applied judiciously cause a quickening in the growth of cotton plants, a larger plant growth, earlier blooming, the production of more and earlier bolls, and an earlier opening of bolls, thus insuring greater production at the first picking. It is probable, however, that much of the apparent increase in earliness is due to a general increase in productiveness.

Most experimenters have reached the conclusion that nitrogenous fertilizers used in limited amounts or in connection with other nutrients will increase earliness to some extent; but if used alone in large quantities, they produce extra vegetative growth and retard fruiting.

Phosphoric acid, if used properly, tends to hasten maturity more than do other fertilizers. Results secured at the Texas Station and reported in *Bulletin* 75 are very striking. Plants on the plots fertilized with acid phosphate were 18 inches high when those on neighboring plots receiving other fertilizers were half the height and had less than half as many squares. Williams, of North Carolina, secured a gain of 12 to 35 per cent at the first picking from the use of phosphoric acid. Potash used on sandy land or on land where cotton is inclined to rust will retard maturity in a sense. It will tend to cause the plants to hold their leaves, remain alive, and open their bolls later. This retarding is really not making the plants later, but it is keeping them alive and allowing them to function more normally. A later but much heavier crop is produced.

White, of the Georgia Experiment Station, found that a complete fertilizer advanced the period of boll opening about 10 days on the average. Williams, of North Carolina, reported a gain of 11 per cent at first picking from the use of 1,000 pounds of slaked lime applied at 4-year periods, when used in connection with a complete fertilizer.

Cumulative Effect of Manures and Fertilizers.—With the limited amount of commercial fertilizer ordinarily used per acre, the cumulative effect from year to year is very slight. In fact, it most frequently happens that, with the light application of fertilizers and poor cultural practices followed, the land is continuously becoming less productive. Where the application is moderately heavy, say 500 pounds or more, the effect the second year, or the accumulative effect, may be considerable. Ordinarily, the effects of nitrogenous fertilizers are not carried over to an appreciable extent. But this may not be the case. At the Mississippi Delta Branch Experiment Station, the effect the second year from 100 pounds of nitrate of soda was sufficient to cause an increase in yield amounting to about 100 pounds of seed cotton. The nitrate was applied to cotton the first year. Dry weather ensuing probably retarded its assimilation. Phosphatic and potassic fertilizers are held in the soil longer than nitrogenous. It is well known that the beneficial effects of barn manures extend over 3 or 4 years. Only about half the effect is secured the first year.

White[1] gives the following summary of experimental work:

> The cumulative effect of manures in the soil is fairly well evidenced in several cases. Nitrogenous manures increased the yield the second season without additional fertilization (Alabama, Arkansas), but not the third season (Alabama). Phosphatic manures increased the yield

without additional fertilization, the second and third seasons (Alabama). The cumulative effects of heavy applications of a complete fertilizer were manifest the second and third years (Georgia). Floats alone gave a greater increase over no manure the third year after application than in the first or second year (Alabama).

Fertilizers in Relation to Cotton Diseases.—Fertilizers are of considerable value in checking the ravages of certain cotton diseases. Potash is almost a specific remedy for black rust. Barn manure or complete fertilizers applied to soil infected with wilt may so stimulate and increase the growth of the cotton plants that they make fair yields before the disease cuts them down. (This applies especially to land not badly infected with wilt.) In general, the same principles apply to the use of fertilizers to lessen the damage done by insect pests. Heavy applications of nitrogen, however, sometimes cause very luxuriant growth and heavy development of foliage. The dense shade thus produced is favorable for fungous growth on the bolls, and much boll rot results.

Maintaining Productivity by the Use of Chemical Fertilizers Alone.—It is a difficult matter to maintain the productivity of cotton soils by the use of chemical fertilizers solely. These supply the common plant nutrients but do not provide for an increase of vegetable matter. If the land is in good condition to begin with, and sufficient fertilizers are used, the vegetative growth of the cotton plants may be sufficient to supply organic matter enough to last for a considerable time.

Judging Fertilizer Needs by the Appearance of Plants.—It is not possible to say just what fertilizers are needed on a particular tract of land from the appearance of the plants growing on it, but a good guess may be made. If plants are vigorous in growth and dark-green in color, there is doubtless plenty of nitrogen in the soil. On the other hand, if the plants are small, pale, and sickly looking, it does not mean, necessarily, that nitrogen is lacking. Faulty nutrition of any sort, or diseases, may have a similar effect on plants. Premature curling or shedding of leaves may indicate black rust and a need for potash. The safest and by far the best method of determining what fertilizers will be helpful and profitable on any particular field is to conduct a fertilizer test in which various fertilizers are tried under uniform conditions.

Suggestive Plan for Fertilizer Test.—Select an acre of land with as uniform and typical soil as is available. Divide this into 10 equal plots. Plow, plant, thin, and handle these plots alike in every way except in respect to the fertilizer applications. Use fertilizers on the plots as follows:

Plot 1. No fertilizer.

Plot 2. Ten pounds of nitrate of soda as a side dressing after chopping out cotton.

Plot 3. Twenty pounds of 16 per cent acid phosphate in the drill before planting.

Plot 4. Ten pounds of kainit in drill before planting.

Plot 5. No fertilizer.

Plot 6. Twenty pounds of acid phosphate in drill before planting and 10 pounds of nitrate of soda as a side dressing after chopping out.

Plot 7. Twenty pounds of acid phosphate and 10 pounds of kainit in drill before planting and 10 pounds of nitrate of soda as a side dressing after chopping out.

Plot 8. Place 300 pounds of barn manure and 20 pounds of superphosphate in drill-row furrows, and bed on them.

Plot 9. Grow and turn under the previous season a crop of soybeans or cowpeas, or grow and turn under a winter cover crop of vetch or other winter legume.

Plot 10. No fertilizer.

Stands should be the same on all plots. All fertilizers used should be weighed carefully, and extreme care taken to avoid mistakes when applications are made. All plots must be picked and weighed separately. The average yield from plots receiving no fertilizer will serve as a check with which to compare yield of plots receiving different treatments. One year's testing will be of help to a grower in deciding what fertilizers he should use, but tests running through several years will give more reliable information. It is better to change the location of the plots in different years and duplicate the series.

Home-mixed Fertilizers.—Home-mixed fertilizers are not so well mixed as the factory product, but there are several advantages to be gained from buying the ingredients separately and mixing them in the proportion wanted. The total cost of the fertilizer will not be so great, and it will be much easier to get just the ingredients needed for the land in question. In many instances, complete fertilizers are bought when only one or two constituents are needed. Home-mixed fertilizer may be built up to suit the particular crop.

Amount of Fertilizers Used by Practical Growers.—Cotton growers use but little barn manure on cotton. Not much is saved on the farms, and a considerable part of this is applied to corn and other crops. Green manures in the form of legumes employed in rotations are used on probably 10 per cent of the land normally devoted to cotton.

As has been pointed out, chemical fertilizers are used much more extensively in the eastern than in the western states of the Cotton Belt, but their use is increasing everywhere (see Table XXIV).

TABLE XXIV.—PERCENTAGE OF COTTON ACREAGE FERTILIZED, AMOUNT OF FERTILIZER USED PER ACRE, AND AVERAGE YIELD OF COTTON PER ACRE, IN 1930,[1] BY STATES

	Acreage ferti-lized, per cent	Average amount fertilizer per acre used, pounds	Average yield lint cotton per acre, pounds	Number acres in cotton, acres	Commercial fertilizer used, short tons
Southeastern states:					
Virginia	95.5	408	228	90,000	17,544
North Carolina	97.0	425	233	1,644,000	338,938
South Carolina	90.9	330	227	2,211,000	331,980
Georgia	95.9	272	190	3,946,000	515,168
Alabama	91.9	262	188	3,820,000	460,334
Florida	91.4	248	232[2]	105,000	11,904
South-central states:					
Mississippi	58.0	220	169	4,296,000	274,120
Louisiana	52.0	180	162	2,125,000	90,450
Arkansas	43.9	185	112	3,985,000	162,152
Tennessee	54.4	218	156	1,252,000	75,101
Missouri	6.9	145	207	377,000	1,885
Southwestern states:					
Texas	7.0	175	108	17,536,000	107,450
Oklahoma	1.9	175	106	4,165,000	7,262
Arizona	...	...	324	212,000	
New Mexico	...	...	333	134,000	
California	...	...	402	273,000	

[1] From U. S. Dept. Agr. *Yearbook*, 1931.

[2] Ten-year average 1919–1928 was 106 pounds per acre.

Profits from the Use of Commercial Fertilizers.—That there is a profit in the use of chemical fertilizers is evidenced by the fact that their use is increasing. The percentage of profit fluctuates considerably, of course, depending on the price of cotton and the cost of fertilizers, but if they are applied judiciously there will usually be a fair profit. Duggar concluded, from a number of experiments in Alabama, that a ton of mixed fertilizer when applied to 4 or 5 acres of cotton would make an increase equivalent to a 500-pound bale of cotton and 1,000 pounds of seed. This seems to us to be a fair estimate for regions where a complete fertilizer is needed. The gain is equal to a 100 per cent profit on the investment at present prices of cotton and cotton seed. If but one or two of the essential fertilizer constituents are needed, the margin of profit will be greater.

Recent unpublished work by Davis, Lovett, and others, of the Louisiana Agricultural Experiment Station, has shown an average annual profit of $13.92 per acre from the use of nitrogen fertilizers on alluvial lands of the Mississippi and Red rivers. This work included three or more tests per year for 5 or more years. This represents an average return of $4.25 per $1 spent for nitrogen fertilizers. The investigators just mentioned secured an average net profit of $16.53 per acre from the use of fertilizers on prairie soils in southwestern Louisiana. This represented an average return of $3.33 for each dollar invested.

Fertilizers Used on Cotton in Foreign Countries.—In the rich Nile Delta of Egypt, commercial fertilizers are used but little. Fertilizer tests made by Hughes,[12] in 1908, showed but slight gains from their use. The annual overflow of the Nile brings much nutrient material. All available barnyard manures and some Koufri manure (soil from old village sites) are used. Considerable berseem clover is grown.

Experiment stations in India have shown that nitrogenous fertilizers increase production and are urging the use of legumes, green-manure crops, and crop rotation. Farmers use a little poudrette, night soil, cattle manure, vegetable waste, wood ashes, etc. Only a small amount of commercial fertilizer is used. With the introduction of better cultural practices will doubtless come more extended use of fertilizers.

The fertility of the soils in China has been maintained by the use of homemade fertilizers, such as night soil, bone meal, ashes, barnyard manures, soybean meal, and talc rock.

Fertilizers are used only in limited areas of Brazil. Much rich virgin soil is still available.

Rotations.—Rotations are advantageous for several reasons: (1) They permit the use of crops that add an extra supply of organic matter to the soil. (2) They allow the use of legume crops, which increase the supply of nitrogen. (3) A change of crops tends to check insect and fungous diseases. (4) Troublesome weeds are held in check more easily. (5) Some crops feed from deeper layers of soil than others; a change may tend to rest some parts or draw most heavily on fresher parts and give better returns in yields. (6) Not all crops require just the same proportion of mineral nutrients. A change may work to advantage.

Rotations Alone Will Not Maintain Yields.—By means of proper rotations the supply of organic matter in the soil can be kept up, and the purchase of nitrogenous fertilizers will be unnecessary. But if the soil is lacking in phosphoric acid or potash, these must be secured from some outside source, as commercial fertilizers or barn manures, and applied to the soil.

Good Rotations to Use.—The rotation most suitable for any particular field or farm is determined largely by the type of farming followed as a main business. If the chief interest is in dairying or in raising stock, and cotton is grown as a secondary crop, rotations that will allow plenty of feed crops should hold first place. If the chief interest is growing cotton, a rotation system that will give cotton crops as often as it is possible for good crops to be made will be most satisfactory.

THREE-YEAR ROTATION FOR GENERAL FARMING

First year. Corn, with cowpeas at laying-by time, or soybeans in drill at planting.

Second year. Oats, followed by cowpeas after harvesting the oats.

Third year. Cotton, with rye and vetch in middles after first picking.

FOUR-YEAR ROTATION FOR A COTTON FARMER

First year. Cotton, with rye and vetch in middles after first picking.

Second year. Corn, with cowpeas at laying-by time or soybeans in drill at planting.

Third year. Oats, followed by cowpeas after harvesting oats.

Fourth year. Cotton, with rye and vetch in middles after first picking.

THREE-YEAR ROTATION FOR A COTTON FARMER

First year. Cotton, with rye and vetch in middles after first picking.

Second year. Corn, with cowpeas at laying-by time, or soybeans in drill at planting.

Third year. Cotton with rye and vetch in middles after first picking.

In some cases, it may be desirable to grow cotton 3 years in succession, with corn and legumes the fourth year. Various combinations can be devised to suit particular conditions.

References

1. White, H. C.: "The Manuring of Cotton," in "The Cotton Plant," *U. S. Dept. Agr. Bull.* 33, 1896.
2. Duggar, J. F.: "The Manuring of Cotton," *Ala. Agr. Sta. Bull.* 107, 1899.
3. Voorhees, E. B.: "Fertilizers," The Macmillan Company, New York, 1916.
4. Blackwell, C. P.: "Fertilizer Experiments with Cotton," *S. C. Agr. Exp. Sta. Bull.* 211, 1922.
5. Ames, C. T.: "Report of Cotton Experiments at the Holly Springs Branch Experiment Station," *Miss. Agr Exp. Sta. Bull.* 192, 1920.
6. Ferris, E. B.: "Cotton Growing in South Mississippi," *Miss. Agr. Exp. Sta. Bull.* 196, 1920.
7. Fraps, G. S.: "Coöperative Fertilizer Experiments with Cotton, Corn, Sweet Potatoes, and Irish Potatoes, 1908–1917," *Tex. Agr. Sta. Bull.* 235, 1918.
8. Cauthen, E. F.: "Comparison of Peanut Meal, Cotton Seed Meal, Velvet Bean Meal, Ammonium Sulphate, and Nitrate of Soda as Fertilizers for Corn and Cotton," *Ala. Agr. Sta. Bull.* 215, 1921.
9. Walker, G. B., Ewing, E. C. and Trotter, I. P.: "Nitrogenous Fertilizer Experiments," *Miss. Agr. Exp. Sta. Bull.* 207, 1922.
10. Cauthen, E. F. and Williamson, J. T.: "Time of Applying Nitrate of Soda to Cotton," *Ala. Agr. Exp. Sta. Bull.* 209, 1920.
11. Williams, C. B.: "Effect of Different Methods of Applying Fertilizers to Corn and Cotton," *Jour. Amer. Soc. Agr.*, vol. 5, p. 3, September, 1913.
12. Hughes, Frank: "Report on the Manurial Trials on Cotton Carried Out during the Season of 1908," *Year Book*, Khedivial Agricultural Society, 1909.
13. Anonymous: "Rates of Applying Nitrogen to Cotton," *Miss. Delta Branch Exp. Sta. Service Sheet* 209, 1937.

CHAPTER XII

COTTON SOILS AND CLIMATE

SOILS

The Cotton Belt of the United States contains a variety of soils. Cotton is cultivated on most of them, especially where climatic conditions are favorable, but the profits secured vary considerably. Plants on the upland sandy soils, for the most part, yield poorly unless fertilized well. Rich clay soils commonly produce good vegetative growth, but plants do not fruit in proportion. This is especially true if the rainfall is heavy. Bottom lands likewise may produce vegetative growth in excess, unless conditions are very favorable. Fungous diseases are apt to be serious if vegetative growth is excessive. The best and safest lands are medium grades of loam.*

The total area of the Cotton Belt in the United States is about 300,000,000 acres. In 1920, 35,878,000 acres were in cotton. This was about 12 per cent of the entire area and about 50 per cent of the acreage in crops. The soils of the belt are such that the acreage in cotton can be increased greatly if the demand for cotton justifies it.

Chemical Properties of Soils.—When agricultural chemists first began analyzing soils it was thought that, by determining the quantity of the various essential soil elements present, and by adding fertilizers to make up deficiencies, soils could easily be made productive. A few years' experimenting revealed that there was no simple relation between the quantity of fertilizer constituents as shown by soil analysis and the yields. A soil might show an abundance of nutrient elements in the chemical analysis and yet produce poorly. They may be present but in a form not available to plants, or some contrary factor may inhibit production.

* This discussion of soils is based largely on information contained in various papers and reports from the Bureau of Soils, U. S. Department of Agriculture.

Physical Structure and Consistence of Soils.—The physical structure and consistence of soils are matters of much importance in the growth of plants. Soils may be loose or compact, hard or friable, granulated or nongranulated and different in many other ways. According to Lyon,[1] granulation is the most important of the favorable structural conditions. It is, in substance, the drawing together of the small soil particles around a suitable nucleus, so that a crumb structure is produced. The small grains cease to function individually but serve in a larger unit.

Granulation in a heavy soil is instrumental in making the soil looser; this allows the air to circulate more freely, permits the excess water to drain out with little hindrance, and makes it possible for water to respond freely to the capillary pull of the plant roots. Various causes influence or bring about the granulation of soils, among them unequal expansion and contraction of different parts of the soil mass, tillage operations, development of plant roots, the burrowing of animals and insects, the action of decaying organic matter, the effect of wetting and drying, the effect of freezing and thawing, and the flocculation of soil substances induced by an application of lime. The colloidal matter in soils, which acts as a binding agent, has a very important relation to granulation.

Soils well supplied with organic matter have a loose structure; they contain large air spaces and possess numerous lines of weakness between particles. They expand and contract freely with change of water content; the dried soil contracts so much that large cracks appear in the surface of the ground. These soils are commonly black. As they become older or the supply of organic matter is decreased, they become heavier, more compact, lighter in color, and less productive.

Soils are said to be in condition of good tilth if they work well and crops grow vigorously on them. Conditions that are most favorable for plant growth are also best for effective plowing and for the development of crumb structure in the soils. On account of the high cohesion and plasticity of heavy soils, the moisture zone for successful plowing is narrow; that is, much of the time it is either too wet or too dry to plow. Their cohesion causes them to be hard to plow when dry, and when they are plowed they break up cloddy. The high plasticity causes the soil to puddle badly if plowed too wet.

Soil particles vary in size from stone pebbles to the microscopic particles of clay less than 0.005 millimeter in diameter. They vary in form, ranging from rounded or nearly spherical to angular and irregular bodies of every possible shape. Sandy soil has relatively large particles, which do not cling together to produce a granular structure. The pore space between individual particles is large, and water passes through freely, but the total percentage of pore space is less than is found in a finer textured soil, like clay. Water enters clay slowly and also moves through it in like manner; but the water-holding capacity of this soil is high on account of its numerous small particles, each of which holds a film of water about itself. Loam is more or less midway between sand and clay. There are some large particles which function separately as in sand. This facilitates drainage. There are also other medium-sized ones which form nuclei around which smaller ones collect, thus helping along soil granulation. The small spaces between the finer particles serve to retain a supply of water. Numerous particles of partly decayed organic matter are present in the loam soil. These serve to hold a supply of water. They also expand and contract considerably, thus moving other soil particles and lightening the whole mass of soil, so that air may circulate more freely and render conditions more favorable for plant growth.

Very sandy soils have a low absorptive capacity and allow water to move through them so freely that the water-soluble nutrient elements are leached out. If rains are not frequent, plants will suffer from lack of water. Heavy clay soils hold moisture so well that plants often suffer from excess of water in the soil or lack of proper aeration for their roots. Loam soils, having characteristics somewhat intermediate between the two soils just mentioned, are medium in water-holding and aeration capacities and are thus better suited to cotton plants.

Soil Regions of the Cotton Belt.—A study of the soils in the regions represented in the Cotton Belt reveals the presence of a large number of types. No attempt will be made in this brief discussion to consider or even mention all the different types.

Important Soils of the Atlantic and Gulf Coastal Plain.—This large group of soils lies largely in the belt designated by Marbut[11] in his latest classification as the "red and yellow soils of the

United States." It extends from southern Virginia southward to southern Florida and along the Gulf of Mexico to the mouth of the Rio Grande River. In North Carolina, it is about 400 miles wide. Toward the Mississippi River, it widens and extends northward to Kentucky. West of the river, it covers much of Arkansas, Louisiana, and Texas. In the soils map (Fig. 48), this area covers the parts indicated as Atlantic Coast Flatwoods, Florida Flatwoods, Middle Coastal Plain, Upper Coastal Plain, Clay Hills, Sand Hills, Black Prairies, Mississippi Bluffs and Silt-loam Uplands, Interior Flatwoods, Interior Coastal Plain,

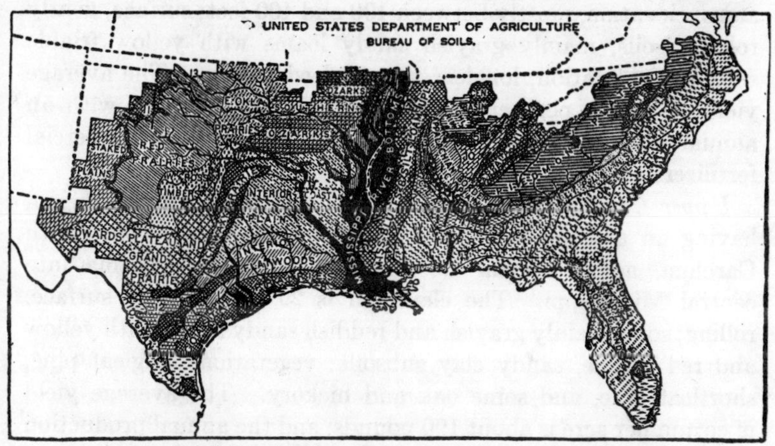

FIG. 48.—Soil regions of the Cotton Belt. (*After Bennett.*)

Black Waxy Prairies, and certain other small districts. The land is rather flat or gently rolling. The soils are mainly sandy with more or less loam or clay in various mixtures. The following discussion of various parts of the Coastal Plain Province is from Bennett.[2]

Atlantic Coast Flatwoods.—This region extends along the coast of the Carolinas and Georgia and over all of the state of Florida except the northwestern part (see Fig. 48). The area is about 21,200,000 acres; elevation is from sea level to 150 feet; surface is flat to undulating with many swamps and much poorly drained land; soils are mainly dark and grayish sands or sandy loams with yellow, gray, and mottled sand and clay subsoils. There is considerable silt loam in northeastern North Carolina. The vegetation is principally open forests of longleaf pine, with gallberry and grassy undergrowth. Fertilizers are used extensively. The average yield of cotton per acre is about 200 pounds of lint, and the average production for the area is about 290,000 bales.*

Florida and South Mississippi Flatwoods.—The area of this region is about 15,000,000 acres. The deep sandy soils are dark and grayish to white; vegetation is longleaf pine with saw palmetto undergrowth. Little cotton is grown.

Middle Coastal Plain.—This region extends across the Carolinas and Georgia into western Florida. Its area is 20,600,000 acres; elevation, mostly between 100 and 400 feet; surface, gently rolling; soils, mainly grayish sandy loams with yellow friable subsoils; vegetation, longleaf pine and wire grass. The average yield of cotton per acre is about 205 pounds of lint, with an annual production for the region of 1,140,000 bales. Commercial fertilizers are used extensively.

Upper Coastal Plain.—This is a comparatively large region, having an area of 28,700,000 acres and extending from South Carolina, across central Georgia and southern Alabama into central Mississippi. The elevation is 20 to 500 feet; surface, rolling; soils, mainly grayish and reddish sandy loams with yellow and red friable, sandy clay subsoils; vegetation, longleaf pine, shortleaf pine, and some oak and hickory. The average yield of cotton per acre is about 190 pounds, and the annual production 1,130,000 bales. Much fertilizer is used.

Clay Hills.—This region extends from western Georgia through Alabama and Mississippi almost to the Tennessee line. The total area is 8,000,000 acres; surface is hilly; some of the land is rough and stony and is known as "white-rock" land in east central Mississippi and in western Alabama; subsoil is stiff clay. The average yield of lint cotton in this region is about 145 pounds, and the annual production is 320,000 bales.

Sand Hills.—This is a narrow soil belt extending along the upper margin of the Coastal Plain from central North Carolina, across South Carolina and Georgia into Alabama; area, 3,400,000 acres; soil, mainly deep, loose sand, grayish at the surface and yellowish beneath; vegetation, longleaf pine and forked-leaf

*** Figures of the average yield per acre of lint cotton are from the four census years 1879, 1889, 1899, and 1909, unless otherwise noted. Figures of average production are for the 5 years 1911–1915.**

blackjack oak. Fertilizer is used extensively. The average yield of cotton per acre is about 180 pounds, and the annual production is 175,000 bales.

Black Prairies.—This crescent-shaped belt, extending from eastern Alabama into northeastern Mississippi, contains about 4,000,000 acres. Its elevation is 200 to 500 feet; surface, gently rolling with some flat "post-oak land"; soils, mainly dark-gray and brown limey clays. The "post-oak" soils are brown sandy loams and clays with reddish subsoils and contain less lime. The yield of cotton per acre averages less than 150 pounds, owing largely to continuous cropping and poor cultural practices. This area was originally considered one of the best in the Cotton Belt. The annual production is 235,000 bales.

Mississippi Bluffs and Silt-loam Uplands.—This division borders the Mississippi bottom lands on the east and extends from Louisiana into Kentucky. The area is 16,800,000 acres; elevation, 100 to 600 feet; surface, level to undulating and badly gullied in places; soils, mainly brown silt loams of loessial origin, which become thinner toward the eastern border. A narrow strip of the same soil occupies Crowley's Ridge, extending from southeastern Missouri to Marianna, Ark.; trees, principally oak, sweet gum, and poplar. The yield of cotton is about 200 pounds per acre, and total production is 586,000 bales annually.

Interior Flatwoods.—The largest area of the Interior Flatwoods extends from the Mississippi bottoms in Louisiana to the Guadalupe River in Texas. A long, narrow, crescent-shaped belt extending from central Alabama into northeastern Mississippi has similar soils and topography. The total area, including 200,000 acres in Alabama and Mississippi, is 13,000,000 acres; elevation, 100 to 500 feet; surface, prevailingly flat with much poorly drained land; soils, mainly gray sandy loams, silt loams, and clays with compact gray and yellow mottled subsoils; trees, longleaf pine in Louisiana and Texas and post oak in central Texas, Alabama, and Mississippi. The yield of cotton is about 175 pounds per acre, and annual production is 366,000 bales.

Interior Coastal Plain.—This region covers a large area of prevailingly rolling country in northeastern Texas, northwestern Louisiana, and southwestern Arkansas. The area is about 28,800,000 acres; elevation, 100 to 500 feet; soils, mainly grayish, brownish, or reddish sandy loams, sands, and clays; trees in the east, longleaf pine, shortleaf pine, and some oak. The average yield of cotton per acre is about 164 pounds, and the total production is 995,000 bales.

Black Waxy Prairies.—There are two main bodies of these prairies in Texas, one extending from the Red River in northeast Texas to San Antonio, and the other from the Brazos River nearly to Laredo. The area of the first belt is 8,400,000 acres, and of the second, 4,900,000; elevation, 100 to 600 feet; surface, flat to undulating with good drainage; soils, black and dark-gray calcareous clays. There is a small belt of similar soils in southeastern Oklahoma and southwestern Arkansas. The average yield of cotton per acre is about 175 pounds, and the average production is 1,320,000 bales.

Less Important Areas of the Coastal Plain.—These include the *Sandy Rolling Lands of Interior Florida*, which contain some red and brown loamy hammock lands, where Sea Island was formerly grown to some extent; *the Appalachian Border of the Coastal Plain* in northern Alabama and northeastern Mississippi, a region of rolling to hilly, grayish sandy loams, with yellow and red subsoils, supporting mixed pine and oak forests; the *Pontotoc Ridge in Mississippi*, which has soils similar to those of the Upper Coastal Plain; and the *Gulf Coastal Prairies* of Louisiana and Texas, which is a flat, imperfectly drained region of black, brown, and gray clays and loams with black, yellow, and gray mottled clay subsoils. The eastern portion of the Coastal Prairies is largely devoted to rice, and the western portion principally to grazing and to corn growing, with cotton ranking second in importance. The *Red Lands of Southwest Texas* produce some cotton in the northern portion. The soils are mainly reddish sandy loams, gravelly loams, and loams with red subsoils.

Alluvial Soil Regions.—These regions include all alluvial soil along the rivers east of the great Plains Region. The soil is "made land," having been deposited by the adjacent rivers, and is naturally productive.

Alluvial Bottoms of the Mississippi and Other Rivers.—This region includes as its principal area the bottoms of the Mississippi River from Cairo, Ill., to the Gulf. It also includes the bottoms of all other rivers within the Cotton Belt, many of which are too small to show on the map. Much of the land is subject to overflow and requires the protection of levees. The area of the district is about 16,500,000 acres; elevation, sea level to 300 feet; surface, level. The soils of the Mississippi bottoms are mainly brown or mottled clays, silt loams, and fine sandy loams, with gray, light-brown, and mottled subsoils. Characteristic trees are cypress, red gum, and oak. Annual production of cotton is about 940,000 bales. The most important cotton-growing section of this region is that known as the Yazoo-Mississippi Delta, in Mississippi, in which the average yield of cotton per acre is about 265 pounds. In the bottoms of the streams east of the Mississippi River the principal soils are brownish loams, silt loams, and fine sandy loams, with yellowish and mottled subsoils. West of the Mississippi the principal soils are chocolate-red, brown, and black loams, silt loams, and clays, usually calcareous.

Mississippi River Second Bottoms and Silty Prairies.—These second bottoms are extensively developed in southeastern Missouri and northeastern Arkansas. They lie above overflow. The important soils are brown and gray silt loams, and fine sandy loams with light-brown, gray, and mottled subsoils. The gray soils are poorly drained. The better-drained soils are extensively used for cotton and give good yields.

Important Soil Regions of the Piedmont, Appalachian, and Ozark Plateaus, Mountains, and Valleys.—These regions are hilly to rolling. The soil is chiefly residual and not especially rich. The parts of most importance for cotton growing are discussed in the following paragraphs.

Piedmont Plateau.—This region extends from New York City southwestward to Alabama, but is planted in cotton only from North Carolina southward; area in the Cotton Belt, 26,700,000 acres; elevation, 100 to 1,500 feet; surface, rolling to hilly; the cultivated slopes often require terracing to prevent erosion; soils, red-clay loams and grayish sandy loams, with red- or yellow-clay subsoils, derived principally from granite, schist, and diorite; trees, largely oak, shortleaf pine, and hickory. The average yield of cotton per acre is about 180 pounds, and the average production is about 1,860,000 bales. Fertilizer is used extensively. In the Carolina state belt portion of the Piedmont, the principal soils are gray silt and slate loams and red-clay loam, with yellow- and red-clay subsoils. Fair to good yields of cotton are made, depending on fertilization.

Less important cotton regions are the Appalachian Limestone Valleys, including the Tennessee River Valley of northern Alabama, and the Central Basin in Tennessee. These are regions of fertile, brown, reddish, and gray loams, silt loams, and clay loams, with reddish and yellowish subsoils, largely derived from limestone and often cherty. A little cotton is grown in the southern portion of the Highland Rim region, which has gray silty and stony soils; and also in the valleys and smooth uplands of the southern Ozarks of Arkansas and eastern Oklahoma, which is a hilly to mountainous region, having brownish or reddish sandy-loam soils, often stony, with red sandy-clay subsoils.

Important Soil Regions of the Western Prairies and Plains.—The lack of sufficient rainfall limits greatly the production of cotton in parts of the Western Prairies and Plains within the Cotton Belt. Their soils are residual and fairly rich.

Eastern Oklahoma Prairies.—This region extends northward from the Red River in south-central Oklahoma and includes most of the eastern half of the state. It is used principally for growing corn and hay and for grazing cattle. The area in the Cotton Belt is about 11,000,000 acres; elevation, 800 to 1,200 feet; surface, gently rolling, with some rough areas; soils, mainly brown, black, and reddish loams, clays, and stony loams with clay subsoils, often of a clay-pan nature; vegetation, prairie grasses with occasional areas of post oak, blackjack oak, and red cedar. Yield of seed cotton in 1909 was 182 pounds per acre; the average production is 350,000 bales.

Red Prairies.—This region extends across western Oklahoma into north-central Texas and is largely devoted to grazing but has recently experienced. extensive development in cotton production; area, 31,700,000 acres; elevation, 1,000 to 2,000 feet; surface, undulating to rolling with many rough eroded areas along the western margin. Soils are mainly red and brown fine sandy loams, silt and clay loams, with red- and brown-clay subsoils, often of a clay-pan nature. Included in this region are small "sand-hill" areas (dunes). The yield of cotton per acre in 1909 was 105 pounds; the average production is 825,000 bales.

Edwards Plateau and Grand Prairie.—This region extends from central Texas westward to the Rio Grande River, but only the eastern portion is used extensively for the production of cotton. The area of the portion east of the Colorado River is 7,400,000 acres; elevation of the eastern part, 1,000 to 2,000 feet; surface, rolling to hilly with some level lowlands; soils, mainly calcareous silt and clay loams, black, brown, gray, and red in color; vegetation, largely prairie and plains grasses with scattered post oak and cedar. The average yield of cotton per acre in the eastern portion is about 115 pounds; the average production is 410,000 bales.

Less important regions are the East and the West Cross Timbers in north-central Texas. The soils of both regions are mainly brown sandy loams, often stony, with reddish and yellowish subsoils; vegetation, mostly post oak and blackjack oak. These regions are largely used for grazing cattle, but they produce some cotton.

Cotton-soil Series and Types.—The soil regions mentioned in the foregoing pages contain many different soil types which have been given definite names. The most important of these, as outlined by Morgan,[3] are as follows:

Coastal Plain
- Norfolk sand and fine sand
- Norfolk sandy loam and fine sandy loam
- Tifton sandy loam
- Orangeburg sand and fine sand
- Orangeburg sandy loam and fine sandy loam
- Greenville clay loam, sandy loam, gravelly sandy loam, and loamy sand
- Ruston fine sandy loam
- Susquehanna fine sandy loam
- Houston black clay, loam, and clay loam
- Houston clay
- Victoria clay, loam, and sandy loam
- Durant fine sandy loam

Piedmont Plateau
- Cecil clay
- Cecil clay loam
- Cecil sandy loam
- Louisa slate loam, fine sandy loam, and loam

Appalachian Region
- De Kalb fine sandy loam
- De Kalb silt loam
- Fayetteville fine sandy loam

Limestone Valleys and Uplands
- Clarksville gravelly loam
- Clarksville silt loam
- Decatur clay loam

Hagerstown loam

Loessial Region (Mississippi Bluffs and Silt-loam Uplands)

Memphis silt loam

River Flood Plains Region

Miller fine sandy loam and clay loam

Trinity clay

Sharkey clay

Ocklocknee fine sandy loam and loam

Congaree loam

Kalmia fine sandy loam

Cahaba fine sandy loam

Great Plains Region

Vernon fine sandy loam, loam, and silt loam

Crawford stony clay

Amarillo loam and silty clay loam

Soils of Other Cotton-producing Regions.—A brief mention of a few of the characteristics of the soils of other important cotton-producing regions of the world may be of some general interest and value.

India.—Smith[8] divides the soils in the more important cotton-producing regions of India into the following groups.

1. The "regur" or "cotton" soils, which are deep, limey, black, brown, or gray sticky clay soils, similar to the "black waxy" soils of the southern states and the "black adobe" of California. They are widely distributed over the large traprock area of central Bombay, northern Hyderabad, Berar, the western portion of the Central India Agency, and most of the peninsula of Kathiawar. These soils also occur in portions of southern Madras and cover much of the alluvial plain in Bombay, extending from south of Surat to Ahmadabad. Owing, in large measure, to more or less continuous cultivation for at least 2,000 years without fertilizer or manure, the yield of cotton per acre is less than 100 pounds of lint.

2. The "red soils," which are derived from crystalline rocks, are mainly of a sandy-clay texture. They cover most of the Madras Presidency, eastern Hyderabad, and Orissa, and though varying greatly in composition, they are, in general, less fertile than the regur soils.

3. The alluvium of the Indo-Gangetic Plain and of the upper valleys of the Nerbudda and the Tapti rivers. These soils consist mostly of reddish, brownish, or yellowish clay, often sandy, especially in the Indus Plain, rich in lime and potash.

Egypt.—Cotton lands in Egypt are limited almost wholly to alluvial soil along the Nile River and in its delta. The silt-laden waters of the river in a conjunction with the berseem clover grown in rotations keep this land very productive.

Central, Western, and Eastern Africa.—Vast areas in the Sudan south of Egypt are suitable to cotton production, in so far as soil conditions are concerned. Lack of labor and transportation facilities stands in the way of their usage at present. It is estimated that there are 25,000,000 acres in Nigeria and Kamerun adapted to the raising of cotton. Other suitable areas are found in Uganda, in East Central Africa, British and German East Africa, Nyasaland, and Rhodesia.

South America.—It is estimated that Brazil has 30,000,000 acres or more of fertile land adapted to cotton raising. Much of this is virgin soil. Argentina has extensive areas. Peru has only limited areas in narrow coastal valleys. This land must be irrigated, but it is productive, and the cotton crop of Peru is considerable.

Asia Minor and Persia, Russia, China and Chosen, Mexico, and the West Indies.—These countries all have lands adapted to cotton growing, and much can be produced when conditions demand it.

Soil and Climatic Adaptations of the Cotton Plant.—It is well known that certain kinds of cotton will yield much better in particular regions than other kinds. Sea Island cotton, for instance, did well in eastern South Carolina, Georgia, and Florida prior to the coming of the boll weevil, but in no other part of the Cotton Belt has it been profitable, and it is not suited to many other cotton districts of the world. Egyptian cotton seems to be particularly adapted to the dry, hot, irrigated regions of Egypt and Arizona. Express, Delfos, and some other long-staple varieties of similar type are prized in the Yazoo-Mississippi Delta but are not at all satisfactory in Texas or in several other parts of the Cotton Belt. Triumph, Lone Star, and Rowden, which are vigorous-growing, big-boll, short-staple varieties, are especially fitted to Texas and Oklahoma conditions.

Very little investigational work has been carried on to determine definitely the reasons for particular cotton adaptations. It is very probable, however, that the adaptations mentioned above are influenced more by the climate, temperature, and rainfall than by soils. The two influences are so combined that it is impossible to attach the proper weight to each without careful research.

Effect of the Soil Fertility on Length and Character of Staple.—This is another subject on which there is a dearth of reliable data. Cotton buyers consider that the character of staple from different soil regions varies considerably, but nothing very definite is known—only their impression without accurate measure.

The average length of the staple of 13 varieties grown on thin hill land by the Mississippi Experiment Station[4] at State College, Mississippi, in 1921, was 0.95 inch. The average length of the same varieties grown the same year on medium-fertile valley land about ¼ mile distant was 1.01 inches. The seed planted in both fields was from the same source. There were many poorly developed bolls on the hill-land cotton. These naturally had shorter and weaker staple. Well-developed bolls from plants grown on unfertile land appear to have practically as long and as strong fiber as similar bolls of the same variety produced on fertile land. The difference is brought about by the relative percentage of good and poor bolls.

In most sections of the Cotton Belt, an application of barn manure or of proper chemical fertilizers invigorates cotton plants and is thought to result in slightly longer and stronger staple. But in Egypt it is believed that a heavy application of barn manure to soil will prolong vegetative growth and injure the character of the lint. Phosphates are thought to hasten maturity and thus improve the staple.

Effect of Soil Fertility on Lint Percentage.—Cotton plants on rich soil grow vigorously and become considerably larger than plants on less fertile land. The seeds are larger and heavier. This results in a lower lint percentage. The average lint percentage of 14 varieties grown on poor soil at Heathman, Miss., in 1922, was 33.4 per cent. The average on richer soil at Trail Lake was 30.5 per cent.

Ewing[5] found that the use of 100 pounds of nitrate of soda on Delta soil at Scott, Miss., decreased lint percentage 0.39 per cent.

CLIMATE

Cotton plants are found growing wild in the tropical and warm temperate zones of both hemispheres. As cotton is a native of warm climate, its growing as a commercial crop is naturally limited to such climates. In the United States, very little cotton is grown north of latitude 37°. This is near the northern boundary of North Carolina. At times, for instance during the Civil War, when the price of cotton was very high, a considerable amount was grown north of this latitude. But it is only during very favorable seasons or under the stimulus of high values that cotton production in this region can be profitable.

The states north of the thirty-seventh parallel have a growing season too short for cotton to mature, and the mean temperature is too low. According to Whitney,[6] the mean temperature for the year is about 15° higher in South Carolina, Georgia, and Mississippi than in Massachusetts, New York, and Pennsylvania. The South has a mean summer temperature about 10° warmer and a winter temperature about 20° warmer than the North. This indicates a longer and warmer growing season in the South. The daily range of temperature is about the same in the two regions, with an increase in both as distance from the seacoast increases.

Whitney,[6] in discussing the climate of the United States, says:

> The mean annual rainfall for the northern section of the United States is about 40 inches, while in the South that amount is exceeded by some 16 or 17 inches. The rainfall in both sections generally increases from the winter, reaching a maximum about the middle of summer, the autumn being the driest period. This larger rainfall and higher temperature of the South give considerably more moisture in the air. The temperature of the dew point in the South is 10 or 12° higher than in the northern states, and a given volume of air contains nearly twice as much moisture. But as the amount of moisture which the air is capable of holding increases with the temperature, the per cent of the saturating quantity or the relative humidity is about the same in the South as in the northern states mentioned. The relative humidity varies somewhat throughout the year, but it is slightly greater during the summer in the South than in the North.

The period of maximum precipitation, as mentioned by Whitney above, applies more particularly to the eastern states. For the central and western states of the Cotton Belt, the heaviest rainfall occurs from March to May (see Figs. 49 and 50). The average annual precipitation for the belt, according to the Weather Bureau, U. S. Department of Agriculture,[7] ranges from 23 inches in western Oklahoma and Texas to 55 inches in eastern North Carolina and 60 inches in southern Mississippi. Throughout much of the area, the range is from 30 to 50 inches (see Fig. 51). The spring rainfall ranges from 6 inches in western Texas to 16 inches in Arkansas and southern Mississippi, being heavier in the Mississippi Valley states than in Texas or the south Atlantic states. Summer rains are moderate in the central states and

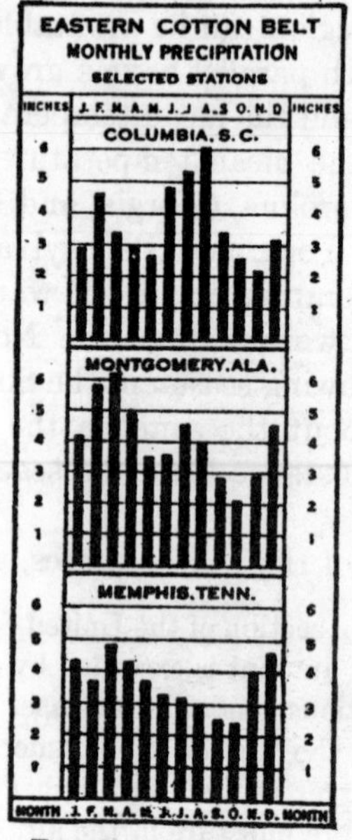

Fig. 49.—Monthly precipitation at selected stations in the eastern part of the Cotton Belt. (*After Weather Bureau, U. S. Department of Agriculture.*)

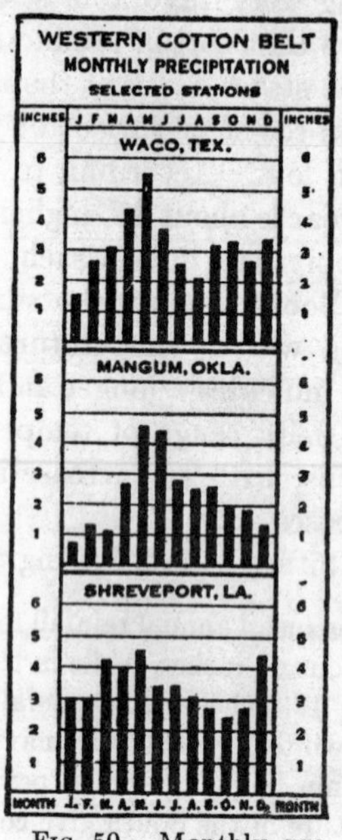

Fig. 50.—Monthly precipitation at selected stations in the western part of the Cotton Belt. (*After Weather Bureau, U. S. Department of Agriculture.*)

heavier toward the southeast and east, reaching a maximum of 20 inches in southern Mississippi and in eastern North Carolina and eastern South Carolina. The Black Prairie of Texas receives only an average of 8 inches of summer rainfall. Autumn is the driest season of the year for the belt, the rainfall for the fall months being less than 10 inches in nearly every region. The Cotton Belt has an average summer temperature of 77° along the northern boundary. This, as pointed out by the Weather Bureau, U. S. Department of Agriculture, appears to be the limit beyond which commercial production of cotton becomes unprofitable. In the southern part of the Cotton Belt, the average temperature is 80 to 85°, and in the Imperial Valley of California it averages 95°. Along the northern margin of the Cotton Belt the last killing frost in the spring is, on the average, about Apr. 10, and the first killing frost in the fall about Oct. 25, the growing season for cotton being about 200 days. In the southern portion of the Cotton Belt the last killing frost comes about Mar. 10, on the average, and the first killing frost in the fall about Nov. 25. This gives a growing season of about 260 days (see Fig. 54). The extra length of frostless season in the southern part of the belt, however, is of very little consequence, on account of the prevalence of the boll weevil. No

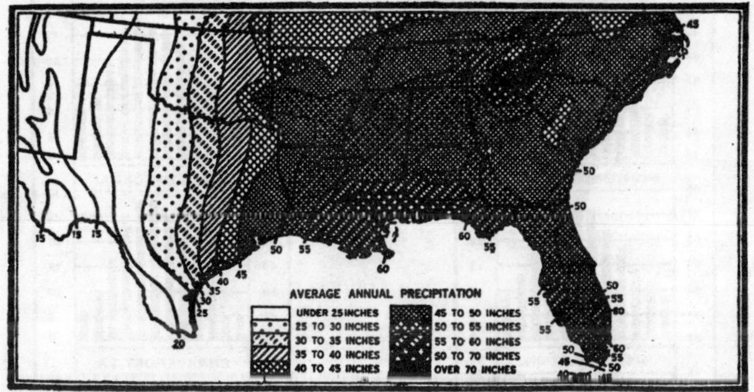

FIG. 51.—Average annual precipitation in the Cotton Belt. (*After Weather Bureau, U. S. Department of Agriculture.*)

cotton is produced after the first of August, as a rule. Larger yields are to be obtained where the growing season is long, but as weevil infestation is heavy during the latter half of the growing season in most areas of the belt, only early-developing varieties can be used advantageously—varieties that will mature bolls beyond weevil damage within a period of 110 days.

Optimum Weather Conditions for Making a Cotton Crop.—Most of the land that is to be planted in cotton is flat broken or broken by listing and bedding prior to the middle of April. On the large cotton plantations, this work starts as soon as the cotton of the preceding crop is picked and continues during the winter and spring season or until completed. There is not enough freezing weather during the winter to impede greatly the progress of this work, but very commonly there is too much rain. Land should be dry enough to crumble freely when turned, and for this condition to prevail rainfall must be only moderate. Occasional rains are needed in April to firm the beds that have been made and to moisten hard, cloddy soil, so that it may be well pulverized in preparation of seed bed.

Frequent warm showers are desirable at planting time. Cotton seed cannot be planted deep, and consequently frquent rains are helpful to keep the surface of the soil moist. Continued

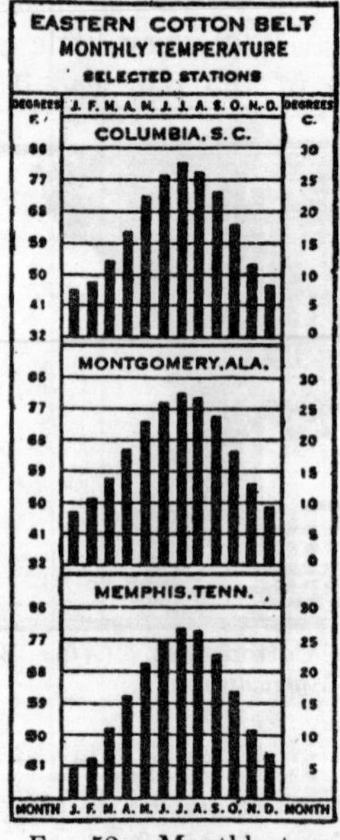

FIG. 52.—Monthly temperature at selected stations in the eastern part of the Cotton Belt. (*After Weather Bureau, U. S. Department of Agriculture.*)

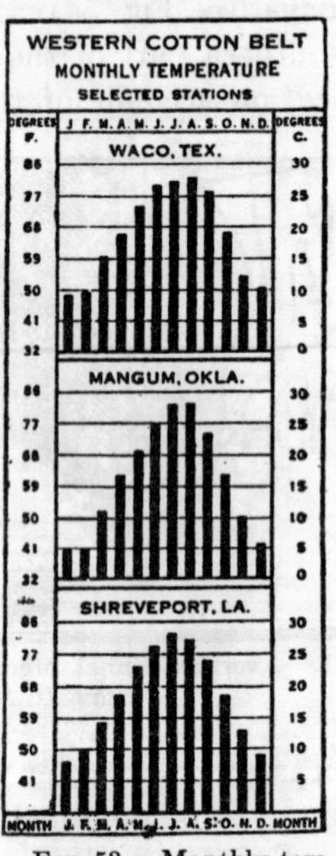

FIG. 53.—Monthly temperature at selected stations in the western part of the Cotton Belt. (*After Weather Bureau, U. S. Department of Agriculture.*)

heavy rains are harmful, especially if accompanied by cool weather. Cotton seed to germinate well must be moist but not wet. If kept wet any length of time, they rot in the ground instead of germinating. If the rainfall is scanty and there are drying winds, the surface of the soil may become so dry that the seed will not secure enough moisture to start growth. After the young plant is above the surface of the soil, cool weather retards its growth and root development; drying winds draw water from its tissues and cause a collapse of a part of its cells, thereby bringing about a disorganization. This is likely to result in a stunting of the plant and a permanent injury to its structure.

From about the middle of May to about the middle of August, the cotton plant is in the period of active vegetative growth. During this time, best growth is made if a warm shower occurs about once a week. Where boll weevils are numerous, comparatively dry weather from the latter part of June until the first part of August is desirable. This retards the growth of plants to some extent, but its beneficial influence in preventing the multiplication of the weevils more than overbalances the harmful

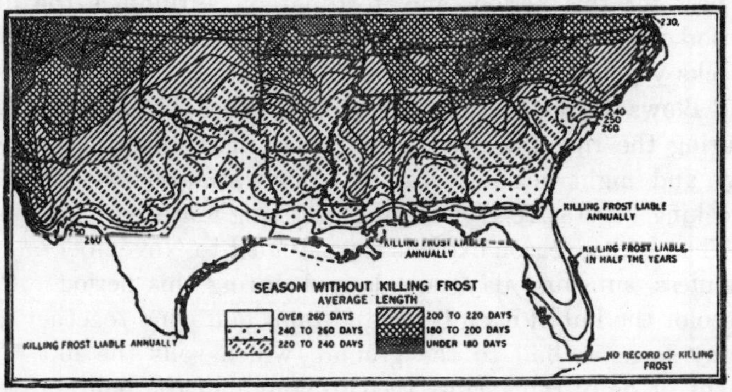

Fig. 54.—Season without killing frost in the Cotton Belt. (*After Weather Bureau, U. S. Department of Agriculture.*)

effect. Rainy weather during the growing period not only leads to worse weevil damage but induces the production of too much "weed," or vegetative growth; it makes proper cultivation impossible, so that grass and weeds have free rein; the cotton plants develop numerous roots near the surface instead of sending them deep into the soil, as they normally do. With only the shallow roots to depend on, the plants suffer from lack of water if dry weather ensues later in the season. Squares, young bolls, and a part of the leaves may be shed as a result. Cotton plants will not make much growth if there is an excess of water in the soil. This excess of water lowers the temperature of the soil, prevents proper aeration, and interferes with the regular upward movement of the capillary water; it is this water that carries to the plant roots their mineral nutrients, both from fertilizers and from the natural soil ingredients.

An abundance of sunshine and warm uniform temperature during the growth period is desirable. Cold periods or wide contrasts between day and night temperatures injure the plants and tend to cause them to check growth and mature prematurely. Rains during the day while blooms are open interfere with pollination and cause some shedding.

During the ripening period, extending from August to October, the fruit that had started to develop comes to maturity, and most of the bolls open. Vegetative growth should cease, and the foods that have been elaborated by the plant and stored in its tissues are used in the production of seed and lint. Dry weather is best for this period, and, fortunately, autumn is the driest period of the year in most parts of the Cotton Belt. Dry weather checks vegetative growth, stimulates fruiting, prevents boll rot, and allows the seed cotton to be picked before the lint is damaged. During the ripening period, the range in temperature between day and night is considerable. This variation also assists in checking vegetative growth and hastening maturity.

The picking season extends from August to November or later. Rainless, sunshiny days are desired during this period. Rains discolor the lint of open bolls, and rains and wind together cause many locks to fall to the ground, which soils the lint. Frequent or daily rains cause seed to sprout in the boll and result in much boll rot, even in open bolls. Prior to the advent of the boll weevil, it was very important that killing frosts should not come until late in November, the late bolls thus being allowed a chance to mature. With weevils prevalent nearly everywhere, few bolls set after the first of September amount to anything. The grade of the lint cotton is somewhat better if frosts are late. The leaves crumble after frosts fall, and some of the pieces lodge in the lint. Early frosts produce much tinged, stained or discolored cotton if there are many late bolls.

Climate of Other Leading Cotton-growing Regions of the World.—The United States leads the world in cotton production. The question is often raised as to whether there is something in the soil or climatic conditions of the Cotton Belt that especially fits it for growing cotton. Several other lands possess as fertile or even more fertile soils. The difference is evidently not due to the soils. The matter of better climatic adaptations probably has some weight. The southern states appear to have a more uniform distribution of rainfall through the year than do most other cotton-growing countries. There is probably some advantage, too, in the gradual change of temperature here from summer to autumn.

West Indies.—According to Mell,[9] from whom this discussion of the climate of foreign countries is largely taken, the West Indies have a temperature that ranges between 77 and 82°, and frosts are very infrequent. A short, wet season begins in April and lasts from 2 to 6 weeks; this is followed by a dry season, during which the thermometer remains about 80°. The heat is very oppressive during July and August, and the summer is dry. The main rainy season begins about Oct. 1 and lasts until December. Following this there is a dry spell that lasts until April. The annual rainfall is 63 inches.

India.—The territory of India is so extensive that there are marked climatic differences in different sections. The climate of a large part of the country is greatly influenced by the two annual monsoons that flow from the northeast and southwest. Such great extremes of moisture and temperature precede or accompany these monsoons that the cotton plants growing in the regions covered are subjected to very hard conditions.

The mean annual temperature of Bombay for the cotton year, from February until June, is 90°. The mean monthly temperature at Calcutta ranges from 60° in January to 85.7° in May. The winter mean is 67.3°; the spring, 83.7°; the summer, 82.5°; and the autumn, 78.5°. The mean annual temperature at Madras is 86.6°; at Benares, 80.3; at Cawnpore, 80°, at Dehra Dun, 70.5°.

The monthly precipitation at Madras is as follows:

	Inches		Inches
January	1.33	July	3.20
February	0.23	August	5.24
March	0.36	September	4.76
April	0.63	October	10.09
May	1.03	November	12.43
June	2.03	December	3.25

The annual precipitation at several stations in India is as shown at top of p. 274.

Some parts of India, especially the northern part, are mainly dependent on irrigation, while others have sufficient rainfall.

Mexico.—Cotton is grown, in small quantities at least, throughout nearly the whole of Mexico. There is an abundance of

	Inches
Bombay	68.73
Tanna	106.16
Dapoolee	134.96
Mahabaleshwar	254.84
Poona	19.02
Dharwar	38.81
Madras	43.58

rainfall along the Pacific coast. Irrigation is possible in much of the interior. The temperature is mild and very uniform over most of the country. At Veracruz, the mean monthly temperatures run as follows:

	Degrees Fahrenheit		Degrees Fahrenheit
January	70.0	July	81.5
February	71.6	August	82.4
March	73.4	September	81.0
April	72.2	October	78.4
May	80.5	November	75.4
June	81.9	December	71.1

Australia.—Frosts in Australia are so light that cotton plants grow on from year to year. The temperature for the cotton months, which are September to June, ranges from 60 to 100°. The mean annual temperature at Sydney is 62.4°; at Victoria, it is 56.8°. The rainfall at Melbourne is 25.66 inches.

Brazil.—The climate of Brazil is in some respects well suited to growing cotton. Mell[9] says:

At the city of Rio de Janeiro, which is situated on the boundary line of the torrid and temperate zones, the average temperature, according to the Annuario de Imperial Observatorio for 1887, is 74.1°F. These figures are the average for 36 years. Only two seasons are known here, summer, or the rainy season, which lasts from October to the end of March, and winter, or the dry season, which lasts from April to September. The average temperature of summer is about 78.8°, and that of winter 69.8°. The highest temperature noted was 99.5 and the lowest 50.3°.

The following meteorological data from selected stations in Brazil are taken from reports made by the Brazilian government.

TABLE XXV.—METEOROLOGICAL DATA FROM CERTAIN STATIONS IN BRAZIL

Stations	Latitude	Altitude, feet	Mean annual temperature, degrees Fahrenheit	Maximum temperature, degrees Fahrenheit	Minimum temperature, degrees Fahrenheit	Annual rainfall, inches
Maranhão, Maranhão	1° 27′	142	81.32	92.84	69.98	96.65
Amarante, Parnahyba	6° 13′	.....	80.78	95.90	64.40	6.30
Bahia, Bahia	12° 58′	210	78.80	88.70	35.78	85.16
Santa Cruz, Rio de Janeiro	22° 56′	85	71.96	97.88	66.56	184.33
Passo-Fundo, Santa Catharina	28° 28′	2,060	62.78	93.92	32.00	
Rio Grande do Sul	32° 00′	52	65.84	90.32	33.80	35.91

Argentina.—The climate of Argentina is very similar to that of the United States, as shown by Mell[9] in a quotation from Barnes of the U. S. Department of Agriculture:

The mean annual temperature of the Argentine Republic is about the same as that of the United States; that is to say, that both countries are included within the limits of similar isothermal lines, from 70 to about 40 in the latter country, exclusive of the Florida peninsula, and also from 70 to about 40 in the former. The average range of the thermometer is therefore about the same. Both are situated geographically and as to range of climate within so-called temperate zones, and, other things being equal, the character and range of productions of the two would be the same.

Climatic conditions vary greatly in different parts of Argentina. There is a range of altitude from sea level to the snow-capped peaks of the Andes. The southerly equatorial current moderates the climate of the southern portion.

Between the thirtieth and thirty-first parallels, the region is subtropical, the temperature ranging from 66° on the eastern plains to 62.5° in Cordoba, and rainfall ranging from 2 inches in San Juan to 47 inches in Entre Rios. In the vicinity of Buenos

Aires, the mean annual temperature is about 63°, with a maximum of 104° and a minimum of 32°; the annual rainfall is 34 inches.

Egypt.—Upper Egypt, or the southern portion of the country, has practically no rainfall, while the region about the mouth of the Nile River has a rainfall of about 8 inches. The mean summer temperature in Lower Egypt ranges from 80 to 90°, and the winter from 50 to 60°. Cotton can be grown only by means of irrigation.

Table XXVI from Mell[9] shows the comparative temperature and rainfall of several of the leading cotton countries of the world.

TABLE XXVI.—COMPARATIVE TEMPERATURE AND RAINFALL OF CERTAIN COTTON-PRODUCING COUNTRIES

Countries	Temperature, degrees Fahrenheit			Annual precipitation, inches
	Mean annual	Mean spring	Mean summer	
United States (Cotton Belt)	63.5	63.5	78.0	50.80
West Indies	79.5	78.9	78.5	63.00
British India	77.8	81.8	86.2	74.22
Mexico (Vera Cruz)	76.6	75.3	82.0	
Australia (Sydney)	59.6			
Brazil (Rio Grande do Sul)	65.6	63.8	74.7	72.36
Argentine (Buenos Aires)	61.2	69.9	73.6	32.06
Egypt, lower (Alexandria)	....	....	76.6	7.51

Figure 55 shows graphically the mean monthly temperature and rainfall at eight different stations in important cotton-producing countries.

Acclimatization.—Cotton plants are very sensitive to climatic or weather changes. A variety moved from one region or country to another rarely produces well for a time and may never do well under the new conditions. A period of dry weather following wet may cause a shedding of squares, bolls, and even leaves. Cold, wet weather during the spring or grass and weeds in the field may prevent proper growth of plants during the forepart of the season, but with the coming of good weather conditions the plants will react and make satisfactory growth and yields, provided they are cultivated properly and the fruiting season is not cut short by boll weevils or frosts. Cook[11] says:

Although many varieties show striking responses to new conditions, with very abnormal behavior in the first years, it is possible, with many varieties, to secure a high degree of adjustment to the new conditions, with a normal and regular behavior of the plants, after a few years of acclimatization or local adjustment, assisted by a proper course of selection.

The range of adaptive possibilities is strikingly shown when a new type of cotton, distinct from any formerly grown in the United States, is able to thrive under many different conditions in widely separated regions. Thus, the Durango, a new upland cotton, introduced only a few years ago from Mexico and acclimatized in Texas, has shown its ability to produce large crops of good fiber over almost the entire range of cotton cultivation in the United States.

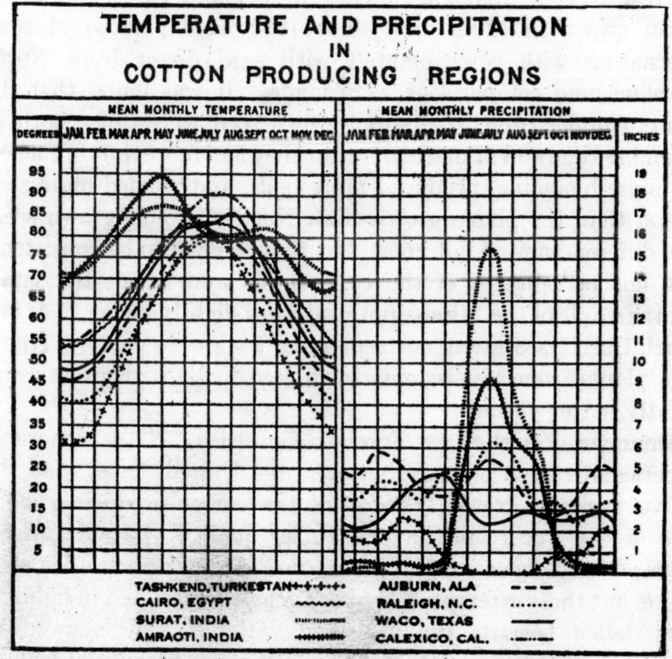

FIG. 55.—The summer temperatures in the various cotton-producing regions are similar, the mean July temperature being between 80 and 90°; and the distribution of precipitation, though varying greatly in amount at other seasons of the year, is similar in showing an October rainfall of 3 inches or less. Heavy rains in the fall interfere with picking and also injure the quality of the lint. (*After Weather Bureau, U. S. Department of Agriculture.*)

Acclimatization Experiments.—Duggar,[10] in 1898, sent seed of a certain strain of King cotton to Goldville, S. C., to Stillwater,

Okla., and to Abbeyville and to Dillburg, Ala. The strain was also planted at Auburn the same year. The following year, seed from the five sources were planted in a comparative test at Auburn, Ala. The differences in yield were not significant. It was thought that one year was not sufficient time to affect the plants enough for differences to show in a later generation.

Winters, of the North Carolina Experiment Station, in 1913, saved seed from a certain plant growing in a field of King cotton near West Raleigh, N. C. In 1914 and 1915, progenies from this 1913 selection were grown at West Raleigh, and the plants self-pollinated each year. In 1916, the seed of the strain were divided, a part being planted at State College, Miss., and the rest at West Raleigh, N. C. In 1917 and 1918, seed of the strain grown in Mississippi since 1916 were planted in rows alternating with rows planted with seed direct from North Carolina, and comparisons were made. It was found that the plants coming from Mississippi-grown seed were taller both at blooming time and at maturity, that they had more fruit branches, did more blooming, produced more bolls, and yielded more seed cotton than the plants grown from North Carolina-grown seed. The differences were not great, but they seemed to be consistent. The lint percentages of the two strains were about the same, but the staple of the Mississippi strain was slightly longer. At the North Carolina Station, corresponding tests were made. There, also, plants from Mississippi-grown seed made best growth and slightly better yields.

American Varieties in Foreign Countries.—Since American varieties of cotton produce better in the United States than the native varieties grown in most of the cotton-producing countries of the world and have a better quality of staple, many attempts have been made to introduce American varieties elsewhere, but these attempts have very generally resulted in failures. They failed because the imported varieties were allowed to become mixed either in the field or at the gin, because of lack of uniform water supply, and because not sufficient chance was given the strains to become acclimated. The usual practice has been to import seed, grow the variety for a few years or until it deteriorated, and then import a fresh supply of seed. Had selections been made and new strains developed from plants that had shown some adaptation to the new conditions, success might have been attained.

India has been importing and experimenting with American varieties since 1790, endeavoring to have them grown there, but over much of the country inferior native varieties are still in use. The Indian Cotton Committee appointed by the Governor General in Council, in its report for 1919, recommended that the work of introducing acclimated American varieties be continued in most sections of the country, but much stress was laid on improving irrigation facilities, establishing seed farms where pure seed of adapted strains could be increased, and improving marketing facilities so that growers of improved varieties, or varieties with better staple, could receive prices in keeping with the quality of cotton produced.

G. A. Gammie, formerly Imperial Cotton Specialist for India, says, "The trials with numerous American cottons show that they all require considerable acclimatization." T. S. Kuo, in the Report of the Cotton Experiment Stations of the Chinese Cotton Millowners' Association for 1920, says with reference to the use of American varieties in China:

> In this country, upland [American] varieties have been tried at many places for many years already. These trials convinced us that they can be grown successfully. All instances of deterioration so far alleged were due to the carelessness of mixing with inferior varieties and lack of selection. We believe those desirable varieties, such as Lone Star, Acala, etc., not only can be quickly acclimatized but will retain their good qualities indefinitely if the line selection of the seed is strictly practiced.

References

1. LYON, T. L., FIPPIN, E. O. and BUCKMAN, H. O.: "Soils: Their Properties and Management," The Macmillan Company, New York, 1918.
2. BENNETT, H. H.: "Soil Regions of the Cotton Belt," in "Atlas of American Agriculture," U. S. Department of Agriculture, part V, sec. A, 1918.
3. MORGAN, J. O.: "Field Crops for the Cotton Belt," The Macmillan Company, New York, 1920.
4. BROWN, H. B. and O'KELLY, J. F.: "Cotton Experiments," *Miss. Agr. Exp. Sta. Bull.* 205, 1921.
5. WALKER, G. B., EWING, E. C. and TROTTER, IDE P.: "Nitrogenous Fertilizer Experiments," *Miss. Agr. Exp. Sta. Bull.* 207, 1922.
6. WHITNEY, MILTON: "Climatology and Soils," in "The Cotton Plant," *U. S. Dept. Agr. Bull.* 33, 1896.
7. Weather Bureau, U. S. Department of Agriculture: "Climate of the Cotton Belt," in "Atlas of American Agriculture," U. S. Department of Agriculture, part V, sec. A, 1918.

8. SMITH, H. D.: "Geography of Cotton Production," in "Atlas of American Agriculture," U. S. Dept. Agr., part V, sec. A, 1918.
9. MELL, P. H.: "Climatology of the Cotton Plant," *U. S. Dept. Agr., Weather Bur. Bull.* 8, 1893.
10. DUGGAR, J. F.: "Experiments with Cotton," *Ala. Agr. Exp. Sta. Bull.* 89, 1898.
11. MARBUT, C. F.: "Soils of the United States," "Atlas of American Agriculture," U. S. Department of Agriculture, part III, 1935.

CHAPTER XIII

COTTON CULTURE

Considering the crop as a whole, it may be said that the cultural practices followed in growing cotton are poorer than those used in growing any other of the important agricultural crops. Morgan[5] mentions five probable reasons for this:

(1) A relatively large percentage of the cotton crop is produced on "one-horse" farms, where thorough plowing and the use of improved implements are impossible. (2) A scarcity of heavy draft animals is often the cause of poor tillage practices even on the large farms. (3) A large percentage of the acreage in cotton is tilled by renters rather than land owners. In most cases little or no supervision of farm operations is given by the land owner, and, as a result, very superficial tillage is practiced. (4) Many unprofitable practices employed in the early days of cotton production in the South have become more or less traditional, being handed down from one generation to the next. (5) Little knowledge of the fundamental principles underlying the growth and nutrition of crops.

Cultural practices vary considerably in different sections, owing to difference in soil, climate, insect pests, etc. In areas such as the Mississippi Delta, Ellis County, Texas, and parts of South Carolina, excellent practices are generally followed.

Disposal of Stalks and Debris.—The first step in the preparation of a good seed bed is the proper disposal of the vegetation on the land. In most cases, cotton follows either corn or cotton, and if the crop has been good there is a considerable mass of old stalks on the lands. Usually it is difficult to turn these under properly with a plow unless they have been cut or broken into short pieces. This is accomplished by running over them with a stalk cutter, the blades of which cut the stalks into short pieces, or breaking them up more or less with a hoe or a club after they have become old and brittle. Where the growth of old stalks is very heavy, a disk harrow is sometimes used in addition to the stalk cutter. In too many instances, old stalks are cut, raked, and burned. This is to be condemned severely, because by such practice the organic matter, which the soil needs badly, is entirely lost. The nitrogen of the stalks, as well as a large part of the mineral nutrients, is also lost. The ashes from the burned stalks are leached out and largely carried away by water from heavy rains.

TABLE XXVII.—TILLAGE PRACTICES WITH COTTON IN NINETEEN REGIONS SURVEYED, SHOWING DATA IN REGARD TO PLOWING (*After Cates*)

	Region surveyed	Fall plowing		Spring plowing		Farmers plowing land		Plows used by farmers			
	County, state, etc.	Farmers practicing, per cent	Depth, inches	Farmers practicing, per cent	Depth, inches	Level, per cent	Into beds, per cent	One-horse, per cent	Two-horse, per cent	Three-horse, per cent	Four-horse, per cent
A	Pemiscot, Mo.	..	....	100	5	60	40	...	76	24	
B	Mississippi Delta	16	4	84	5	24	76	...	76	88	16
C	Robeson, N. C.	4	7.5	96	6.5	56	44	36	64		
D	Mecklenburg, N. C.	52	6.5	48	5.5	80	20	4	88	8	
E	Barnwell, S. C.	..	....	100	6	100	...	52	48		
F	Pike, Ga.	56	6.5	44	5.5	80	20	12	88		
G	Tift, Ga.	16	7.5	84	7	100	...	4	92	4	
H	Giles, Tenn.	12	5	88	5.5	100	...	...	96	4	
I	Bulloch, Ga.	...	....	100	7	96	4	32	68		
J	St. Francis, Ark.	...	....	100	4	4	96	56	44		
K	Ellis, Tex.	4	6	96	4	4	96	...	4	..	96
L	Chambers, Ala.	...	....	100	5.5	72	28	28	72		
M	Johnson, Okla.	8	4	92	4.5	60	40	...	60	36	4
N	Jefferson, Fla.	8	10	92	4.5	52	48	48	52		
O	Lincoln Parish, Louisiana	...	....	100	4.5	28	72	32	68		
P	Lavaca, Tex.	68	5	32	5	...	100	...	80	..	20
Q	Houston, Tex.	8	7	92	4	8	92	...	96	..	4
R	Monroe, Miss.	4	3.5	96	5	20	80	12	84	..	48
S	Bexar, Tex.	92	6	8	7	92	8	4[a]	12	36	48

[a] Five-horse plow.

Plowing by Flat Breaking.—According to data collected by Cates,[1] of the U. S. Department of Agriculture, a little more than one-half of the cotton land in the United States is plowed broadcast, or broken flat (see Table XXVII, columns 7 and 8). The work is done with one-, two-, three-, or four-horse turning plows or with heavy disk, sulky, or gang plows, such as are used in breaking land in the Corn Belt. The type and size of the plow used seem to depend largely on the section of the country and the topography of the land. On the level lands of Texas, Okla-

TABLE XXVIIIA.—COTTON CULTURE IN THE UNITED STATES

State	How land is plowed	When land is plowed	When cotton is planted	How planting is done	Seed used per acre	Delinted seed used[a]	Spacing in row	How thinning is done
California	Broadcast	Fall, reworked with disk in spring	Apr. 10 to 25	Disk planters	12 to 30 lb.	Very seldom	10 to 12 in. mostly	With hoes
Arizona	Broadcast	Winter and early spring	March and early April	With 2-row planters mostly	About 22 lb.	About 30 %	14 to 16 in.	With hoes
New Mexico	Broadcast	January and February	Apr. 15 to May 15	With 2-row planters mostly	15 to 40 lb.	None	6 to 12 in. mostly	Blocked with hoes
Texas	Broadcast and by bedding	Fall to spring	Apr. 15 to May 1	Single- and double-row planters	½ to 1 bu.	Very little	8 to 12 in.	With hoes mostly
Oklahoma	Broadcast mostly; bedding in places	Fall and winter	Apr. 15 to June 1	Mostly 2-row planters	½ to ¾ bu.	None	Generally 6 to 8 in.	With hoes mostly
Arkansas	Bedding mostly	Winter and spring	Apr. 20 to May 10	1-row planter	1 bu.	Very little	2 plants each 15 in.	With hoes
Missouri	Bedding mostly		Apr. 20 to May 10	1-row planter mostly	1 to 1½ bu.	None	1 to 3 plants each 10 to 15 in.	With hoes
Louisiana	Bedding mostly	Fall, winter, and spring	Apr. 1 to May 1	1-row planter	1 bu.	Very little	2 plants every 15 in.	With hoes
Mississippi	Bedding mostly	Fall and early spring	Apr. 10 to May 10	1-row planter	1 to 1½ bu.	Very little	2 plants every 15 in.	With hoes
Tennessee	Broadcast or bedding	Spring mostly	May 1 to 12	1-row planter	1 to 2 bu.	Very little	10 to 15 in.	With hoes
Alabama	Bedding mostly	Spring	Mar. 20 to May 10	1-row planter	1 to 2 bu.	Rarely used	12 to 18 in.	With hoes
Florida	Broadcast	December and January	Mar. 20 to April 10	1-row planter	1 to 2 bu.	None	12 to 18 in.	With hoes
South Carolina	Broadcast and later bedding	Fall and spring	Mar. 15 to May 10	Largely 1-row planter	1 to 2 bu.	Rarely	2 or 3 plants, hoe's width	With hoes
North Carolina	Mostly broadcast, and bedding later	Winter and spring	Apr. 1 to May 1	Mostly 1-row planter	1 to 2 bu.	Very little	9 to 15 in.	Mostly blocked out with hoes
Virginia	Broadcast and then bedding	Spring	May 1 to May 15	1-row planter	1 bu.	None	8 to 12 in.	With hoes

[a] This information was obtained in 1932 from a questionnaire sent to agronomists and others familiar with cotton growing in the various states of the Cotton Belt. At present (1937), more delinted seed, especially mechanically delinted, is used in some states.

TABLE XXVIII*B*.—COTTON CULTURE IN THE UNITED STATES

State	Implements used in cultivation	Duration of cultivation period	Number of times cultivated	Practical value of tractors in growing cotton	Percentage of cotton poisoned for weevils	Value of boll-weevil poisoning
California	Chisel and duckfoot cultivators	Until blooming	Every 2 or 3 weeks	Helpful on large areas	None	
Arizona	Single-row mule and 2-row tractor	Until plants are too large	5 to 7	Practical if acreage is sufficient	None	
New Mexico	Largely 1-row riding cultivator	Until plants are too large	3 times dry land; 6 to 8 times on irrigated	Not practical in New Mexico	None	
Texas	2- and 4-horse cultivators mostly	Until July 15	3 to 4	Valuable in southwest Texas	Never more than 33 %	Helpful if done properly
Oklahoma	1-mule and 2-row cultivators	Until about Aug. 1	3 to 5	Practical only in west Oklahoma	Less than 1 %	Will pay in some cases
Arkansas	1- and 2-mule cultivators	Until about Aug. 5	6 to 8	May be practical on large farms	Less than 1 % in 1932	Profitable under some conditions
Missouri	1-row cultivators mostly	Until about last of July	5 to 7	Not practical in Missouri	None	
Louisiana	1- and 2-mule cultivators	Until Aug. 1	7 to 9	Not practical on most farms	Not more than 25 %	Helpful if properly done
Mississippi	1- and 2-mule cultivators	Until about Aug. 1	6 to 8	Not practical on most farms	Not more than 15 %	Helpful if applied right
Tennessee	1-mule cultivators mostly	Until about Aug. 1	4 to 6	Not practical on most farms	Almost none	
Alabama	70 % one-mule plows	Until July 15 or Aug. 1	6 to 8	Not of practical value	20 to 30 %	Ordinarily profitable
Florida	1-mule cultivators mostly	Until July 15	6 to 8	Not practical in Florida	5 %	Questionable except when well handled
South Carolina	1-mule cultivators mostly	Until July 15	6 to 8	Zero	30 %	Profitable
North Carolina	1- and 2-mule cultivators	Until Aug. 1 to 15	6 to 8	Practical only on large coastal-plain farms	1 %	Profitable if handled properly
Virginia	1- and 2-mule cultivators	Until blooming	4 to 6	Probably not practical in Virginia	Less than 5 %	Boll weevils not often serious

homa, and Mississippi, large plows and heavy teams or tractors are used, whereas on the uplands of the Carolinas, Georgia, Alabama, and Mississippi smaller turning plows are in use, many being one-horse. Flat breaking is advantageous in that none of the soil is left unbroken, but it is more expensive than breaking by bedding.

Plowing by Bedding.—Plowing by listing, or bedding, is extensively practiced in many parts of the Cotton Belt, even in regions where good cultural practices are followed. It is somewhat more speedy than flat breaking. Fields are covered more quickly, and it is not necessary to rebed or lay off rows prior to planting.

There are two general methods of bedding land: One is to use a heavy middle burster or lister requiring a heavy team of four or six mules. This method is used in Texas and in the Mississippi Delta and is becoming fairly general. The lister is run in the ridge of the old row, and a new bed is made at each "through."

The second method is to use a turning plow of one-, two-, or three-horse size, various sizes being used in different sections. A furrow is made through the old middle, and then one or two furrow slices are thrown toward it from each side. If but one furrow slice is thrown, there is a considerable strip, or "balk," left in the middle between the new ridges. This is thrown out with a middle burster.

On land that has been flat broken, low beds for planting are sometimes made with small disk harrows or with cultivators with disks or shovels on them set to make a ridge. When large listing plows are used, such as the Avery "Uncle Sam" four-mule middle burster (Fig. 56), good work is done. Apparently, the yield is much the same whether the land is bedded or broken broadcast, but probably more labor is required in cultivating the crop if the land is not flat broken. Land put into beds during the fall or winter dries out earlier in the spring, warms up sooner, and is in condition for planting at an earlier date. This makes an earlier crop possible. An early crop gives best returns in boll-weevil districts.

Time of Plowing.—According to the survey made by Cates,[1] most cotton land is plowed in the spring (Table XXVII). This practice varies considerably with the locality and the season, some winters being much more favorable for work than others. Where conditions will permit it, there are several advantages to be gained from fall or winter plowing: (1) It gives the planter a chance to get ahead with his work, so as to be able to plant his crop more quickly when suitable weather comes in the spring. (2) Vegetation turned under early has time to decay and to benefit the following crop. (3) Fresh soil brought up from the subsoil has time to weather, and its important chemical elements

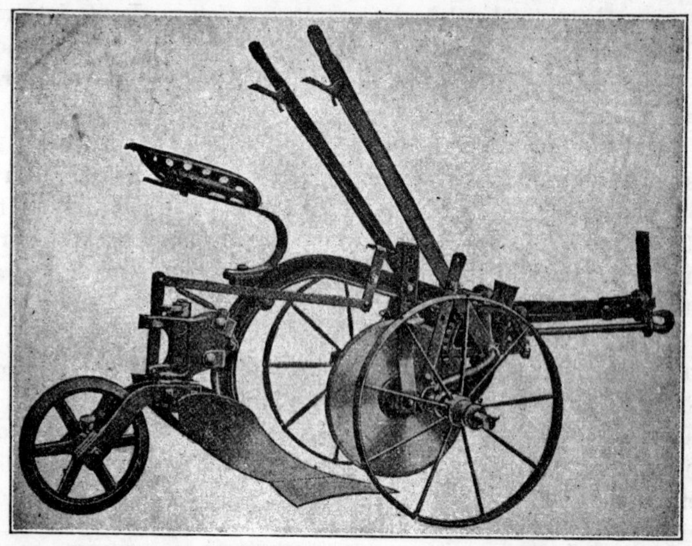

Fig. 56.—Uncle Sam Middle Burster. A heavy double-mold board plow used in bedding cotton land. (*Courtesy B. F. Avery and Sons.*)

become available to plants. (4) Insect pests are controlled in a measure.

In hilly regions where the land erodes badly, fall plowing is not advisable, nor is it possible if cover crops are grown.

Depth of Plowing.—Breaking for cotton is 3 to 8 inches deep, but far too much land is broken only 3 or 4 inches (see Table XXVII). Breaking is commonly somewhat deeper when done in the fall. Sandy soil is usually plowed more deeply than clay, since the plow runs more easily. Clay should be plowed to a depth of 6 to 8 inches for the best results. A deep layer of cultivated soil holds more moisture, and plants on it grow more vigorously. It is not advisable to plow deeply late in the spring. Sterile soil from the subsoil brought to the surface late in the spring does not have time to disintegrate but bakes, forms clods, and is troublesome during the season. If shallow plowing has been practiced, it is considered best to increase the depth gradually, say an inch a year, until a satisfactory depth is attained. Data secured by Cates[1] failed to show any correlation between depth of plowing and yields.

Subsoiling.—"Subsoiling" is a term applied to the loosening of the subsoil without bringing it to the surface. It is accomplished by running an ordinary turning plow and following this, in the same furrow, with a special subsoiling plow which has no mold board and does not displace the soil. In a few experiments, subsoiling has shown profitable results, but on the whole it has not paid, and the process is to be considered of doubtful value.

The Seed Bed.—For cotton, a firm but well-pulverized seed bed is always desirable. This may be secured by thoroughly pulverizing the soil some weeks before planting, letting it rest until a good rain has fallen to firm or settle it, and then freshening the surface by a light harrowing or dragging.

Land that is listed or bedded in the fall or winter will require freshening in the spring before planting. This is commonly accomplished by rebedding or by running over the bed with a deeply running harrow, such as a two-horse disk or a disk cultivator. After the beds have been freshened, they are pulverized and dragged down to a height of 6 inches or less with a spike-tooth harrow or a drag. Beds that have been made recently may be pulverized and dragged down to the desired height with a spike-tooth harrow or drag. Cotton planted late is sometimes planted on the level. In parts of Texas and Oklahoma and in a few other of the drier districts of the Cotton Belt, cotton is planted in lister furrows.

There is a considerable advantage to be gained in most places by planting cotton on beds or low ridges. Young cotton plants are tender and will not thrive if the land is wet or the weather cool. The top part of a ridge becomes dry sooner than level land and, being drier and presenting more surface to the sun, becomes somewhat warmer.

Time of Planting.—The best date for planting cotton naturally varies in different places on account of climatic differences. Table XXVIII gives planting dates for various parts of the Cotton Belt. These range from Mar. 1 for southern Texas to

June 1 for the more northern regions, the average date being about Apr. 15.

TABLE XXVIII.—TILLAGE PRACTICES WITH COTTON IN NINETEEN REGIONS SURVEYED, SHOWING DATES AND METHODS OF PLANTING (*After Cates*)

Region surveyed (see Table XXVII)	Planting date		Farmers[a] planting, per cent			Average distance apart				Planters used by farmers, per cent		
	Average	Range	Level	On beds	Listing	Rows, feet	Hills, inches	Average space per hill, square feet	Seed per acre, pecks	One-horse	Two-horse: Two-row	Two-horse: One-row
A	Apr. 21	Apr. 10–May 10	..	100	..	3.5	16	5	4	60	40	
B	Apr. 11	Mar. 25–Apr. 30	16	84	..	4	17	6	4	96	4	
C	Apr. 14	Apr. 8–Apr. 20	8	92	..	4	13	4	4	100		
D	Apr. 23	Apr. 15–May 5	..	100	..	3.5	12.5	3.5	4	100		
E	Apr. 11	Mar. 25–Apr. 30	..	100	..	4	15.5	5	4			
F	Apr. 15	Apr. 1–May 1	16	84	..	3.5	13	3.5	5	100		
G	Apr. 4	Mar. 15–Apr. 20	28	64	8	4	16.5	5.5	2.5	100		
H	Apr. 15	Mar. 10–Apr. 30	20	80	..	3	13	3.5	5.5	100		
I	Apr. 1	Mar. 15–Apr. 15	4	84	12	4	16.5	5.5	3	100		
J	Apr. 25	Apr. 1–May 20	4	96	..	3.5	14	4.5	5.5	100		
K	Apr. 13	Apr. 1–May 15	8	92	..	3	13	3.5	2.5	...	4[b]	96
L	Apr. 11	Mar. 25–May 1	..	100	..	3.5	13.5	4	5	100		
M	Apr. 21	Apr. 1–June 8	16	64	20	3.5	15.5	4.5	2.5	64	4	32
N	Apr. 1	Mar. 1–Apr. 15	8	92	..	4	13	4	4	92	4	4[c]
O	Apr. 18	Apr. 1–May 15	..	100	..	4	14	4.5	4	100		
P	Mar. 25	Mar. 1–May 1	..	100	..	3.5	13	4	5.5	8	..	92
Q	Apr. 16	Mar. 30–May 30	4	96	..	3.5	12.5	3.5	3	92	..	8
R	Apr. 15	Apr. 1–May 1	4	96	..	3.5	15	4.5	4	96	..	4
S	Mar. 25	Mar. 1–Apr. 15	40	20	20	3.5	18.5	5.5	3	...	4	96

[a] All planted with drill.
[b] Three-horse one-row planter.
[c] Hand planter.

Ewing, of Scott, Miss., planted cotton at 5-day intervals from Mar. 25 to about the middle of May for a series of years. The optimum date was found to be between Apr. 10 and 20, with but little difference in yields secured from plantings made at different times within those dates. Plantings made before or after those dates gave considerably less yield, and the farther removed from the date the lower the yield. These experiments were conducted in the central part of the Mississippi Delta. Boll weevils were plentiful the latter part of the season.

The optimum date for planting in any particular regions varies in different years, on account of seasonal weather variation. This variation may be as much as 10 days or 2 weeks. Cotton should not be planted until danger of frost is over and the land has become warm. If planted too early, the seed may not germinate because of the coldness and wetness of the soil. Or, if the plants come up and cold weather follows, they become stunted, and many die from fungous diseases. Where boll weevils are plentiful, it is advisable to plant as early as a good stand of healthy plants can be secured and kept. Where there are no boll weevils, planting may be later, and good yields may be secured provided weather conditions during the latter part of the season are favorable. Later plantings require less cultivation than early ones.

Methods of Planting.—In most cases, cotton is planted in drills with a one-horse planter, which opens a small furrow, drops the seeds, and covers them. Some planters drop bunches of seed in hills the desired distance apart. If the seed bed is in proper condition and the planter handled carefully, this may be a satisfactory way of planting. If planted in hills the right distance apart, cotton is more easily thinned properly. Frequently good stands are ruined by careless thinning. In some areas, however, where the death of young cotton plants from seedling diseases has reached serious proportions, hill planting may not be a good practice. If one or more plants in the hill develop a disease, it may spread and destroy all the plants in the hill. If the whole hill is lost, that makes a considerable gap in the row. In such areas, drill planting is apt to result in better stands. The writer used hill-drop planters when he first began experimental work at Baton Rouge, La., but found it necessary to discard them for the drill planters. King and Loomis,[10] of the Sacaton, Ariz., Field Station, had a similar experience with the hill-drop planters. In an experiment that they conducted, 3.2 per cent of the plants were lost in the drilled rows and 16.8 per cent in the hill-drop rows.

A few farmers drop seed by hand, and some others use two- or four-row planters. Planting by hand is not economical if planters are available. The larger planters may not regulate the depth of planting so well as the smaller ones, but they may be more economical to use in large-scale operations.

Cotton is sometimes checked, or planted in hills, so that it may be plowed in two directions. This practice reduces the necessary amount of hoeing to a minimum and makes it possible fairly easily to control troublesome weeds, such as coco grass (*Cyperus rotundus*). But the yield from checked cotton is not so great, as a rule, as from drilled cotton that is closely spaced and well cultivated.

A cotton planter, to be satisfactory, must be so constructed that it will drop seed regularly; will press the soil around the seed; and will plant the seed at a uniform depth. A planter

Fig. 57.—A good one-row cotton planter. (*Courtesy B. F. Avery and Sons.*)

with a small shovel to open a furrow, a roller behind to press the soil down on the seed, and a lever by means of which depth of planting can be regulated closely is the most satisfactory (see Fig. 57).

Depth to Plant.—The cotton plant is a native of warm regions, regions considerably warmer than the Cotton Belt during the spring, the time the young plants are getting a start. Being adapted to warmer regions, they are naturally weak and delicate in temperate climates. On account of their lack of strength, they cannot get out of the ground if planted very deeply or if the surface of the soil has a hard crust. Seed should never be placed deeper in the soil than is necessary for them to secure moisture sufficient for germination. A depth of an inch is sufficient usually, for rains are rather frequent over much of the Cotton Belt, frequent enough to keep the soil moist near the surface. "Barely cover the seed" is a rule followed by many.

During periods of drought or in regions of light rainfall, seed, of course, must be planted considerably deeper.

Quantity of Seed to Plant.—It is the common practice to plant cotton seed at the rate of a bushel an acre. This is much more than would be needed if every seed would make a large plant. But they do not all make plants. Some fail to germinate; others germinate, but the plants never get above the surface of the soil; and many others die of disease or are accidentally destroyed. To be sure of a good stand, it is necessary to plant a large quantity of seed and then thin out plants not wanted. More seed is needed on stiff clay land or on old land low in organic matter where the soil tends to bake than on sandy soil or on new lands. If a planter that drops seeds in hills is used, a smaller quantity can be planted with safety, but it is never advisable to use less than ½ bushel per acre.

Planting seeds are sometimes delinted by running them through a machine that removes the linters or a large part of the short fibers and fuzz on the seed; or they may be delinted thoroughly by the use of strong acids. Acid delinting is somewhat tedious and dangerous if done without special equipment. Delinted seed pass through the planters more freely, plant more regularly, and germinate a few days sooner if there is a shortage of moisture in the soil; but if a period of cold wet weather follows planting, the delinted seed are more inclined to rot before germinating.

Vitality of Cotton Seed.—There seems to be some difference in different varieties of cotton in respect to the time seeds retain their vitality. In experiments conducted by the writer, the big-boll varieties, especially ones with large seed, gave a lower percentage of germination when the seed was more than a year old. Less than 5 per cent of the large seed that were 3 years old germinated, whereas seed from some of the small-seeded varieties gave a germination of 85 per cent. As was shown in Chap. IX, storage conditions have much to do with length of time that the seeds retain their vitality. If the moisture in the seed is reduced below 8 per cent and is kept low, they retain their vitality well. Seed 10 years old have been known to give 90 per cent germination.

The use of seed more than 1 year old for field planting is sometimes advocated, and there are good reasons. Certain fungous organisms that cause diseases of mature plants are present in the seed but will not retain their vitality more than 1 or 2 years at most. Plants coming from seed more than 2 years old will then be less subject to fungous diseases. There is still another reason for favoring old seed. Weaker seeds which would probably produce weaker and lower-yielding plants lose their vitality in a year or two and are thus eliminated.

Field tests made by Duggar failed to show any consistent gain in favor of the old seed. Tests made by the writer in the Mississippi Delta gave somewhat better yields for both 2- and 3-year-old seed. Apparently, old seed may be planted to an advantage, but planting should never be done until a germination test has been made.

Cotton Spacing.—The subject of cotton spacing, or the number of plants to be left to the acre, is of much importance to cotton growers, and it has received considerable attention from experimenters. Results obtained in different places have varied considerably, owing largely, no doubt, to the varying conditions under which the experiments were conducted.

Width of Rows.—The width of row now in use varies from 3 to 4 feet, but 3½ feet is the most popular (Table XXVIII). It is generally considered best to make the row the width that will be most convenient for cultivation and to regulate the number of plants to the acre by the spacing within the row.

Distance in the Drill.—The spacing of cotton plants is influenced by the prevalence of boll weevils more than by any other factor. The best distance for plants on a certain type of land before the weevils came is not necessarily the best distance now. Brown[7] made a review of all the more important cotton-spacing experiments carried on during the last 35 years. In his summary of the pre-boll-weevil experiments, he says:

> In the sixty-four distance-in-the-drill-spacing experiments listed in the foregoing pages, the rows varied in width to some extent, and several different drill distances were used, some more often than others. The character of the soil and the climate varied in different places also. This makes an accurate summary or set of averages impossible, yet a survey of the results shows that 12-inch spacing, in 3½ to 4-foot rows, was first in yield in twenty-four of the tests given. This was a distance used more often than any other, but it averaged best. Boll weevils did little or no damage in any of the experiments listed. On the less productive land the closer spacing gave better yields in the majority of cases.

On the richer lands wide spacing gave best yields in a number of cases, but the results varied widely, owing probably to differences in rainfall, length of fruiting season, etc.

In his review of the experiments made under weevil conditions, Brown[7] says:

In the eleven different tests conducted under moderate to heavy boll-weevil infestation, nine gave best yields for a spacing of 12 inches or closer. From this it seems fairly certain that, under heavy boll-weevil infestation, cotton should be spaced closely. The only question remaining to be decided is just how close it is to be spaced—how much closer than 12 inches. While leaving the rows unthinned gives the best yields in some cases, this practice is uncertain, and in no case will it be practicable on account of the difficulty in removing grass and weeds. There must be room between plants or between small clusters of plants for a hoe to work if grass and weeds are to be removed easily and effectively. This requires a space of 10 or 12 inches, and often more is taken. Hoes are 7 or 8 inches wide, and it is impossible to get a cotton chopper to hit within an inch of the place at which she is looking. Too often she is not looking at the row at all.

During the last few years, several additional spacing experiments have been conducted. Brown and Cotton,[11] in work conducted at Baton Rouge, La., obtained best yields on both alluvial and bench land from two stalks every 20 inches, but there was no significant difference between any of the spacings from 10 to 30 inches. Four-foot rows were used. This was on rather rich land in a humid area. McNamara,[12] on the Texas blacklands at Greenville, obtained best yields from unthinned rows and from 6-inch spacing. These experiments were conducted under droughty conditions. Other experiments have been conducted by Reynolds of Texas, Ware of Arkansas, Tisdale of Alabama, McKeever of California, and others. Most of the tests failed to show significant differences in yields from spacings such as are ordinarily used in cotton culture. In the main, the unthinned cotton gave lowest yields, but close spacing gave better yields than wide spacing. Close spacing reduced boll size and had a tendency to lower the number of locks per boll.

Prior to the coming of boll weevils, the fruiting season for cotton plants was long. Except in the northern edge of the Cotton Belt, there was time enough for plants to grow large, to bear bolls to the outer end of later branches, and to utilize much space in the row. Under such conditions, especially on rich land, it was not essential to have the plants closely spaced in the row. They could grow large, utilize all the space in the row, and produce as good yields as, or possibly even better yields than, if more plants were in the row. Under heavy weevil infestation, the fruit must be set in a short space of time—a month or a little more. There is not much time for plants to make growth during the fruiting period; so they are still rather small when this period is over. This makes it necessary for them to stand close together in the row if all space is used to the best advantage.

If weevil infestation is heavy, it is not practicable to leave plants close enough together in the row for maximum yields if only one plant is left in a hill. Leaving two or more plants in a hill a hoe's width apart obviates this difficulty. On poorer soils, better yields may be obtained usually if three or four stalks are left in a hill. On rich soil, one plant every 12 inches is sufficient.

Under ordinary plantation conditions, cotton plants are rarely spaced close enough for best results. Lengthy skips are frequent, and hills are left too far apart by the choppers.

Effect of Spacing on Comparative Yield, First Picking.—In the spacing experiments in which comparative yields from different pickings are recorded, the closely spaced plots yielded the highest percentage of their crop the first picking. Duggar,[6] in Alabama, secured results as follows: 42 per cent of the whole crop was obtained the first picking from plants spaced 12 inches; 38 per cent, from plants spaced 18 inches; 30 per cent, from plants spaced 24 inches; and 26 per cent, from plants spaced 36 inches.

Close spacing is conducive to earlier maturity, or at least to more open bolls at an early date. If there are more plants on the ground, there is a higher percentage of primary bolls, or bolls springing directly from the main stalk or from fruit branches on the main stalk. Bolls on these parts develop earlier than bolls on secondary branches; and if the percentage of these is higher, an earlier crop is produced.

Number of Stalks to the Hill.—Several tests have been made to determine whether or not there is an advantage in leaving more than one stalk of cotton in a hill. Balls and Holton,[8] in some careful tests with Egyptian cotton in Egypt, secured 10 per cent greater yield regularly from two stalks in a hill as compared with one stalk. Each hill had 0.34 square meter space. This represented 12,000 hills to the acre.

Lee, in northern Louisiana, secured better yields from two stalks in a hill at all distances except 8 inches, one stalk to the hill leading at this distance. Rast, in *Bulletin* 115 of the Georgia College of Agriculture, says that in tests made in Georgia two stalks to the hill yielded more than one for all distances greater than 14 inches.

Ayres, in cultural experiments on rich bottom lands in Arkansas, secured best yields the first picking, and also the best total yield from two plants to the hill. His plants were spaced 18 inches in the row, the rows being 4 feet apart. Walker and Ayres, at the Mississippi Delta Experiment Station in 1919, secured better yields from two plants in a hill for 12, 16, 24, and 28 inches. With 8-inch spacing, one plant to the hill led two by a margin of 50 pounds of seed cotton per acre.

From the tests made it appears that for distances that are practicable for cultivation—12 inches or more—two plants to the hill will give better yields than one.

Redding,[3] of Georgia, has shown that the nearer the area allotted to each plant approaches a square the greater its yield will be, comparatively. Probably maximum yields are to be secured from one plant in a hill, but the difficulties in the way of cultivation when plants are spaced closely enough for best yields make this type of spacing unsatisfactory.

Spacing for Different Varieties.—Very little work has been done to determine the difference between different varieties in regard to space required for best yields. Duggar[6] found that Welborn, a cluster variety, gave its maximum yield when spaced 1 foot in the drill, whereas Peeler, a spreading variety, yielded best where the plants stood 2 feet apart in the row. Four-foot rows were used in both cases. In work carried on by Walker[9] at the Mississippi Delta Experiment Station, Trice, a dwarfish variety, gave its best yield when spaced at 6 inches. Lone Star and Columbia, larger growing varieties with heavy foliage, yielded best at 12 and 18 inches, respectively. Results by Pittuck, of Texas, and others show no consistent differences. It is reasonable to conclude, however, that a vigorous, rank-growing variety will be able to use to advantage more space than a weaker, smaller growing variety. In the case of varieties with heavy foliage, crowding the plants may result in much boll rot during a rainy season. The dense shade produced is favorable to fungous growth on the bolls.

Single-stalk Cotton Culture.—The essential feature of the single-stalk system of cotton culture is the suppression of vegetative branches by delaying thinning until the plants are 10 or 12 inches tall. When plants are allowed to crowd each other in the row for a time, the development of vegetative branches is prevented to some extent. It has been thought by some that the late thinning fosters the development of fruit branches on the lower part of the plant, that the plants having few vegetative branches fruit earlier, and that the crop yield is larger because more early plants can be grown on the same area.

Apparently the advocates of the single-stalk system have not distinguished between the effect of the delayed thinning and that of close spacing. Close spacing gives all the beneficial results claimed for the single-stalk system; and, as shown by the work of Ayres, of the Arkansas Experiment station, McClelland, of Georgia, and Brown,[7] of Mississippi, cotton thinned at the regular time, when the plants are 3 to 5 inches tall, will yield more than cotton thinned late, provided the same number of plants is left in both cases. Delaying the thinning has a tendency to make plants tall and "whiplike" and hinders the development of fruit as well as vegetative branches. The bulk of experimental evidence indicates that cotton plants should be thinned as early as it is safe to thin them, that is, as soon as the danger of losing a stand from cold weather, damping-off fungi, etc., has passed, and before the plants are stunted by undue crowding.

Cultivation.—Intertillage of cotton has for its purpose the destruction of weeds, the loosening of the soil so that it may be properly aerated, the aiding of water penetration on tight soils, and the placing of the soil in proper condition for plant nutrients to be rendered most available. Work by Cates[1] showed a close relationship between cotton yields and the number of cultivations the crop received. The average yield of 258 farmers who worked their cotton four or five times was 763 pounds of seed cotton, whereas that of 171 farmers who cultivated six or seven times was 907 pounds, and that of 20 farmers who cultivated eight or nine times 1,012 pounds.

Broadcast Tillage.—Under certain conditions, broadcast tillage with a light harrow or weeder may be very helpful and economical. It may be used to break the crust after a rain or to kill numerous young weeds. A hard crust may do much harm in preventing young plants from breaking through the soil. But it is only rarely that broadcast tillage can be used satisfactorily. Since cotton is usually planted on beds and covered very lightly, any harrowing that will tear up the crust satisfactorily is apt to tear up too many young plants or drag the seeds out of the row if they have not yet germinated.

The middles between the rows will not be cultivated to any extent unless the ridge is torn considerably. If there is much trash or litter in the soil, the harrow will catch it, drag it out, and tear up cotton plants; or if the stand of plants is poor, the harrow will destroy too many, especially in spots where there is none to spare.

First Plowing of Rows.—The purpose of the first plowing is to loosen the surface of the soil and to destroy weeds. If the land has been well plowed in the preparation of seed bed, it is not necessary to cultivate deeply; but if the soil has not been well plowed, or if heavy, packing rains have fallen, deep cultivation will be helpful. Deep cultivation will not injure the cotton plants while small. It is important to plow close to the plants, so that there will not be a wide strip left to be hoed; but care must be taken not to throw much dirt to the row, or the small plants will be covered. All that are covered are killed. It is thus necessary to use plows with small shovels or to use fenders on plows with large shovels. A spring-tooth cultivator (Fig. 58) is satisfactory for shallow cultivation, whereas a side harrow, a Perry cultivator, or a heavy plow with small shovels is better for deep cultivation.

If fields become very weedy or grassy owing to continued wet weather which prevents plowing or to the prevalence of very troublesome weeds, such as coco or Bermuda grass (*Cynodon dactylon*), it may be necessary to use a scrape or turning plow. The use of these implements for barring off cotton rows is not a good practice ordinarily, but it is effective in the destruction of a heavy growth of weeds when nothing else will serve. Weeds must be kept down at all hazards. If the scrape or turning plow is held so that the side next to the row does not run deeply, little harm will be done.

Probably slightly better work can be done with one-horse than with two-horse plows, but the use of two-horse plows is much more economical in so far as saving labor is concerned. The negro laborers prefer to work one mule; and since with one mule a family can easily cultivate as much cotton as they can pick, there seems to be no reason for using two. The survey made by Cates[1] showed that one-horse implements are used very generally in the cultivation of cotton in the United States.

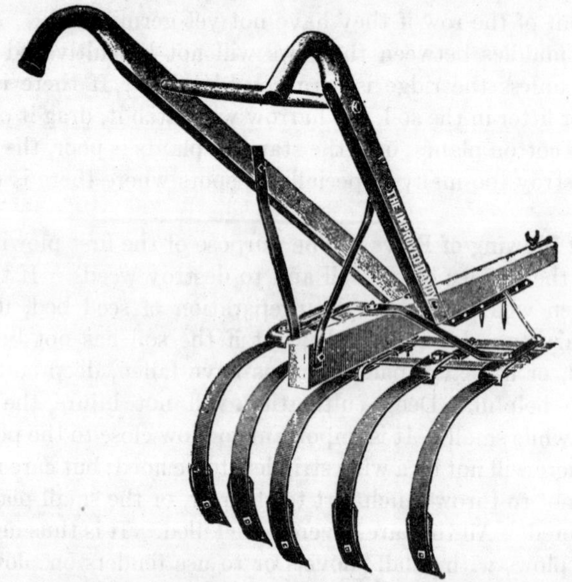

Fig. 58.—Spring-tooth cultivator. A good implement to use in plowing small cotton. (*Courtesy Moline-Hooper Company.*)

Hoeing.—The first hoeing that cotton receives is frequently spoken of as "chopping." The laborer scrapes, or scalps, all the land in the row left unplowed and takes out unnecessary plants, leaving two or more in hills or bunches the desired distance apart. A machine with revolving blades has been made to chop out cotton, but it is doubtful if it will come into general use. It is not satisfactory for use where stands are "skippy" or where the surface of the rows is not even. One or more later hoeings are given, the number being determined by the amount of weeds not destroyed by the plow. At later hoeings, the number of plants in a hill is commonly reduced to one, but better yields would be secured if two were left. It seems to be a very common mistake for hoe hands to cut out too many plants when thinning cotton. They leave the hills too far apart and usually but one tender plant in a hill. Many plants are destroyed accidentally in later hoeing. Careless plowing covers up a good many more; and if the season is unfavorable while the plants are small, many die from disease. The ultimate result is a poor stand. If more plants are left in the chopping process, the final stand is better and the yields greater.

Hoeing is expensive tillage, but apparently there is no way to avoid it. On account of the weakness of young cotton plants,

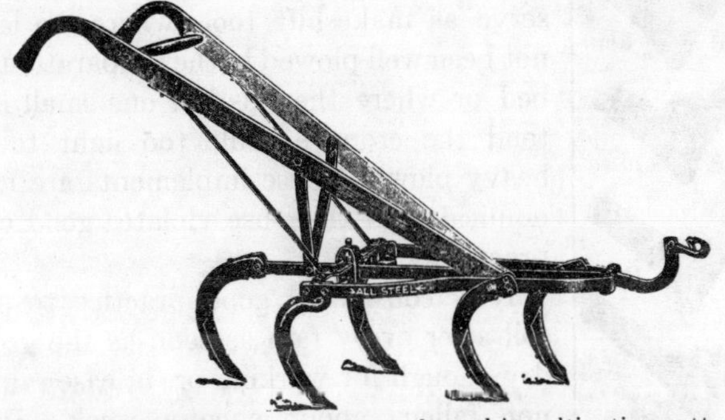

FIG. 59.—A five-shovel cultivator much used in cultivating cotton.

it will always be necessary to plant an extra amount of seed to insure a good stand. There will then be plants to remove, and this can be done only with hoes. Drilled cotton will yield more than checked cotton. If it is drilled, the weeds in the row can be removed only with hoes.

Second and Subsequent Cultivations.—The purpose of the second and later cultivations is similar to that of the first. If the row has been barred off at the first cultivation, it is essential that soil be thrown back to the row at the second. It is well, too, for a small amount of soil to be thrown to the row at each subsequent plowing until the plants are large enough to shade the ground in the row. This tends to keep down grass and weeds and to supply a mulch between plants. The second cultivation may be deep, but later ones should be shallow. It is very necessary, however, that all cultivations be deep enough to destroy grass and weeds, as the elimination of these is more important than anything else. Cultivators with three to five medium-sized shovels are best for the second cultivation (Fig. 59). For later cultivations the same implements may be used, but the shovels should be replaced by small buzzard-wing sweeps (Fig. 60). A double shovel with two large sweeps or a two-horse cultivator with sweeps (Fig. 61) is satisfactory for later cultivations. In a few regions where intensive cultivation is not needed, as in some of the semiarid regions of Texas, two- or four-row cultivators are used. Small turning plows or stocks with one large shovel are used in later cultivations by many small cotton farmers. These may serve as makeshift tools where the land has not been well plowed in the preparation of seed bed or where there is but one small mule to tend the crop—a mule too light to pull a heavy plow. These implements are not economical, and their use violates good cultural practices.

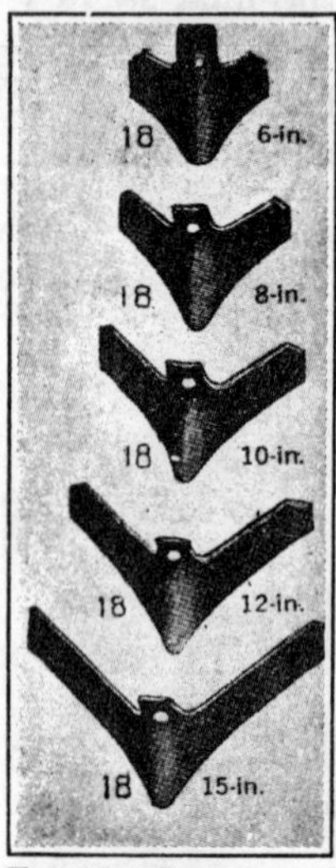

Fig. 60.—Sweeps used in cultivating cotton.

It is considered good practice to stir the soil after every rain as soon as the ground is dry enough for working or, in case rains have not fallen, about once a week. It often happens that continued wet weather prevents plowing for days or even weeks; consequently, the number of cultivations is lessened. The ground should not be plowed while wet. Cultivation should be continued until plants are mature or have put on a good crop of bolls. In regions where boll weevils are numerous, it is advantageous to cultivate late in order to keep plants growing and blooming. As long as there are enough flower buds for the weevils, they will not attack young bolls seriously. In the central part of the Cotton Belt, "laying by" takes place, or cultivations cease, about the first of August. Farther south, the date is earlier.

Depth of Cultivation.—It is not possible to say just how deep cotton should be cultivated, but the following rule should serve as a guide. Cultivate as shallow as is consistent with the securing of the main objects of tillage, that is, just deep enough to form a suitable soil mulch and to destroy weeds. The first cultivation may be deep, medium, or shallow without injury to the plant. If the seed bed has been well prepared before planting, however, deep cultivation is not needed, nor is it desirable. It is never desirable when it can be avoided. Plants get only a limited amount of nutrients from the upper few inches of the soil that is stirred. The deeper the soil is stirred the deeper the

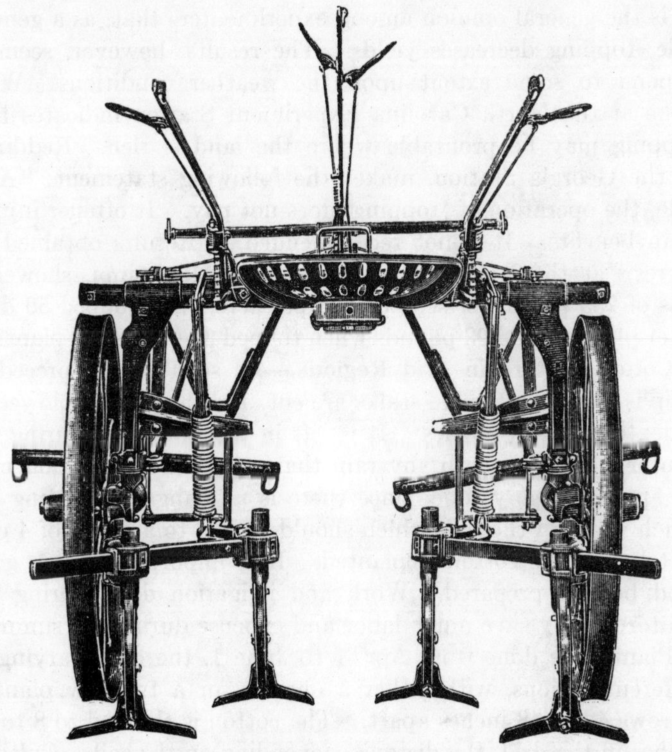

Fig. 61.—Planet, Jr., pivot-axle two-horse riding cultivator properly equipped with sweeps for shallow cultivation of cotton.

plant roots must go to obtain nutrient materials and moisture. It is necessary that cultivation be deep enough to destroy vegetation. Under ordinary conditions, 1½ to 2 inches is deep enough to meet the requirements of good tillage, though there are conditions that make deeper plowing necessary. If, owing to unfavorable conditions, weeds and grass have become too large for shallow cultivation to be effective, then it is necessary to plow deeper. Shallow cultivation, except when adverse conditions demand that it be deep, is to be recommended for cotton.

Topping Cotton.—Some farmers practice cutting off the tip of the main stem, or the tip of the main stem and of the main branches, a few weeks before bolls begin opening. The theory is that the topping checks vegetative growth and increases fruiting, but it is very doubtful if this belief is well grounded. It is the general opinion among experimenters that, as a general rule, topping decreases yields. The results, however, seem to depend to some extent upon the weather conditions. Work done at the North Carolina Experiment Station indicates that topping may be profitable where the land is rich. Redding,[3] of the Georgia Station, makes the following statement: "As a rule, the operation of 'topping' does not pay. It oftener injures than benefits. It is not recommended." Results obtained by Ayres,[4] at the Delta Branch Station of Mississippi, showed a loss of 233 pounds of seed cotton per acre when topped 50 days after planting and 93 pounds when topped 60 days after planting.

Cotton Culture in Arid Regions.—As soon as the preceding crop is removed, the old stalks are cut, and the ground plowed to a depth of 6 to 8 inches. It is left in this condition during the winter so as to absorb any rain that may fall. Irrigation may be applied at any time, since there is no danger of getting too much water in the soil, which should be wet to a depth of 4 or 5 feet before the cotton is planted. It is important that a good seed bed be prepared. Work and irrigation done during the winter usually save much labor and expense during the summer.

Planting is done from Apr. 1 to June 1, the date varying in different regions, with either a one-row or a two-row planter, in rows 42 to 48 inches apart. The cotton is thinned to 8 to 14 inches in the drill, the distance depending on the soil. Cultivation does not differ much from cultivation in other sections. It should begin early and continue late, the primary objects being to keep down weeds and conserve moisture. It is always necessary to stir the soil after each irrigation. All cultivations are shallow, since the soils tend to dry as deeply as stirred.

It is not possible to say just when or how cotton should be irrigated, since this is determined by soil and climatic conditions. Water should be turned on when the plants show the need for it. Sometimes it is necessary to irrigate when the plants are very small, whereas in other cases irrigation may be delayed for a time. Need of water is indicated by a wilting of the plants during the hot part of the day, especially when the wilting begins in the forenoon. The plants should not be allowed to suffer for water after blooming begins, for it is then that they need their maximum supply. Irrigation should be continued until most of the bolls are mature; if it is not continued, the quality of the lint may be poor. Much care should be taken with late irrigation, since too much moisture may cause the bolls to rot.

Cotton Culture in Foreign Countries.—Cotton in Egypt is grown principally under irrigation and is largely cultivated by hand. A primitive plow is used in many instances for breaking the land and for making crude ridges. The land is usually broken two or three times. After the ridges are formed with the plow, the hoe is used for breaking the clods and completing the ridges. They are made about a month before time of planting. Ten to twenty seeds are planted in each hill. The hills are not placed on top of the ridge but 3 inches down on the south side, so as to protect the seedling plant from the north wind. They are about 12 to 16 inches apart, and the rows 24 to 30 inches. The ground is dry when the seed is planted but is watered soon after. Planting is done in March or April. About 3 weeks after the cotton is planted it is hoed, the plants being thinned to one or two in a hill, the weeds destroyed, and the soil stirred. The fields are watered at this time, again a little later, and again about the end of May. Watering after May is governed by the supply, which is limited. The use of artificial fertilizer is not general.

Cotton culture in India, although it is now making rapid strides forward, is still very primitive. A large part of the work is done by hand. Cultural methods vary in different sections. In the Punjab, a large part of the cotton is grown under irrigation. Seed are sown by hand, either broadcast or in drills, between April and June. The land is plowed once before planting, and the plow is run through the cotton 2 or 3 months after it is planted. Picking begins about the first of September. Cultivation in the United Provinces is about the same as in the Punjab. In the Bombay Presidency, cotton culture is probably further advanced than in any other part of India. Most of the land is plowed. Seed are planted in lines with drills, and rotations are being followed to some extent. Planting dates range from June to September. Practically no commercial fertilizer is used.

Cotton culture in China is intensive, although very primitive, most of the work being by hand. The planters are men of small holdings with not enough capital to buy artificial fertilizer or the better implements, were they available. All natural manures are used. Two-handled wooden plows pointed with iron or steel are used for plowing the land. A rather peculiar system of rotations is generally followed. In May, just before the wheat is ready to harvest, cotton seed is planted among the wheat stalks and covered with a light layer of soil. The wheat is cut about the end of May, and the young cotton plants left standing. When they are about 6 inches high, the ground is hoed and the dirt pulled to their roots. The plants are left 7 to 8 inches apart. At intervals during the growing season, weeds are removed from among the plants by hand. The rainfall is ample, although not always well distributed.

South America seems to be a land of great possibilities as a cotton-producing country, but it is lacking in development. Cultural methods in most parts are of the most primitive type. There is very little seed-bed preparation, as the fields are usually not plowed. It is a common practice to burn off the grass and rubbish and then drop the seed in holes made with hoes. Cultivation is practically all with hoes. Planting is from the middle of December to June 1, the time varying with the section. There are few large plantations. Small fields are cultivated by the different farmers. Agricultural machinery, fertilizers, and scientific rotations are hardly known except in Peru, Argentina, and parts of Brazil.

References

1. CATES, H. R.: "Farm Practice in the Cultivation of Cotton," *U. S. Dept. Agr. Bull.* 511, pp. 1–62, 20 figs., 1917.
2. DUGGAR, J. F.: "Southern Field Crops," The Macmillan Company, New York, 1911.
3. REDDING, R. J. and KIMBROUGH, J. M.: "Cotton Culture," *Ga. Agr. Sta. Bull.* 75, 1906.
4. AYRES, W. E.: "Cotton Experiments," *Miss. Exp. Sta. Cir.* 42, 1921.
5. MORGAN, J. O.: "Field Crops for the Cotton Belt," The Macmillan Company, New York, 1920.
6. DUGGAR, J. F.: "Cotton Experiments," *Ala. Agr. Exp. Sta. Bull.* 107, 1899.

7. BROWN, H. B.: "Cotton Spacing," *Miss. Agr. Exp. Sta. Bull.* 212, 1923
8. BALLS, W. L. and HOLTON, F. S.: "Spacing Experiments with Egyptian Cotton," *Phil. Trans. Roy. Soc. London*, ser. B, vol. 206, pp. 103–180, 1912.
9. RICKS, J. R. *et al.*: "Cotton Experiments 1916," *Miss. Agr. Exp. Sta. Bull.* 178, 1916.
10. KING, C. J. and LOOMIS, H. F.: "Agricultural Investigations at the United States Field Station, Sacaton, Arizona, 1925–1930," *U. S. Dept. Agr. Cir.* 206, 1932.
11. COTTON, JOHN R. and BROWN, H. B.: "Cotton Spacing in Southern Louisiana in Relation to Certain Plant Characters," *La. Agr. Exp. Sta. Bull.* 246, 1934.
12. MCNAMARA, HOLMER C.: "Cotton-spacing Experiments at Greenville, Texas," *U. S. Dept. Agr.* Dept. *Bull.* 1473, 1927.

CHAPTER XIV

FUNGOUS, BACTERIAL, AND OTHER DISEASES OF COTTON

The damage to cotton plants caused by fungous and bacterial diseases results in an annual loss of millions of dollars in the value of the cotton crop. In the Cotton Belt of the United States, the annual loss is from 20 to 30 millions, or about 4 per cent of the value of the crop. The damage varies greatly with the locality and the season, usually being much greater during years of heavy rainfall. In certain areas, the damage is so great that cotton cannot be grown profitably under common methods of procedure.

The cotton plant is a native of tropical and subtropical regions and is naturally adapted to a hot climate, but most of the commercial crop is grown in temperate regions with hot summer weather. The young plants are very tender and easily injured by cold weather. A temperature a few degrees below normal seems to lessen their resistance to diseases, making them much more susceptible to injury by fungi or bacteria. Plants under cultivation are usually much larger but tender and more susceptible to disease than their wild relatives. Diseases, also, have a much better chance to spread in cultivated fields, mainly because of the large number of plants growing near together.

The widespread and continued movements of modern trade and commerce serve to scatter cotton diseases as well as many others, unless careful preventive measures are used. Cotton seeds are being moved freely from locality to locality, from state to state, and even from one country to another.

A certain amount of disease is to be found in any cotton field at any time, but in most instances diseases are held in check by natural conditions. Wilt, anthracnose, angular leaf spot, and root knot are the most troublesome.

Fusarium Wilt (*F. vasinfectum* Atk.).—Cotton wilt, a disease known also under the names of "black heart" and "frenching," has long been known by cotton growers. In the United States, it is to be met with in all the southern and eastern states and is most prevalent in eastern South Carolina, southwestern Georgia, and southeastern Alabama (Fig. 62). New areas are being invaded each year. Many foreign countries also record the disease.

Cotton on sandy or sandy-loam soils is most commonly affected, but the disease is not unknown on clay and other soils. Probably all sandy soils of the Cotton Belt will be infected even-

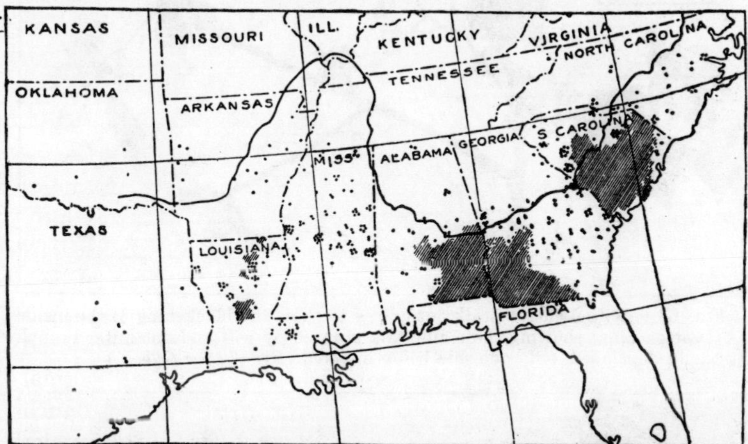

Fig. 62.—Map of the southeastern states, showing the geographical distribution of cotton wilt in 1921. The shaded portions show where the wilt is general, while the dots indicate scattered localities where the disease is known to occur. (*After Gilbert.*)

tually. Gilbert[1] estimates that the annual loss in the United States due to cotton wilt is $10,000,000. It is a difficult matter to estimate the exact damage, for many plants infected with wilt show no external evidence, but their yield is necessarily reduced more or less. Other plants are stunted, and their yield is reduced considerably, but they do not die outright.

The first external evidence of the wilt disease is usually a yellowing of the edges of the lower leaves and the area midway between the main ribs of the leaf. This yellowing spreads widely in the leaf, and the part first discolored soon turns brown and becomes shriveled and somewhat curled. The shriveled leaves later fall, leaving stem and branches nearly bare. Often the disease does not cause a shedding of leaves but does cause a stunting or dwarfing of the main stem of the plant. One or

Fig. 63.—Wilted cotton stalk cut across to show the blackening of the inside (*A*); cotton plant showing the main stalk stunted by wilt and one lower branch normally developed, only one side being infected (*B*). (*After Gilbert.*)

Fig. 64.—Cotton wilt; wilted plant on right, resistant plant on left. (***After W. A. Orton, Bull.* 333, *Bur. Plant Industry, U. S. Department of Agriculture.***)

more of the lateral branches outgrows the main stem (Fig. 63). In some cases, all the leaves on an apparently healthy plant suddenly wilt, and many fall (Fig. 64). If the interior of the stem, root, or even the petioles or leaf blades (in later stages) of a wilt-diseased plant is examined, the woody parts are seen to be brown or darkened (Fig. 63*A*).

Atkinson[2] observed cotton seedlings growing on the black-belt soils of Alabama with indications of wilt when but a few days old. Ordinarily, the disease does not do much damage until the

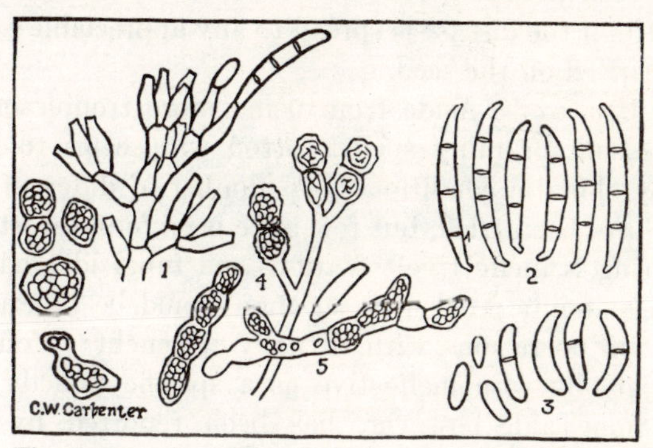

FIG. 65.—The cotton-wilt fungus (*Fusarium vasinfectum* Atk.); (1) conidiophore; (2) macroconidia; (3) microconidia; (4) chlamydospores; (5) bits of swollen hyphæ. (*After Gilbert.*)

plants are about a foot high. In cases of light infection, the plants in many instances attain full size.

The Causal Organism.—The cotton-wilt disease was first identified and the causal fungus described by Prof. G. F. Atkinson,[3] in Alabama, in 1891. The fungus, *Fusarium vasinfectum* Atk., has hyphae 2 to 4μ in diameter and yellowish at maturity; conidia are born singly; microconidia, oval, continuous; macroconidia, falcate, two to three septate, 1 to 2 by 2 to 4μ. Ascigerous stage has not been identified. Chlamydospores are also formed (Fig. 65). Fungus lives saprophytically in soil once infected and enters cotton plants through the roots. Inside the plants the hyphae grow freely throughout the vessels and water-carrying ducts, apparently forming a poisonous chemical which causes the wilting of the plant.

The organism causing okra wilt was once considered identical with the organism just described, but it is now thought to be different.

After soil is once infected, the wilt-producing organism may live in it saprophytically for years and spread to some extent by natural growth. If susceptible varieties of cotton are grown on the soil, spread by natural growth is more rapid. There are numerous ways whereby uninfected soil may become infected if near soil contains the wilt organism, for instance by the transference of soil from one field to another on plows and other implements and on the feet of men and animals, by water during floods, and by infected manure that is hauled out and spread on the farm. It is doubtful if the disease is spread to any appreciable extent by infection carried on the seed.

Control Measures.—Aside from being more troublesome during wet seasons or rainy spells, cotton wilt seems to be little affected by climatic conditions. A number of different control measures have been tried, but few have been found worth while. Late planting is ineffective. Various soil fungicides have been applied in amounts much heavier than would be practicable to use in farming operations, with virtually no benefit. Commercial fertilizers are likewise ineffective as a specific remedy for the disease. Some aid, however, has been reported by Young, Janssen, and Ware,[25] in Arkansas; and by Neal,[26] in Mississippi, from the use of a liberal application of a complete fertilizer and from fertilizers containing potash.

Stable manure has been found to be of some help on land that is not heavily infected. The manure probably does not affect the fungus but does stimulate the plants to better growth and thus increases their resistance. Fulton[4] reported a loss twice as great from unmanured plots as from plots that received 30 loads of manure per acre. The disease also spreads more rapidly in the unfertilized plots. Plowing under cowpeas and other leguminous crops had a similar, though less marked, beneficial effect. The writer has secured considerable benefit from the use of nitrate of soda or barn manure on wilt-infected soil at the Mississippi Experiment Station, but the wilt there was not severe.

The burning of diseased plants has been recommended by some as a control measure. This is hardly worth while except to retard the spread of the disease when it is just starting in a new area. Robbing the soil of the much needed organic matter that would come from the decay of the cotton stalks is a weighty reason for not burning them.

Since the wilt fungus can live in the soil saprophytically for a number of years, it is a difficult matter to eradicate it by crop rotation. Short rotations do no good. Keeping cotton off the land 8 or 10 years has been found to lessen the soil infection but does not eradicate the organism altogether. Since root knot caused by nematodes commonly occurs with wilt and increases its severity, rotations may be helpful. Nematodes may be so reduced that susceptible crops can then be grown profitably by growing, for 2 or 3 years, crops immune to nematode attack.

For wilty soil infected with nematodes, Cauthen[5] recommends the following rotation for Alabama:

First Year.—Plant corn and between the corn rows or hills plant Iron or Brabham cowpeas. Where early autumn pasture for cattle is desired, velvet beans may be planted with the corn and grazed while green and in time to sow a fall grain crop.

Second Year.—Plant oats; after the grain is cut for hay or seed, plant the stubble in Iron or Brabham cowpeas for hay or seed. Follow this with some winter grain for a cover crop.

Third Year.—In the spring, plow under the cover crop and plant some wilt-resistant variety of cotton.

Wilt-resistant Varieties.—For fields heavily infected with cotton wilt, the only control measure of much value is the use of wilt-resistant varieties. Different varieties, and even plants of the same variety, differ greatly in their resistance. Some are nearly immune, while others are very susceptible. By selecting, for a series of years, immune plants in fields where the disease was prevalent, it has been possible to develop certain valuable wilt-resistant varieties (Figs. 66 and 67).

The first wilt-resistant variety developed was Dillon. This came from some selections made in a badly wilt-infected field of Jackson Limbless cotton growing near Dillon, S. C., in 1900. Selections were made by a representative of the U. S. Department of Agriculture. The strain isolated proved to be resistant to wilt and continued to be one of the most resistant of varieties; but it was objectionable on account of its cluster habit, late maturity, and bad picking qualities.

On rich soil, plants belonging to the Dillon variety grew tall, with two or three long vegetative branches originating near the base; fruit limbs were very short; bolls in clusters, small, slender, somewhat pointed, and difficult to pick; seed, small and fuzzy; lint, 7/8 inch; lint percentage, 37 to 38.

A number of other wilt-resistant varieties have been developed, Dixie being the most prominent in the early history of the work. The original Dixie strain is not grown to any extent at present, but there are a number of selections from it and from hybrids of this variety in use, and there are also several other good wilt-resistant varieties. Among the other varieties may be mentioned Dixie-14, Dixie Triumph, Tri-Cook, Cook-307-6, Super Seven, Rhyne Cook, Covington-Toole, Council Toole, Lewis-63, Watson's Long Staple, and Clevewilt.

FIG. 66.—Typical plant of Dixie wilt-resistant cotton. (*After Gilbert.*)

There are several other varieties that possess a moderate degree of resistance or are wilt-tolerant. In this class may be mentioned Arkansas Rowden-40, Cleveland-54, Miller-642, Express-121, D. & PL.-4-8, and Lightning Express.

Descriptions of the leading wilt-resistant varieties may be found in Chap. III.

General Characteristics of Wilt-resistant Varieties.—Most of the wilt-resistant varieties that have been developed belong to the Dixie or Peterkin type, with tall, vigorous-growing plants. The resistant Cook selections are marked exceptions to the general rule.

The extra-early varieties like King, Simpkins, and Trice are very susceptible to wilt. Cleveland and certain strains of Cook

FIG. 67.—Resistant and susceptible varieties of cotton growing on wilt-infested soil. Note the vigorous growth and healthy appearance of the resistant varieties at the right and left as compared with the scant growth and sickly appearance of the two susceptible varieties in the center. (*After Neal and Gilbert.*)

are more or less wilt-resistant, but several other of the big-boll varieties, as, for instance, Lone Star and Russell, are very susceptible. Most Sea Island varieties are susceptible, but Egyptian varieties, when grown in the United States, appear to possess some resistance.

No variety of cotton is resistant to wilt if the soil is infested with nematodes, which cause the root-knot disease of cotton. The wilt fungus enters wounds in the roots made by the nematode worms. Before cotton can be grown successfully on such land, it is necessary to eradicate the worms or reduce the nema-

TABLE XXIX.—STUDY OF WILT RESISTANCE
(*Mississippi Experiment Station*, 1917)

Variety	Pounds of seed cotton per acre	Per cent lint	Pounds of lint per acre	Character of staple	Length of staple, in inches	Value of lint per pound, in cents	Value of lint	Value of seed at $70 a ton	Total money value per acre	Per cent of uninfected plants	Per cent of plants infected and slightly damaged	Per cent of plants killed or badly injured	Rank in money value
Trice[a]	612	29.9	182.9	Good	1 1/16 full	33½	$ 61.27	$15.05	$ 76.32	3.2	16.5	80.3	11
Improved Dixie	837	30.2	252.7	Good	1	31	78.33	20.47	98.85	55.7	15.4	28.9	8
Wannamaker-Cleveland[b]	906	35.6	322.5	Good	1 full	31½	101.58	20.44	122.02	37.3	30.8	31.9	3
Tri-Cook	1,030	33.8	348.1	Good	1	31	107.91	23.87	131.78	67.1	24.6	12.3	1
Cook-307-6	767	33.6	257.7	Fair	⅞	30	77.31	17.85	95.16	63.2	31.9	4.9	9
Express-350[a]	782	28	218.9	Good	1 3/16	39	85.37	19.79	105.16	26.7	20.0	53.3	6
Dix-Afifi	767	29.9	229.3	Fair	1 3/16	38½	88.28	18.83	107.11	79.2	13.3	7.5	5
Lewis-63	1,030	32.9	338.8	Fair	1	30½	103.33	24.22	127.55	88.7	4.6	6.7	2
Covington-Toole	713	34.3	244.5	Fair	⅞	30	73.35	16.41	89.76	81.5	8.8	9.7	10
Council-Toole	844	29.9	252.3	Good	1	31	78.21	20.72	98.93	77.6	12.5	9.9	7
De Soto	937	29.9	280.1	Fair	1	30½	85.43	22.99	108.42	76.8	12.3	10.9	4

[a] Susceptible. [b] Somewhat resistant.

tode infestation by growing for 2 or more years other crops that are immune or resistant to the nematodes.

As has been shown, different varieties of cotton differ considerably in their ability to resist the wilt disease. Brown and Ames[6] found by cutting into the stems and roots of cotton plants late in the season that a large percentage of the plants of all varieties, resistant as well as nonresistant, were infected (Table XXIX). Although infected, the resistant varieties were able, in most cases, to withstand the attack and gave no external evidence of disease.

Cauthen,[5] in his experiments with wilt-resistant varieties of cotton grown at a number of different places in Alabama, had from none to 82.5 per cent of his plants die, the percentage depending on the variety and the location (Table XXX).

TABLE XXX.—PERCENTAGE OF PLANTS DEAD OR NEARLY DEAD UP TO THE LAST COUNT OF THE SEASON

Location in Alabama	Name of varieties tested										
	Year	Dillon	Modella	Cook (check)	Wood	Covington-Toole	Tri-Cook	Cook-307-6	Cook (check)	Dixie	Dix-Affi
Loachapoka	1911	1.4	2.9	22.1	10.5	5.2	6.0	0.7	15.1	3.7	
Loachapoka	1912	3.0	28.0	35.2	17.2	25.1	12.9	12.4	39.4	4.9	
Brundidge	1912	3.7	10.2	68.2	11.6	10.8	6.6	10.2	33.7	3.2	
Liverpool	1913	7.4	29.2	68.0	27.5	9.6	19.0	28.4			
Notasulga	1913	3.5	21.3	43.9	19.5	7.4	12.0	14.3	....	18.7	
Tuskegee	1913	3.0	6.2	43.5	7.7	3.5	5.2	5.1	24.5	2.9	
Brundidge	1913	15.5	35.4	56.3	20.1	34.8	9.5	10.1	33.2	20.0	
Auburn	1914	....	7.0	21.6	14.6	4.6	0.9	6.1	3.0	2.7	
Notasulga	1914	....	13.8	42.4	14.4	4.9	2.4	1.2	46.5	14.4	
Liverpool	1914	....	1.6	8.0	1.4	1.1	0.6	0.0	4.9	7.7	0.0
Brundidge	1914	....	34.3	82.5	21.8	8.6	12.5	10.2	71.2	26.1	
Lowndesboro	1914	....	2.9	53.3	11.1	7.6	3.5	5.4	15.1	6.8	0.0
Notasulga	1914	....	10.1	45.2	21.5	16.8	7.0	10.4	26.6	4.5	
Notasulga	1915	....	14.8	3.0	8.3	6.3	10.1	1.8	46.5	1.2	
Notasulga	1915	....	9.8	18.9	13.2	15.7	3.7	16.3	18.9	7.1	
Greenville	1915	....	8.3	33.1	20.5	5.4	4.8	13.5	29.9	3.9	
Average percentage of loss	....	5.4	14.7	40.3	15.1	10.5	7.3	9.3	29.2	8.5	

Lewis and McLendon[8] report from 5 to 96 per cent healthy, resistant plants in their variety tests at Americus, Ga. These figures were probably based on the condition of plants as determined from external appearance (Table XXXI).

TABLE XXXI.—VARIETY TESTS, AMERICUS, GA.

Variety	1914				1915				1916					
	Pounds seed cotton per acre	Per cent lint	Per cent healthy stalks	Rank in yield lint	Pounds seed cotton per acre	Per cent lint	Per cent healthy stalks	Rank in yield lint	Pounds seed cotton per acre	Per cent lint	Pounds lint per acre	Per cent healthy stalks	Value per acre	Rank in value
Improved Dixie	1,326	37	96	2	2,030	36	96	2	1,080	35	378	80	$ 96.66	3
Lewis-63	1,326	37.5	96	1	2,100	36.5	96	1	891	36	321	80	81.30	6
De Soto	1,309	35	95	3	1,960	35	94	5	891	35	312	81	79.77	7
Council-Toole	.....	....	..	..	1,925	37	95	3	945	37	350	75	87.85	5
Modella	.....	....	..	..	1,925	36	94	4	837	36	301	74	76.28	8
King	374	36	10	6	665	35.5	20	12	324	36	117	10	29.61	16
Bank Account	.....	....	..	..	.....	....	..	..	216	34	73	8	18.86	17
Simpkins	.....	....	..	..	.....	....	..	..	324	35	113	10	28.93	15
Triumph	.....	....	..	..	.....	....	..	..	486	36	175	50	44.33	13
Half and Half	612	40	18	5	350	39	10	13	567	40	227	43	55.60	12
Cook	680	38.5	15	4	1,050	38	83	11	915	39	357	70	88.14	4
Cleveland	.....	....	..	..	.....	....	..	..	378	38	144	40	35.82	14
Dix-Afifi	.....	....	..	..	1,925	31	95	9	1,215	32	389	85	125.92	1
Express	.....	....	..	..	1,750	30	85	10	756	31	234	70	71.82	9
Curtis Long-staple	.....	....	..	..	.....	....	..	..	720	30	211	70	69.59	10
Webber	374	32.5	10	7	840	30	24	14	162	32	52	5	16.72	18
Toole	.....	....	..	..	1,820	36.5	80	6	675	38	257	30	63.94	11
Pulnott	.....	....	..	..	1,750	37	80	7	1,215	36	437	85	110.55	2
Cook Alabama-307	.....	....	..	..	1,645	37.5	90	8						

Yields of Wilt-resistant Varieties.—Under favorable conditions, wilt-resistant varieties yield comparatively well, a bale to the acre being not uncommon on productive land. On land badly infected with wilt, they yield considerably more than nonresistant varieties, provided they are not hindered by boll weevils and are

TABLE XXXII.—STUDY OF WILT RESISTANCE, MARIANNA, ARK., 1933

Variety name and strain number	Total plants wilted, per cent	Earliness,[a] per cent	Lint, per cent	Lint yield, pounds	Staple length, thirty-seconds	Bolls to the pound[b]	Acre value of seed and lint	Time required to pick a bale, days
Ck-Trice-304	32.3	85.3	33.0	249	32	89.3	$28.19	9.00
Missdel (wilt-resistant)	25.2	88.7	32.6	246	35	72.5	29.65	7.41
Missdel-3	48.1	84.6	32.2	162	36	84.7	20.17	8.77
Missdel-4	51.8	92.4	33.9	190	34	91.7	22.50	9.00
Delfos 668	26.5	89.6	31.2	290	35	82.0	35.20	8.77
Ck	28.9	92.6	32.7	220	33	90.9	25.97	9.26
Dixie-Triumph	3.8	81.5	33.9	332	32	80.6	37.22	7.94
Dixie-Triumph-6	10.7	84.4	35.4	277	33	76.9	32.10	7.25
Clevewilt-3079	9.6	86.4	34.5	234	34	84.7	27.53	8.13
Clevewilt	12.1	86.9	35.9	279	33	85.5	31.96	7.94
Ck	26.4	91.2	33.3	211	32	90.1	23.72	9.00
Dixie-14	4.1	81.2	36.2	322	32	78.7	35.39	7.25
Half and Half	64.9	83.1	42.9	193	25	79.4	19.27	6.17
Wilson Type Big Boll	22.6	87.9	33.0	253	31	82.6	27.74	8.13
D. P. and L.-11	33.4	91.2	41.1	293	34	88.5	33.57	7.25
Ck	32.4	91.7	33.0	199	32	94.3	22.44	9.52
Dixie-14-1	8.3	86.9	34.9	262	32	79.4	29.37	7.58
Arkansas Lone Star-2124	14.5	92.8	35.1	245	34	76.3	28.55	7.25
Arkansas Rowden-2088	7.8	90.6	34.0	215	32	65.4	23.90	6.41
Stoneville-3	35.3	94.3	37.4	255	32	93.5	27.99	8.33
Ck	27.6	91.3	33.9	225	32	91.7	25.43	9.00
Arkansas Rowden-40	9.0	86.5	34.3	239	33	71.4	27.86	6.94
Stoneville-5	33.9	93.6	37.8	251	33	90.1	28.91	7.94
Arkansas Rowden-1049	11.4	88.3	34.2	221	32	64.1	24.86	6.29
Qualla	55.2	67.6	38.8	166	32	62.5	18.11	5.38
Ck	25.3	82.9	33.6	233	32	90.9	26.20	9.00

[a] Based on perentage of total crop harvested at first picking.

[b] Seed cotton.

allowed sufficient fruiting period to produce a full crop. Ware and Young,[27] in Table XXXII, give the comparative values and other characters of wilt-resistant and nonresistant varieties grown on artificially inoculated soil at Marianna, Ark., in 1933.

Lewis and McLendon[8] find that, as a rule, resistant varieties yield better than nonresistant when grown on infected Georgia soils, in the absence of boll weevils (Table XXXI).

TABLE XXXIII.—STUDY OF WILT RESISTANCE MISSISSIPPI EXPERIMENT STATION, 1919

Variety	Pounds of seed cotton per plot	Pounds of seed cotton per acre	Lint, per cent	Pounds of lint per acre	Length of lint, inches	Value of lint per pound, in cents	Value of seed at $75 a ton	Value of lint	Total money value	Rank in money value
Tri-Cook	22.50	703.1	34.3	241.2	$\frac{15}{16}$	41½	17.32	100.10	117.42	7
Cook-1110	20.75	648.4	39.4	255.5	⅞	41	14.73	104.75	119.48	5
Wannamaker-Cleveland	20.75	648.4	34.6	224.3	⅞	41	15.90	91.96	107.86	10
Council-Toole	26.50	828.1	31.7	262.5	$\frac{15}{16}$	41½	21.21	108.94	130.15	2
Express-432	22.00	687.5	31.0	213.1	1 ⅛	50	17.79	106.55	124.34	3
Dix-Afifi	19.75	617.2	26.0	160.5	1 $\frac{3}{16}$	60	17.13	96.30	113.43	8
Express-350-64	27.25	851.5	26.9	229.0	1 ⅛[a]	55	23.34	125.95	149.29	1
Cleveland × Covington-Toole	25.00	781.2	30.2	235.9	$\frac{15}{16}$	41½	20.45	97.90	118.35	6
Lewis-63	25.75	804.7	29.6	238.2	$\frac{15}{16}$	41½	21.24	98.85	120.09	4
Improved Dixie	12.75	398.4	34.1	135.8	⅞	41	9.85	55.68	65.63	13
Express × Covington-Toole	21.75	679.7	30.6	208.0	1 $\frac{1}{16}$	45	17.69	93.60	111.29	9
Covington-Toole	16.00	500.0	32.4	162.0	$\frac{15}{16}$	41½	12.68	67.23	79.91	12
Trice (Arkansas Station)	22.75	710.9	28.4	201.9	$\frac{15}{16}$	41½	19.09	83.79	102.88	11

[a] Full.

TABLE XXXIIIA.—AVERAGE RESULTS OF WILT-RESISTANT COTTON VARIETY TESTS IN ALABAMA, 1934–1936

Variety	Yield, pounds lint cotton per acre, average 3 tests, 3 years (1934–1936)						
	Auburn	Prattville	Aliceville	Gastonburg	Monroeville	Brewton	Headland
Cook-307 (C. Rhyne)	451	493	472	292	482	417	591
Cook-307 (B. Rhyne)	448	481	459	270	471	403	552
Cook-307-92 (Smith)	467	474	441	307	457	395	520
Cook-1006	468	463	458	331	454	352	510
Clevewilt (C. Rhyne)	471	459	451	276	456	379	537
Cook Wiregrass	440	466	427	272	428	375	592
Cook-27-54	412	476	445	271	456	375	551
Local Variety	442	437	451	262	448	361	545
Cook-144 (Smith)	468	477	459	310	435	297	470
Clevewilt (Coker)	452	435	429	246	441	344	559
Dixie Triumph (Marrett)	433	465	409	282	447	333	470
Dixie Triumph (Wann.)	431	433	415	281	429	286	482
Delpress-3	402	418	386	286	375	240	409

Brown and Ames,[6] in Table XXIX, give comparative yields of resistant and nonresistant varieties grown in east-central Mississippi on soil with moderate wilt infection and medium-heavy weevil infestation. The wilt-resistant varieties led in yields under conditions prevailing.

In Table XXXIII, Brown and Anders[7] show yields under heavy boll-weevil infestation. The soil was the same as that used in the previous experiment (Brown and Ames[6]). In the latter experiment, an early variety, Express-350-64, made a better yield than any of the wilt-resistant varieties. Similar results were secured other years when weevil infestation was heavy.

An earlier wilt-resistant variety of cotton is needed. The ones grown at the present time are too late to be profitable if weevil infestation is heavy.

In Table XXXIII*A*, Tisdale, of the Alabama Experiment Station, gives comparative-yield data on wilt-resistant varieties grown in Alabama.

Verticillium Wilt (*V. albo-atrum* Reinke and Berth.).—The verticillium wilt of cotton is similar to the fusarium wilt, discussed above, in its general effect on the plants, but it is not so prevalent or so harmful. It causes a mottling, turning brown, wilting, and shedding of leaves and a discoloration of wood of stem and roots like the fusarium wilt, but the plants are not dwarfed and are rarely killed. The only sure way of distinguishing the two diseases, however, is to grow the causal organism in pure culture in a laboratory.

Verticillium wilt of cotton was first found in the United States in California in 1927. It has since been found in Arkansas, Mississippi, Oklahoma, Tennessee, and Virginia.

No control measures are known.

Root Knot of Cotton (*Heterodera radicicola* (Greef), Muell).—Root knot is caused by a nematode, a small eel-like worm which enters the roots of cotton plants, multiplies greatly, and causes swellings, or galls; later, the nematodes go into a resting condition in a thick-walled cyst. This disease is to be found in most areas where cotton wilt prevails, being prevalent on the sandy coastal plain from Virginia to Texas. Although root knot may be met with in different soils, it is rarely found in any but light, sandy soils. Root-knot troubles with certain crops have been reported from almost all countries of the world. Probably few cotton-growing regions are free from the disease.

Root knot is second in importance to wilt in damage done to the cotton crop. Gilbert estimates the average annual loss to the Cotton Belt in the United States to be around 200,000 bales of cotton and 100,000 tons of seed. A further loss is attributable

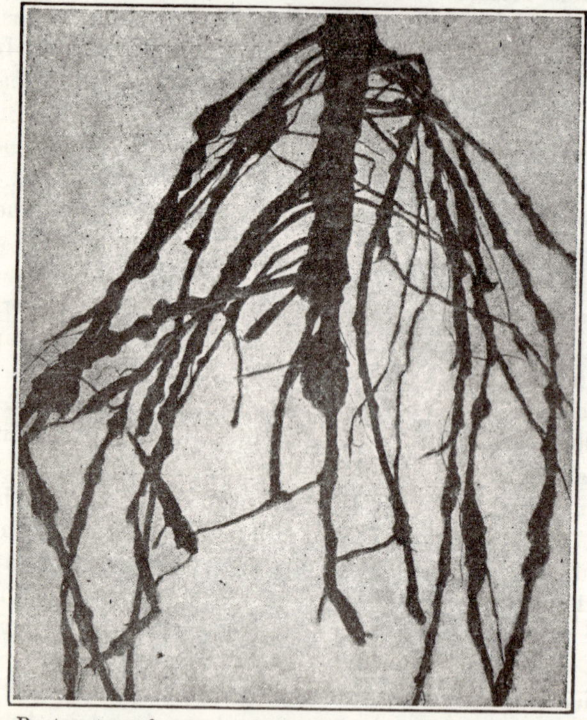

FIG. 68.—Root system of a cotton plant severely attacked by root knot. (*After Gilbert.*)

to the disease, as cotton plants are more susceptible to wilt if infected with nematodes.

Symptoms of Root Knot.—This disease can be identified most easily by examining the roots. If they show numerous more or less irregular knots or swellings ranging from the size of a pinhead to ½ inch in diameter, root knot may be suspected (Fig. 68). If the interior of the knots is examined, small pearly-white, rounded bodies about the size of a mustard seed may be seen. These are the female nematodes (Fig. 69*A*). A microscopic examination is necessary to show the presence of the male worms. They are very slender, narrowed toward either extremity, and about 1 to 1½ millimeters in length (Fig. 69*B*).

Nematode-infected plants usually appear dwarfed or stunted and are a pale, yellowish-green color. During hot, dry weather, or even when there is sufficient moisture in the soil, infected plants tend to wilt more than normal ones do. If all the plants in the field are diseased, their abnormal appearance, as described above, may not be observed. Mild attacks are often unnoticed,

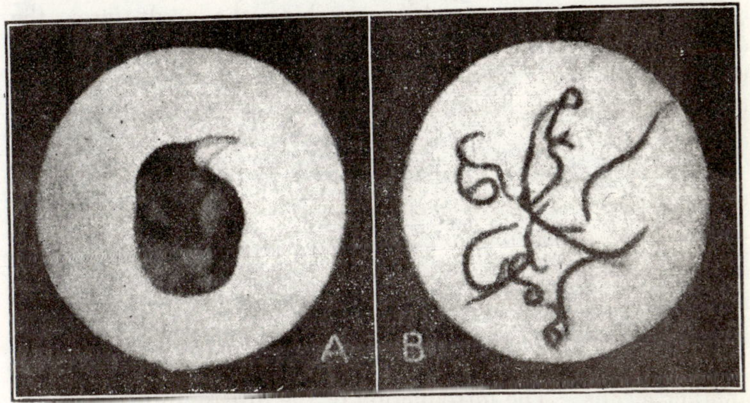

Fig. 69.—Adult female nematode full of eggs (*A*); young male and female nematodes (*B*) (both illustrations greatly enlarged). (*After Gilbert.*)

but severe attacks may kill practically all the plants in the field.

Life History of the Causal Organism.—Hatching from an ellipsoidal egg, 67 to 128 by 30 to 52μ, according to Bessey,[16] is a slender, cylindrical larva 375 to 500μ in length and about 12 to 15μ in diameter. The larva moves through the soil with considerable activity and, although making no growth, may remain alive there for several months. On finding a suitable root, it seeks the growing point and enters, boring its way in by the aid of a spearlike tongue. Once inside, the young nematode ceases its activity and begins to enlarge and to absorb nourishment from the root. The presence of the worm in or among the cells of the root of the plant stimulates them to active multiplication, and the abnormal growth known as a "gall," or "root knot," is produced. It requires about 4 weeks for the female to reach full development and begin laying eggs. It is then flask-shaped, 400 to 1,300μ in length, and 270 to 500μ in diameter

(Fig. 69). Five hundred or more eggs are produced by one female. The larvae produced by these drill their way out of the eggs, in some cases even before the eggs leave the mother's body.

From external appearance, the larvae that produce males cannot be distinguished from the ones that produce females when they enter the plant roots. Inside the root they enlarge a little; later there develops inside the old skin a slender, much elongated worm having the same general shape of the larva. This is the adult male (Fig. 69). It is 1,200 to 1,500μ in length and 30 to 36μ in thickness. After reaching the adult form, the elongated male breaks out of the old skin and moves through the tissues of the plant root in search of a female. The adult male probably takes but little food and after meeting a female soon dies.

Control Measures.—Uninfected soil may become infected in various ways. Through their own activity, nematodes travel through the soil a few feet in the course of a year. Bessey[16] estimates that the movement may be as much as 6 feet. The transference of infected soil from one field to another by the feet of animals or men, on plows, or on other farming implements serves to spread the disease. The flow of drainage water is another means. The introduction of the disease into new localities is very often brought about by the use of infected nursery stock, cabbage, tomato, strawberry plants, etc.

Root-knot nematodes have been found attacking more than 400 different species of plants. This list includes many of our leading cultivated plants as well as several species of weeds. On the other hand, there are several important cultivated crops that are not attacked by nematodes. This makes it possible to control them fairly easily by planning rotations in which resistant crops are grown 2 years in succession in a 3- or 4-year rotation. Corn, wheat, rye, oats, sorghum, most grasses, velvet beans, Brabham cowpeas, Iron cowpeas, Laredo soybeans, and peanuts are immune or nearly so and may be used in rotation with cotton.

Various chemicals, such as sulphur, formaldehyde, carbon bisulphide, calcium carbide, potassium sulphocarbonate, and ammonium sulphate, have been applied to the soil in the effort to eradicate nematodes. Some of these were found to be effective if applied in sufficient quantity but were not of practical value for field use. Chemical fertilizers that will stimulate plants to increased growth and vigor may be helpful as a control measure. The plants may be able to make satisfactory growth in spite of the influence of the parasites.

Cotton Anthracnose. [*Glomerella gossypii* (South.) Edg. (*Colletotrichum gossypii* South.)].—Cotton anthracnose, or pink boll rot, has probably been present in cotton fields for many years. The disease was first described by Miss E. A. Southworth,[9] of the U. S. Department of Agriculture, in 1890, and the causal organism named *Colletotrichum gossypii.* Atkinson,[10] of Alabama, working independently, published a paper in 1891 in which he verified Miss Southworth's work. Edgerton,[11] of Louisiana, discovered what he interpreted to be the perfect stage in 1909 and proposed the name *Glomerella gossypii* (South.) Edgerton. Edgerton's discovery has not been verified except that Shear of the U. S. Department of Agriculture obtained the perfect stage of the fungus in pure culture. Barre,[12] of South Carolina, has contributed much to our knowledge of the disease, especially of the growth of the causal organism in cotton bolls and seed and of control measures.

Cotton anthracnose is probably widespread in most of the cotton-growing regions of the world. Edgerton[13] mentions it as being reported from Trinidad, British Guiana, West Indies, and possibly India and Natal. In the United States, it has been reported in most of the cotton-growing states.

The annual loss due to cotton anthracnose amounts to millions of dollars. It causes damage by killing young seedling plants before they get aboveground or soon after, thus making stands poor. It also causes many bolls to rot, ruining them altogether; others are infected, and the lint is damaged but not destroyed. The damage to bolls is considerably greater during years when there is much rain, especially during late summer and fall; it is greater in more humid regions, as in countries near the seacoast and in the Mississippi Delta; and on very rich land where plants become large. The damage is worse on varieties with large leaves. The heavy foliage shades the lower parts of the plants more completely, thus making conditions favorable for fungous growth.

The amount of damage varies greatly in different years, in different localities, and with the condition of the seed planted.

Estimates made by different observers vary all the way from 6 to 60 per cent. Under certain conditions, 25 to 50 per cent of the bolls may be rotted or less seriously damaged, but probably, on the average, the loss is not more than 1 per cent of the value of the crop.

Symptoms of the Disease.—Anthracnose attacks all parts of cotton plants except the roots of large plants. In case the seeds planted are infected, the fungus may attack the sprouts of the germinating seed and prevent further growth (Fig. 70). The failure of a large percentage of the cotton seeds that are planted to produce seedlings is doubtless to be attributed largely

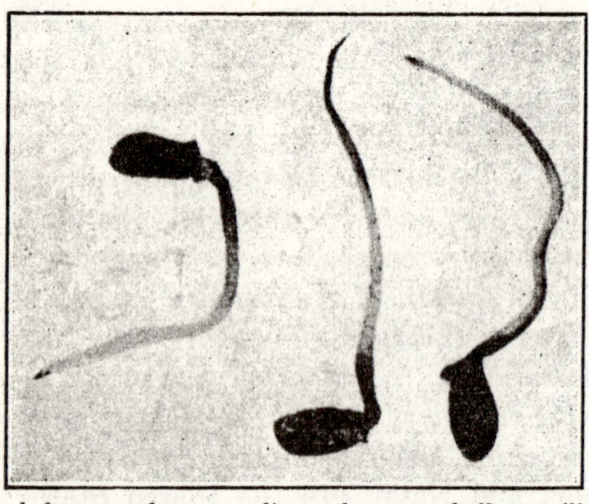

Fig. 70.—Seed from anthracnose-diseased cotton bolls sterilized externally and germinated in sterile Petri dishes. The fungus grew out from underneath the seed coats and attacked the sprouts in the black areas. (*After Gilbert.*)

to this disease. During cool weather, young seedlings only a few inches high are often attacked and injured or destroyed. The stem may be attacked either below or a short distance above the surface of the ground, the injury causing a "damping-off." It is rather difficult to distinguish this trouble from a damping-off caused by other organisms. Anthracnose, according to Atkinson,[2] causes tissue to redden somewhat and shrink in longitudinal lines; while the "sore-shin" attacks caused by Rhizoctonia produce well-defined ulcers, or diseased depressions of a brownish color, on the stems of seedlings just below the surface of the soil. These injuries are frequently outgrown unless the disease has penetrated the stems too deeply. On the cotyledons of young seedlings, anthracnose produces irregular diseased areas around the edges.

Anthracnose sometimes attacks the stem of plants that are mature or nearly so. The infected areas are at first dull red, then black, and finally become pink if spores are produced. Stem infection usually occurs in places where plants have been injured or in leaf scars, the fungus living saprophytically in part. Stem infection sometimes causes the breaking down of branches and the dropping of bolls. As a rule, injury to stems is not serious on upland cottons. The stem lesions. however may serve as a source of boll infection.

FIG. 71.—Anthracnose of cotton. (*After R. W. Harned.*)

The anthracnose fungus is, according to Atkinson,[2] frequently found on leaves of mature cotton plants, being more liable to develop on sickly or injured leaves. Converse[14] found that 90 per cent of the nectar glands of leaves of diseased plants contain the cotton-anthracnose organism. This is probably an important source of infection for bolls.

Anthracnose disease is most characteristic on cotton bolls of mature or nearly mature size, and it does most damage to the bolls. On the bolls the disease first appears as a small dull-red spot or spots. As the spots enlarge, the tissue darkens; different spots may become confluent, covering one-fourth to one-half the boll. A little later, with the production of spores, the spot becomes grayish in the central or older part; and if the weather is moist, spores appear in such abundance that the center of the spot becomes pink (Fig. 71). As the pink area enlarges, it continues to be surrounded by a rather well-defined dark border. Fungus threads enter the boll, arrest its growth if not full size, and cause a darkening and rotting of the parts within. Diseased bolls frequently become darkened and hardened throughout and never open; others crack, but the lint in the locks is weak and discolored and does not become fluffy (Fig. 71). In many instances, a part of the locks open normally, apparently, but the lint and seed in these are, in most cases, infected with the disease-producing organism.

The Causal Organism.—Glomerella gossypii (South.) Edg. has a richly branched septate mycelium, usually hyaline but sometimes somewhat darkened. It grows in and between host cells, causing collapse, loss of chlorophyll, and browning. Mycelium grows throughout diseased bolls, entering seeds and lint fibers.

Conidia are formed in acervuli, subtended by stromata; acervuli, erumpent; conidiophores, 12 to 28 by 5μ; conidia, irregularly oblong, hyaline, or flesh-colored in mass; setae, single or tufted, dark at base and colorless above, straight, rarely branched; setae increase in number with the age of the acervulus, and conidia are occasionally found on them. Conidia in germination usually develop one septum, sometimes two septa, and produce dark chlamydospores. Acervuli are common on bolls; less so and smaller on leaves and stems.

The perithecia, as found on cotton bolls in the field by Edgerton in Louisiana, were usually entirely imbedded, with the beaks only protruding, and were often numerous and crowded. Perithecia are described as being subglobose to pyriform: 80 to 120 by 100 to 160μ; beak, 0 to 60μ long; asci, numerous, clavate, 55 to 70 by 10 to 14μ; spores, elliptic, hyaline, rarely curved, 12 to 20 by 5 to 8μ; paraphyses, long and slender, very abundant (Fig. 72).

How Cotton Plants Become Infected with Anthracnose.—Barre[12] has shown that spores and filaments of the fungus may remain alive on old stalks and bolls in the field for a year but not longer. When bolls were buried, they lived only about 7 months. On the outside of seed, spores retained their vitality for 9 months; but on the inside, spores or fungus hyphae lived as long as 3 years. Edgerton[13] found many spores on cotton seed, estimating the number as high as 800,000 per seed from apparently healthy cotton. Barre[12] found that many seeds which looked sound on the outside contained fungous filaments and spores capable of infection.

It is likely that diseased cotton seeds are the chief source of infection for the plants of a given crop. The organism in the

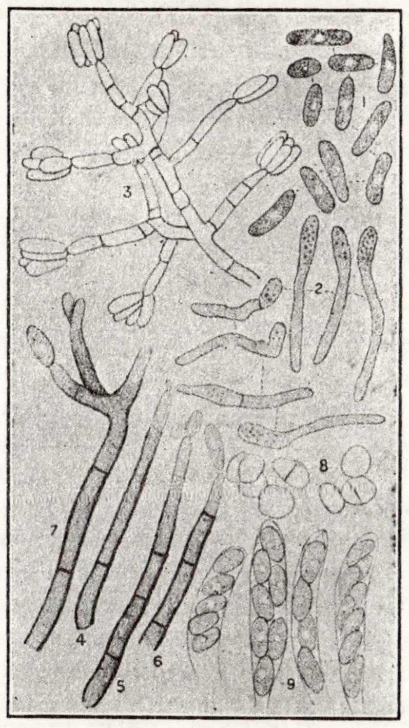

FIG. 72.—Glomerella gossypii: (1) spores, (2) spores germinating, (3) spore formation on mycelial threads, (4, 5 and 6), setae with spores in different stages of development, (7) a branched seta, (8) yeastlike cells formed in Elfring's nutrient solution, (9) asci and ascospores. (*After Edgerton.*)

seed attacks the germinating seeds and, if the weather is too cool for the young plants to grow rapidly, causes the death of many of them even before they get aboveground. The disease spots on ones that do survive constitute a source of infection for other parts of the plant later. Infection is probably spread by water, especially the splashing of raindrops. Barre[12] and Edgerton[13] have both shown that a high percentage of boll infection results from spraying suspension of spores on cotton bolls;

the younger the boll the higher the percentage. Table XXXIV, from Edgerton, brings out this point.

TABLE XXXIV.—INFECTION EXPERIMENTS WITH BOLLS OF DIFFERENT AGES

Age of bolls, days	Number of checks	Per cent of checks diseased	Number of inoculations	Per cent of inoculations diseased
4	13	0	27	77.8
7	..	....	41	95.1
11	..	....	38	81.6
16	1	0	25	24.0
21	3	0	1	0.0
26	12	8.4	17	11.8
30	4	50.0	4	0.0
32	4	25.5	7	14.3

Control Measures.—Planting only seed from disease-free fields, the use of 2-year-old seed, the use of crop rotations, and fall plowing are control measures that have proved effective and of practical value. If it is not possible to gather seed from disease-free fields, picking cotton from disease-free plants is a help. Since the fungus is not able to live on old plant debris for more than a year, a 2-year rotation in which some other crop is alternated with cotton serves to kill out the disease in the field. Running over the old stalks with a stalk cutter and turning them under thoroughly by fall plowing is helpful, because the organism lives a much shorter time when buried in the soil.

Other methods of control, such as sterilizing seed by the use of concentrated sulphuric acid and mercuric chloride, soaking seed in hot water for a few minutes, drying them artificially to a moisture content of 2 per cent, using gases on them, and spraying plants with Bordeaux mixture and other sprays, have been tried. Most of these measures were effective to a large extent, but they did not eradicate the disease entirely and are not suitable for practical use.

Varietal Resistance.—Some varieties, such as Toole, Dixie, Dillon, Cleveland, Russell, Express, Rowden, and Truitt, are generally considered somewhat resistant to anthracnose; Cook, Half and Half, Acala, Lone Star, Wilds, and Triumph are regarded as susceptible. Inoculation experiments carried on by

Edgerton[16] showed that there is practically no difference between different varieties in actual resistance after inoculation. Barre[12] also found that no variety possesses marked resistance. It is probable that the habit of some varieties makes them more susceptible to natural inoculation than other varieties.

Rust, or Black Rust.—Rust, or black rust, also known under the names of "yellow leaf blight" and "potash hunger," is due to unsatisfactory cultural conditions and is termed a "physiological trouble." It is very common on the poorer soils throughout the Cotton Belt and causes damage ranging from practically none to nearly total. Gilbert[17] estimates the average damage for the entire Cotton Belt to be from 4 to 5 per cent.

Symptoms.—Plants affected with the black rust fail to make normal growth, being small and of a sickly green color. Toward the middle of the season, the leaves take on a mottled appearance, yellow spots appearing over areas farthest removed from veins. These spots enlarge and become brownish, and later the whole leaf becomes somewhat blackened, curled up, and ragged. Most, if not all, of the leaves are shed prematurely; consequently, the bolls fail to develop properly. Many remain small and fail to open; others that do open have lint of very inferior quality. The blackening of the leaves is due largely to attacks of Alternaria and other saprophytic fungi.

Cause.—Black rust of cotton is caused by improper soil conditions. No disease-producing organism initiates the trouble, but certain fungi attack the weakened or dead plants. Faulty soil conditions responsible are:

1. Lack of humus, or vegetable matter, in the soil.
2. Lack of potash.
3. Improper drainage.

Continuous cropping to corn and cotton without keeping up the supply of organic matter in the soil has put many fields in such a condition that cotton rusts badly. Rust also frequently develops on only the lowlands of a field or only after a period of continued heavy rains.

The growing and plowing under of green manures, such as vetch, rye, cowpeas, velvet beans, and soybeans will usually check black rust. Ricks[21] and others have shown that a liberal application of barnyard manure or of kainit is effective in checking rust. Muriate of potash and other carriers of potash are also helpful.

Magnesium Deficiency.—In some areas in the southeastern states, there is a deficiency of magnesium in the soil. This deficiency may cause a reddening or discoloration of the leaves of cotton plants and some reduction in production. Application of magnesium salts in the fertilizer or the use of dolomitic limestone corrects the deficiency.

Angular Leaf Spot (*Pseudomonas malvacearum* E. F. S.).—Angular leaf spot is the only bacterial disease of cotton of eco-

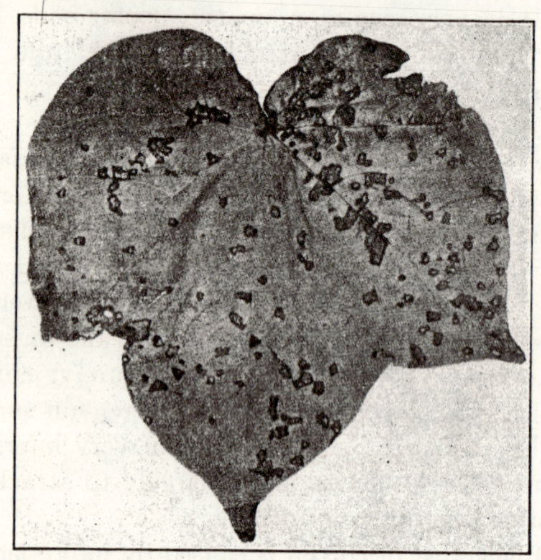

Fig. 73.—Angular leaf spot of cotton. (*After Dr. E. F. Smith.*)

nomic importance. Various names have been applied to this disease, as "bacterial blight," "bacterial boll rot," "black arm," according to the part of the plant attacked.

This disease is widely distributed in the cotton-producing states of the United States and has been reported from Natal, the West Indies, and from several other cotton-growing districts. Leaves are attacked very generally, but the chief damage is to the bolls. A very small percentage of the bolls in a field of cotton are commonly infected, but an infection as high as 60 per cent has been reported. Rainy weather favors the spread of the disease. A tropical storm in June, 1934, caused such a spread of the disease that the cotton plants in southern Louisiana shed most of their leaves.

Symptoms.—Angular leaf spot is most noticeable on the leaves, but it also attacks cotyledons, stalks, branches, bracts, and bolls.

On the leaves, the disease first causes dark-green angular spots on the undersurface. These later show on the upper surface and become reddish brown. The brown spots rarely cross the veins. This is the cause of their angular shape (Fig. 73), but several spots may coalesce more or less along either side of large veins, making zigzag streaks. If the leaf is badly infected, it may curl up and drop off.

When the disease attacks the stem or branches, it causes a blackening of the parts affected, and it is then called "black arm."

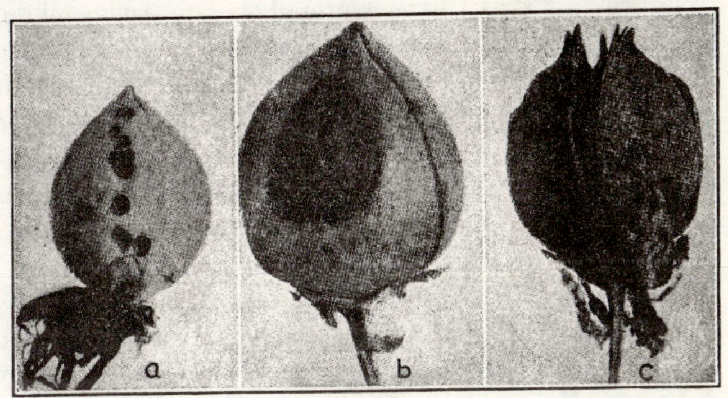

FIG. 74.—Bacterial boll-rot: (*a*) early; (*b*) intermediate; and (*c*) late stages. (*After Gilbert.*)

This, according to Gilbert,[17] is worse on Egyptian and Sea Island cottons, which seem to be especially susceptible to the disease.

On the bolls the first signs of disease are small, dark-green, water-soaked, roundish spots. These spots gradually enlarge and turn black in the center as the tissue is killed. Two or three locks or all the locks in the boll may be so injured that they fail to open (Fig. 74).

Causal Organism.—The angular-leaf-spot disease of cotton is caused by a bacterium (*Pseudomonas malvacearum* E. F. S.), which is described by Faulwetter[18] as follows:

It morphologically resembles the other yellow, one-flagellate rods causing diseases of walnut, bean, hyacinth, citrus, etc. It is in the form of short rods, usually occurring in pairs or chains of three and sometimes more. The flagellae have been observed in young cultures. One, sometimes branched at the end, or occasionally two flagellae occur at one end, making possible very active motility. No endospores, granules or capsules have been demonstrated. Moderate growth takes place on all the sugar agar slant media.

Successful inoculations were made by E. F. Smith by spraying a suspension of young agar culture of the organism on cotton leaves and bolls. Numerous successful inoculations have also been made by Faulwetter, Rolfs, and others.

Fig. 75.—Phymatotrichum omnivorum on roots of cotton. (*After B. M. Duggar.*)

Initial infection in the spring probably results largely from bacteria that have wintered over on seeds and on lint attached to seeds. Primary infection is on the cotyledons of young seedlings; other parts of the plant are infected from these. According to Faulwetter,[18] infection is spread very largely by rain, and, in case of storms, by rain and wind.

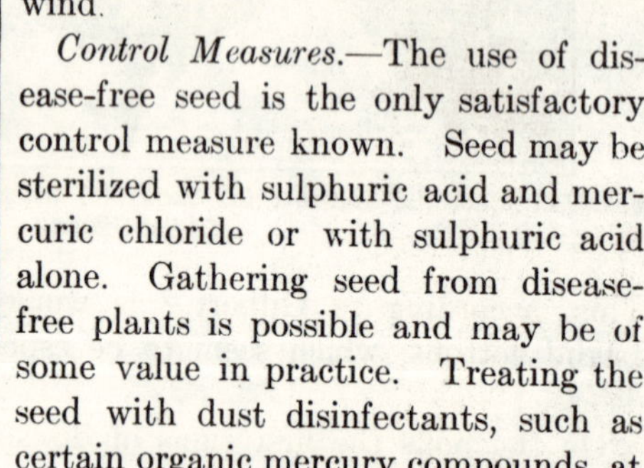

Control Measures.—The use of disease-free seed is the only satisfactory control measure known. Seed may be sterilized with sulphuric acid and mercuric chloride or with sulphuric acid alone. Gathering seed from disease-free plants is possible and may be of some value in practice. Treating the seed with dust disinfectants, such as certain organic mercury compounds, at the rate of 2 to 4 ounces of the dust to a bushel of seed has given good results in the control of the angular leaf spot, or bacterial blight, and also of some other seed-borne diseases.

Texas Root Rot (*Phymatotrichum omnivorum* (Shear) Duggar). A form of cotton root rot known commonly as "Texas root rot" causes much damage to the cotton crop in Texas and smaller losses in Oklahoma, Arizona, New Mexico, and California. The disease has not extended east of the Mississippi River. In Texas, the annual loss has been estimated to be as high as $20,000,000. This disease has been studied by Atkinson,[2] Shear,[20] Pammel,[19] McNamara, Peltier, Taubenhaus, Neal, King, and others.

Symptoms.—The general characteristics of the disease, as described by Pammel,[19] are as follows: The first indication is the sudden wilting of a single plant or a number of them; this is likely to occur in May or June. During the last of June or the first of July, a large number of wilted plants may be seen, later forming irregular dead patches. The plants wilt most frequently on hot days following rain. Continuous dry weather tends to hold the disease in check.

If the roots of dead cotton stalks or of those wilting are examined, brown threads of fungus will be found closely surrounding the taproot and some of the lateral roots (Fig. 75). Small wart-like bodies, or sclerotia, are found in numerous places on the taproot and lateral roots. The roots shrink and decay. The fungus mycelium in younger stages is white and may be found on roots of plants that appear healthy.

Causal Organism.—The fungus causing Texas root rot is *Phymatotrichum omnivorum* (Shear) Duggar. Shear[20] describes it as follows:

> It is a facultative parasite infesting the soil and attacking the roots of a great variety of plants, and causing serious damage to cultivated crops, such as cotton, alfalfa, cowpeas, beets, and fruit trees. Few crops except grasses and grains are free from it. No fructification is definitely known. Sterile mycelium is a dirty yellow, sometimes whitish when young or growing in cultures or in the vessels of vascular bundles of plants; hyphae forming strands and spreading from them, producing a rather dense arachnoid layer on the surface of the host and bearing one to four branches arising and growing at right angles from the same point near the ends; diameter 3 to 5,μ tapering toward the ends.

Recent work has shown that the small sclerotia serve as a holdover stage for the fungus and are one of the principal means of propagating the disease. The sclerotia in the soil are known to retain their vitality for 3 or 4 years, and they may live for a much longer period.

Control Measures.—Since this is a soil-borne disease, great precaution must be used to prevent the transference of inoculated soil to disease-free areas. Soil is often transferred to carry legume inoculation, and this practice may serve to spread the disease.

The use of immune crops, such as grains, 2 years in succession in a 3-year rotation is a control measure that has been recommended, but it fails in some cases, as shown by the work of

Scofield[24] at San Antonia, Tex. Deep plowing, especially subsoiling, has given some measure of control. Recently, King[28] secured splendid control on irrigated land by placing 15 tons of barn manure or 30 tons of green alfalfa hay in deep furrows and then bedding and planting on this layer. The method as he has summarized it is as follows:

> It consists in burying liberal quantities of organic manures in deep furrows in infested areas during the winter, irrigating to encourage rotting, and planting cotton over the decaying material in the spring. The control is more efficacious with successive treatments, and should be continued over a period of years in order to build up and maintain an abundant microfloral population in the soil. The root rot fungus does not thrive in the presence of great activity on the part of saprophytic organisms, and the cotton plants frequently escape infection.

Diplodia Boll Rot (*Diplodia gossypina* Cooke).—Diplodia boll rot, according to Edgerton,[13] causes considerable damage to the cotton crop in Louisiana, being worst on the alluvial soils and bluff lands. It is estimated that the average damage for the state as a whole is 2 per cent. No reports have been published giving losses elsewhere. Probably they are not serious.

Diseased bolls turn black and, after a short time, become thickly covered with pycnidia, which exude quantities of black spores, making the whole surface of the boll appear smutty. The contents of the boll become black and rotten.

Edgerton[13] found that inoculations were effective only when made in wounds on bolls; that spores in pycnidia on old diseased bolls retained their vitality for 9 months; and that the disease was not seed-borne. Deep fall plowing which will bury all the old diseased boll is probably a satisfactory control measure.

The causal fungus, *D. gossypina*, was described by Cooke, in 1879, on cotton grown at Bombay, India.

Cotton Diseases of Lesser Importance.—In addition to the more important cotton diseases that have been discussed, there are several others of minor importance that should be mentioned.

Fusarium Boll Rot.—A boll rot caused by a species of Fusarium (determined tentatively by Edgerton[13] as *F. roseum*) does some damage to cotton bolls. This organism attacks only bolls that have been injuried in some way and lives, in the main, as a saprophyte, but at times, after getting a start in the boll, it grows into living tissues and may also damage the lint in mature bolls.

This disease resembles anthracnose considerably in that spots on the bolls at certain stages are coated with pink spores; but the spot is not slimy, as in anthracnose spots, and the spores are a lighter color. Infection is carried over winter by the seed.

Other Boll Rots.—Boll rots are caused by Rhinotrichum, Olpitrichum, Volutella, Sclerotium, Botryosphaeria, Schizophyllum, and several other fungi, but the damage done by these is not of importance.

Areolate Mildew.—This disease is also called "frosty mildew," from the frosty appearance of the underside of the leaves. It is caused by *Septocylindrium areola* (Atk.) Ehrlick and Wolf, a fungus first described by Atkinson, in Alabama, 1890. This disease is widespread, being especially common the latter part of the season on rank plants growing on fertile, moist soil. Although it causes some shedding of leaves, the economic importance is slight.

Leaf Spot (*Mycosphoerella gossypina* (Atk.), Earle; *Cercospora gossypina* Atk.).—This disease is common on cotton, but it is of slight economic importance, since it attacks only weakened or old leaves. Small red dots first appear on the leaves. These enlarge centrifugally, the center becoming brown and, later, white if spores are produced. In the last stages, the leaves are ragged and full of holes.

Rust.—A true rust of cotton caused by *Kuehneola gossypii* (Lag.), Arth. has been reported from Florida, Cuba, Porto Rico, and British Guiana. Apparently, this is of little economic importance.

Cluster Cup Rust (*Aecidium gossypii* E. and E.).—This is another true rust, which has been found on cotton in Florida, Mexico, and Texas. The most serious outbreak on record occurred in Texas in 1917. Plants were partly defoliated, and the loss in infected fields was estimated at 20 per cent.

Smut (*Doassansia gossypii* Lag.).—This disease has been found attacking cotton leaves in Ecuador and in the West Indies.

Red-leaf Blight.—Frequently when cotton is grown on poor soil, the leaves tend to turn red late in the season, resembling the autumnal coloration of woody plants. This is a physiological trouble caused by lack of nutrient elements in the soil.

Damping-off (*Pythium de baryanum* Hesse and other fungi).—During seasons when the weather is too cool for young cotton plants to make proper growth, they are apt to be attacked by certain fungi that cause damping-off, or a soft rot of the stem at the surface of the ground or below.

Sore Shin (*Corticium vagum; Rhizoctonia*).—This disease attacks young seedling plants during cool, wet periods and may damage the stand seriously. The disease is characterized by the presence of dark rusty-brown cankers on the stem at or below the surface of the soil. These may penetrate so deeply that they cause the plant to bend over and die.

Crown Gall [*Pseudomonas tumefaciens* (S. and T.), Stev.] is mentioned by Stevens and Hall[22] as doing some injury to cotton.

Crazy Top (*Acromania*).—This disease, or disorder, is named "crazy top" on account of the abnormal branching and fruiting of the plants. It has been reported in California and Arizona. The fruiting branches in the upper part of the plants are frequently replaced by vegetative branches producing an abnormal appearance. There is a general shedding of flower buds and young bolls during a part of the season, but late in the season many bolls may be set near the end of the branches. The flowers and bolls are small and more or less malformed. Both Egyptian and upland cottons are affected.

Hope[29] found that the incidence of crazy top was related to the checking of growth, due to water shortage, and the resumption of growth when abundant moisture was restored. Proper irrigation will prevent the trouble.

References

1. Gilbert, W. W.: "Cotton Wilt and Root-knot," *U. S. Dept. Agr. Farmers' Bull.* 625, pp. 1–21, 15 figs., 1915.
2. Atkinson, G. F.: "Diseases of Cotton," *U. S. Dept. Agr. Bull.* 33, pp. 278–316, 8 figs., 1896.
3. Atkinson, G. F.: "Some Diseases of Cotton," *Ala. Coll. Sta. Bull.* 41, 1891.
4. Fulton, H. R.: "Cotton Wilt," *La. Agr. Exp. Sta. Bull.* 96, pp. 1–15, 3 figs., 1907.
5. Cauthen, E. F.: "Wilt Resistant Varieties of Cotton," *Ala. Agr. Exp. Sta. Bull.* 189, pp. 65–88, 3 figs., 1916.
6. Brown, H. B. and Ames, C. T.: "Cotton Experiments, 1917," *Miss. Agr. Exp. Sta. Bull.* 184, pp. 1–28, 2 figs., 1918.
7. Brown, H. B. and Anders, C. B.: "Cotton Experiments 1919 and 1920," *Miss. Agr. Exp. Sta. Buil.* 187, pp. 1–32, 4 figs., 1920.

8. LEWIS, A. C. and MCLENDON, C. A.: "Cotton Variety Tests for Boll Weevil and Wilt Condition in Georgia," *Ga. State Board Entomology Bull.* 46, pp. 1–36, 1917.
9. SOUTHWORTH, E. A.: "Anthracnose of Cotton," *Jour. Myc.*, vol. 6, pp. 100–105, 1891.
10. ATKINSON, G. F.: "Anthracnose of Cotton," *Jour. Myc.*, vol. 6, pp. 173–178, 1891.
11. EDGERTON, C. W.: "The Perfect Stage of Cotton Anthracnose," *Mycologia*, vol. 1, pp. 115–120, 1909.
12. BARRE, H. W.: "Cotton Anthracnose," *S. C. Exp. Sta. Bull.* 164, pp. 1–22, 10 figs., 1912.
13. EDGERTON, C. W.: "The Rots of the Cotton Boll," *La. Exp. Sta. Bull.* 137, pp. 1–113, 13 plates, 1912
14. CONVERSE, O.: "Report of Botany Division," *S. C. Agr. Exp. Sta. 32d Ann. Rept.*, 1919.
15. EDGERTON, C. W.: "Experiments on Varietal Resistance to the Bean and Cotton Anthracnose," *La. Exp. Sta. Bull.* 155, pp. 1–24, 2 figs., 1916.
16. BESSEY, E. A.: "Root-knot and Its Control," *U. S. Dept. Agr. Bur. Plant Ind. Bull.* 217, pp. 1–89, 23 figs., 1911.
17. GILBERT, W. W.: "Cotton Diseases and Their Control," *U. S. Dept. Agr. Farmers' Bull.* 1187, pp. 1–32, 18 figs., 1920.
18. FAULWETTER, R. C.: "The Angular Leaf Spot of Cotton," *S. C. Agr. Exp. Sta. Bull.* 198, pp. 1–41, 4 figs., 1919.
19. PAMMEL, L. H.: "Cotton Root-rot," *Tex. Agr. Exp. Sta. Bull.* 7, pp. 1–30, 16 figs., 1890.
20. SHEAR, C. L.: "New Species of Fungi," *Bull. Torrey Botan. Club*, vol. 34, pp. 305–317, 1907.
21. RICKS, J. R.: "Cotton Experiments 1913," *Miss. Agr. Exp. Sta. Bull.* 164, pp. 1–24, 1914.
22. STEVENS, F. L. and HALL, J. G.: "Diseases of Economic Plants," pp. 1–507, The Macmillan Company, New York, 1921.
23. COOK, O. F.: "A Disorder of Cotton Plants in China: Club-leaf or Cyrtosis," *Jour. Heredity*, vol. 11, p. 3, 1920.
24. SCOFIELD, C. S.: "Cotton Root-rot in the San Antonio Rotations," *Jour. Agr. Research*, vol. 21, pp. 117–125, 1921.
25. YOUNG, V. H., JANSSEN, GEORGE and WARE, J. O.: "Cotton Wilt Studies, IV. Effect of Fertilizers on Cotton Wilt," *Ark. Agr. Exp. Sta. Bull.* 272, 1932.
26. NEAL, D. C.: "Cotton Wilt: A Pathological and Physiological Investigation," *Miss. Agr. Exp. Sta. Tech. Bull.* 16, 1928.
27. WARE, J. O. and YOUNG, V. H.: "Control of Cotton Wilt and Rust," *Ark. Agr. Exp. Sta. Bull.* 308, 1934.
28. KING, C. J.: "A Method for the Control of Cotton Root Rot in Irrigated Southwest," *U. S. Dept. Agr. Cir.* 425, 1937.
29. HOPE, CLAUDE, KING, C. J. and PARKER, ORLAN.: "The Effect of Crazy Top Disorder on Cotton Plants and Its Control by Irrigation Management," *U. S. Dept. Agr. Tech. Bull.* 515, 1936.

CHAPTER XV

COTTON INSECTS

The cotton plant has several different means of attracting insects. On the midrib on the underside of each leaf there is a nectary, a shallow cup-shaped structure that often contains a droplet of sweet fluid; at the base of each bract on the outside of the square is another nectary; inside the flower near the base of the petals are others. Many insects visit the plants to get the nectar. Others come to the cotton flower to gather or feed on the pollen. The tissues, or substance of seeds, leaves, stem, and roots of cotton plants, are rich or moderately rich in nutrient materials. These invite various insects to feed on the plants and, in some instances, to live within their tissues.

Of the many insects that visit cotton plants, a few are beneficial in that they foster pollination or prey on harmful insects. A comparatively small number of others are injurious to the plants. The greater number of the insect visitors may be classed as neutral, being of no help to the plants but doing no harm. Of the harmful insects the boll weevil, the pink bollworm, the cotton leaf caterpillar, and the bollworm are the most destructive. These will be discussed in some detail, and control measures mentioned. Several other insects of lesser importance will be considered briefly.

With the extension of cotton culture in the different cotton-producing countries of the world and with the transference of cotton seed for planting purposes has come the spread of harmful insects. Many countries have had no quarantine measures until recently and did not realize the need of them until too late.

In the United States, the cotton leaf caterpillar and the bollworm were the only insects that did any considerable damage prior to 1892, and they appeared in serious outbreaks only occasionally. In 1892, the boll weevil crossed the Rio Grande River into Texas and has since spread over practically all the Cotton Belt east of New Mexico, damaging the cotton crops to the extent of hundreds of millions of dollars. Within the last two decades, the pink bollworm has been introduced into and has spread over much of the cotton-producing region of Egypt and Brazil. This worm has been found in Texas, Arizona, New Mexico, Louisiana, Georgia, and Florida, but it has been either eradicated or placed under control in the areas where found.

THE MEXICAN BOLL WEEVIL

The Mexican boll weevil (*Anthonomus grandis* Boh), which is by far the most destructive insect that has ever attacked the cotton plant in the United States, is a native of Central America or Mexico. All evidence indicates that it has fed on the wild tree cottons of those regions for many years. Very little is known concerning the boll weevil in Mexico, except that it was generally distributed and did considerable damage many years previous to its entrance into Texas in 1892.

Boll weevils were found near Brownsville, Tex., in 1892, and since that time they have spread over nearly all of the American Cotton Belt. They did considerable damage in Texas in 1894 and entered Louisiana in 1903, Mississippi in 1907, and Alabama in 1909. The spread throughout the eastern Gulf states was rapid. The boll weevil is known as a cotton pest from Mexico southward to Costa Rica and in Cuba. Other cotton-growing districts are free from weevil infestation. Figure 76 shows in detail the spread of the boll weevil over the Cotton Belt.

Anthonomus grandis thurberiae Pierce, a variety of boll weevil distinct from the common weevil but belonging to the same species, was found feeding on the wild cotton, *Thurberia thespesioides*, in the mountains of southeastern Arizona, in 1913. In 1920, it was seen on cultivated plants north of Tucson. Since that date, the infestation has spread. In 1932, this weevil was known to occur on cultivated cotton in parts of five counties in Arizona and infested some 12,000 acres. Strict quarantine regulations are being enforced to restrict this weevil to its present location.

Description and Life History of the Boll Weevil.—The adult boll weevil is a small grayish or brownish weevil, about ¼ inch in length and varying from ⅛ to ⅓ inch, with a breadth one-third of the length. The size varies considerably, being determined

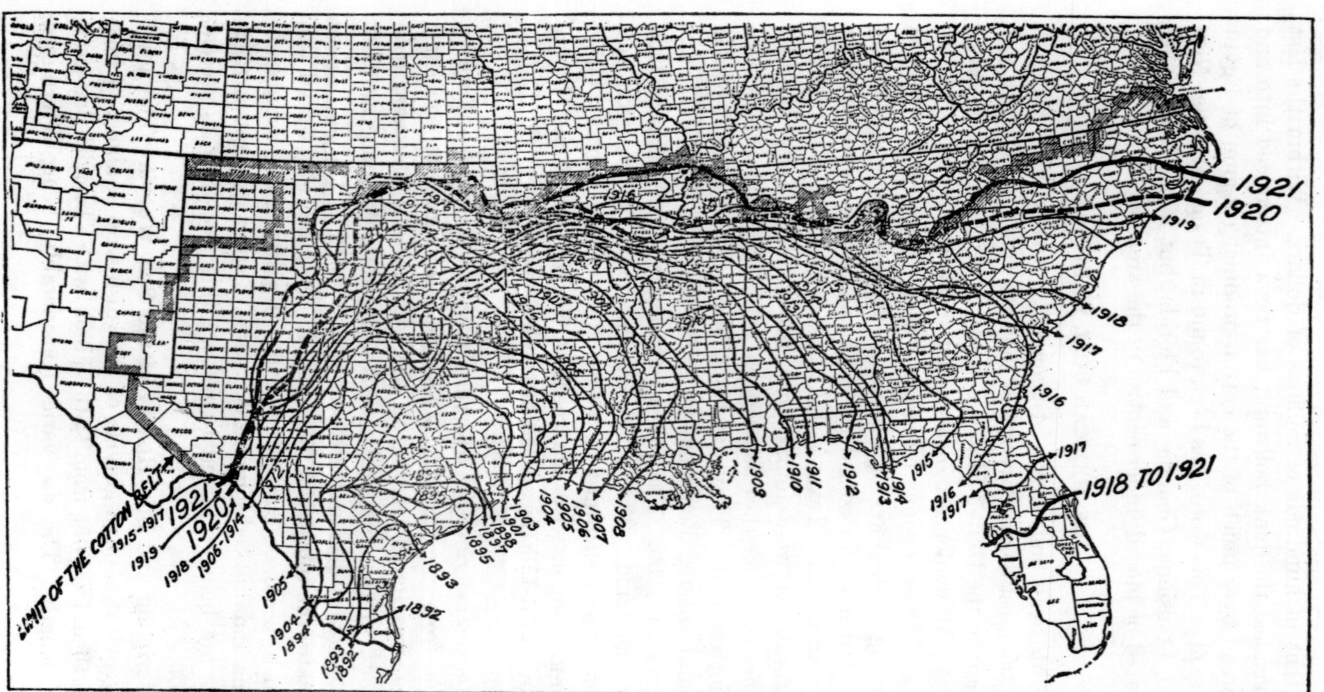

FIG. 76.—Map showing dispersion of the Mexican cotton boll weevil in the United States from 1892 to 1921. (NOTE: The outer limit of the Cotton Belt advances or recedes slightly from year to year and in the map is not absolutely accurate for 1921, as shown by the fact that in a few places the boll-weevil line is a little in advance of the indicated limit of cotton culture. (Bureau of Entomology, U. S. Department of Agriculture.)

by the amount of food supplied the developing larva. Weevils that develop in bolls, where food is abundant, are considerably larger than ones that grow in squares. The color is largely

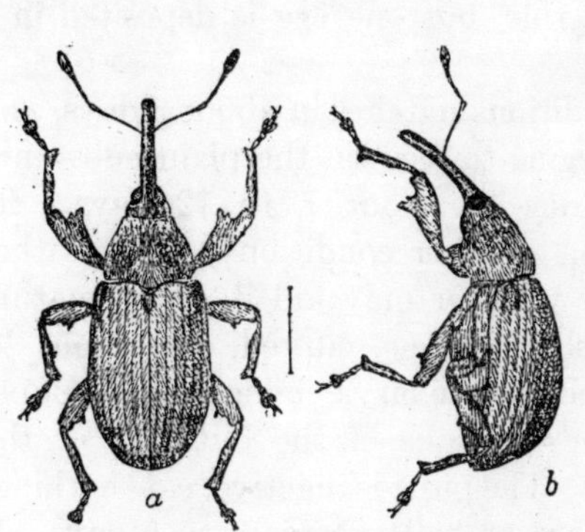

FIG. 77.

FIG. 78.

FIG. 77.—Much enlarged picture of adult boll weevil, showing hairy covering, snout, antennae, and the two small teeth on projections on the front leg. Line indicates true size. (*After W. D. Hunter, U. S. Department Agriculture.*)

FIG. 78.—Square showing position of boll weevil egg puncture near base of flower bud. (*After Smith.*)

dependent on the age of the weevil; newly hatched ones are yellowish brown, while older ones become grayish and darker. Figure 77 shows the shape and general appearance of the adult weevil. The ridged or lined appearance of the wing covers and the two projections at the outer end of the first joint of the front legs are distinguishing marks helpful in determining whether or not a specimen is a boll weevil. Anyone who has once seen boll weevils will have but little trouble identifying them. In a cotton field the presence of weevils is indicated by the flaring of the bracts of cotton squares and by an excessive amount of square shedding. If the squares when cut open are found to contain a white curved grub which has eaten out the inside of the flower bud, there is not much doubt that boll weevils are present.

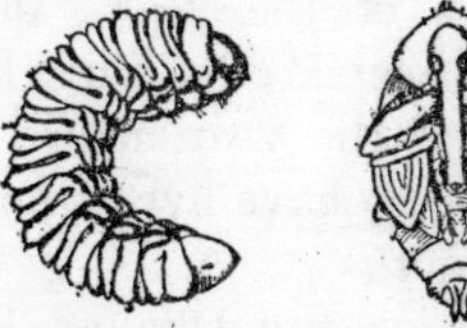

FIG. 79.—Cotton boll weevil; larva at left, pupa at right; about three times natural size.

The boll weevil passes through the winter, or hibernates, in the adult, or winged, stage. Egg laying does not begin until cotton squares have formed. Eggs are deposited in small pits or openings which the female makes with her snout. They are usually placed near the base of the flower bud and inside the corolla (Fig. 78). A hibernated female, as shown by Smith,[1] lays, on the average, 76 eggs, but some under observation laid as many as 218 eggs. As a rule, but one egg is deposited in a square.

The egg under normal conditions hatches in about 3 days, and the larva, or young grub, begins to feed on the plant substance about it. It reaches naturity in about 7 to 12 days. Its development is dependent on weather conditions; cool weather retards growth, but hot dry weather may kill it. The mature larva is white with brownish markings, curved, and about ½ inch in length (Fig. 79). The mature larva changes into a pupa, a form corresponding to the chrysalis of the butterfly or the cocoon of the moth (Fig. 79). The pupa is inactive, eats nothing, and remains within the square or boll. After 3 to 5 days, an adult weevil hatches from the pupa. The young adult begins to produce a new generation in 5 to 7 days after emergence. The condition of the weather has an important bearing on the length of time required for a boll weevil to pass through its life cycle. The time required varies from 18 days to 4 weeks or longer, being longer during cool weather or toward the end of the season. Warm weather favors rapid development, but hot, dry weather increases mortality greatly.

The length of time that the mature weevils live varies with the season of the year. In the summer, according to Hunter,[2] most weevils do not live longer than 50 days. During the cooler part of the year, many live as long as 6 months, and one is known to have lived from December to the following October—11 months.

Fenton and Dunnam,[11] in their studies of the biology of the boll weevil at Florence, S. C., noted:

> The average longevity of over-wintered weevils in hibernation cages before cotton was available for food was 5.65 days. In the late fall weevils may live as long as 29 days without food or water. In field cages the average length of life before square formation was 8.13 days, and after square production it was 19.32 for males and 15.99 for females. The maximum longevity was 82 days for males and 81 for females. In these same cages the pre-oviposition period averaged 7.21 days; oviposition period, 12.66 days; the average number of eggs deposited per female

was 81.21, and the daily rate of egg deposition, 6.42. The maximum fecundity in field cages was 440 eggs. The average period of development in squares in the field was 18.3 days and in bolls 32.31 days.

The average longevity of over-wintered females mated the previous fall was 79.2 days, the preoviposition period 7.9, the oviposition period 64.8, the fecundity 90.5 eggs, and the average daily rate of oviposition 1.40 eggs.

Dispersal by flight was influenced by temperature to a slight extent after it had already begun, more weevils being in flight when the temperature was high than when it was lower. Dispersal was correlated with the interrelation between weevil population and square production by the cotton plant, or, in other words, with the percentage of infestation. Trap-crop records indicate that weevils were flying from field to field at least as early as July 13, 1925, and July 22, 1926.

The average winter survival at Florence, South Carolina, for four seasons (fall of 1922 to spring of 1926) was 3.27 per cent, the maximum being 11.05 per cent in 1923, and the minimum 0.35 per cent in 1924. In 1925 and 1926, 3.17 per cent of weevils survived in field cages as compared with 4.31 per cent in the woods under conditions otherwise the same.

The shelter giving the most protection was piled cotton stalks, the survival in this material for the period 1924 (fall) to 1926 (spring) averaging 5.64 per cent. The survival in other shelters was as follows: Cornstalks, 4.03 per cent; pine straw, 3.56; Spanish moss, 2.63; sawdust or shavings, 1.37; oat straw, 0.72; and cage only, 0.43 per cent.

For four years (1923 to 1926) 21.26 per cent of the weevils surviving emerged in March, 41.78 in April, 29.19 in May, 7.76 in June, and 0.11 in July. The percentage of emergence in all hibernation cages at date of first square formation ranged from 99.1 in 1923 to 90.02 in 1925.

The greatest mortality in fallen squares was due to heat, averaging 41.18 per cent in 1925 and 25.73 per cent in 1926. The mortality caused by parasites averaged 6.75 per cent in 1925 and 7.84 per cent in 1926. Predators caused death rates of 0.59 and 0.03 per cent; proliferation, 3.22 and 1.0 per cent; disease or unknown causes, 0.93 and 4.11 per cent.

Food of the Boll Weevil.—The favorite food of the boll weevil is pollen in the unopened flower bud of the cotton plant. This it obtains by making a number of punctures through the corolla of the unopened flower. The upper part of the bud may show a dozen or more punctures with loose pollen scattered around them. Weevils that appear in fields before squares have formed attack the young leaf buds to some extent, and in the latter part of the season, after unpunctured squares have become scarce, they attack young bolls. Even bolls of full size, but with tender walls, may be punctured.

Apparently, the young pollen grains of the medium-sized flower bud are the favored food of the weevil larvae. The mother weevil deposits the egg inside the bud among the anthers of the flower, so that the first food of the young grub may be pollen. Later in the season, after squares have become scarce, a good many young bolls are punctured, and eggs are deposited in them. The developing larva either ruins the boll completely or damages it badly.

The cotton plant constitutes almost the sole food of the boll weevil. In 1913, a boll weevil was observed at Victoria, Tex., feeding on *Hibiscus syriacus* (Howe[3]), a plant botanically related to cotton. Since that time cage experiments and field observations have been conducted which go to show that the weevil can live for a time at least on plants closely related to cotton. The most important of these seem to be the various wild species of Hibiscus and okra and the wild cotton of Arizona. Occasionally, eggs have been deposited on Hibiscus, and one of the research men working at the Boll Weevil Laboratory at Tallulah, La., reared seven weevils in Althea buds. When confined in bottles, adult weevils feed rather freely on the kernels of cotton seed and have been observed, under the stress of starvation, to eat various other substances, such as apples and bananas.

Nature and Amount of Boll-weevil Damage.—Both the adult weevil and the larva damage cotton plants. The adult feeds on young leaf buds and punctures both squares and young bolls for feeding and egg laying. The larvae feed on the inside of squares and bolls. The squares are shed or fail to develop flowers. Young bolls are commonly shed, and older ones have one or more locks ruined. Some locks that open have the lint so stained or damaged that the grade of the whole lot of lint cotton is lowered.

The damage done by boll weevils varies largely in different localities and even in different fields of the same locality. The cause of much of this difference is difficult to explain. For some reason, the weevils fail to appear in certain fields in sufficient numbers to do much damage until many bolls have been put on and a good crop made. The amount of damage done in different years varies much also, owing to differences in the weather. Rainy or cloudy weather during the period the plants are blooming most, or during the latter part of June, in July, and the first half of August, is conducive to heavy weevil damage. Hot, dry weather during the same period serves to hold the weevils in check on account of the heavy mortality of the larvae. Some squares become so dry while still clinging to the plant that the larva or pupa within perishes, and others that fall on the hot ground become so heated that these immature stages are killed. A warm or open winter is favorable to weevils, because a high percentage, comparatively, survive the winter and are ready to multiply in the spring. Low winter temperatures, however, are not so effective in control of the weevil as once thought and certainly not so effective as dry weather during the summer. The writer has seen a heavy initial infestation in central Mississippi following a winter during which the temperature dropped to zero at two different times.

The damage, as was pointed out above, varies greatly in different localities, ranging from less than 5 per cent to practically total loss. The Bureau of Crop Estimates of the U. S. Department of Agriculture, in the fall of 1920, estimated that the average annual loss for the American Cotton Belt for the 4 years just preceding was $300,000,000. This was direct loss due to the nonproduction of seed and lint. The usual loss when cotton values are lower is not quite so great, but it is enormous. In addition to the direct loss, there are others of immense importance occasioned by the damage to the cotton crop, as fall in price of land, depreciation in value of gins or oil mills which have to close or do a limited business, depreciation in railroad values, etc. However, figures given on the lowering of values of any particular crop may be somewhat misleading. As a rule, an estimate is made of the percentage of the crop destroyed, and then values are reduced in proportion to this percentage. This is not quite fair, for, if a portion of the crop is destroyed, the value of the rest will usually be higher.

Movement of Weevils.—The boll weevil after landing in a cotton field shows but little disposition to leave if squares are plentiful. It flies for short distances but is not apt to fly to another field until the seasonal migration period in August, unless food becomes scarce. When the migration period comes on, the weevil instinctively takes to wing and by a series of short flights may travel a hundred miles or more. The movement is usually in the direction of the prevailing wind, but it may be in any direction. In the fall, beginning about the time of the first frost, there is also a movement to hibernating quarters. This flight is usually short and is terminated when a hibernating place is found. In the spring, there is a movement from the hibernating quarters back to the cotton fields, the flight continuing until the cotton plants are found (see Fig. 76 for the spread of weevils over the Cotton Belt).

Hibernation.—By "hibernation" is meant the act of passing through the winter period in a dormant, or quiescent, state. With the coming of cool weather in the fall of the year, boll weevils begin to seek shelter. Many enter cracks in the ground or hide under grass, weeds, or other trash in the cotton fields. Others fly to fence rows, grassy ditch banks, or the woods for shelter. The festoons of Spanish moss hanging from limbs of trees so abundantly in sections of the cotton states afford splendid winter quarters. Most weevils that attempt to pass the winter in the cotton fields fail to survive. Apparently, woods, Negro cabins, or haystacks afford the safest places, for infestation in cotton fields early in the season is usually heaviest near such places.

During the winter period the weevils take no food and are practically inactive. When warm days come, they may move about feebly. In the parts of the belt farthest south, the hibernation is shortest, and there is more moving about during the period. In northern Florida, in 1918, Smith[1] found that weevils did not enter hibernation until about Dec. 7 and began coming out on Feb. 10. Farther north, the period lasts from the middle of November to the first of March. In Texas, more than 50 per cent of the weevils emerge in March; in Louisiana, about 25 per cent. Emergence begins in March or earlier in nearly all states and continues until about the middle of July. It comes in waves, which are influenced by temperature and rainfall.

In a series of hibernation experiments carried out in Texas and Louisiana, from 1906 to 1911, the average survival was 6 per cent. This probably represented about the percentage of survival in nature for the localities where the experiments were conducted. In colder regions, the mortality is greater.

Control Measures for Boll Weevils.—Boll weevils reproduce with such rapidity that the offspring from a single pair, if all survived, would total several millions in one season. Fortunately, several natural agencies tend to hold them in check. The most important of these are hot, dry, summer weather and insect parasites that prey on the weevil larvae. There are also fairly satisfactory artificial means of control.

Natural Means of Control.—As was pointed out above, hot, dry weather causes heavy mortality of the immature forms of boll weevils. When infested squares fall on dry ground and land where the sun's rays will reach, they become so heated that the immature weevils within are killed. Others perish in squares which dry up while still clinging to the plants. The immature weevils are not able to escape from the squares.

A very large percentage of hibernating boll weevils perish. Cold weather during the winter is responsible for a part of the mortality.

There are also a number of small insects that prey on the boll weevil. These serve as another natural method of control, and one of much importance. Forty-five species of weevil enemies have been identified. A large number of these are parasites which deposit eggs on the larvae or pupae within the cotton square. The young of the parasite feed on the substance within the body of the immature weevil and eventually destroy it. Boll-weevil parasites are much more plentiful in certain localities than in others. This probably is the reason why the weevil damage is much less in some regions than in others with apparently the same climatic conditions. Hunter[2] states that in one instance, in 1907, in a field near Robson, La., the mortality due to parasites was 77 per cent and in a field at Victoria, Tex., 61 per cent. Studies in various parts of Mississippi, made by Allen of the Mississippi State Plant Board, showed an infestation ranging from 40 per cent downward. These parasites have doubtless been in the country for a long time. Prior to the coming of the boll weevil, they fed on other closely related weevils, which were not very abundant. The presence of large numbers of well-fed boll-weevil larvae invited them to make a change. There is a possibility that in the course of time parasites may become so plentiful that they will help greatly as a control measure.

In addition to the parasites just mentioned, there are other insects that feed on the boll weevil, the most important being ants. Twelve species of ants are known to attack the weevil. They are the small brownish or yellowish ants which run about over the ground and plants in cotton fields. They gnaw into the infested squares, both when hanging on the plants and on the ground, and soon destroy the helpless grub inside. Hunter[2] reports that in some cases half of the immature stages in fields have been destroyed by ants.

Birds render some service by catching adult weevils.

Plant cells, in the effort to heal the wound made by the weevil in puncturing the squares or bolls to deposit an egg, may exert such pressure on the egg or young larva that it is crushed. This is spoken of as "proliferation."

Artificial Control.—The control of boll weevils by artificial measures is a very difficult matter. The young develop inside the flower bud, where they cannot be reached with a poison, and the adults are so protected by the foliage and flower parts that they cannot be collected to any advantage or poisoned with a contact poison. The Bureau of Entomology of the U. S. Department of Agriculture and different experiment stations in the Cotton Belt have labored long and hard to find some control measures that will be of practical use in the cotton fields. Hunter and Coad[4] give a good digest of the results obtained from various boll-weevil control experiments and make recommendations based on the best information available at present. They say:

> There are the direct and indirect methods [of control], both of which are of vital importance. The importance of indirect methods is often more difficult for the farmer to appreciate than that of the direct; but, in reality, successful weevil control cannot be accomplished unless full advantage is taken of every possible method, and the campaign must be based upon a combination of the different methods rather than concentrating all efforts on direct control.
>
> The farmer is aided in his fight against the weevil by a number of important natural factors which tend to reduce the possible weevil damage. Some of the more important of these which must be taken into consideration in planning a fight on the weevil are as follows:
>
> 1. The weevil is practically dependent on cotton for reproduction.
> 2. The mortality of the weevil during the winter is very high.
> 3. Hot, dry weather during the summer exercises a tremendous control upon the weevil stages, while moist, cloudy weather removes the control and greatly accelerates multiplication.

4. The weevil is attacked by many different species of insect enemies.

5. The emergence from hibernation quarters during the spring is slow and prolonged until well into the summer.

6. Early in the season, on account of comparatively lower temperature, the development of the weevil is much slower than during the midsummer months.

7. The cotton plant produces many more squares than it can carry to maturity as bolls. This surplus is shed by the plant throughout the season, under normal conditions about 60 per cent of the fruit being shed.

8. Up to a certain point, weevil puncturing of fruit does not reduce the cotton crop, because large numbers of forms would be shed normally.

9. The weevil has a decided tendency to seek moisture wherever it may be found on the surface of the plant.

Direct Control by Poisoning with Calcium Arsenate.—In the years which have elapsed since the advent of the boll weevil into the United States, every conceivable means of direct control of the weevil has been tried repeatedly. Owing to the peculiarities of the weevil attack which have been mentioned, most of these attempts have been unsuccessful. Some methods were found which would control the weevil, but these were either impracticable or too expensive for use on a commercial scale. During comparatively recent years, however, a method of poisoning has been developed which has proved very successful. This consists of treatment of the plants with powdered calcium arsenate by a specialized method.

It has long been known that poisoning the boll weevil is possible to a certain extent, but on account of the peculiar habits of the insect it was difficult to develop methods for the application of the poison.

The first question which occurs to the cotton farmer contemplating poisoning is whether it will pay to do so. From the following, any cotton grower should be able to determine the question for himself:

It will pay to poison—

If the weevils are really injuring your crop seriously; and if your land is sufficiently fertile to yield at least one-third bale per acre with weevil injury eliminated; and if your farming organization is such that you feel assured that the poison application will be made at the right time and in the right manner; and

If you are willing to spend the full amount necessary to provide an adequate amount of dusting machinery and poison.

You should not poison if the cost of the calcium arsenate, the cost of the labor to apply it, and the depreciation on the dusting machines will total more per acre than the current value of 100 pounds of seed cotton.

Hand guns should be figured as depreciating 50 per cent in a season and the larger machines about 25 per cent.

The next important question is that of the dusting machinery which should be used. Extensive experience has shown that it is impossible to get satisfactory results by using makeshift devices to apply the poison, and the only safe procedure is to provide an ample supply of the specialized dusting machinery which is now on the market for the treatment of cotton for the control of the boll weevil. Machines of various types, prices, and capacities are now being manufactured which meet the requirements and circumstances of almost all classes of growers. The following is a brief description of the several types and their uses:

The hand gun is the smallest type of cotton-dusting machine, and, as the name implies, must be carried and operated by the laborer. These machines are generally quite unsatisfactory, owing to their necessarily frail construction and laboriousness of operation. The selling price ranges from $12 to $25 each. *They should be used only when no other machine is suitable.* Not more than 8 acres should be allotted to one hand gun, and it has generally been found inadvisable to attempt the treatment of more than 25 acres of cotton in one organization with hand guns.

The one-mule machine is the smallest of the traction type of dusters. It is a one-wheel, one-mule machine, which the operator handles as he would handle a walking cultivator or any other walking implement. The machine has two nozzles and will treat either two or three rows of cotton at a trip, thus covering from 15 to 20 acres in a night of operation. It should be allotted not more than 60 acres of cotton throughout a season. This machine is now selling at from $100 to $125.

The cart machine is a two-wheel two-mule machine, which straddles a row of cotton. It has three nozzles and will cover from 25 to 30 acres of cotton in a night of operation. It should be allotted not more than 100 acres of cotton for treatment through the season, and is the type most suitable for large farms. This machine is now selling from $250 to $400.

In the early stages of the dusting work an engine-power machine was tried and a few of these are still in use, but it has been found that they are generally too complicated for satisfactory operation except by expert labor. Still other types of machines to suit different conditions are in the process of development, but the present supply will meet almost any conditions fairly well.

The following condensed rules have been prepared for the guidance of those planning to poison:

Use only pure calcium arsenate in the form of a dry powder.

Apply this only in the dust form.

Purchase this to conform to the following specifications:

Not less than 40 per cent total arsenic pentoxide.

Not more than 0.75 per cent water-soluble arsenic pentoxide.

Density not less than 80 or more than 100 cubic inches per pound.

Use only dusting machinery especially constructed for cotton dusting.

Poison only when the air is calm and the plants are moist. Practically, this means making only night applications.

Use about 5 to 7 pounds of calcium arsenate per acre for each application.

Start poisoning when the weevils have punctured from 10 to 15 per cent of the squares.

Keep your cotton thoroughly dusted until the weevils are under control. This usually means about three applications at the rate of one every 4 days.

Then stop poisoning until the weevils again become abundant.

If the weevils become abundant early enough to injure your young bolls, make one or two more applications late in the season.

If you have a heavy rain within 24 hours after dusting, repeat this application immediately.

Do not expect to eradicate the weevils. Poisoning merely controls them sufficiently to permit a full crop of cotton, and you can always find weevils in the successfully poisoned fields.

Keep your cotton acreage low and do everything possible to increase your yield per acre, as it costs just as much to poison one-quarter-bale-per-acre cotton as a bale-per-acre cotton.

Always leave an occasional portion of a cut unpoisoned for comparison with the adjoining poisoned tract. This will show how much you have increased your yield by poisoning.

Indirect Means of Control.—In addition to the use of poison, there are numerous other practices which tend to reduce the weevil injury, some of which are of general value, while others can be used only locally. The following pages list a few of the more important of these measures. Even when poisoning is practiced, the most thorough attention should be given to the indirect means of control, since they reduce the amount of poisoning which might be necessary and increase the profit which may be secured.

Fall Destruction of Infested Plants.—One of the most important steps towards reducing the weevil infestation, when it can be practiced, is the destruction of the cotton plants in the early fall, before the weevils have hibernated. To be of the greatest value, however, the plants must be completely destroyed by fire or plowed under deeply before the first killing frost, and this limits the use of this control measure to the districts where conditions are such that the entire cotton crop can be picked in time to permit such an early plant destruction.

For many years preceding the development of the calcium arsenate method of control, removal of the cotton plants from the field as early as practicable in the fall was advocated by this department as the most important step in controlling the weevil. The purpose of this operation is to destroy as many as possible of the immature forms of the weevil still remaining in bolls and squares. These immature forms, if undisturbed, transform into weevils which live over winter and lay eggs the following spring. Fall destruction of plants likewise eliminates hibernating places of the weevil in the field.

Grazing.—In some districts where it is impossible to practice fall destruction of the plants, somewhat the same results can be accomplished by grazing the field with cattle, sheep, or goats. This is only a local measure, however, since the supply of livestock in regions where the bulk of the cotton crop is produced is not sufficient for the purpose. Even when poisoning is practiced, fall grazing is still advisable, and no danger need be apprehended of poisoning the stock, since there is rarely sufficient poison on the plants to injure stock, even immediately after application, and, furthermore, a considerable period of time generally elapses between the last poison application of the season and the earliest grazing.

Sprout and Volunteer Cotton.—Considerable local difficulty in the control of the boll weevil is experienced in southern Texas and occasionally in Louisiana, owing to the prsence of stumpage or sprout cotton. Sprout plants are sometimes encouraged, because they produce a small but very early crop. This may have been defensible before the advent of the boll weevil, but at present the practice is undoubtedly the worst that could possibly be followed. The weevils seek out these plants in the early spring and produce progeny much earlier than they otherwise could, and these progeny infest the planted cotton at an abnormally early date. Volunteer cotton causes the same results over a considerable portion of the Cotton Belt. The cotton seed scattered about seed houses and gins and along roadsides frequently produces plants which furnish early-season breeding places for the weevil. Needless to say, all such plants should be destroyed.

Destruction of Weevils in Hibernation.—It is often possible for the farmer to reduce considerably his spring weevil infestation by proper winter clean-up measures around his fields. The weevils will hibernate successfully in any trash or rubbish, and it is a very good practice to burn over or clean up any such situations around the cotton field during the winter, especially the fence rows and ditch banks.

In addition, much can be accomplished by the elimination of the hibernation quarters. Especially along the more northerly portions of the weevil zone, the most successful hibernation is largely confined to the timbered areas, and as a result serious weevil injury is experienced only in the fields adjoining such timber. Under such conditions it is of the utmost importance to plan all clearing operations so that the open areas for cultivation are consolidated into as large tracts as possible, and thus increase the amount of land which is sufficiently distant from timber to suffer a minimum amount of weevil injury.

Locating Fields to Avoid Weevil Damage.—Nearly every farmer who has been raising cotton for a few years in the presence of the boll weevil knows that there are certain fields on his place where the weevils always appear first and in greatest numbers. With this information as a basis, it is some times possible to reduce the damage by refraining from planting cotton in such fields and planting the more distant fields. This practice, however, is advisable only when no attempt is made to control the weevil by poisoning. The fields adjoining timber where the weevil infestation is heaviest are usually the new lands of the place, and are thus the most fertile and capable of producing the best cottom crop if the weevil injury is eliminated. Furthermore, when such fields adjoin hibernation quarters, the weevils concentrate on them instead of scattering over larger areas, and they serve to a certain extent as trap crops, making it possible to poison the weevils on these fields and thus prevent their spread over the remainder of the crop.

Procuring an Early Crop of Cotton.—The foregoing facts relative to the life history, hibernation, emergence, and multiplication of the weevil show very plainly the importance of producing the cotton crop just as early in the season as possible. In reality, the production of cotton in the presence of weevils is nothing more or less than a race between the setting of bolls on the plant and the multiplication of weevils, and everything possible should be done to aid the cotton plants in winning this race. The following are some of the more important steps which may be taken.

Early Removal of Plants and Preparation of Land.—The first step in producing an early crop is the removal of the plants, so that the land may be plowed during the fall and winter and the seed bed thoroughly prepared. Just how much can be done is, of course, a problem for the individual farmer to determine and depends largely upon labor and weather conditions, but the importance of a well-prepared, solid seed bed can hardly be overestimated. Furthermore, unfavorable weather conditions shortly before planting often prevent plowing at that time, and early preparation does away with the risk.

Use of Early Varieties of Cotton.—One of the most important steps which have been taken to reduce the weevil damage has been the development and introduction of varieties of cotton which mature their crops earlier in the season than those varieties which were planted before the weevil invasion. The variety to be planted in order to obtain a profitable crop under weevil conditions will depend upon a number of factors as well as on the severity of the infestation. The soil, climate, and other factors must be considered. In many localities it is extremely important to select varieties which are resistant to diseases. The first effect of the boll-weevil invasion was to force the abandonment of the longer-staple and large-boll varieties of cotton and the adoption of small-boll, early varieties of very short staple, such as King and its derivative, Simpkins. During recent years, however, numerous other varieties have been developed to the point where they are sufficiently early to mature a crop in the presence of the weevil, and these are being rapidly adopted.

The long-staple cotton situation is particularly interesting. The upland long staples, which were cultivated when the weevil arrived, were slow-maturing, non-prolific varieties with a very thin boll wall, and were thus subject to a maximum amount of weevil damage. The weevil soon eliminated practically every one of these varieties, and for some time it appeared that the production of long-staple cotton would be practically prohibited. The situation, however, has been met by the development of several long-staple varieties which are prolific and reasonably early. . . .

Fertilizers.—An important step in procuring an early crop is the use of commercial fertilizers. In many large areas in the Cotton Belt the land is not impoverished to the extent that it actually needs fertilizers under normal conditions, but it has been demonstrated many times by the different experiment stations in the South that the maturity of cotton frequently can be hastened materially by the use of fertilizers. On impoverished soils, moreover, fertilizers containing a high percentage of nitrogen give increased yields under boll-weevil conditions.

Other Cotton-dusting Machines.—In addition to the types of dusting machines mentioned by Hunter and Coad[4] on a previous page, there are some others of more recent development that are of some importance.

The Saddle Gun.—This machine is made to fasten to a saddle and to be used from the back of a mule. It is similar in construction to the hand gun described by Hunter and Coad[4] but is somewhat larger, holds more poison, and has two nozzles. With this machine two rows may be covered at a trip. It is less laborious to use than the hand machine and will cover twice as much acreage in a given length of time. It is tiresome to use, however, and does not make a very effective cloud of dust. It may be used on a small acreage fairly satisfactory if a larger machine is not available.

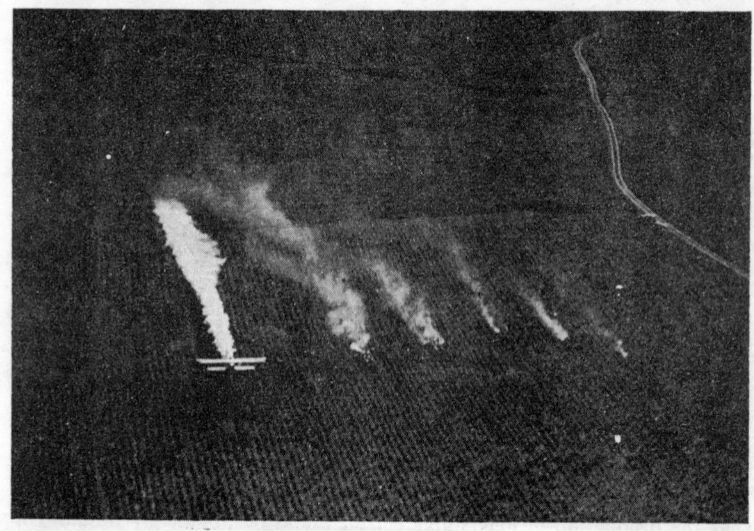

Fig. 79A.—Types of cotton dusting machines in use. From left to right, airplane, power duster, cart traction duster, one-mule machine, saddle gun, hand gun. (*Courtesy R. C. Gaines, Bureau of Entomology U. S. Department of Agriculture.*)

Power Dusters.—Some of the power dusters made within the last few years have proved to be very satisfactory. A light gasoline engine suitable for mounting on a cart-type duster has been developed. This engine is simple and inexpensive to operate, and the power that it furnishes supplies an air blast sufficient to make a good cloud of dust. With this machine five rows may be covered satisfactorily at one trip.

Another type of power duster being used to some extent has but a single large nozzle. The nozzle is either stationary or is made to oscillate back and forth to distribute the dust. A strong air blast carries the dust into the air and across 20 or more rows of cotton. It is possible with this machine to dust a cut by driving along the edge or by driving through it about every 25 rows. This machine is speedy, but it probably does not give a very even distribution of the poison.

Dusting with Airplanes.—In some of the Delta areas of Mississippi and Louisiana, a considerable acreage of cotton is dusted from airplanes for boll-weevil control. The dust is placed in a large hopper in the body of the plane. Air currents carried into this hopper by funnels opening forward force the dust out beneath the machine. It is here dispersed by the air current generated by the plane as it moves onward at a rapid pace. A dust cloud forms in a long uniform band behind the plane and gradually settles to the ground beneath, covering the cotton plants very uniformly and thoroughly.

The airplane method of dusting is not very practical except in areas where there are large fields of cotton or many adjoining small fields. With planes the dust may be applied quickly, only a few seconds being required to dust an acre uniformly and economically. Certain companies own a number of airplanes and charge a moderate sum per acre for protecting the cotton or for giving it a definite number of applications of poison. It may cost the grower even less to hire the planes than it would to dust the cotton himself.

How the Boll Weevil Ingests Poison.—When calcium arsenate was first used for poisoning boll weevils on cotton, it was thought that the weevils got the poison by drinking the dew on the plant that had been treated. A chemical examination of dew, however, showed that there was practically no poison in it; furthermore, it was shown that poison on plants under a shelter where no dew formed on them destroyed the weevils placed on the plants.

Observations made by Grossman[13] and others showed that the weevil in crawling over a plant frequently touched the tip of his proboscis to the surface, by this means collected some of the poison on his mouth parts if the surface of the plant was covered with poison, and was apt to swallow some of it when he started feeding. Grossman[13] gathered further evidence by catching weevils and dipping the tip of their snout into the dust and into poisoned molasses. Weevils so treated had a much higher mortality than untreated weevils.

Florida Method of Boll-weevil Control.—A method of control somewhat different from the one outlined in the preceding pages has been devised by Professor Smith of the State Plant Board of Florida. A short time after all boll weevils have come out of hibernation, all the squares on the cotton plants are removed by hand and destroyed. The plants are then dusted with calcium arsenate, especially the buds at tip of stem and branches. The immature weevils are destroyed by the square removal, and adult forms are poisoned. This method of control seems to have given good results in Florida, but it is doubtful if it will be satisfactory farther north where weevil emergence is much later.

Use of Poisoned Molasses for Weevil Control.—In South Carolina and in other areas, a mixture of 1 pound of calcium arsenate, 1 gallon of water, and 1 gallon of cheap molasses is used in poisoning the boll weevil. The mixture is applied by brushing the leaf bud at the tip of the stem and the upper part of the stem with a homemade mop or brush which has been dipped into the solution. It can be applied rapidly by simply dragging the brush over the top of the plants. One to two gallons will treat an acre. The first application is made before the squares on the plants are large enough for the weevils to puncture. This application destroys most of the hibernated weevils, and no other treatment may be needed. Usually, however, weevils increase in number later, and some planters continue the poisoned-molasses treatment, whereas others use calcium arsenate dust. The advantage of the poisoned-molasses method is that it is cheap, especially in so far as the chemical is concerned, and no special equipment is needed for applying the mixture.

Numerous experiments carried on at the Boll Weevil Laboratory at Tallulah, La.,[12] from 1921 to 1929, showed that the use of the molasses-calcium arsenate mixture under Louisiana conditions gave some increase in yield over untreated checks but that it did not give so large an increase as calcium arsenate dust. In one set of 16 tests, the molasses mixture, first applied before squaring and then continued at 4- and 7-day intervals until 7.4 applications had been made, gave an increase in yield of 5.7 per cent over the untreated checks. In 14 experiments with calcium arsenate dusting, carried on under similar conditions, except that only 5.6 applications were made, the dusted plots made 24.2 per cent more seed cotton than the untreated checks.

The molasses mixture used as a presquare treatment gave only a 2.4 per cent gain. When used as a presquare treatment and followed by dusting, a 21 per cent gain was secured. Calcium arsenate dusting instead of the molasses presquare treatment on the same date and calcium arsenate dusting on all the other dates the same as above gave an average gain of 30.4 per cent over the check.

In boll-weevil experiments conducted by the laboratory at Florence, S. C., early applications of the molasses mixture gave only 5 per cent increase in yield of seed cotton. An early application of the mixture followed by dustings gave 39.2 per cent increase. The dustings without the early application of the mixture gave an increase of 29.0 per cent.

Objections to the Use of Calcium Arsenate.—Serious drawbacks to the use of calcium arsenate in boll-weevil control are:

1. It costs so much to buy the poison and machines and to hire the labor necessary to apply the poison that it is not recommended for land that will not make one-third of a bale of cotton to the acre. Much of the land of the Cotton Belt does not yield that amount.
2. The work must be done at night largely, which makes it disagreeable and costly.
3. During rainy seasons, when the control is most needed, it is a difficult matter to keep the poison on the plants, and the ground is so soft that the machines will not run well.
4. Lice usually multiply on plants that have been poisoned, causing a partial defoliation and a lowering of the grade of the lint cotton produced.
5. The gain in yield is very frequently not sufficient to offset the expense of the poisoning.
6. In some cases, the use of heavy applications of calcium arsenate to cotton interferes with other crops grown subsequently on the same soil.

Importance of Cultural Methods.—Under the boll-weevil regime, it is very important to have a full stand of cotton plants and to space them closely—close enough so that all the space between plants in the row will be covered early in the season. With wide spacing, in course of time the plants may become large enough to cover all the space in the row and yield in proportion to the space occupied; but if weevils are plentiful, they prevent the formation of fruit the latter part of the season, and, consequently, there will not be so much fruit put on as there would be in a row containing more but smaller plants.

It is also necessary to have land well drained, so that the crop can be planted early and cultivated properly. Cultivation should be frequent and thorough enough to keep down all grass and weeds and to allow the plants to grow freely. One cultivation a week is a good rule. Cultivation should be continued late in order to keep the plants growing and forming squares until there is a good crop of large bolls. As long as there are plenty of squares for the weevils to attack, they will not bother the bolls seriously. It pays to continue plowing as long as there are numerous bolls liable to weevil damage. In the central part of the Cotton Belt, cotton should not be "laid by" before the first of August.

Ineffective Methods of Control.—On account of the severity of boll-weevil damage and the difficulty in controlling it, a great many different methods of control have been proposed, many of them of no value or impracticable. The Entomological Laboratory at Tallulah, La., has tried out a great variety of control measures that have been proposed. The following discussion from Hunter and Coad[4] is based largely on results secured at that laboratory.

Late Planting.—Late planting is foremost among the futile means of control. At various times it has been suggested that late planting, especially if following early fall destruction, would so lengthen the hibernating period that no weevils would survive. Very numerous experiments in the field and in cages have proved that the weevils in considerable numbers are able to survive from any reasonable time of early destruction in the fall to beyond the date in the spring when any return whatever could be expected from planting cotton, even if the weevils were entirely eliminated. In a field experiment in Kerr County, Texas, the plants were removed very thoroughly early in November. Neither stumpage nor volunteer plants were allowed to grow during the winter. No other cotton was planted within 9 miles. On the experimental field, planting was deferred until June 10. In spite of this fact, weevils appeared as soon as the plants were up, and multiplied so rapidly that the production was not sufficient to warrant picking. Similar experiments under different conditions by the State Crop Pest Commission of Louisiana agree in every way with those of the Bureau of Entomology in Texas.

Weevil and Square Collection.—The possibility of weevil control by hand picking of the adults in the early spring and of the infested squares later in the season has been thoroughly tested on numerous occasions. Undoubtedly, this method is efficient when practiced with sufficient thoroughness, but numerous attempts to carry it out on a practical scale have shown that the labor difficulties are almost always prohibitive. This work is of value only comparatively early in the season and thus falls at the same period when there is a very heavy demand for the labor for other purposes, and it is generally impossible to collect the weevils or infested squares without neglecting other more important work. Consequently, this procedure can be recommended only under rare conditions, when the infestation is not excessively heavy and an abundance of cheap labor is available.

Many attempts have been made to collect the weevils by means of mechanical devices. Hundreds of such devices have been tested and all are to be condemned. They do not collect an appreciable number of weevils unless they are so violent in the agitation of the cotton plant that they are actually injurious to it.

Trap Rows.—The idea of attracting weevils to a few early plants or trap rows has frequently been advanced. Practical experience, however, has shown that the only possibility of success in such a procedure lies in the use of entire fields adjoining hibernation quarters, the fields to be poisoned later. The use of a few rows as a trap crop has been found to be absolutely valueless.

Attraction to Lights.—Many insects more or less resembling the boll weevil are attracted to lights Many attempts have been made to destroy the cotton pest by taking advantage of this supposed habit. The boll weevil, however, is not attracted to lights. Numerous tests have been made in which many thousands of other insects were collected around strong lights in cotton fields, but not a single boll weevil was found, in spite of the fact that there were multitudes of these pests in the fields surrounding the lights.

Chemical Treatment of Seed.—Any money expended by the farmers in attempting to destroy the boll weevil by soaking the planting seed in chemicals in the hope of making the plants that are to grow from them distasteful or poisonous to the insects would be entirely wasted. The same remark applies to the various proposed treatments of the plants or soils with chemicals which are supposed to be taken up by the plants to the detriment of the weevils feeding upon them.

Topping of Plants.—The topping of plants is sometimes recommended for fields infested with boll weevils. The practice generally results in more harm than good, since it removes a portion of the plants upon which the weevil is most dependent for food during the latter part of the season, and furthermore practically always produces an exceedingly dense foliage growth which greatly reduces the sun control of the weevil stages and promotes such dangerous diseases as boll rot.

Sweetened Poisons.—Many attempts have been made to make poisoned substances attractive to the weevils by introducing sweets and other ingredients. Some known sweets, such as honey, have a slight attraction for the weevil, but not enough to assist in practical control. Numerous tests of such sweetened mixtures have been made and it has always been found that, though they may have a slight value, results are far inferior to those which can be obtained by applying dry calcium arsenate under the same conditions.

Contact Poisons.—Poisons designed to kill the boll weevil by suffocating them have been proposed. They cannot, of course, be effective against the immature weevils within the cotton fruit. Normally, also, the adult weevils are found inside the bracts of squares, where they cannot be reached by sprays. Numerous chemicals have been found which, if placed directly on the weevil, will cause immediate death, but this does not mean that the chemicals are of the slightest value when applied in the field. In the first place, they are nearly always exceedingly injurious to the cotton plant, and furthermore, when applied to the plant under field conditions, do not come in contact with the weevils sufficiently to kill any appreciable number. In spite of the numerous chemicals tested, not a single contact poison has been found to have any practical value in field use against the weevil.

Repellents.—In the same way it has been claimed that numerous chemicals, fumigants, etc., have a repellent value against the boll weevil. Almost every conceivable compound has been tested for this action, and not a single one has been found which had the slightest repellent action against the weevil.

Other Proposed Remedies.—Many other remedies have been suggested for the weevil. Literally, hundreds of these have been carefully investigated, and it has been found that the claims of their advocates were based on faulty observations and careless experiments. The claims made at different times of the repellent power of tobacco, castor bean plants, and pepper plants against the boll weevil have no foundation whatever. In fact, none of these plants has the least effect in keeping weevils away from cotton.

Effect of the Boll Weevil on Cotton Production in the United States.—Since 1892, the year that boll weevils first entered the United States, there has been a gradual upward trend in the amount of cotton produced. The increase has been marked by fluctuations, for example the low-crop years of 1921, 1922, and 1923, the very high production of 1926, and the low production of 1934 and 1935, due to the reduced acreage. Hinds,[5] as a result of his studies, estimated that the boll weevil has reduced yields in fields invaded 37 per cent on the average. This reduction has been offset by greater acreage in the western part of the Cotton Belt and better cultural practices resulting in better yields in other places. Good yields may be secured by the use of poison even if weevils are present, but here the greater yield is obtained at the expense of greater effort.

In some respects, the boll weevil has been a blessing in disguise to the cotton growers. It has forced them to use better cultural methods in growing cotton, and these are being used on other crops as well. It has caused many to keep more livestock and to grow other crops in addition to cotton, thus following a better balanced system of agriculture.

THE PINK BOLLWORM

The pink bollworm (*Pectinophora gossypiella* Saunders) was described by Saunders, of England, in 1842, from specimens sent to him from India, where the insect was observed doing damage in cotton fields. India or other parts of southern Asia were probably the original home of the pink bollworm, but it is now found, according to Hunter,[14] in East Africa, West Africa, Egypt, Angola, Nigeria, Sierra Leone, Sudan, Zanzibar, India, Bengal, Ceylon, Burma, Siam, China, Chosen, the Philippines, Hawaii, Porto Rico, Santo Domingo, Haiti, Mexico, and Australia. In

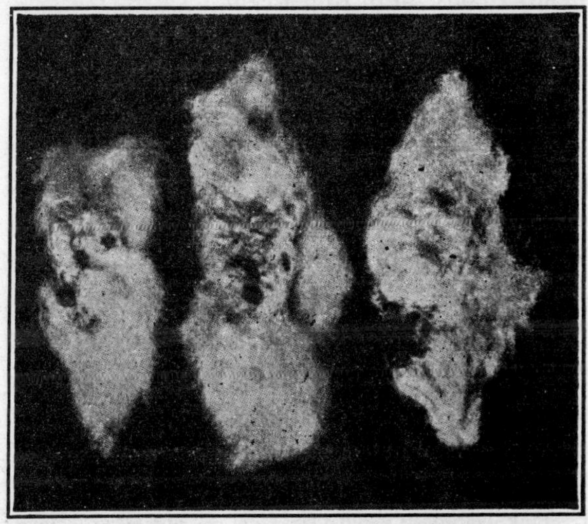

Fig. 80.—Cotton locks showing typical injury by the pink bollworm (*Pectinophora gossypiella*). (*After Hunter.*)

the United States, specimens have been found in Arizona, New Mexico, Texas, Louisiana, Georgia, and Florida, but infestations here are believed to have been eradicated or placed under control.

The pink bollworm was introduced into Egypt in 1906–1907 in cotton sent from India for ginning. It was carried from Egypt to Brazil in cotton seed in 1911–1913 and to Mexico in the same way in 1911. In 1916, 446 carloads of cotton seed were shipped from Mexico to oil mills in 11 different places in Texas. A part of this seed contained pink bollworms, and as a result infestation was started in the region about Hearne and about Beaumont, two of the places that had received seed. Infestations found in other parts of Texas are traceable to other sources.

Damage.—The pink bollworm ranks at least second among destructive cotton insects. The larvae enter seeds, eat their kernels, and render them practically valueless. The lint on the infested seeds fails to reach full development, being short and kinky (Fig. 80). A part or all of the locks of many bolls are practically ruined. Bolls formed early in the season are infested much less than later ones. Gough examined 106,400 bolls in Egypt one season. Infestation varied as follows:

July bolls, less than 10 per cent.
August bolls, 10 to 25 per cent.
September bolls, 25 to 75 per cent.
October bolls, 75 to 89 per cent.

Fig. 81.—Pink bollworm; adult, larva, pupa, and egg. (Figure reduced slightly in reproduction but cut shows insects enlarged about two and one-half times.) (*After Hunter.*)

It is estimated that there is a loss of 17 per cent of the Egyptian crop. In Brazil, the estimate of damage runs from 30 to 66 per cent of the crop; in Mexico, in 1917, the estimated damage was from 30 to 50 per cent.

Appearance and Life History.—The pink bollworm is the larval stage of a small dark-brown moth. The different stages of the insect are described by Hunter[6] as follows:

The pink boll worm has four stages, namely, egg, larva, pupa, and adult, or moth. The moth [Fig. 81] resembles somewhat the common clothes moth of this country. From tip to tip of the extended wings it measures from ⅗ to ⅘ inch. It is of a dark-brown color, the fore wings ending in a rather sharp point. The hind wings are somewhat broader than the fore wings and end in an even sharper point. The eggs are very small objects, somewhat oval, about $\frac{1}{25}$ inch long and $\frac{1}{50}$ inch broad. The surface is white and finely wrinkled. The larva [Fig. 81], when first hatched, is glassy white with light-brown anterior markings. It grows rapidly and when mature measures nearly $\frac{1}{2}$ inch in length. It is cylindrical, white, with the dorsal side strongly colored with pink. The pupa [Fig. 81] is about $\frac{2}{5}$ inch in length, reddish brown, with the posterior end pointed and ending in a hook-like process.

The small, pink corn worm (*Pyroderces rileyi* Walsingham) is sometimes mistaken for the pink bollworm. It is of much the same size and color and is frequently found in cotton bolls, but it enters only old decaying or diseased bolls. It does not enter perfect seed and does no damage to cotton.

According to Busck,[10] who has spent considerable time studying the pink bollworm in Mexico and in Hawaii, the female pink bollworm lays eggs singly or in small groups on the green bolls or in the flowers. One moth will lay about 100 eggs. These hatch in 4 to 12 days. The larva immediately bores its way into the boll and attacks the seed. As soon as the contents of one is eaten, it enters the one next above. As a rule, the larva remains within the lock, but it may drill a hole through to an adjoining lock and enter. It never leaves one boll to attack another. When full grown, the larvae frequently web together two seeds which they have hollowed out, thus making a cell of some size. These double seeds are characteristic of the work of the pink bollworm, and they furnish a quick method of determining whether or not a lot of seed is infected.

The larva develops in 20 to 30 days during the summer, but later in the season the period is considerably longer. Gough reports larvae in Egypt remaining alive in a quiescent condition for over 2 years. Busck found larvae alive in cotton bales 18 months after having been placed in the bales. Following the larval stage is a pupal stage, which lasts 10 to 20 days. The moth that hatches from the pupa lives but a comparatively short time, the majority, according to Busck, living only 14 to 20 days.

The pink bollworm has been reported as feeding on several different species of plants, but more recent studies have indicated that it uses only cotton or some closely related species in the same family of plants.

Control Measures.—Although the pink bollworm has a number of insect enemies, apparently none of them is very troublesome, since they have but little influence as control measures. According to Hunter,[6] the most important one is a small mite, *Pediculoides ventricosus* Newport.

Since the insect winters in cotton seed, it is possible to fumigate the seed or to sterilize with heat sufficiently great to kill the insect but not destroy the vitality of the seed. Collecting all undeveloped or defective bolls left in the field by the pickers is also an aid in control.

Several indirect methods of control, such as are used in boll-weevil districts, as early varieties, early planting, good culture, are helpful.

As soon as the pink bollworm was found in Texas, the Federal Horticultural Board of the U. S. Department of Agriculture started active operations to eradicate it. A large force of entomologists was assembled, and much scouting was done to determine the limits of the infestation. Laborers went over the infested fields and some adjoining ones, collecting and destroying with kerosene all trash and all portions of cotton plants. The cotton picked from the infested fields and near-by ones was also handled so as to prevent any infestation's coming from it. The planting of cotton in the areas about the regions where the bollworms were found was prohibited by the governor.

To safeguard against further introduction of the pest into the country, the Department of Agriculture put forth several regulatory measures, some of which were:

1. The exclusion from the United States of cotton seed from all foreign countries except the Imperial Valley of Lower California, and exclusion also of seed from Hawaii.
2. Regulation of the entry of cottonseed products from all foreign countries and from Hawaii.
3. Regulation of entry and disinfection of all imported cotton and cotton waste and also burlaps that have been used as wrapping of foreign cotton, including such material from Hawaii.

COTTON INSECTS OF LESSER IMPORTANCE

Although the boll weevil and pink bollworm are the cotton insects that do by far the most damage, there are several others that injure cotton plants more or less which deserve mention.

The cotton leaf worm; the southern grassworm, bollworm, and red spider; and the cotton aphis, or louse, are the most important of these.

The Cotton Leaf Worm or Cotton Caterpillar (*Alabama argillacea* Hübn.).—The cotton leaf worm, also sometimes called "army worm," eats leaves, flowers, young bolls, and other tender parts of cotton plants. It probably ranks third among American cotton insects in respect to destructiveness. It has been known in the United States since 1793 and was for a good many years considered a very serious cotton pest. Its seriousness was largely due to the fact that there were no satisfactory control measures prior to 1871. About that time it was discovered that it could be poisoned effectively by the use of paris green or other arsenical poisons.

Since 1890, the cotton leaf worm has appeared rather sporadically, but it would have done considerable damage certain years had it not been controlled. As a rule, it has appeared during the latter part of July or during August, so late in the season that it did but little damage and in many instances was a distinct help in that plants were defoliated and checked in their growth, thus robbing the boll weevil of food the latter part of the season.

The cotton leaf worm is thought to be a native of Central and South America. It probably migrates northward each spring or early in the summer. It flies freely, having been reported some years as far north as Canada.

Life History.—The eggs of the cotton leaf worm are deposited singly on the underside of leaves. They are pale green in color and not difficult to distinguish. They hatch in 3 to 20 days, the time depending on temperature, producing a larva that feeds voraciously on the younger leaves. The larvae (Fig. 82) are about 1½ inches in length when grown and somewhat varied in coloring, some being yellowish green without prominent stripes, whereas others have a black stripe down the back with a fine central yellow stripe. Each segment has four black dots above. The larva when mature webs itself in a leaf rolled up on one side and there pupates. Other larvae frequently eat away the leaf, leaving the pupa (Fig. 82) hanging by a thread. The larval stage lasts from 10 days to 3 weeks. From the pupa there hatches a tawny-colored moth with a wing spread of a little more than an inch. The pupal stage lasts 1 to 4 weeks.

Control Measures.—The cotton leaf worm is easily controlled by poisoning with powdered lead arsenate or calcium arsenate.

Fig. 82.—Cotton leaf worm: stages and work. (*After Pierce.*)

The poison dust is most effective if applied when the plants are moist and is best applied with dusting machines, such as are used for boll-weevil poisoning, at a rate of 4 or 5 pounds per acre. Another method of applying the poison which is fairly satisfactory is known as the "pole-and-bag method." In applying the poison, the operator rides a mule between rows of cotton, carrying in front of him a pole about 4 feet long, to each end of which is attached a mulsin bag containing poison. The bags are agitated sufficiently as the mule walks along to cause the poison to shake out slowly. By weighing the bags before and after a known area

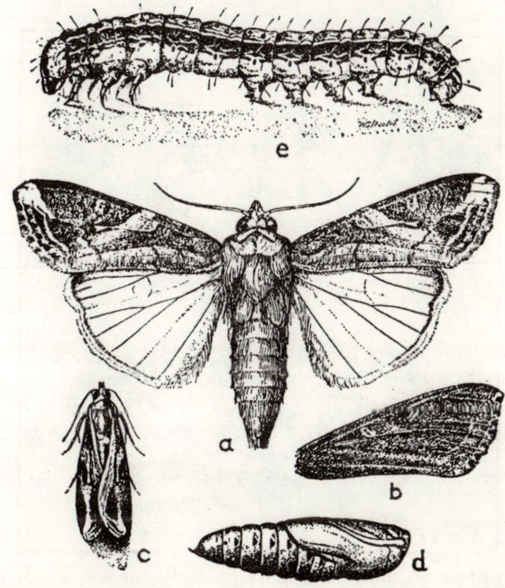

FIG. 83.—Southern grassworm, or fall army worm: (*a*) adult male moth; (*b*) right front wing of female moth; (*c*) moth in resting position; (*d*) pupa; (*e*) full-grown larva. *a, b, d, e,* about twice natural size; *c*, about natural size. (*Walton and Luginbill.*)

has been covered, the operator may know just how much poison is being applied. Poisons should be applied before defoliation becomes heavy.

Southern Grassworm.—The southern grassworm (*Laphygma frugiperda* S. and A.) (Fig. 83), also called "fall army worm," is another insect that at times does some damage to cotton plants. The larvae appear in numbers and feed on many different plants, including grasses and cultivated plants. They gnaw the stems and eat the growing tips of cotton plants. They enter the ground to pupate.

The southern grassworm may be controlled with poisons. Methods such as are used in poisoning the leaf worm are satisfactory.

The Cotton Bollworm.—The cotton bollworm (*Chloridea obsoleta* Hübn.) (Fig. 84) destroys cotton squares and bolls by eating their interior. In some localities, especially in Texas, the damage done is considerable. This insect is also known as the "corn-ear worm" and the "tomato fruit worm," because it attacks corn and tomatoes.

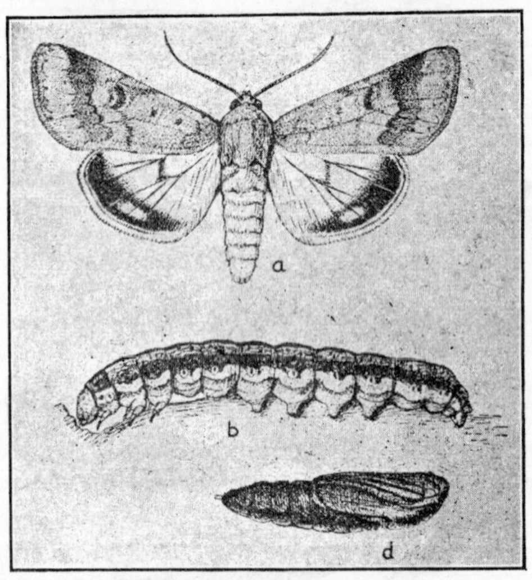

Fig. 84.—Bollworm: (*a*) moth, or adult; (*b*) larva, or worm; (*d*) pupa. About two-thirds natural size. (*Howard.*)

Eggs are laid on leaves. After hatching, the young larva feeds on the surface of the leaves for a short time and then enters a square or boll. After destroying the contents, it seeks another, so continuing until several have been ruined. When full grown, it enters the ground to pupate. A moth hatches from the pupa in about 2 weeks. This insect passes the winter as a pupa in the soil.

But little effort is made to control the cotton bollworm. Fall plowing and frequent plowing of growing crops are helpful, as the pupae may be destroyed in the soil. When the attack is serious, poisoning may be resorted to, the same plan being followed as was given for the cotton leaf worm.

Cotton Square-borer.—The cotton square-borer (*Uranotes melinus* Hübn.) (Fig. 85) is described by Pierce[7] as follows:

Frequently, the squares are bored by small, oval, flattened, pea-green, velvety haired larvae, known as square-borers. Each larva can destroy

Fig. 85.—Cotton square-borer: (*a*) adult butterfly, top view; (*b*) same from side, with wings closed; (*c*) larva, or borer, from side; (*d*) pupa; *a–d*, somewhat enlarged. (*Howard.*)

many squares. The larvae transform into small pupae, and when mature these become the dainty little blue swallow-tailed butterflies so often seen in the cotton fields. It is seldom necessary to take active measures against them.

Cotton Red Spider.—The red spider (*Tetranychus telarius* L.) on cotton is not a spider but a small mite (Fig. 86) which attacks cotton, beans, corn, tomatoes, and many other plants both wild and cultivated. It is rather generally distributed throughout the southern states, but it is most prevalent in the eastern states of the group. It is usually brick red in color and about 1/50 inch in length, being so small that it is difficult to see it without a magnifying glass. Numbers of red spiders together give the cotton leaf the appearance of rust, and the disease is sometimes wrongly called "rust."

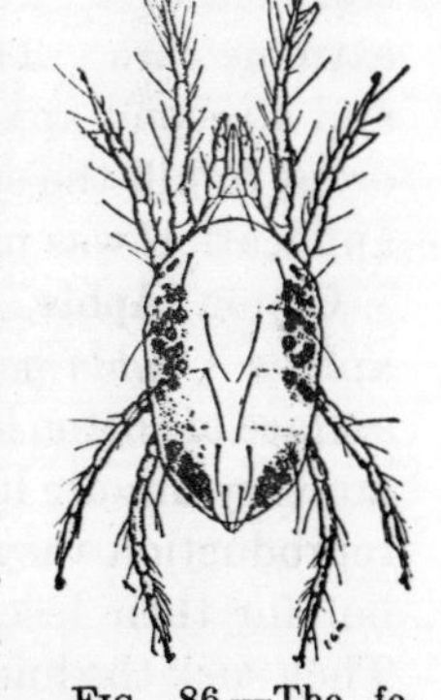

Fig. 86.—The female red spider. Highly magnified (*McGregor and McDonough.*)

As a rule, red spiders become very numerous only in spots in fields. They rarely spread over more than a few acres but may do much damage in the area covered. They collect on the underside of cotton leaves, suck juices from them, and cause them to redden, curl up, and fall. Frequently the entire plant dies. McGregor[8] estimated that in South Carolina, in 1912, the loss occasioned by this pest amounted to two-fifths of the crop over an area of 20,000 acres.

Red spiders spread over a field, normally by crawling from plant to plant. The spread is often hastened by rains' washing them from the plants and transporting them some distance. Winds, animals, etc., may also assist in their dispersal. They go through the winter on weeds, such as sow thistle, Jerusalem oak, evening primrose, or the violet—plants that hold their leaves during the winter.

Since the red spider must have green leaves on which to winter, one important control measure is to destroy such plants as were mentioned above before the winter season. Outbreaks may sometimes be checked in the early stages by pulling up and carefully removing all infested plants and burning them. In case the infestation is widespread and serious, it is advisable to spray. Several sprays are satisfactory. Potassium sulphide (1 ounce to 2 gallons of water), lime-sulphur (homemade or commercial), and kerosene emulsion (prepared according to the usual formula) are recommended by McGregor.[8] Dusting with finely powdered flowers of sulphur is also a good control measure. To be effective, the spraying must be done with extreme care. The material must reach the underside of leaves, and a second application must be made a week after the first in order to kill the spiders that were in the egg stage when the first application was made.

Cotton Aphis, or Cotton Louse.—The small green lice, or aphids (*Aphis gossypii* Glover) (Fig. 87), found on the lower surface of cotton leaves are known to every cotton planter. They are very minute in size, but because of their extraordinary rate of reproduction they sometimes do considerable damage. Were it not for their insect enemies, they would do enormous damage. They suck the juices from the leaves, causing a curling or dwarfing of the leaves and sometimes death. Young plants, during cool weather, are frequently damaged considerably. Plant lice frequently become very numerous on cotton that has been dusted with calcium arsenate to control boll weevils. According to studies made by some of the men working at the Boll Weevil Laboratory at Tallulah, La., the white color of plants dusted with the arsenate attracts the winged forms of lice to the plants, and once there they multiply rapidly, since the poison has a tendency to destroy such predators as lady beetles and some of the parasites of the lice. The lice cause a shedding of leaves and often a blackening of the entire plant, including the surface

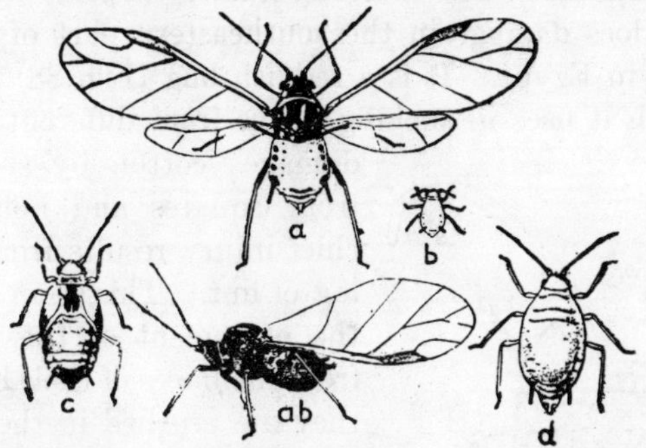

FIG. 87.—Cotton aphis: (*a*) winged female; (*ab*) dark female, side view; (*b*) young nymph, or larva; (*c*) last stage of nymph; (*d*) wingless female. All much enlarged. (*Chittenden.*)

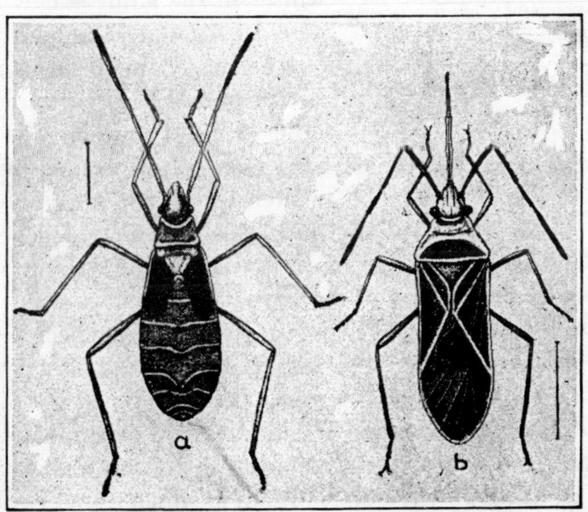

FIG. 88.—Cotton stainer: (*a*) fourth-stage nymph, or pupa; (*b*) adult. Enlarged. (*Insect Life.*)

of the lint in the open bolls. The blackening is due to a mold that grows on the honeydew excreted by the lice.

Aphid infestation on cotton may be controlled by the use of nicotine sulphate. This chemical may be mixed with calcium arsenate or with hydrated lime at the rate of 6¼ pounds to 100 pounds of the carrier. The mixture is applied with dusting machines at a rate of 6 to 9 pounds per acre. The dusting should be done when the air is calm, as the nicotine fumes must hang around the plants for a time to have satisfactory results.

Cotton Stainer.—The cotton stainer (*Dysdercus suturellus* H. Schf.) does damage in the southeastern part of the United States and in Egypt. It is a reddish bug (Fig. 88) with a long beak, which it uses in sucking juices from different plants. It damages cotton by sucking juice from squares and bolls, but the chief injury results from the staining of lint. This stain comes from the excrement of the insects and from the juices of their bodies when they are crushed in the picking or ginning process.

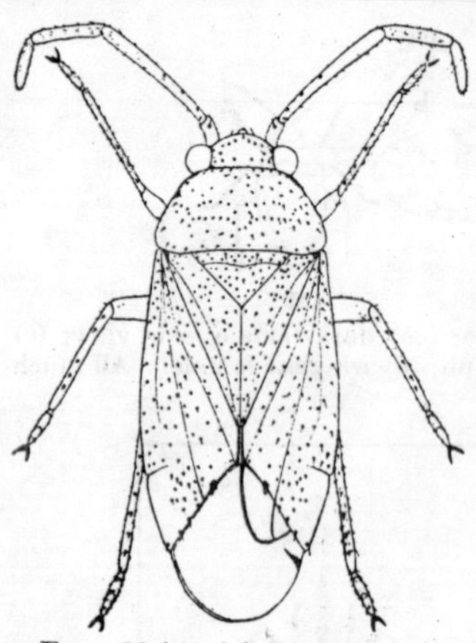

FIG. 88*A*.—Adult cotton flea hopper. Greatly enlarged. (*After Reinhard.*)

The cotton stainer passes the winter in the adult stage on weeds or old cotton stalks about the cotton fields. Control measures consist in burning old cotton stalks or other trash in which the insects may have taken refuge during the winter or in trapping them with bunches of cotton seed placed about the field. After the bugs have taken refuge in these bunches, they may be collected and destroyed with hot water.

Cotton Wireworms.—The larvae of different species of click beetles (*Monocrepidius vesperinus* Fabricius and other species), commonly known as "wireworms," feed on the roots of cotton plants and do considerable damage. They are mentioned by Pierce[7] as being especially troublesome in South Carolina. They are best controlled by a crop rotation, in which a small grain crop followed by cowpeas or some other legume is included.

Cotton Flea Hopper.—The cotton flea hopper (*Psallus seriatus* Reut.), sometimes called "cotton hopper" or "cotton flea," is another insect that causes considerable damage to cotton plants. In some localities, the damage is serious in certain years. The trouble has been most marked in the coast counties of Texas but has been prevalent in other parts of Texas and in other states.

The insect is a small green bug not more than ⅛ inch in length (see Fig. 88*A*). It has a wide distribution in the United States, having been found as far north as Minnesota and as far west as California. It breeds on several species of goatweed (Croton), on horsemint, and on many other weeds.

The flea hopper damages cotton principally by puncturing young squares. Punctured squares turn brown or black and fall from the plant. Another characteristic of flea-hopper damage is the abnormal growth and branching of affected plants (see Fig. 88*B*). After the hoppers have left the plants, new squares may be put on, and a crop of bolls produced. In many cases, the loss from the hopper damage is considerable.

As a control measure it is recommended that weeds near the cotton field be destroyed, especially Croton, or goatweed. Dusting cotton plants with finely ground sulphur at the rate of 10 pounds per acre when the air is calm is fairly effective in destroying the hoppers. Three or four applications may be required to give satisfactory control.

Fig. 88*B*.—Cotton plant deformed by the attacks of the cotton flea hopper. This plant shows late-season recovery as a result of which a few bolls were produced at the top of the plant. (*Courtesy of K.P. Ewing, U.S.D.A.*)

Tarnished Plant Bug.—The tarnished plant bug (*Lygus pratensis* L.) is a small yellowish-brown to blackish bug, variegated with shades of yellow, red, and black (see Fig. 88*C*). The nymphs, or young bugs, are green. The bug feeds on a large number of plants, weeds, fruit trees, truck crops, etc. It damages cotton in much the same way as the flea hopper and may be controlled by the same measures.

Rapid Plant Bug.—The rapid plant bug (*Adelphocoris rapidus* Say) belongs to the same family of insects as the cotton flea hopper and the tarnished bug and has similar habits. On cotton it punctures squares and young bolls, causing them to shed. The adult bug is dark brown above and edged on each side with yellow. The nymphs are pale green marked with red (see Fig.

88*D*). This insect may be controlled by dusting plants with sulphur, as was recommended for the flea hopper.

Thrips.—Thrips are tiny, slender, black, or tinged with yellow or red insects which are found in the flowers and leaf buds of many

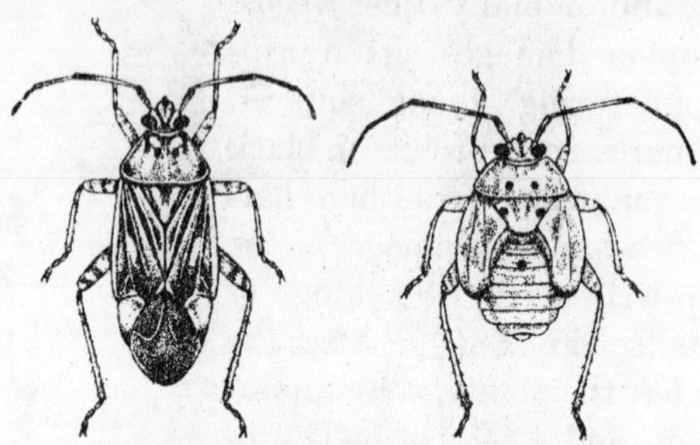

Fig. 88*C*.—The tarnished plant bug: adult bug at left; last stage of nymph at right. Nearly 4 times natural size. (*Orton and Chittenden.*)

plants. They are so small that it would take a dozen of them placed end to end to measure an inch. They are found very commonly in the terminal bud of young cotton plants 2 to 4 weeks of age. They damage the embryonic leaves in the bud, making them curled, ragged, and irregular when they unfold.

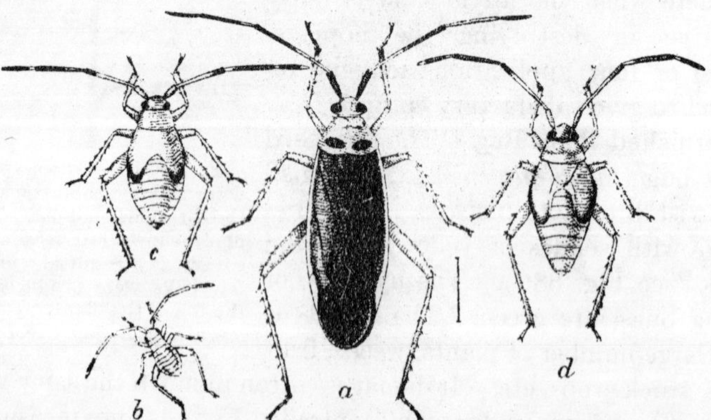

Fig. 88*D*.—Rapid plant bug: (*a*) mature bug; (*b*) young nymph; (*c*) fourth-stage nymph; (*d*) fifth-stage nymph. Enlarged to 4 times natural size. (*Sanderson.*)

This injury retards the growth of the new leaves and stunts the plants to some extent. If growing conditions are favorable, the plants soon outgrow the injury, and no permanent harm results. No practical control measures have been developed.

Corn Root Aphis.—The aphis that frequently attacks corn roots (*Aphis maidiradicis* Forbes) is sometimes found on the tap-root of cotton. A 3-year rotation consisting of corn, cotton, oats and cowpeas is a satisfactory control measure.

Miscellaneous Insects.—Several other insects feed on or live about cotton plants. Among these may be mentioned cutworms, which burrow into the soil during the day and then come out at night and cut off young plants; May beetles, which in the adult stage attack young cotton plants; grasshoppers, which eat cotton leaves rather seriously in Texas and occasionally in other regions; cotton bugs (several different species of sucking bugs, "squash" bugs, "stink" bugs, "plant" bugs, etc.), which attack cotton squares and bolls; flower beetles, frequently found in blooms of cotton, sometimes eating petals, stamens, and pistils; moth stalk borer, a caterpillar which tunnels through the heart of cotton stalks; snowy tree cricket, which makes egg punctures in stems; sharpshooters, which puncture stalks and lay eggs in leaves; ants, which destroy some boll-weevil larvae but are harmful in that they assist harmful aphids; lady-bug beetles, both the larvae and adults of which feed on aphids and are very helpful in holding them in check; and many parasites that prey on other insects mentioned.

References

1. Smith, George D.: "Studies in the Biology of the Mexican Cotton Boll Weevil on Short-staple Upland, Long-staple Upland, and Sea Island Cottons," *U. S. Dept. Agr. Bull.* 926, 1921.
2. Hunter, W. D.: "The Boll-weevil Problem with Special Reference to Means of Reducing Damage," *U. S. Dept. Agr. Farmers' Bull.* 848, 1917.
3. Howe, R. W.: "Studies of the Mexican Cotton Boll Weevil in the Mississippi Delta," *U. S. Dept. Agr. Bull.* 358, 1916.
4. Hunter, W. D. and Coad, B. R.: "The Boll-weevil Problem," *Farmers' Bull.* 1262, 1922.
5. Hinds, W. E.: "Boll-weevil Effect upon Cotton Production," *Ala. Agr. Exp. Sta. Bull.* 178, 1914.
6. Hunter, W. D.: "The Pink Boll Worm with Special Reference to Steps Taken by the Department of Agriculture to Prevent Its Establishment in the United States," *U. S. Dept. Agr. Bull.* 723, 1918.
7. Pierce, W. D.: "How Insects Affect the Cotton Plant and Means of Controlling Them," *U. S. Dept. Agr. Farmers' Bull.* 890, 1920.
8. McGregor, E. A.: "The Red Spider on Cotton and How to Control It," *U. S. Dept Agr. Farmer's Bull.* 831, 1917.

9. Hunter, W. D.: "The Cotton Worm or Cotton Caterpillar," *U. S. Dept. Agr. Bur. Entomology Cir.* 153, 1912.
10. Busck, August: "The Pink Bollworm, Pectinophora Gossypiella," *Jour. Agr. Research*, vol. 9, No. 10, 1917.
11. Fenton, F. A. and Dunnam, E. W.: "Biology of the Cotton Boll Weevil at Florence, South Carolina," *U. S. Dept. Agr. Tech. Bull.* 112, 1929.
12. Harned, R. W. and Loftin, U. C.: "A Survey of Methods Now Used for Boll Weevil Control, with Comments on Certain Methods," mimeographed report by Division of Cotton Insect Investigation of the Bureau of Entomology and Plant Quarantine, U. S. Department of Agriculture, 1935.
13. Grossman, Edgar F.: "How the Boll Weevil Ingests Poison," *Fla. Agr. Exp. Sta. Bull.* 192, 1928.
14. Hunter, W. D.: "The Pink Bollworm with Special Reference to Steps Taken by the Department of Agriculture to Prevent Its Establishment in the United States," *U. S. Dept. Agr. Dept. Bull.* 1397, 1926.

CHAPTER XVI

COTTON HARVESTING

Cotton harvesting, or "picking," as it is commonly termed, is the most expensive single operation connected with the production of cotton. At present, cotton is mostly picked by hand, a boll at a time. This makes the harvesting slow, tedious, and expensive. The difficulty of getting cotton picked is the chief limiting factor in its production in many regions. This is especially true in sections of the American Cotton Belt where labor is scarce or in sparsely settled regions of the world, such as parts of South America and Africa. No doubt, in the future, satisfactory mechanical cotton pickers will be made. Although a number have been invented, none is in common use today.

Cost of Harvesting.—Since cotton is picked by hand almost altogether, the cost of harvesting is governed largely by the wages paid day laborers in the particular region. The wages paid vary considerably in different localities, usually being higher near towns or cities or near regions where there are factories, sawmills, etc. If the market value of the lint cotton is good, the growers can afford to pay more for picking and do so in order to save as much of their cotton as possible and to get it out early that the grade may be good. If labor is scarce, the competition for pickers is greater, and consequently higher wages are paid. Wages paid during the first part of the season, while the cotton is green and heavy and has not yet been picked over, are lower than later in the season after the seed cotton has become lighter in weight, the bolls less plentiful and harder to pick free from hulls and trash.

In the United States prior to the World War and during the depression of 1929–1935, the value of cotton was low, much of the time being worth less than 10 cents a pound. During that time, the price paid for picking was usually from 40 to 60 cents for 100 pounds of seed cotton. Since the war, cotton values have been considerably better. The price paid for picking has been from 75 cents to a $1.25 per hundred during the first part of the season and, in some localities, $2 or more during the latter part. During the latter part of the season, the days are shorter, bolls less plentiful, burs rotten, the weather disagreeable, and the ground often wet or muddy. Picking cotton under such conditions is worth considerably more.

Time of Harvesting.—The time for beginning cotton picking varies somewhat with the season, being as much as 2 weeks earlier some years than others, and varies much with the latitude. In southern Texas, picking begins about the first of August and continues until the crop is all gathered, which is usually in November or December. In states farther to the north, the date for beginning picking is successively later until, in the most northern states, picking does not begin until late in September.

The date picking is begun depends to some extent, also, on the variety grown. The bolls of some of the large-boll varieties do not open quite so early as those of some of the early small-boll varieties. Hot, dry weather during July and August may cause cotton plants to check their vegetative growth and open their bolls prematurely. The defoliation of plants by insects may also cause premature opening. Although this is, of course, not desirable, since the cotton from the prematurely opened bolls is not of good quality, it does advance picking dates.

In the southern part of the Cotton Belt, cotton picking is usually finished in November or early in December. In the most northern states of the belt, there is usually cotton in the fields to pick after Christmas; and if the crop is heavy and the weather during the fall and winter unsatisfactory for picking, there may still be some cotton in the fields at plowing time in the spring.

Picking Cotton by Hand.—A person must have considerable skill at the work to be a good cotton picker. He must be able to get his hands on the bolls quickly, remove the cotton from the whole boll in one or two quick grasps, and avoid getting an excessive amount of dead leaves and trash in the seed cotton. Picking cotton is very tiresome work for inexperienced persons—and they do not get much cotton picked, either. In picking, it is necessary to stoop or bend over almost continuously for the greater part of the day. Early in the season the weather is much too hot for comfort, and late in the season it is too cold—pickers suffer from cold fingers. Dragging or carrying a heavy bag of cotton is tiring also.

Much of the cotton in America is picked by negro women and children. They are trained from childhood, becoming skillful and so accustomed to the work—perhaps having their muscles hardened to it—that they do not complain of being tired. They usually work together in small groups, gossip, joke, sing, and apparently have a good social time. Negro men do not like to pick cotton so much as the women do; and since on most plantations there is other heavier work to do, the men are relieved

FIG. 89.—Picking cotton by hand in Mississippi.

from a part of the picking. If a family tends a few acres as its special crop, the men in the family usually help pick.

In certain sections of the South, there are many white tenant farmers whose families work in the fields and pick cotton. The white people usually work more energetically and are more concerned about getting their work completed. In Texas and in some border states, Mexicans are brought in to pick cotton. In parts of the Southwest, Indians do considerable cotton picking.

The picker carries a long sack (see Fig. 89), which is held in place by a strap over the shoulder. The sack has a wide opening at a place conveniently reached by the right hand. The lower end of the long sack drags on the ground. This relieves the picker of the necessity of supporting the weight of the cotton while picking. The bag is dragged down the row as the picker moves from plant to plant. A bag containing 50 to 75 pounds of cotton may thus be handled with reasonable ease.

The amount of cotton that a picker can gather in a day depends on several factors—skill and energy, the abundance of open cotton bolls, size of the bolls, the picking qualities of the variety, etc. For average cotton, 100 pounds a day is a fair day's work, especially when this amount is picked day after day during the picking season. In good-picking cotton, where bolls are very plentiful, considerably more than this has been picked. Some persons have picked as much as 500 pounds in a day. Such rapid picking is, however, not satisfactory, because when an effort is made to pick such a large quantity, some cotton is left on each stalk, and much trash is gathered with the cotton.

Picking the Crop Grown.—It is sometimes estimated, on the basis of the size of the rural population of the South, that a crop of 40,000,000 bales could be picked. Such opinions do not, as a rule, indicate much experience in cotton fields on the part of the estimator, and leave out of consideration several factors. They base their conclusions on picking weights obtained under the most favorable conditions and on everybody's picking cotton every day. As a matter of fact, more than half of the rural people do no picking. A good average picker will not average more than 100 pounds a day. Since the women do not pick cotton on Saturday or Sunday, that leaves only 22 or 23 days in the month for picking. An allowance of 7 or 8 days a month for bad weather, sickness, attendance at funerals in the neighborhood, etc., leaves only 15 days for work. This means a bale a month and, as the season lasts about 3 months, three bales per picker for the steady pickers. Under the conditions prevailing at present, any great increase in amount of cotton produced would entail some difficulty in getting the crop harvested unless much of the cotton were snapped or mechanical harvesters were used.

Picking American-Egyptian and Sea Island Cotton.—American-Egyptian and Sea Island cottons have a long, silky staple and are used for making fine fabrics. They must be picked and ginned very carefully to be satisfactory. The grower cannot obtain a commensurate price for a low grade.

These cottons are hard to pick. It was thought for a while that the expense of picking the American-Egyptian was going to be so great that it would prevent its cultivation. The bolls are small, not easily handled, and must be picked free from trash. It has been found, however, that a laborer can pick about half as much of this cotton as of upland varieties and that, if he is paid twice as much per hundred, he is satisfied.

Sea Island cotton is picked even more carefully than Egyptian. After it is picked, it is gone over by hand, and leaves, trash, and defective locks are removed. It is dried in the sun for a day or two and then stored in bins for 5 or 6 weeks before ginning.

Harvesting Bolly Cotton.—In the northern part of the Cotton Belt, frosts commonly come in the fall before a portion of the bolls on the cotton plants have opened. These frost-bitten bolls contain cotton of inferior quality, but it is of some value. The unopened bolls are gathered and put through a gin with a huller attachment which bursts the bolls and removes the hulls. The lint ginned from such bolls is known as "bollies," or bolly cotton.

Snapped Cotton.—In parts of western Texas and Oklahoma and in a few other localities, considerable cotton is harvested by snapping, that is, gathering by hand the burs, seed cotton, and all. This method is used when there is a scarcity of labor or when there is so much cotton to be picked that it is difficult to harvest it in any other way. A laborer may be able to gather more pounds of lint cotton—even twice as much under some conditions. This snapped cotton is taken to a gin equipped with an extractor especially designed for handling it. The lint cotton ginned from snapped bolls ordinarily contains more trash than picked cotton, and the fiber may be damaged somewhat by the beating that it gets in the cleaning process; consequently, the product is of lower grade. If the cotton is snapped, some 700 pounds more to the load must be hauled to the gin than would be required in the case of picked cotton. This extra weight consists mainly, of course, of burs; and since the burs are rich in potash, considerable soil nutrient is removed from the farm when they are hauled away. The ginning charge for snapped cotton is about twice as great as for picked cotton. If all features are considered, there is not much difference in the total cost per bale. The following table given by Jones, Hurst, and Scoates[2] shows comparative costs of the two methods of harvesting cotton in Lubbock County, Texas, in 1927.

TABLE XXXIVA.[a]—APPROXIMATE COST OF HARVESTING AND GINNING PICKED, SNAPPED, AND SLEDDED COTTON IN LUBBOCK COUNTY, TEXAS, 1927

Method of harvesting	Pounds required to make a 500-pound bale, approximate	Harvesting cost		Ginning cost		
		Per cwt.	Per bale	Per cwt.	Per bale	Total
Picked.......	1,400	$1.50	$21.00	$0.40	$ 5.50	$26.60
Snapped.....	2,100	0.75	15.75	0.50	10.50	26.75
Sledded......	2,900	.....	2.55	0.50	14.50	17.05

[a] Requirements and cost for picking, snapping, and sledding cotton in west Texas and Oklahoma, Bureau of Agricultural Economics, U. S. Department of Agriculture, *Preliminary Report.*

Cotton Picking in Foreign Countries.—In several of the foreign countries where cotton is grown, such as China, India, and Egypt, labor is very abundant, and there is no trouble in getting cotton picked. In China, it is a common practice to gather the unopened bolls, store them, and pick out the seed cotton by hand after they have opened. This practice entails extra labor, but it gives an excellent grade of staple, since the lint is not exposed to the weather and is relatively free from leaf fragments and trash. In Egypt, much care is exercised in picking cotton in order to produce staple that will grade high; but in India, this does not seem to be the case. Much of the cotton there is of a very low grade.

Mechanical Cotton Harvesters.—It has been the hope of cotton growers for many years that a machine would be made to pick cotton. Inventors have labored over the problem and devised many machines. A cotton picker was patented in the United States in 1850, and since that time about 800 other patents have been granted. All these machines pick more or less cotton, and some of them have been fairly efficient; but none of them has been economically successful to a degree that would warrant their common use.

It is a difficult matter to devise a machine to pick cotton, for several reasons. The cotton bolls do not all open at the same time. The first ones that open must be picked before later ones open to prevent the staple of the early ones from being damaged by exposure to the weather. The plants and late bolls must not be injured while gathering in the early bolls. The seed cotton must be picked with a minimum of trash and leaf fragments. The bolls are high and low on plants and distributed throughout a considerable extent of space. Some bolls do not open well. The machine must gather the cotton more cheaply and more rapidly than it can be picked by hand. One writer has said that making a machine to pick cotton is much like devising a machine to pick strawberries or raspberries. The machine must be so constructed that it will get all of the berries, but it must not mash any of them.

Notwithstanding the difficulty of the problem, it seems reasonably certain that some sort of mechanical cotton picker will be perfected in the future which will pick at least a part of the cotton and be of economic value, especially under favorable conditions, in regions where labor is scarce.

The cotton-harvesting machines that have been devised fall into four general classes: the stripper, the vacuum-suction, the mechanical finger, and the whirling spindle.

Stripper, or Cotton Sled Harvester.—The simplest type of cotton stripper, or sled harvester, consists of little more than a large box with the front end removed. A V-shaped slot sawed in the bottom at the front end guides the cotton plants to a narrow slit which extends across the bottom of the box. The slit continues between two parallel two-by-fours which reach from the floor of the box to near the top at the back. The box is mounted on low runners and is dragged straddle of the row. The stalks enter the front opening of the slit, and as the box moves forward the stems are dragged through the opening, but the bolls, being too large to pass through the slit between the parallel two-by-fours, are pulled off.

Another form of stripper is the finger type. This consists also of a large box open at the front end and supported by runners or wheels. Instead of a triangular slot which the plants enter first, there are eight or more fingers consisting of parallel bars of iron about 2 feet long and ½ inch square, fastened securely in the back and sloping downward in front of the box (see Fig. 89*A*). The tip of the bars is pointed and bent upward slightly. A man stand-

FIG. 89*A*.—Homemade finger-type cotton sled or stripper. (*Courtesy H. P. Smith.*)

FIG. 89*B*.—Stripping cotton in northwest Texas. (*Courtesy Texas Agricultural Experiment Station.*)

ing in the box forks back into the box the stripped bolls which collect on the fingers.

After the growers had begun using these crude strippers rather extensively, the farm-machinery manufacturers became interested and put several commercial strippers on the market. These were built mainly on the slot-stripper principle.

FIG. 89C.—Cotton stripper equipped with bur extractor. (*Courtesy H. P. Smith.*)

The cotton strippers gather cotton rapidly and cheaply but are adapted to rather small plants only (see Fig. 89*B*), and they, according to Jones, Hurst, and Scoates,[2] lose from 4 to 16 per cent of the cotton. H. P. Smith, of the Agricultural Engineering Department, Texas A. & M. College, has devised and built several stripper machines, some of which contain bur extractors and cleaners and are a marked improvement over the original type of stripper, both in efficiency and in the quality of cotton obtained.

Vacuum Type of Mechanical Cotton Pickers.—The pickers of the vacuum type are not altogether mechanical, because it is necessary for the operator to touch each cotton boll with a suction nozzle. This type of machine is simple in construction. It consists of a small engine, mounted on a truck and so connected that it drives a fan which draws the seed cotton through a tube, or set of tubes, into a hopper. Several smaller tubes are joined to the large ones. The operator touches the tip of the open boll, and the seed cotton is quickly sucked in and carried to the storage receptacle. Some of the machines of this type have a nozzle on the end of the tube which contains revolving

Fig. 90.—Indiana (Stukenborg) electric cotton picker in use. (*Courtesy Indiana Silo Company.*)

brushes, wheels, or some sort of mechanical device to catch the exposed cotton and assist in pulling it from the boll. This may remove from the bolls cotton that the simple suction would not remove.

Illustrations of the suction type of mechanical cotton pickers are to be found in the Worswick-Haardt cotton picker, invented by J. E. Worswick, Montgomery, Ala.; the Thurman vacuum cotton harvester, invented by John S. Thurman and manufactured by the Vacuum Cotton Harvester Company of St. Louis, Mo.; and the Indiana (Stukenborg) electric cotton picker, invented by L. C. Stukenborg of Little Rock, Ark., and manufactured by the Indiana Silo and Tractor Company, Anderson, Ind. The last-mentioned machine has nozzle attachments at the end of the tubes and is illustrated in Fig. 90.

Objections raised against the vacuum type of pickers are that they are expensive, that there is much hand labor involved

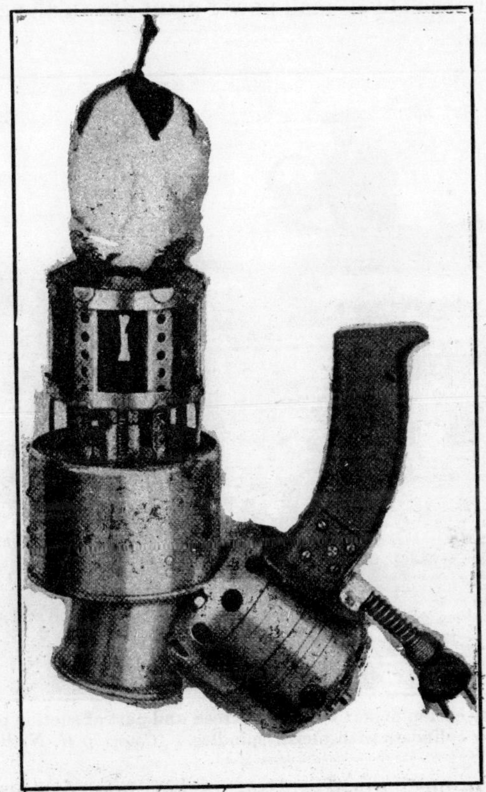

FIG. 91.—Staso mechanical cotton picker. (*Courtesy Staso Mechanical Cotton Picker Company.*)

in moving the nozzle to each boll, and that the suction draws in an extra amount of foreign matter.

Mechanical-finger Type of Cotton Pickers.—The third type of cotton pickers mentioned above is designated as the mechanical-finger type, because the essential part of the machine consists of rapidly moving fingers, covered with hooks or teeth, which quickly pull the cotton from the open boll (see Fig. 91). The nozzle containing the fingers must be moved by the operator to each boll that is picked. From the nozzle the cotton is carried by a series of conveying teeth through the barrel of the machine and automatically discharged into a funnel-shaped canvas sack which is attached to the back end of the machine. The fingers are driven by a small electric motor attached to the machine. Power is furnished by a small power plant drawn along the cotton row by a mule.

FIG. 92.—Berry cotton picker with wire screen and part of suction pipes removed to show cylinders with picker spindles. (*Courtesy H. N. Berry.*)

The Staso mechanical cotton picker manufactured by the Central Commercial Company, of Chicago, Ill., is an illustration of the type of picker just described.

The Lowry cotton picker, invented by George A. Lowry, of Boston Mass., works on much the same principle but is constructed differently.

Objections similar to the ones raised against the vacuum type of picker are made against the mechanical-finger type.

Whirling-spindle Type of Cotton Pickers.—Mechanical pickers of the fourth, or whirling-spindle, type are altogether automatic. The machine is driven straddle of a row, and numerous whirling, needle-like spindles move through the cotton plant as the machine passes over it. When a spindle touches a lock of cotton, the lint clings to it and becomes twisted around it in such a way as to draw the cotton from the boll. After the spindles move out of the plant, the cotton is stripped from them by mechanical means and carried to a container.

The Berry cotton picker is a machine of the type just mentioned (see Figs. 92 and 93). It is an invention by the late H. N.

Fig. 93.—Berry cotton picker complete. Insert, H. N. Berry, inventor.

Berry, of Greenville, Miss., and contains a number of improvements over other machines of this type that had been constructed previously. It has been demonstrated that this machine will pick cotton rapidly; get most of the open bolls; collect very little, if any, more trash than a person picking by hand; and damage plants but very little. It may be driven at a rate of 5 miles an hour, picking a row as it goes. It has picked a bale of cotton in 3 hours. One man can operate the machine. It occurs to the writer that this machine or one of similar type will be a success and will probably do much to revolutionize cotton growing.

On account of its apparent value, space is taken to describe the machine in some detail.

The picking and power mechanism of the Berry cotton picker is mounted on an arched framework supported by four wheels similar to heavy automobile wheels. The arched framework permits the picker to straddle the row of cotton that is to be picked (see Fig. 92). The machine is driven by a four-cylinder motor similar to the motor used in a light truck. This engine furnishes motive power for propelling the machine and also power for driving the picking mechanism.

The picking mechanism consists of two cylindrical revolving drums which are 12 inches in diameter and about 3 feet long. Each drum contains more than 500 radiating, whirling steel needles. The drums are between front and rear wheels, one on each side of the row, and near enough together that the tips of the radiating needles almost meet in the center of the plant that is being picked (see Fig. 92). The distance of the drums from the ground is regulated by changing the height of the rear wheels in respect to the arch. The radiating steel needles are tapering in shape, about ¼ inch in diameter and 9 inches long. The drums are geared directly with the driving mechanism which turns the traction wheels, so that, as the machine moves forward, the drum revolves into the center of the plant at the same speed. Thus the needle remains almost stationary in the plant, in respect to movement in line with the row of cotton, and so avoids damage to the plants. While the needles are within the plant, they whirl on their own axis at a rate of about 600 revolutions per minute. If a lock of cotton comes in contact with one of these whirling needles, the loose fibers are immediately firmly twisted around the needle, and the lock drawn from the boll. Since a large part of the needles move through the space within the limits of a single plant, and since the limbs and bolls of the plant move to some extent, there is not much chance for a boll to be missed. After the needles have passed through the space within the plant, they stop whirling for a moment, then start whirling in the opposite direction. This tends to unwind the lint that is wound around the needles and facilitates the removal of the locks of cotton. Each needle passes through an aperture in a movable bar which serves as a stripper. As the needle is unwinding the lint about it, the bar moves outward so as to push the cotton from the needle. The movement of the stripper bars is controlled by eccentric tracks above and below each drum. After being dropped by the needle, the cotton is picked up by air blast and carried to a receptacle in the rear of the machine.

The Rust cotton picker, another machine of the whirling-spindle type, has received considerable publicity the last year or two. It works on much the same principle as the Berry picker but has spindles on only one side of the plant. The plant is compressed into narrow space so that the spindles may reach all the bolls. This machine is drawn by a tractor and is more simple in construction than the Berry picker and less expensive, but its efficiency might be improved.

References

1. O'KELLY, J. F. and HULL, W. W.: "Effects of Delayed Harvesting by Picking and Snapping on Six Varieties of Cotton," *Miss. Agr. Exp. Sta. Bull.* 316, 1936.
2. JONES, D. L., HURST, W. M. and SCOATES, D.: "Mechanical Harvesting of Cotton in Northwest Texas," *Tex. Agr. Exp. Sta. Cir.* 52, 1928.

CHAPTER XVII

COTTON GINNING

Before cotton fibers can be used in spinning or making cloth, or even before they can be marketed advantageously, they must be removed from the seed by some kind of ginning process. During early times, and even as late as 1793, the date of the invention of the saw gin, a considerable quantity of lint cotton was picked from the seed, or ginned, by hand. This was a very slow and laborious process, but enough lint could be secured by a family in this way, when they were not otherwise employed, to supply the needs of the home. During pioneer days it was the custom in many homes for the family to occupy themselves thus while seated around the fireside in the evening.

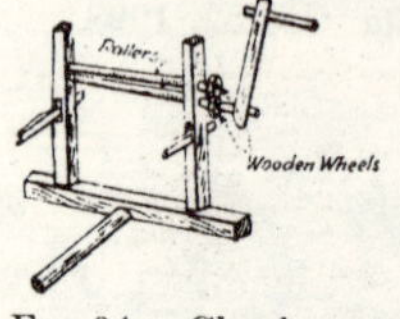

FIG. 94.—Churka cotton gin.

"Churka" Gin of India.—The first recorded gin made was the small roller, or "churka," gin, used in India many centuries ago. This machine consisted simply of two upright pieces of wood, made secure at the base in some way, between which were two small, horizontal, parallel wooden rollers so adjusted that they barely touched (see Fig. 94). Attached to one of the rollers was a hand crank, by means of which the rollers were turned. The seed cotton was fed between the rollers by hand. The lint was caught by the rollers and drawn between them and thus separated from the seeds, which would not pass between the rollers. This type of gin will not work well except with the smooth-seeded varieties like Sea Island. Five pounds of lint cotton was a fair day's work with one of these primitive gins. They are still used to a limited extent in China and India.

Some of the primitive gins of a slightly different construction had fluted wooden rollers, and others had one small iron roller and a larger wooden roller.

Roller Gins.—A number of different kinds of roller gins were made prior to 1793. These were improvements over the primitive roller gin of India but were fashioned after it. None of them proved to be of great economic value. According to Miller,[1] the following persons invented improved gins: Krebs, of Mississippi, in 1772; Burden, of South Carolina, in 1777; Eve, of Augusta, Ga., in 1790; and Pottle, of Georgia.

Fig. 95.—A typical 20-stand roller gin. Lint is shown coming over the rollers and brushes and falling to the floor. (*After Townsend.*)

The first roller gins were driven by hand power. Later ones made use of horsepower or water power. The present-day gins use, of course, steam, gas, or electric power.

The modern roller gin is small and simple in construction. The essential parts consist of a roller covered with leather—walrus hide being preferred, or some other covering that is firm and of such texture that cotton fibers will cling to it freely—and a large knife. The knife is set so that its edge is parallel to the surface of the roller and barely touches it. The locks of seed cotton fall or are fed by hand on one side of the roller, which picks them up and carries them to the knife. The lint clinging to the roller passes under the knife and is stripped from the seed, which cannot pass under the knife. The lint falls on one side of the roller, and the seed on the other (see Fig. 95). Some of the more recent types of roller gins have a mechanical feeder, a contrivance by means of which the seed cotton is carried to the roller regularly and in small bunches. Some also have an extra roller and an agitator for moving locks and partially ginned seeds.

Roller gins are used in America for ginning Sea Island and American-Egyptian cotton. They are used in Egypt exclusively and to a limited extent in India and China. Roller gins are considered better than saw gins for ginning cotton with very long staple. The saws may cut the fiber. Objections sometimes made to roller gins are that the lint that comes from them is so packed or wadded up in parts that it makes a poor sample and makes neps, which bother in spinning; that the lint contains too many pieces of broken seed (true if the gin is not adjusted and run properly); and that the ginning is slow and expensive. A single gin will yield 50 to 80 pounds of lint cotton per hour.

Saw Gins.—Prior to 1793, it had been demonstrated conclusively that cotton could be grown easily in the South. England needed cotton but did not want it in the form of seed cotton, because she had no good means of separating lint and seed. The South had plenty of fertile land and labor and needed a profitable crop to grow. Great need for a machine that would separate the fiber from the seed quickly and economically was therefore felt. It is true that some crude cotton gins had been made but they were not satisfactory.

Eli Whitney.—It has often been said that necessity is the mother of invention. Thus the great need for a machine that would separate lint cotton from the seed rapidly and thoroughly led to the construction by Eli Whitney of the saw gin, a machine that will do more than the early cotton growers imagined a machine could ever do.

Eli Whitney was born at Westboro, Mass., in 1765 and graduated from Yale University in 1792. He possessed much skill in the use of tools, being "a natural-born mechanic." It is related by Scherer[2] that at the age of twelve he took his father's watch to pieces but possessed enough ingenuity to put it together again, his father being none the wiser.

After graduation from college, young Whitney accepted a position as teacher in a school in South Carolina. Going south on a boat, he chanced to meet Mrs. Greene, widow of Gen. Nathaniael Greene of Revolutionary fame. Mrs. Greene became interested in Whitney and invited him to visit her plantation near Savannah, Ga. While there he saw negroes picking lint cotton from the seeds with their fingers, and he also heard certain visitors at the plantation, who were prominent men of the South, mention the great need of a machine that would gin the Green Seed cotton, which was the short-staple cotton with fuzzy seeds and which produced considerably more cotton than the smooth-, or black-seeded, sort like Sea Island.

Whitney at once became interested in the problem and fitted up a workshop in the basement of Mrs. Greene's house, where, working with only the plantation tools available, he constructed the first model of his gin in 10 days. First he placed small, sharp hooks over the surface of a board and drew unginned cotton over them. From this came the idea of making a wooden cylinder with hooks attached. With the addition of another revolving cylinder, to which were attached brushes for pushing the cotton from the hooks, the gin was essentially complete.

The first model of the gin was so promising that Phineas Miller, manager of the plantation belonging to the Greene estate, became interested and agreed to furnish the money needed to perfect this new invention and to put it on the market. In May, 1793, Whitney returned to New England, where he secured better tools and materials to complete his machine. (see Fig. 96). On June 20, 1793, he applied for a patent, which was granted him on Mar. 14, 1794. George Washington, then President, signed the papers.

The model presented contained a wooden roller into which were driven spikes. The spikes were bent and sharpened to form hooks. Seed cotton was placed in a hopper, or "gin breast," as it is now called, in contact with the roller containing spikes. As the cylinder turned, the hooks passed between strips of iron at the back of the hopper. These strips were so placed that the hooks would pass between them in their circuit, carrying some lint cotton which they had engaged in their movement through the hopper but too close to permit cotton seeds to pass. The lint was in this way pulled from the seeds. Just back of the roller containing spikes was another roller, to which were attached four strips of brushes (see Fig. 96). This roller was set so that the brushes would remove the lint from the hooks, being geared to revolve in the opposite direction from that of the hooks and considerably faster. Although revolving in the opposite direction, the two cylinders moved in the same direction where they came in contact.

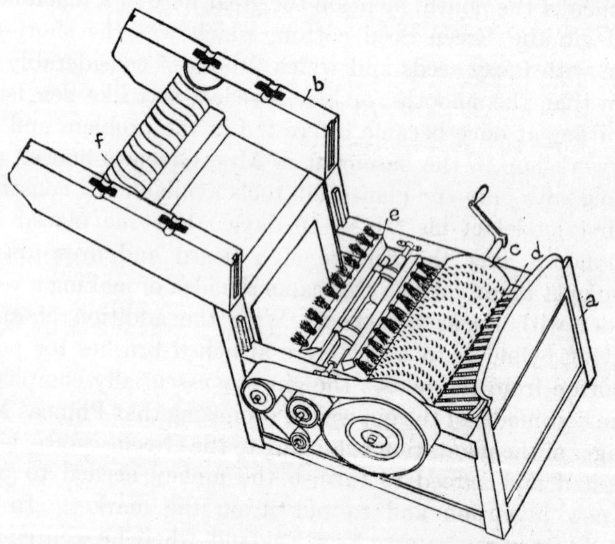

Fig. 96.—Original model of Whitney's cotton gin. (*After Tompkins.*)

An Improved Saw Gin.—In May, 1796, just 2 years after Whitney obtained his patent, Hogden Holmes, of Augusta, Ga., obtained a patent on an improved cotton gin. The improvement over Whitney's gin consisted in the substitution of circular saws for the rows of hooks. This was really an important improvement and one for which Holmes deserved credit, although but little has been given him.

Antagonism at once sprang up between Whitney and Holmes. Whitney became involved in much litigation with Holmes and others in defense of his patent rights. Eventually, South Carolina voted Whitney a royalty of $50,000 for his invention; North

Carolina gave about $30,000; and Tennessee about $10,000 Most of this money was expended in lawsuits.

Whitney's Business Policies.—It was Whitney's plan to maintain a monopoly of cotton ginning. He refused to sell his gins but built them at various places and asked a toll of 1 pound of lint cotton out of every 3 ginned. Where gins were built by others, Whitney sought to levy a tax of $200 on each gin. These methods of securing remuneration for his invention were not popular with the cotton growers and brought him into disfavor in the eyes of many, although he was really a benefactor of the public and deserved honor and financial reward.

FIG. 97.—Eli Whitney, inventor of the cotton gin.

Value of the Cotton Gin.—The following paragraph taken from an address made by Judge Johnson, of South Carolina, in the lawsuit brought by Whitney and Miller to recover from the state of South Carolina $50,000 appropriated by the legislature for the purchase of the right to use the gin in the state gives a good picture of conditions in the South at the time the cotton gin was invented and its beneficial effects.

The whole interior of the southern states was languishing, and its inhabitants emigrating for want of some object to engage their attention and employ their industry, when the invention of his machine at once opened views to them which set the whole country in motion. From childhood to age it has presented to us a lucrative employment. Individuals who were depressed with poverty and sunk in idleness have suddenly risen in wealth and respectability. Our debts have been paid, our capital has increased, and our lands have trebled in value. We cannot express the weight of obligation which the country owes to this invention; the extent of it cannot be seen.

Fig. 98.—Old plantation gin house and screw. (*After Tompkins.*)

Improvements in the Saw Gin.—It was at once very evident that the circular saw as designed by Holmes was an improvement over the spiked cylinder of Whitney, and it was used, even by Whitney. Gins have been enlarged and adapted to the use of horse, water, or other power, but the essential principles of the machine have remained the same even to this day. The large amount of machinery found in modern ginneries is used largely to clean seed cotton, carry it to the saws, and convey the lint cotton away from the saws.

Old-time Plantation Gin House and Screw Press.—Before the Civil War, a gin house, with a screw press for baling cotton, was a familiar sight on nearly every large plantation in the South (see Fig. 98). The ginning was done during the winter or at times when there was no work to do in the fields. At such times there was plenty of labor available, both man and mule. The plantation manager even had trouble in finding jobs enough to keep his hands employed during the off season of the year; consequently there was little need or desire for labor-saving contrivances at the gin. The seed cotton was carried to the gin by hand and fed into it by hand; the lint was carried from the gin to the press in baskets, emptied into the press box, and tramped down by men; and so on.

Power to drive the gin and to turn the press screw was furnished by mules. Two mules were hitched to each of two large levers which were fastened to a big vertical wooden shaft. This shaft bore a large wooden cog wheel which engaged another wooden cog wheel on a horizontal shaft. A belt running from the horizontal shaft to the gin carried the motive power (see Fig. 98).

When sufficient lint cotton had been packed in the press box by tramping to make a bale, pressure was applied by means of the large wooden screw above the box (see Fig. 98). The screw was raised and lowered by large levers to which mules were hitched. The bale was covered with jute or gunny sack and bound with two hemp ropes.

The Modern Saw Gin.—Modern saw gins contrast greatly with the gin just described. In them every known contrivance is used to avoid hand labor. Electric, steam, or gas engines furnish motive power. The seed cotton is drawn out of the grower's wagon bed by suction and carried mechanically through cleaner, distributor, and feeders to the saws. From the saws the seed is carried by mechanical conveyors to the grower's wagon, a railway car, or a storage bin. The lint is carried by current of air to condensers, which condense it and drop it into the press box. There it is packed by mechanical trampers and pressed into a bale by hydraulic or other power. About all that has to be done by hand is to place the bagging and ties that are to cover the bale in the press box, fasten the ties, and roll out the bale of cotton.

Modern ginneries contain from two to eight gins, or "gin stands," as they are commonly called (Fig. 99). Where there are more than four stands, they are arranged in two rows, commonly parallel. The smaller gins are mostly private gins and are found on plantations, while the larger ones are chiefly public gins.

Elevators.—As was mentioned previously, at modern ginneries the seed cotton is drawn by pneumatic elevators from the wagon that hauled it to the gin. These are made of heavy sheet metal and are in the form of tubes 10 to 12 inches in diameter; the outer two joints are movable and telescopic, so that different parts of a wagon bed may be reached. Properly connected tubes carry the seed cotton onward to bins, distributor, cleaner, or feeder, just as desired. The suction is produced by rapidly revolving fans so placed as to draw the air through the tubes.

Fig. 99.—Four-stand gin complete. (*Courtesy Continental Gin Company.*)

Cleaner and Separator.—In most gins, the seed cotton from the wagon is first carried to a cleaner or separator (Fig. 99). Here the cotton is first thrown forcibly against a wire screen; part of the sand and dirt contained passes through the screen and is thus separated from the cotton. The cotton is next taken up by spiked rollers or by agitators of some kind, which tear apart the large, somewhat dense bunches and throw the locks against wire screens which permit dirt to fall through.

Distributor.—From the cleaner the seed cotton is delivered to a large belt which travels over the top of all the gin stands (Fig. 99). Attached to this belt are fingers which assist in moving the cotton to the feeders above the stands. Each stand is regularly supplied with all the cotton that it can handle. Any excess that the belt may have is dropped to the floor where the belt turns (see Fig. 99) and is known as the "overflow." This cotton is taken up by the elevator near by after all the cotton has been removed from the wagon.

In some gins, a pneumatic distributor is used. This delivers the cotton to a cleaner separator above each gin.

Feeder.—Over each gin stand there is a device for supplying the gin saws below with cotton regularly and in small quantities. This is known as a "feeder." The feeder commonly contains a spiked roller (Fig. 100), which picks up the cotton and so drops

Fig. 100.—Combined feeder and cleaner. (*Courtesy Continental Gin Company.*)

it that it will fall to the saws. In many cases, the feeders have cleaning devices in connection, as shown in Fig. 100.

Gin Breast.—From the feeder the seed cotton falls directly into the gin breast, or "roll box," as it is sometimes called (Fig. 101). Here the cotton comes in contact with the rapidly revolving saws. When the saws strike the mass of seed cotton, they cause it to move and soon form a cylindrical mass or roll which revolves rapidly. In this manner, new bunches of cotton are brought against the saws almost continuously.

The front of the gin breast is movable. If it is so set as to lessen the size of the opening at the lower part of the breast, the seeds are held inside the breast longer and ginned closer.

Some gins are built with a double breast, into the outer compartment of which fall hulls, bolls, etc. These come in contact with the saws before they reach the regular roll of cotton. Gins of this construction are especially valuable for ginning bolly cotton or cotton that contains many burs, pieces of bolls, trash, etc.

Saws.—The gin saws are steel disks 10 or 12 inches in diameter (Fig. 101). The saw teeth on the circumference of the disks are made somewhat hook-shaped, so that they will have maximum strength but so that they will cut the cotton fibers a minimum amount. For good ginning, the saw teeth must be kept sharp. Even under ideal conditions the teeth cut the fibers to some extent. The damage done is multiplied if the cotton is damp or if the saws are run too fast. A proper speed for 10-inch saws is 400 revolutions per minute; 12-inch saws, on account of their greater circumference, need not run so fast to accomplish the same results. A single gin contains about 70 saws.

FIG. 101.—Transverse sectional view of saw gin. (*Courtesy Continental Gin Company.*)

Ribs.—On the side of the gin breast next the saws is a phalanx of bars of iron, the gin ribs (Fig. 101). These bars fit together closely and firmly at each end but have a narrow slit between each two in the middle part. The saw projects through this narrow slit and pulls the lint cotton through; the slit is too narrow for the seeds to pass.

Some gins, like certain of the huller type which are designed for removing burs, broken bolls, etc., have two sets of ribs.

Mote Board.—The mote board is a movable board under the brushes of a gin. By moving this board, an air space beneath the gin is enlarged or diminished. Proper adjustment of the mote board assists in removing particles of seed and foreign matter from the lint cotton.

Gin Brushes.—Just back of the gin saws is a revolving cylinder which bears numerous rows of bristles, or "brushes," as they are called (Fig. 101). The brushes are set so as barely to touch the saws and are geared to run about three times as fast as the saws. By this means they remove the lint from the saw teeth, although they move in the same direction as the saws. The brushes, when running, generate a current of air which assists not only in removing the lint from the saws but also in driving it toward the condenser near the press.

In some gins of recent manufacture, an air blast, instead of brushes, is used to remove the lint from the saws (see Fig. 102).

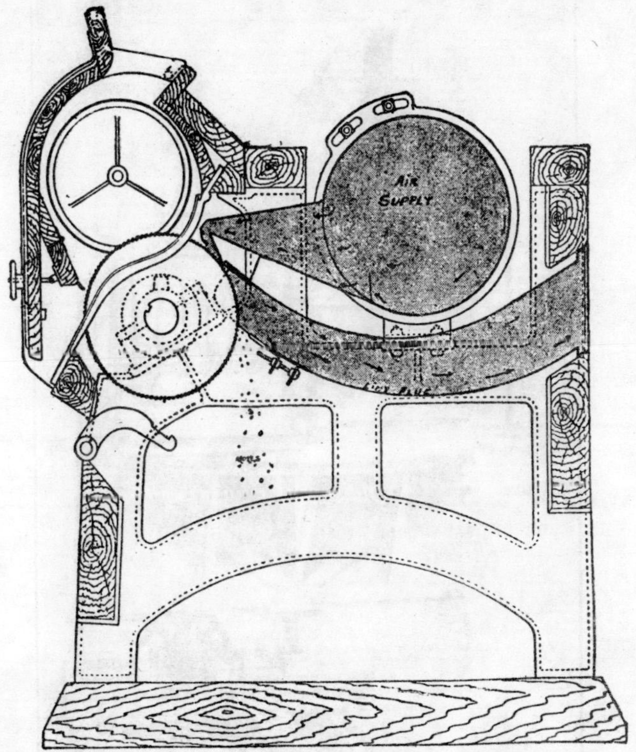

FIG. 102.—Air-blast method of removing lint cotton from gin saw. (*After Taylor.*)

This method seems to be fully as satisfactory as the brush method.

Condenser.—The lint removed from the saws is carried by air blast to the condenser, which consists essentially of a large roller covered with wire screening, and a smaller wooden roller. The lint is blown against the screened roller, which allows the air to pass through but stops the lint, which collects on it. The rollers so turn as to press the accumulated lint between them, thus forming a fairly compact layer of cotton "bats."

Press.—From the condenser, the cotton drops into the press box, over the bottom of which has been spread bale bagging.

FIG. 103.—Revolving double-box press with 5-inch screw and plain steam-cylinder tramper. (*Courtesy Continental Gin Company.*)

As the cotton collects in the box, it is packed every minute or so by mechanical trampers. After enough has accumulated to make a 500-pound bale, an amount estimated by the pressman, heavy pressure is applied by hydraulic, screw, or other power. Just before it is applied, a piece of bagging is placed over the upper part of the mass of cotton in the box. While the cotton is under pressure, the bagging is fitted over the sides of the bale (the bagging is usually not long enough to cover the sides completely), and six iron ties are put around it and fastened. Pressure is now released. The cotton expands somewhat but is held firmly in a compact bale by the bagging and ties. The bale is now complete. In most ginneries, a double press box, like the one shown in Fig. 103, is used. This permits the gin to run continuously, since one box may be filling while the cotton in the other is being compressed and tied.

Cotton Bales.—Before lint cotton can be handled conveniently or marketed, it must be made more compact and tied in a package that can be handled satisfactorily. Various kinds of packages, or types of bales, have been made and are being made by different countries at present.

In America, cotton was first packed in a large sack as a package for the market. This sack was commonly suspended beneath a hole in the gin-house floor, through which lint cotton was pushed. The cotton was packed in the sack by two men who trampled it. When the sack was packed as full as possible, the open end was sewed. These sacks weighed 200 to 300 pounds, and the cotton was compressed to a density of about 5 pounds per cubic foot.

These bags of cotton were bulky, requiring much space in ships. There was at once a demand for a package that would not take so much room in proportion to its weight. The demand brought into use the screw press, such as has been described in the preceding pages. The screw-press bales varied in weight considerably but were intended to weigh 500 pounds. Their density was about 8 pounds per cubic foot.

About 1870, the metallic screw and hydraulic press came into use. By means of these the bale was given a density of about 12 pounds per cubic foot. The size of the press box was standardized to make a bale of uniform size—27 inches thick and 54 inches long. The width varied somewhat, depending on the compression of the bale, but was about 45 inches. The average weight was about 500 pounds. This bale was the same as the modern flat bale which is turned out by ginneries all over the South (Fig. 104).

Round Bales.—Different machines have been devised for making round bales, but such bales are not made extensively. The Bessonette bale, which is the type mostly used, is made by winding the bat around a central core. As the lint is wound, two rollers press against the roll, thus packing the fibers closely. Wound in this way, the fibers adhere. The result is a very compact cylinder of cotton which will not unwind or expand. No ties are needed to keep it compressed. The finished bale is about 25 inches in diameter, weighs 250 pounds, and has a density of about 33 pounds per cubic foot. After completion, the bale is

FIG. 104.—American flat bale.

covered with duck or hemp cloth, which is sewed on. A lint sample is taken while the bale is being made. The covering o the bale is not cut for sampling.

Bales Made by Foreign Countries.—The cotton bale made i India varies in size considerably, but the average weight is abou 400 pounds. It is entirely covered with bagging and well bounc with ties, which run around it spirally. It is a neat, well-pro tected package.

The Egyptian bale, when shipped, is said to be in better condi tion than that from any other country. It weighs from 70 to 800 pounds and is compressed to a density of about 30 pound to a cubic foot. It is completely covered with canvas an securely bound by numerous iron bands.

Compress.—With the increase in cotton production came fu ther demand from railroads and ship lines for a more compac bale. This demand brought the bale compress into use. The compress is constructed much like a gin press except that it is much stronger and can exert more pressure.

When a flat bale is put into a compress, the ties are removed, and usually patches, pieces of bagging, are put over the holes in the old bagging and over the sides of the bale that had not been covered with bagging. A pressure of 120 pounds or more per square inch is then applied. This reduces the bale to a thickness of just a few inches and a density of 50 pounds or more per cubic

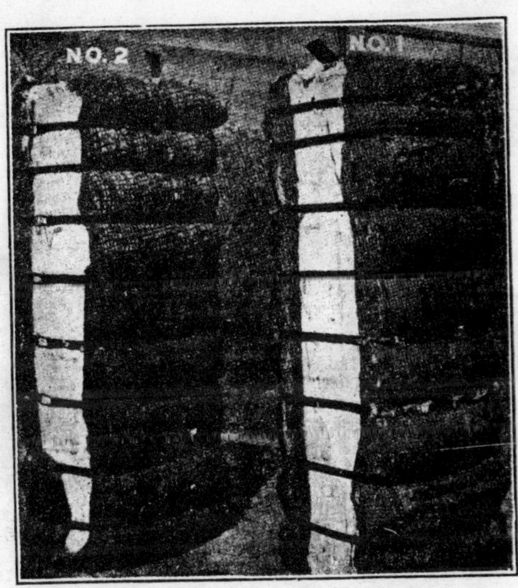

Fig. 105.—Compressed bales.

foot. The ties are then replaced, and usually two others, which are made by splicing pieces cut off the old ties, are added. When the pressure is removed, the bale expands somewhat and becomes less dense, but the density remains above 22½ pounds per cubic foot. Bales are handled so rapidly at the compress that not much pains can be taken to give them a neat appearance; consequently, the compressed bales have irregular shapes and a ragged appearance (Fig. 105). Double the number of bales may be placed in a car if they have been compressed. Compresses are to be found in the South in all towns of any size and at all concentration points for cotton. A charge of 55 to 75 cents a bale is made for compressing cotton.

Gin Compress.—In order to avoid the expense of hauling bales to a compress and the extra labor of handling them there, a gin compress has been constructed. This is similar to the regular baling press at the gin and is used instead of it, but it is built much stronger, so that more pressure may be applied to the bales. This system is surely more economical than the one in vogue, but it is not used extensively. Its adoption would call for new presses at the gins and the doing away with the compresses, in

Fig. 106.—Plantation gin, showing loads of cotton waiting to be ginned.

which much capital is invested, and would require certain changes in marketing methods.

Criticism of American Bales.—The American cotton bale when it arrives in foreign markets is very ragged in appearance. The bagging has been cut at several places to draw samples; some of the cotton is exposed and dirty; and the whole bale has perhaps been damaged by exposure to the weather. The bad condition and appearance are due largely to our sampling and marketing systems. Almost every dealer that handles the bale feels that he must have a new sample from the bale, and usually he cuts the bagging at a new place to get the sample. The producer is far removed from the consumer. He feels that the consumer will never know where the bale came from and that the price he receives for his cotton does not depend to any great extent on how he handles the bale.

Seed-cotton Driers.—Seed cotton that is damp or wet will not gin satisfactorily. If very damp, some of the fibers will be cut by the saws or packed into small dense tufts. The saws may become so clogged that the gin will not run at all. In some areas, especially in the Gulf Coast region, there is so much wet cotton that a special effort is made to dry it before ginning. Some growers spread the seed cotton on sheets in the field and expose it to the sun for several hours. Others spread it out in a thin layer in large wire-bottomed trays and leave them out in the sun for a day or more. Persons using this method either have a large building near by into which the trays may be moved in

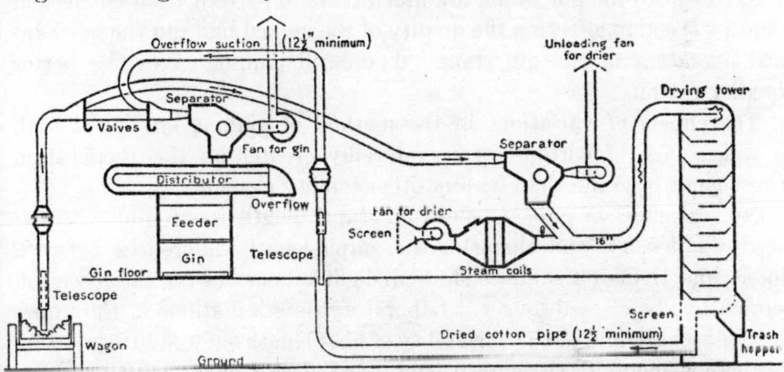

FIG. 106*A*.—A complete layout for cotton gin and two-fan vertical drier, model A, with tower on ground level. A one-fan drier installation is made by connecting both separators to one oversize suction fan. (*After Bennett.*)

case of rain, or they have canvas to spread over the trays to keep off the water.

None of the methods just mentioned is rapid or very satisfactory. Some artificial driers dry the cotton by passing it through a large cylinder or tower through which is forced a strong current of heated air. One of the most successful artificial driers is the one designed by Bennett,[4] of the Bureau of Agricultural Engineering, U. S. Department of Agriculture. The general arrangement of this drier is shown in Fig. 106*A*. With this machine, the damp seed cotton is treated with a continuous current of hot air at the rate of 40 to 100 cubic feet for each pound of cotton. The cotton is exposed to the drying process for different periods ranging from 15 seconds to 3 minutes; the temperature of the drying air is 150 to 200°F.

Important Features Relative to Gin Operation.—The men in charge of the Government Experimental Gin at Stoneville, Miss., have made various studies relating to the operation of gins. The following paragraphs prepared by Bennett give a summary of results obtained:

There were only small effects on the lint quality and the ginning capacity caused by variations of gin-saw speed 100 revolutions per minute above or below the manufacturer's recommended speed. Lowering the speed 100 revolutions per minute improved the quality slightly, but raising the gin-saw speed 100 revolutions per minute did not change the quality appreciably.

Changes in seed-roll density, caused by changes in the rate of feeding seed cotton to the gin stand, are much more important than changes in gin-saw speed in affecting the quality of the ginned lint and the mechanical operation of the gin stand. Loose-roll ginning gives the better quality cotton.

The effects of variations in the method of ginning employed with a wide range of cottons were materially greater on the preparation component of grade than on any other quality element.

No real effect of gin-saw speed on staple length was found. As the seed-roll density was changed, the staple-length differences between loose- and tight-roll samples showed slight tendencies for the loose-roll samples to be classed longer. Laboratory determinations of the upper quartile length and of the variability of fiber length showed little relation of these elements to the gin-saw speed or the seed-roll density.

Grade differences due to variations in gin-saw speed averaged less than one fifth of a grade improvement for the low speed. The change due to increasing the speed was negligible. Grade differences due to change in seed-roll density amounted to nearly two thirds of a grade in favor of the loose roll.

Using prices for cotton of like grades and staples prevailing in the Memphis market during the 1932 crop season as a basis of monetary-value computations, the use of a loose seed roll instead of a tight seed roll with a group of cottons averaging $1\frac{3}{16}$ inches in staple length from seed cottons averaging 14.1 per cent in moisture content showed net benefits of approximately $4.50 per bale, or 13 per cent; and almost $1.50 per bale, or 4 per cent, with a group of cottons averaging $1\frac{5}{32}$ inches in staple length from seed cottons having 10.2 per cent moisture content.

The amount of reduction in value as a result of using a tight seed roll for short-staple cottons (about 1 inch) from seed cottons of relatively high moisture content and of substantially lower moisture content averaged 95 and 62 cents per bale, respectively.

Ginning capacity and lint turn-out for the group of seed cottons giving 1-inch staple length and having 9.6 per cent moisture content were generally slightly less with the low saw speed and often were greater with the high speed. Ginning capacity for an outfit of four 70-saw gins was increased 3 bales per 8-hour day by increasing the gin-saw speed 200 revolutions per minute, and 11 bales by using a tight seed roll instead of a loose seed roll. Lint turn-out showed no definite or consistent relationship to seed-roll density.

Power requirement, though showing some tendency to increase with gin-saw speed, was not materially affected by changing the gin-saw speed 100 revolutions per minute above or below the manufacturer's recommended speed. The power requirement was increased 50 per cent, however, by varying the seed-roll density from loose to tight.

Energy consumption per unit weight of seed cotton ginned was not appreciably affected by variations in gin-saw speed or seed-roll density. The interrelationship of ginning time and power requirements caused the difference in energy consumption to be negligible, although indications were that for these small differences the high gin-saw speed and loose seed roll showed slight advantages.

Number of Gins in the Cotton Belt.—As shown by Hathcock,[5] there has been a gradual decline in the number of cotton gins in the United States since 1912. In that year, there were 28,358 gins, whereas in 1925 there were but 18,262, and of this number only 15,482 were active. In 1933, the number of active gins had been reduced to 13,543. This decline in number was largely due to the abandonment of many small plantation gins and the construction of large public gins. Although the number is smaller, the larger gins are better equipped, and the crop can probably be ginned in less time than formerly. It has been estimated that there are sufficient gins to gin the whole American crop in 30 days if all were kept going at full capacity.

The total investment in cotton gins amounts to approximately $200,000,000.

Cost of Ginning.—The charges for ginning are usually 30 to 45 cents per 100 pounds of picked seed cotton, and in addition a charge of $1.25 to $1.50 per bale is made for the bagging and ties that are put around the bale. The total cost is from $5 to $7. The charges for ginning snapped or sledded cotton, as was mentioned in Chap. XVI, range from about $10 to $15 a bale.

References

1. MILLER, T. S.: "Cotton Trade Guide and Student's Manual," Steck, Austin, Tex., 1915.

2. SCHERER, JAMES A. B.: "Cotton as a World Power," Frederick A. Stokes Company, New York, 1916.
3. TAYLOR, FRED *et al.*: "Cotton Ginning Information for Farmers," *U. S. Dept. Agr. Farmers' Bull.* 764, 1916.
4. BENNETT, CHARLES A. and GERDES, FRANCIS L.: "The Vertical Dryer for Seed Cotton," *U. S. Dept. Agr. Misc. Pub.* 239, 1936.
5. HATHCOCK, JAMES S.: "Practices and Costs of Cotton-gin Operation in North-central Texas 1924–25," *U. S. Dept. Agr. Tech. Bull.* 13, 1927.

CHAPTER XVIII

COTTON CLASSING

The staple of different kinds of cotton varies greatly in length, strength, and character. Different lots of cotton vary much in respect to color and amount of trash and foreign matter in them. There are other differences also that are very evident to persons engaged in the cotton trade but not easily understood by those not well acquainted with the staple. Before cotton can be marketed satisfactorily, the different kinds must be named and designated in such a way that the seller may be able to explain just what kind of cotton he has to offer and the buyers may understand what is offered and be able to say what is the market value of cotton of such a quality. This need of a standardizing system for use in cotton marketing brought into existence the science, or art, of cotton classing. It is more of an art than a science, perhaps, because it is mastered more by practice than by study. While it is true that many cotton sales are made on the basis of samples or on type, there is nevertheless always use for terms in designating market quotations and various business transactions.

A few years ago, very few persons except the cotton buyers in the larger markets knew anything about cotton classing, or grading and stapling cotton. The local buyers had a general notion of grades; knew clean, white cotton when they saw it; and in buying were careful to give themselves a good wide margin in price paid. They did not give the real market values of definite grades. The grower did not know what grade of cotton he had for sale; consequently, the price received depended more on the competition of buyers than on the merits of the cotton. In far too many instances the grower did not get the real worth of his cotton.

More recently much more interest has been taken in cotton classing. The subject is taught in most of the agricultural colleges in the South, in special cotton-classing schools, and in institute courses for farmers. County agents can secure sets of

Official Cotton Standards. Growers may then examine or study these and by means of them get some notion of cotton grades.

History of Cotton Classing.—Names for different grades of cotton were first used in Liverpool about 1800. Since that time, there has been in use at all times some sort of system of cotton classing, but, as a rule, it has been open to many objections. Until very recently not only did the markets in different countries have different standards, but even the standards varied in different cities in the same country. Liverpool used one system, New York another, and New Orleans still another. This made trading between the different markets inconvenient. As a rule, the same, or about the same, grade names were used in the different markets, but they did not in all cases represent the same quality of cotton.

Grade names were used in America, to some extent, during the first half of the nineteenth century, but it was not until the latter half that a systematic effort was made to get the matter on a proper basis. At a meeting of representatives of the American Cotton Exchanges at Augusta, Ga., in 1874, the standards used by the New York Exchange were adopted for use in all the exchanges of the country. After the adoption of the standards, many of the exchanges made no systematic effort to preserve their samples; consequently, different standards were soon in use in different places, and much confusion resulted.

In 1909, because of complaints made by cotton growers, the United States government began working toward the standardization of cotton classing in this country. A committee consisting of expert cotton classers, prominent cotton growers, dealers, and manufacturers was appointed by the Secretary of Agriculture to confer with the government classers. After due deliberation and study of grade types from about 30 American cotton exchanges and types submitted by individual members of the committee, the committee made a unanimous report and submitted a set of nine standards for grade which it had prepared. Copies of this set were prepared and temporarily placed in cotton exchanges, in colleges, and in offices of other organizations. Within 6 months after their promulgation, sets had been purchased by interested parties in 21 states and also in England, Germany, and Mexico.

The use of the first standards prepared was not compulsory. They did not meet with universal favor, because the trade considered them too high, especially for the lower grades.

With the passage of the United States Cotton Futures Act in 1914 it became necessary to require the use of Official Government Standards. In order to have a set of official standards that would be thoroughly acceptable to the trade, some of the leading cotton classers of the country were called to Washington to work with the department experts in the preparation of a new set of standards that would meet the objections raised against the old set. A set of new standards was thus prepared, and on Dec. 15, 1914, the Secretary of Agriculture gave notice that he had on that day established and did thereby promulgate standards for nine grades of white cotton designated as Middling Fair, Strict Good Middling, Good Middling, Strict Middling, Middling, Strict Low Middling, Low Middling, Strict Good Ordinary, and Good Ordinary, to be known as the Official Cotton Standards of the United States. The changes made in the standards made them acceptable to all interests in this country, and their use became general.

In 1915, the Secretary of Agriculture issued an announcement in which were defined certain kinds and conditions of cotton that were not tenderable on future contracts. Standards for color were established and promulgated in 1916; standards for American-Egyptian, in 1918; standards for Sea Island, in 1918; and standards for length of staple, in 1918. In July, 1922, the Official Government Standards were again revised, and certain changes made by a committee representing all branches of the cotton trade.

A numerical system was introduced in the revised standards to designate the different grades, No. 1 being assigned to Middling Fair, No. 2 to Strict Good Middling, and so on, higher numbers being given to the lower grades in numerical order. Additional designations for color were established.

On Mar. 4, 1923, the United States Cotton Standards Act was passed by Congress. This gave legal force to the Official Cotton Standards, established under the United States Cotton Futures Act, in all spot-cotton transactions in interstate and foreign commerce. Their use was to be required after Aug. 1, 1923. This requirement made it very necessary that Liverpool and other foreign markets have the same set of standards as the United States. The desirability of this uniformity had been

TABLE XXXV.—GRAPHIC ILLUSTRATION OF THE UNIVERSAL STANDARDS FOR THE GRADE OF AMERICAN UPLAND COTTON AS PROMULGATED AUG. 20, 1935, EFFECTIVE AUG. 20, 1936

Gray	Extra white	White	Spotted	Tinged	Yellow stained
		MF			
		SGM			
GMG	*GMEW*	**GM**	*GMSp.*	**GMT**	*GMYS*
SMG	*SMEW*	**SM**	*SMSp.*	**SMT**	
	MEW	**M**	*MSp.*		*SMYS*
MG	*SLMEW*	**SLM**		**MT**	*MYS*
	LMEW	**LM**	*SLMSp.*	**SLMT**	
			LMSp	**LMT**	
	SGOEW	**SGO**			
	GOEW	**GO**			

Symbols in heavy type denote grades for which practical forms or grade boxes of the revised standards will be prepared for public distribution. The grades indicated by symbols in italics are descriptive grades and will not be represented by grade boxes. Middling Fair White cotton is that which in color, leaf, and preparation is better than Strict Good Middling. Gray cotton is that which is more gray in color than that in the boxes for White cotton; Spotted cotton is that which in color is between the White and the Tinged; Yellow Stained is that which is more yellow in color than the Tinged; and Extra White is that which is whiter in color than the White grades. On and after Aug. 20, 1936, the grades shown above the horizontal line will be deliverable on futures contracts made in accordance with sec. 5 of the United States Cotton Futures Act. Those below the line will not be deliverable on such futures contracts.

recognized by all the markets for some time, and several unsuccessful efforts had been made to bring about the use of uniform standards.

With the passage of the act mentioned above, the matter of universal standards was again taken up. In June, 1923, there was a conference in Washington of Officials of the U. S. Department of Agriculture and representatives of the Liverpool Cotton Association, the Manchester Cotton Association, the Syndicat du Commerce des Cotons du Havre, and the English Federation of Master Cotton Spinners' Associations. The representatives from Liverpool held proxies from the associations at Bremen, Rotterdam, Ghent, Barcelona, and Milan. After discussion and the submission of recommendations by the visiting representatives, the official standards for grade and color for upland cotton, with certain modifications, were adopted by all the associations represented. The Liverpool representatives asked for certain changes in the white standards for Good Middling, Strict Middling, Middling, and Strict Low Middling, and it was the wish of the foreign delegates that the name "Universal Standards" be used instead of "Official Cotton Standards of the United States."

A second conference was called during July, 1923, at which the associations mentioned above were represented and, in addition, American merchants, exporters, spinners, and growers. This conference agreed on certain changes. The Secretary of Agriculture published an order putting into effect, on Aug. 1, 1924, the changes suggested by this last conference. Other standards not affected by the changes made in the conference were repromulgated, and the name Universal Standards was recognized as an alternative name.

With the exception of the establishment and subsequent modification of standard boxes for five grades of Extra White cotton, the standards for upland grades remained as promulgated on July 30, 1923, until a general revision was officially approved on Aug. 20, 1935, to be effective on Aug. 20, 1936. The reasons for the establishment of the new set of grade standards are given in *Service and Regulatory Announcement* 150, U. S. Department of Agriculture, as follows:

> In recent years various requests have been received that the grade standards be so modified as to make them more representative of the grade characteristics of the bulk of the cotton now produced and to increase their usefulness to buyers and sellers in the United States and abroad. These requests have come from organizations of producers, merchants, and manufacturers in the United States and from merchants and manufacturers in foreign countries. . . . The experience of the Bureau of Agricultural Economics in the cotton standardization work had clearly established the need for a general revision, as hereafter explained.

The Bureau of Agricultural Economics, in its effort to determine to what extent the standards should be modified, made a survey of the color and grades of bales of the American crops for the years 1928–1934 inclusive, as shown in the regular reports made by the bureau on the quality of the American crop. Further studies were made of the color and grades of cotton in important spot markets, but only a few bales of Blue Stained, Yellow Stained, Light Yellow Stained, or even Yellow Tinged cotton were found. It was noted further that very little cotton in these markets was classed as Extra White.

The *Regulatory Announcement* 150, quoted above, continues:

It was apparent from the study as a whole that the cotton graded against the standards for White cotton in recent years was considerably whiter than much of that represented in the old universal standards. Furthermore, since little or no Blue Stained, Yellow Stained, or Middling Fair White cotton had been produced, it was virtually impossible to procure cotton of these descriptions for the preparation of boxes to represent the grades.

In view of these facts it was decided early in 1935 that the preparation of a tentative set of revised standards should be undertaken. In it the grades for Blue Stained cotton were eliminated and the grades for Yellow Stained were made descriptive. The grades for White cotton were shifted to include whiter, and the more creamy samples in the higher grades for White cotton were eliminated, since creamy bales could not be found in quantities sufficient to make copies of the old standards. The highest grade for White cotton, No. 1 or Middling Fair, was made descriptive, as were the grades for Extra White. The latter was increased in number from five to seven. The new Tinged grades did not contain the deeper colored cottons of the old series, and the changes in the White and in the Tinged standards resulted in excluding from the descriptive standards for Spotted, much of the Light Tinged cotton heretofore classified as Spotted. No. 2 Yellow Tinged, or Strict Good Middling Yellow Tinged, was also eliminated in the revision.

The revision mentioned above reduced the number of recognized grades from 37 to 32 and the number of standards represented in grade boxes from 25 to 13. The grades now officially recognized and the grades for which grade boxes may be obtained are shown in Table XXXV.

Cotton Standardization.—A system of cotton classing or standardization must take into account many qualities of cotton and express in some way the degree in which any particular quality is manifested. These qualities cover such attributes as amount of trash, waste, or foreign matter in the cotton; the color of the fiber; body of fibers; uniformity of fibers; damage to fibers in ginning; and mixture of different kinds of cotton in a bale.

Each of these different qualities will be considered in some detail later.

Classification According to Grade.—The Universal Standards, which superseded the Official Cotton Standards of the United States, contain nine grades of white cotton, which are named: Good Ordinary, Strict Good Ordinary, Low Middling, Strict Low Middling, Middling, Strict Middling, Good Middling, Strict Good Middling, and Middling Fair, the last named being the highest. Each standard represents a full grade. The range of these grades is wide enough to include practically all White cotton, especially all that is harvested under ordinary conditions. Middling is the basic grade, or the grade on which market quotations are based. The relative value of grades above or below middling varies in different seasons, the value depending largely on supply and demand.

Other grade names are used for cotton that is tinged, stained, or discolored in some way. These will be discussed in a later paragraph.

Factors That Determine Grade.—The grade of cotton is determined by the amount of leaf fragments, trash, and foreign impurities in it; by its color, whether white, tinged, stained, spotted, blue, or gray; and by the preparation or ginning, whether smooth and well ginned or containing neps and gin-cut staple.

Foreign Impurities.—After a cotton boll opens, it dries out rapidly, and all its parts, except the seeds, soon die. The bracts, the small leaflike structures at the base of the boll, become very dry and brittle. When a boll is picked, these bracts are apt to be broken more or less, and small pieces carried away with the locks of cotton. This probably constitutes the chief foreign impurity found in lint cotton. During the early part of the picking season, it is about the only one. After frosts come, the large leaves of the plant are brittle, and fragments of them become mixed with the cotton when it is picked. During the latter part of the season, a considerable amount of cotton falls to the ground. If this is picked up, it is apt to carry with it leaf fragments, pieces of stems, trash of all kinds, sand, dirt, and dust. The addition of this cotton lowers the grade considerably and is one of the main reasons why cotton from late pickings is of low grade. Leaf fragments of pinhead size, if numerous, are considered more objectionable than larger ones, because they are harder to remove from the cotton in manufacturing processes.

Cotton exposed in the open boll collects a certain amount of dust and sand blown about by the wind. Bolls near the ground collect a considerable quantity of sand and dirt splashed on them by beating rains. Locks picked up from the ground carry still more. When in excessive quantities, the sand and dirt contained may be considered apart from the grade and cause the cotton to be penalized in consequence.

There are always some small, undeveloped seeds in the seed cotton. A part of these pass between the gin ribs with the lint and constitute an impurity known as "motes." Certain years when the weather is very dry or when the cotton plants are defoliated by insects or otherwise, these immature seeds may be rather abundant. The little stem at the small end of the seed may be broken off in ginning; or if the roll is too tight and saws run fast, there may be fragments of hulls from cut seeds in the lint. All these impurities, of course, lower the grade of cotton.

In parts of the Cotton Belt, there is considerable snapped or bolly cotton. The cotton is not picked in the usual way, but unopened bolls, open bolls, burs, and all are gathered. This cotton is ginned on a gin that will tear the bolls open and take out the lint cotton. While such cotton is of value, a large portion of it is low in grade because of the presence of much trash, boll fragments, etc.

The classification of cotton may be raised one or two grades by ginning it on one of the improved gins now in use which have special cleaner devices.

Color.—Although there are separate grades for colored cotton or cotton that is "off" in color, the characteristic of color has a bearing on grading and is considered in the classing of White cotton. According to the Universal Standards, to grade Good Middling or above, a sample must be a bright creamy or white color and be free from any discoloration. Below this grade, some latitude is allowed in range of color. The cotton may be creamy, white, or "dead" white. In the lower grades, grayish, dingy, or slightly reddish colors are allowable.

Ginning.—The quality of the ginning also has a bearing on the grade of a sample of cotton, but it is not of primary consideration. If the ginning is very bad, the cotton may be considered as irregular and penalized accordingly. Bad ginning adversely affects the grading in that it makes cotton look stringy, rough, and tangled.

If cotton is fed to the gin breast too rapidly, or if the saws are run too fast, or if the cotton is damp when it is ginned, the lint is apt to be gin-cut or show neps. Gin-cut cotton has a "chewed-up" and "stringy" appearance. Neps are small masses of densely matted and tangled fibers. They appear as small white masses. When they are present in quantity, they are hard to remove and show in the finished yarn if they have not been removed.

If the seed cotton is stored a few weeks before ginning, it dries out thoroughly. It then gins better, and there is probably less gin cutting of fibers. As has been mentioned in Chap. VII, some authorities are of the opinion that the luster of cotton is improved by storage before ginning, but this has not been demonstrated conclusively.

Description of Grades.—The different grades typified by the Universal Standards are different, but the difference is mainly in the degree in which the various qualities appear. Consequently, it is impossible to draw sharp lines separating them or to give delimiting descriptions. Samples of cotton must be seen and studied to get a definite conception of the grades.

Good Ordinary.—A sample of Good Ordinary cotton contains a considerable quantity of leaf particles, and it may have also a sprinkling of motes, seed fragments, and occasional gin cuts. It may contain 1 per cent or less of sand, dust, or dirt. This grade is considered White cotton, but it has some tinge or stain and is usually more or less smoky and discolored.

Strict Good Ordinary.—Strict Good Ordinary differs from the preceding grade chiefly in having a lesser amount of the imperfections named.

Low Middling.—Low Middling appears to be considerably whiter than the first grade mentioned, but there are indications of tinge and stains in spots. There is also a considerable quantity of leaf and motes and perhaps some gin cuts and neps.

Strict Low Middling.—Strict Low Middling is intermediate between Middling and Low Middling in characteristics.

Middling.—Middling cotton is white and nearly free from gin cuts and neps, but it contains some pieces of cotton seed and a medium amount of fairly large pieces of leaf. Middling cotton may have the color of the higher grades, but the leaf particles and other impurities that it contains hold it to its grade. Middling is the basic grade in respect both to market price and to the measuring of other grades. Different samples of Middling cotton may vary in a way. One sample may rank high in respect to color but low in respect to quantity of leaf and motes. Another may be from cotton that was so carefully picked that it contains but little trash but be held down by having too much color or discoloration for the higher grades. Within certain limits, the grade is determined by the average of the qualities of the sample. Of course, a sample that is badly off in respect to color, gin cutting, or some other characteristic could not be classed as Middling cotton even if it had excellent other qualities.

Strict Middling.—The Strict Middling grade is, of course, one grade better than Middling, but it is difficult to say just wherein it is better. In general, it is a slightly better looking cotton and has better color and less leaf and trash. This grade is very popular with the American spinner of staple cotton.

Good Middling.—Good Middling cotton has a good creamy-white or white color, is free from neps and gin cuts, and contains only a few foreign particles and but little leaf trash. Cotton that is picked early in the season before it has been rained on should grade Good Middling or possibly better if it is well ginned and the picker was careful. Very little of the commercial crop grades higher than Good Middling unless special effort is made in picking.

Strict Good Middling.—Strict Good Middling cotton has a bright creamy or white color, has no neps or gin cuts or broken seeds, and has fewer pieces of leaf than are to be found in Good Middling.

Middling Fair.—Middling fair is the highest grade represented in the Universal Standards, and it is rare in commercial trade. The cotton is of a bright creamy or white color, fluffy, and well ginned. About the only imperfections to be found are a very few leaf fragments.

Grades above and below the Official Government Grades.—The two grades Strict Middling Fair and Fair have been mentioned for cotton of a higher grade than Middling Fair, the highest grade represented in the Government Standards. These grades are more theoretical than real, since such cotton never appears on the market and is rarely seen. They are not now recognized.

Three or four different grades for cotton of lower grade than Good Ordinary have been listed. These are, in order of rank from highest to lowest, Strict Ordinary, Ordinary, Low Ordinary, and Inferior. These grades are not used much by the trade. There is a considerable quantity of this very low-grade cotton, but it is largely called "dogs," or "dog tail," and is sold on basis of sample or type.

Half and Quarter Grades.—Prior to the issuance of the Government Standards, Middling Fair, Good Middling, Middling, Low Middling, and Good Ordinary were considered as full grades, whereas the grades between these, or the grades designated with the prefix "Strict," were regarded as half grades. All these grades were made full grades by the Cotton Futures Act of 1914. Since that time, the term "half grade" has largely passed out of use in this country.

When grading cotton, the classer frequently handles samples that are a little too good for a certain grade but hardly good enough for the next higher. These, being between the half grades formerly in use and whole grades, have been termed "quarter grades" and designated by the names "fully" or "barely." "Fully," when used before the name of a certain grade, indicated a quarter above; and "barely," when used in this way, indicated a quarter grade below the regular grade. The terms "full style" and "shy" express about the same as "fully" and "barely," respectively. Although these quarter-grade designations are not sanctioned in the present standard classification, they are used more or less commercially.

According to Palmer,[4] it is the present custom of the government classers to give cotton of intermediate grade the classing of the lower grade unless there is some special reason for classing it otherwise.

Boxes of Official Standards.—The expert cotton classers of the U. S. Department of Agriculture have prepared boxes of cotton samples illustrating eight of the nine official grades of White upland cotton (Fig. 107). The standard for Middling Fair is descriptive only. Each large box contains twelve fair-sized samples of cotton chosen to illustrate the variation allowable within one grade (Fig. 108). The samples are sewed in the

Fig. 107.—Government classers preparing boxes of cotton samples which illustrate the official standards. (*Courtesy G. L. Meloy.*)

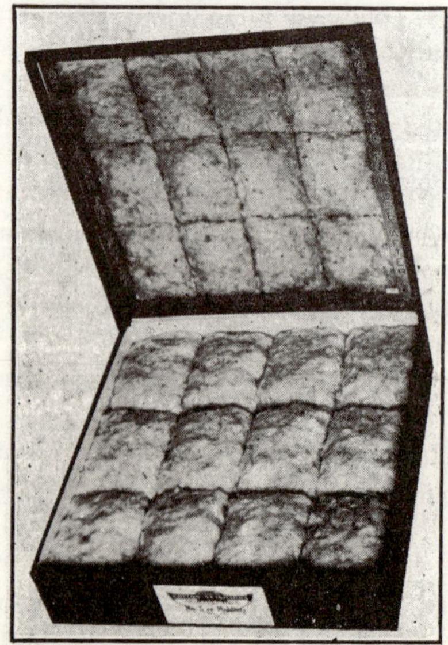

Fig. 108.—Box containing twelve cotton samples which show the range within one grade of the official standards.

box so that they cannot be removed or tampered with, and a full-size photograph showing the appearance of the samples when issued is pasted on the inside of the lid of each box. By a comparison of the samples with this picture, the disturbance of leaf particles, etc., may be detected. The box has a hinged lid which, when closed, will keep out dust and light.

These standards are very useful in giving a beginner in the art of cotton classing a notion of the different grades, and they help keep experienced classers from changing their ideas. Sets are sold by the government for a moderate sum and are in use in agricultural colleges, cotton-classing schools, cotton exchanges, and the sample room of many cotton buyers.

To show their color satisfactorily, the standards should be placed in a room well lighted by skylights. If this is not available, a room lighted by windows on the north side is next best. Direct sunlight should never be allowed to fall on the samples. The true color cannot be judged in sunlight, and the sun bleaches out color.

In using the standards for comparison, the classer rolls the sample he is examining into a shape resembling that of the standard samples and then holds his sample against the side of the standard boxes until he finds one that matches his. If the boxes are tilted at an angle of about 45 degrees, the samples may be inspected more easily, and there is less likelihood that the student will hold his sample over the box of standards. It is best not to hold a sample over the box of standards, because dirt and fragments of foreign material are apt to fall from it. A very little foreign material dropped on the surface of a standard will ruin it. The standard boxes should be kept closed except when in actual use. When opened, the lid should be lifted slowly, since rapid opening or closing may so fan the air that particles on the surface of the samples will be moved.

Classification on Basis of Color.—The amount of discolored cotton harvested varies considerably in different years, depending largely on the time killing frosts occur. If they are early, the quantity is large. Earle and Taylor[1] have estimated that, on the average, one-fifth of the crop cannot be classed as White cotton. Since the advent of the boll weevil, the average percentage is higher. During some years the percentage of discolored cotton runs as high as 35 or more. From this it may be seen that the discolored cottons are of considerable importance.

The discolored cottons have, of course, been handled by the cotton trade for years, but no distinct classes were established for all the different kinds. They were classed as White cottons but designated as "off color," "spotted," "tinged," or "stained," etc. The recent official classification of the U. S. Department of Agriculture recognizes six groups of upland cotton on basis of color. These are, as shown in Table XXXV, Gray, Extra White, White, Spotted, Tinged, and Yellow-Stained. Minor color variations occur in each of these divisions. These variations are probably more marked in the White cotton than in any other. This division, as was brought out in preceding paragraphs, contains cotton that is not altogether white but is more or less tinged with different colors.

Gray cotton is grayer in color than any cotton found in the boxes of White cotton; the Extra White is whiter in color than that found in the White grades. Spotted cotton is lighter than tinged cotton but has too much yellow color in parts to be called white. Tinged cotton has a light yellowish or faint reddish color throughout. Such cotton bleaches and dyes well and consequently is of considerable value. Yellow-Stained cotton is similar to Yellow-Tinged, except that it is darker in color. The difference is largely in degree of color. Frost on unopened bolls is the principal cause of the yellow discoloration.

The discolored cottons are considered inferior to White in value. As may be seen from Table XXXV, a large part of the grades of these cottons are not tenderable on future contracts.

Grades of Discolored Cotton.—The discolored cottons are first classed according to color, as Gray or Spotted. Next they are examined with reference to presence of foreign matter and to ginning preparations. On the basis of their appearance in respect to these qualities, each is given a specific grade classing, as, for instance, Good Middling or Middling. None of the discolored cottons is graded higher than Good Middling, and none is graded lower than Low Middling. The complete range of grades is shown in Table XXXV.

Grades of Sea Island Cotton.—Sea Island cotton is considerably different from upland cotton and so requires a different system of grading. It has a very long, fine, silky staple and is ginned on roller gins, which give the lint cotton a tufted or wadded appearance very different from lint from saw gins. Crushed seeds are also more common than in the lint from saw gins. The lint of Sea Island resembles that of upland in color but is somewhat more creamy.

Official standards for grades for Sea Island cotton were established in October, 1918. These were based on the commercial grades of Sea Island cotton in use at that time and are designated numerically in order from highest to lowest as Nos. 1, 2, 3, 4, 5, and 6. Half grades, intermediate between these, are recognized and designated by the fraction ½, as Nos. 3½, 4½, etc. The one set of grades covers, or takes cognizance of, color, foreign impurities, and ginning qualities. Boxes of samples illustrating the different grades of Sea Island cotton were prepared by government classers and were available by purchase from the Department of Agriculture for a time, but they have been discontinued because of the scarcity of this variety of cotton.

Grades of American-Egyptian Cotton.—American-Egyptian cotton is different from upland cotton and also somewhat different from Sea Island. It is ginned on roller gins; and as the seeds are somewhat more fuzzy than those of Sea Island, it is, according to Palmer,[4] harder to gin and gives a more crinkled or crimped sample of lint, which contains more crushed seeds and is darker in color. The highest grades of American-Egyptian cotton are of a creamy color, whereas the lower tend to be darker. American-Egyptian is derived from cotton grown in Egypt, some varieties of which are brown in color.

Official Standards for American-Egyptian cotton were established by the government in 1918. The grades recognized are designated by the numbers 1 to 5 inclusive, the lowest number being the best grade. Half grades, as in Sea Island cotton, are designated by the fraction ½. Cotton inferior to grade 5 is designated "below grade 5." Descriptive standards only are used, but the descriptive standards for grade have the same legal status as grades represented in physical form.

Classification According to Length of Staple.—If the premium paid for the extra length of long-staple cotton is considered, the classification for length of staple appears to be a matter of considerable importance. Less than 20 per cent of the cotton produced in the United States, however, is long-staple cotton. A large part of the cotton grown is sold in primary markets without regard to its staple length.

Cotton with a staple of $1\frac{1}{16}$ inches in length or less is generally considered short-staple cotton, or short cotton, but is sometimes referred to as "medium-staple" if the length is 1 or $1\frac{1}{16}$ inches. Staple $1\frac{1}{8}$ inches in length or longer is called "long-staple," or "staple," cotton. Cotton $1\frac{1}{8}$ to $1\frac{3}{16}$ inches is sometimes called "benders," and cotton with a staple length more than $1\frac{1}{4}$ inches is called "extra-staple" cotton. At present, there is a tendency to restrict the term "short cotton" to cotton less than an inch in length. Most of these terms are used without close discrimination.

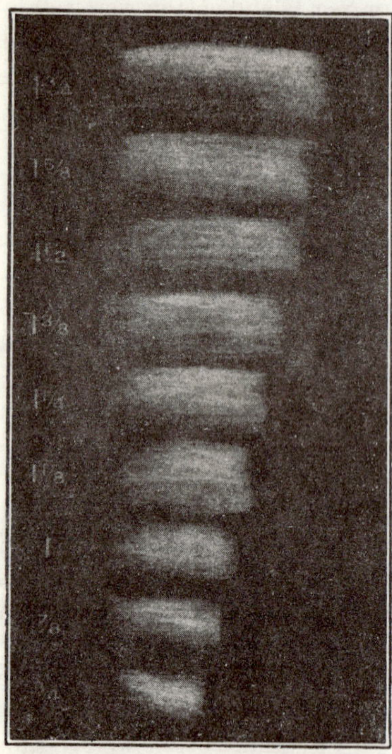

Fig. 109.—A photographic representation of the official cotton standards of the United States of those lengths of staple for which types are available for distribution, each respective length as shown being obtained from the original type bale. (*Courtesy G. L. Meloy.*)

Official Standards for Length of Staple.—Official standards of the United States for length of staple were established in October, 1918. These have been revised somewhat at different times since that date. At present (1937), the standards in physical form include American upland cotton of the lengths $\frac{3}{4}$, $\frac{13}{16}$, $\frac{7}{8}$, $\frac{29}{32}$, $\frac{15}{16}$, $\frac{31}{32}$, and 1 inch, $1\frac{1}{32}$, $1\frac{1}{16}$, $1\frac{3}{32}$, $1\frac{1}{8}$, $1\frac{5}{32}$, $1\frac{3}{16}$, $1\frac{7}{32}$, $1\frac{1}{4}$, $1\frac{9}{32}$, $1\frac{5}{16}$, $1\frac{11}{32}$, $1\frac{3}{8}$, and $1\frac{1}{2}$ inches; and American-Egyptian cotton of the lengths $1\frac{1}{2}$, $1\frac{9}{16}$, $1\frac{5}{8}$, and $1\frac{3}{4}$ inches. Descriptive staple standards for American upland cotton are in use for the lengths $1\frac{13}{32}$, $1\frac{7}{16}$, $1\frac{15}{32}$, $1\frac{17}{32}$, $1\frac{19}{32}$, and upward in gradations of thirty-seconds of an inch. American-Egyptian cotton of the lengths $1\frac{17}{32}$, $1\frac{19}{32}$, $1\frac{21}{32}$, $1\frac{11}{16}$, $1\frac{23}{32}$ inches; and longer than $1\frac{3}{4}$ inches is designated descriptively in gradations of thirty-seconds of an inch.

A difference in staple length less than $\frac{1}{32}$ inch is hardly measurable in practical cotton stapling. In practice, lengths are more commonly expressed with fractional units, none of which is smaller than $\frac{1}{16}$ inch, as $1\frac{1}{16}$, $1\frac{1}{8}$, $1\frac{1}{4}$. In case the length cannot be measured in even sixteenths, it has been the custom of some classers to add the suffix "full" to the nearest sixteenth. This is equivalent to adding $\frac{1}{32}$, since $1\frac{3}{16}$ full means $1\frac{7}{32}$. This intermediate length is also expressed by giving both the even numbers nearest and connecting them with "to"; for example, $1\frac{7}{32}$ may be expressed as "$1\frac{3}{16}$ to $1\frac{1}{4}$." The intermediate length $1\frac{7}{32}$ may also be indicated by the expression "not less than $1\frac{3}{16}$ inches or over $1\frac{1}{4}$ inches."

These samples show actual length of fibers without reference to character or grade. For exact measurements, the temperature must be 70°F. and the relative humidity of the air 65 per cent.

The length of the staple is sometimes expressed in millimeters. This is especially true in European countries.

Determining Staple Length.—Much practice and skill are required to determine the length of cotton fibers, or, in other words, to "staple" cotton. The fibers vary in length, their length varies with the moisture of the air, and they are much tangled, bent, and twisted in the cotton sample. Some care must be taken in drawing out the fibers to be measured in order to have a representative sample; and after they are drawn and arranged parallel, there may be a question as to the exact length of the "pull," since some fibers are slightly longer than others. In fact, within narrow limits, scarcely any two fibers are exactly of the same length. In measuring, the stapler must use his judgment in striking an average.

An experienced classer is so adept at stapling cotton that he can judge the length of the staple in a measure simply by grasping a sample and pulling it apart. For more accurate work, he "pulls" the staple after a plan similar to the following one outlined by the specialists in the U. S. Department of Agriculture.

Detailed Description of Method of Pulling Staple.—Grasp in the two hands a tuft of cotton of a size convenient for the purpose (about $\frac{1}{4}$ ounce). Hold it firmly between the thumb and forefinger of each hand, with the thumbs placed together, the fingers turned in toward the palms of the hands, and the middle joints of the second, third, and fourth fingers of each hand touching the corresponding joints of the fingers of the other hand, so as to give a good leverage for breaking.

Pull the cotton slowly apart with about the same leverage of each hand on the joints of the fingers, separating the tuft into two parts.

Discard the part remaining in the right hand.

Grasp with the thumb and forefinger of the right hand the end of the tuft of the cotton retained in the left. The point of pressure on the cotton in the left hand is just below the joint of the thumb and at the nail joint of the forefinger.

With the right hand draw a layer of fibers from the cotton held in the left hand.

Retain in the right hand the layer so drawn.

Repeat this operation four or five times, placing each successive layer directly over the fibers previously drawn. Be careful that the ends of all the layers are even with one another between the thumb and forefinger of the right hand.

After discarding the cotton in the left hand, hold the fibers thus obtained between the thumb and forefinger of the right hand and smooth them with the thumb and forefinger of the left hand.

Place these fibers on a flat horizontal surface with a black background.

Block off the ends of the fibers with a cotton-stapling rule, so as to indicate the length of the bulk, or body, of the fibers.

Then measure the distance between the blocked-off ends.

If preferred, the left hand may be substituted for the right and the right for the left, as the case may be, throughout the process here described.

Classification According to Character of Staple.—Character of staple is a matter of much importance to spinners and manufacturers of cotton goods, yet it is not at all understood by cotton growers and but vaguely grasped by many cotton buyers. Character has to do with the strength, body, uniformity, drag, elasticity, and other qualities of cotton fibers. Most of these attributes are rather intangible and hard to measure. There are no standards with which to make comparison. The classer is guided by his training and experience, as he interprets the "break" and "drag" of the fibers when he pulls them apart and when he interprets the "feel" of the fibers.

Strength.—The strength of cotton fibers or yarns may be determined fairly accurately by means of a fiber-testing machine, such as was described in Chap. VII. The strength of fibers may be determined roughly by a cotton classer by breaking a small tuft when the ends are clasped firmly between the thumb and forefinger of both hands.

Strength is a desirable quality in cotton fibers, since the strength of yarns and the quality of the goods made from the yarns depend in large measure on the strength of the fibers.

Body.—By body of cotton is meant the relative solidity or density of a mass of the fibers. The fibers appear to fit or cling together closely and to make a firm sample. Good soil usually produces better bodied cotton than poor soil. The cotton from poor soil tends to be light and fluffy, which are not qualities of good body. Varieties that produce only a limited number of bolls to the plant also tend to have better bodied cotton than the more prolific varieties.

Uniformity of Fiber.—To the spinner there is no quality of cotton of more importance than the uniformity of the fibers. If they vary in length, there will be much waste. If they vary in diameter or twist, the yarns will not be uniform. Mixed varieties, as would be expected, produce staple of various lengths, and unselected strains frequently contain many plants that have mixed staple on individual seeds.

The uniformity of fiber can be judged to best advantage by combing out the lint of different plants in the field. Uniformity in sample of ginned cotton can be judged to best advantage by making a large number of "pulls" from a single sample.

Drag.—"Drag" is a term used to indicate the resistance offered by fibers in being pulled apart or drawn past one another. Good drag is considered a desirable spinning quality, because fibers with good drag cling together well in spinning.

Silkiness, or Fineness.—Silkiness, or fineness, is a desirable quality in cotton that is to be used in making fine fabrics. Such cotton feels very soft and smooth. As a rule, the fibers are long and of small diameter.

Drawing Sample from the Bale.—Where gin compresses are in use or in gins where round bales are made, lint samples are taken while the bale is in the press; therefore it is not necessary to cut the bagging to draw them. Most flat bales are cut and sampled one to several times. Frequently, each successive buyer wants a new sample.

There is no special art in sampling cotton, but the operator should know how to do it and understand the importance of correct procedure. The sample should be representative of the entire bale, and to insure this it is necessary to cut each side of the bale.

Miller[2] gives the following instruction for sampling a bale of cotton:

> The cut should be semicircular in form, made with a sharp knife and deep enough to cut through at least an inch of cotton in depth. With the fingers dipping into the opposite end of the cut from the sampler, he should gradually work them under the cotton as he draws his hands toward himself, at the same time tearing it up until he has a sufficiently large sample loosened on one side of the cut. Beginning again in the cut at the opposite side from himself, he should draw the sample as he brings his hands forward. The cotton drawn should be a sample of about the same thickness all the way across the cut. Keep the sample intact; that is, do not draw small quantities repeatedly and attempt to make a sample by combining them.

Hooks, instead of the fingers, may be used in pulling out the sample. Drawn samples that are to be preserved should be rolled and wrapped securely with heavy wrapping paper. A numbered tag is fastened to the bale, and a card bearing the same number is enclosed in the sample.

Classing Samples.—As was shown previously, it is necessary to have proper light to judge the color of cotton satisfactorily. Rough classification may be made in the open air, but for careful work the samples are carried to a cotton-sample room, where the right kind of light may be had (Figs. 107 and 111). With his back to the light, the classer holds up the sample, noting color, foreign matter, etc. The sample is then opened between successive layers, and a mental average taken of the appearance of the whole sample. The grade given the sample is on the basis of this average. In case the sample shows that the two sides of the bale are different, the grade is based on the appearance of the poorer half.

Licensed Classifiers.—Under sec. 3 of the United States Cotton Standards Act approved Mar. 4, 1923, it is provided that the Secretary of Agriculture may, upon presentation of satisfactory evidence of competency, issue to any person a license to grade or otherwise classify cotton and to certificate the grade or other class thereof in accordance with the Official Cotton Standards of the United States.

Application for a license to classify cotton is made to the chief of the Bureau of Agricultural Economics. Upon presenting satisfactory evidence of his competency, the payment of a license fee of $10, and agreeing to comply with the terms and regulations of the Cotton Standards Act, the applicant may be given an official license to classify cotton for a period of one year. This license may be renewed at a cost of $5 annually.

For the classification and certification of any cotton or samples, a licensed classer may charge fees as follows:

1. If the classification is with respect to grade only, at the rate of 15 cents a bale.
2. If the classification is with respect to staple only, at the rate of 15 cents a bale.
3. If the classification is with respect to any other single quality, at the rate of 15 cents a bale.
4. If the classification is with respect to two or more of the qualities specified in paragraphs 1, 2, or 3, at the rate of 30 cents a bale.

Licensed Cotton Samplers.—The reliability of the classification of any cotton depends on the drawing of the sample from the bale as well as on the accuracy in grading and stapling. Unless the sample is authentic and thoroughly representative, the classing is of little value. To make available the means for assuring the regularity and representative nature of samples drawn for classification, the Secretary of Agriculture is authorized to issue sampler's licenses to qualified persons. Application for license is made to the chief of the Bureau of Agricultural Economics.

References

1. Earle, D. E. and Taylor, Fred: "Classification of American Upland Cotton," *U. S. Dept. Agr. Farmers' Bull.* 802, 1917.
2. Miller, T. S.: "Cotton Trade Guide and Student's Manual," Steck, Austin, Tex., 1915.
3. Hubbard, W. Hustace: "Cotton and the Cotton Market," D. Appleton-Century Company, Inc., New York, 1923.
4. Palmer, Arthur W.: "The Commercial Classification of American Cotton," *U. S. Dept. Agr. Cir.* 278, 1924.
5. Coman, J. G.: "Cotton Classing and Marketing," *Clemson Agr. Coll. Bull.* 4, 1913.

CHAPTER XIX

COTTON MARKETING

As is well known, millions of bales of cotton are produced every year. Nearly all this cotton is hauled to the market and sold. The consumers live all over the world, a large number in regions far removed from the fields where the cotton is grown. Much of the product is spun and woven into cloth in mills thousands of miles from the cotton fields. An army of men is required to buy this cotton and distribute it to the manufacturers who need it for making cloth or for other purposes, and millions, even billions, of dollars are required to finance the operations incident to this great volume of business. It should be the object of marketing to convey the commodity from the producer to the consumer by the shortest route and at the least possible cost, so that the producer may receive a satisfactory price for his cotton and the consumer get the goods he buys at a reasonable figure. From data that will be given later in this chapter it appears that the difference between the price the consumer of cotton goods pays and the price the producer of the raw cotton gets is too great, being in a ratio of about 5:1. More than 75 per cent of the increase comes after the cotton has reached the large concentration points, like New Orleans.

Brief History of the Cotton Trade.—Prior to and during the eighteenth century, the manufacture of cotton goods in England was carried on by spinners and weavers in their homes. The goods made were of the rougher sort. The raw cotton was obtained principally from the West Indies. The system of marketing was very simple, the weaver being buyer, manufacturer, and seller. When the cloth was made, he carried it to the town market to find purchasers. As the quantity of goods manufactured increased, there was need for middlemen or merchants to deliver material to the spinners and weavers and in return receive cloth for sale in the market. About 1760, small factories sprang up. Since the quantity of raw material consumed by these was considerable, it came to be the business of certain men to buy and furnish the manufacturers with raw material. Thus special cotton merchants, or factors, came into existence. London and Liverpool were the chief import centers for cotton coming from the West Indies. Importers sold cotton to factors, who, in turn, sold it to the manufacturers.

Some cotton was grown in most of the southern English colonies in North America from early times, but it was not until near the close of the eighteenth century that export trade began. There is a record of eight bags of cotton being shipped to England in 1784. At first the English spinners were prejudiced against American cotton. They had not tried it and imagined that it was inferior, but a trial soon convinced them that it was even superior to cotton from India. In 1791, America exported 189,-500 pounds of cotton to England. France soon began to want American cotton. In 1801, she imported 844,728 pounds, and by 1806 the amount was increased to 7,082,118 pounds.

In 1787, the first cotton mill in America was erected in Massachusetts. This gave an impetus to home demand for raw material. The War of 1812 stopped the cotton trade with England and for a time depressed the industry. After the war, trade was revived, the demand became good, and cotton commanded a fair price. The saw gin and slave labor made the production of a quantity of lint cotton an easy matter. Transportation was poor, however. The cotton had to be hauled in wagons to river landings or export ports for transportation.

Prior to the Civil War in America, most of the American cotton was sold through commission men, who were situated in the export cities of the South and who were, in most instances, the only intermediaries between the producers and the importers in European countries. The commission men handled cotton just as any other article of commerce. Market conditions were not satisfactory, because of lack of ready means of communication and a central point from which quotations might be derived. There were no cotton exchanges, and hence no future markets.

At the beginning of the Civil War, America was producing more than half of the world's supply of cotton. At the close of the war, she was producing practically none. Prices advanced to extreme figures; some Middling cotton is reported to have sold for $1.89 per pound, and the average price was probably about $1 a pound. No regular shipments to New York were made during the latter part of the war. The only cotton received was some that was captured by the Federal government through its blockade of southern ports or that was confiscated in the South. These lots of cotton were, in some instances, bought by speculators, and considerable profit was made. This led to much speculation. Speculators were so anxious for cotton that they began to buy "cotton to arrive." This was probably one of the steps that led to the development of the cotton exchanges, where "futures" are bought and sold.

In 1866, the Atlantic cable was laid. This brought the American and European markets into close contact and stimulated trade. Soon afterward, the Liverpool and New York cotton exchanges were established, and the cotton trade began to follow a system similar to the present one. The commission men who previously had received cotton at every port were displaced by local cotton buyers, who established themselves in all towns where cotton was ginned or assembled in any quantity. There is some advantage to the grower in the new plan. The local cotton buyer examines individual bales and is more inclined to pay the grower just what each is worth than the former commission merchant, who figured on a rather wide margin.

The distribution of American cotton to various countries of the world is shown by the following table, which is taken from one prepared by the Liverpool Cotton Association.

TABLE XXXVI.—DISTRIBUTION OF AMERICAN COTTON

Years	Per cent exported to					Per cent taken by American spinners, North and South
	Great Britain	France	North Europe	Other ports	Total	
1851–1856	52.70	13.13	5.86	5.91	77.60	22.40
1870–1875	49.30	6.79	10.22	2.94	69.25	30.75
1890–1895	37.28	7.59	16.06	6.02	66.95	33.05
1910–1915	25.45	6.72	18.22	9.98	60.37	39.63
1920–1923	13.50	5.58	13.51	13.27	45.86	54.14

This shows that during the last 70 years the relative amount of American cotton taken by Great Britain has gradually declined, whereas the relative amount taken by American spinners has increased greatly. Great Britain is taking about one-half as many bales as she was taking 50 years ago, whereas the American mills are using several times as many.

Several of the cotton-producing countries of the world are competitors for the European trade. India and Egypt are the leading rivals of the United States. India produces a great deal of cotton, but the quality is mostly rather poor, and a considerable amount is used for home consumption. Egypt produces more than a million bales of cotton annually. This is of excellent quality, and most of it is exported. The production of cotton in Egypt is at present limited by the lack of more land suitable for cotton growing.

Present System of Marketing.—As has been said in previous paragraphs, the machinery of cotton marketing in use today is extensive and complex. There are three large cotton exchanges on which contracts for futures, or future delivery of cotton, are bought and sold. These exchanges keep in closest possible touch with the growth and market conditions of cotton in all parts of the world. The trading in the exchange market is influenced by the data gathered. Results of the trading taking place are telegraphed to all parts of the world and become the basis of market quotations everywhere. Prices prevailing in "spot" markets, that is, markets to which a large volume of cotton is shipped for sale, are governed very largely by the quotations from the exchanges that deal in future contracts, or futures.

There are primary markets at nearly every crossroad hamlet in the South, larger interior markets in the larger towns and cities, consuming markets in cities where there are cotton mills or near to them, and export markets along the seacoast and on the Mississippi River. There are numerous buyers, brokers, commission men, and cotton dealers of various kinds. Warehouses are numerous, large, and mostly well constructed and well managed. Most of these units of the marketing system will be discussed in the following pages.

Primary Markets.—In nearly every small town there is at least one cotton buyer. In most instances, this is the country merchant. Census Bureau figures of 1920 show the average size of improved farms in the South to be 47.3 acres, including only land under cultivation. From this it may be seen that the average size of the farm is small. Most of the cotton produced is grown by small farmers, many of whom have to have some financial assistance in making their crop. This assistance is commonly given by the country merchants, the local banks, or cotton factors.

Formerly, many cotton factors in the larger towns entered into business relations with farmers and loaned them money to use in "making a crop." They, of course, charged a good rate of interest on the money loaned and made a profit on the cotton after it was sold to them. In many instances, the factors operated with very little capital of their own. They secured money from banks or other sources by means of the paper that the farmers had given them and did not pay the growers for their cotton that was consigned to them until after it was sold. More stringent financial regulations now in force have interfered with their system of carrying on business, and as a consequence they are passing. To be sure, there are factors who are financially strong and are doing a legitimate business. They have some arguments in their favor, which if put into force may result advantageously to the farmer. The classers they send out are well qualified. They handle a large quantity of cotton, so they can assemble large lots of even-running grade, or staple, and can pool the whole and sell it to good advantage. Their storage facilities in the larger markets are excellent, and insurance rates are low.

The merchant furnishes the farmer with necessary supplies—food, everyday clothing, medicines, etc.—for his family and also his tenants if he has any. The merchant is secured, in that the farmer must give him the first chance to buy the cotton when the bales are put on the market. This may appear to be a fair system, but it is one that has frequently wrought much hardship on the poor, hard-working farmer in the hands of a "Shylock" merchant.

Better means of transportation have increased competition among merchants, so that profits are not so excessive as formerly. Too, the prices paid for cotton are better than formerly. Competition between buyers is so keen that much cotton is handled on a very narrow margin. This has been brought about by better marketing conditions; an increase in number of buyers, especially itinerant representatives of large cotton dealers; and the development of cooperative marketing associations among the growers. Probably the worst defect of the present system is to be found in the lack of careful grading and stapling of cotton in some of the smaller markets. Insufficient premium is paid for the better grades and staple lengths to encourage the growers to produce them. Nor is any difference in character recognized. Practically as much is paid for cotton of an inferior character as for good cotton.

FIG. 110.—Cotton buying on the street. (*Courtesy G. L. Meloy.*)

Trading in the Primary Market.—There are usually one or more cotton gins at a primary cotton market, which is a small town, frequently a county seat. After getting a bale of cotton picked, the farmer may load it into his wagon early in the morning, haul it to town to have it ginned, and sell the bale while there. If there is no debenture against it, the owner may haul it to the local warehouse, have it sampled, get a receipt, and carry the sample around to the different buyers in town for bids. Or if there is no public warehouse in the town, he may haul the bale to the place where he can meet buyers and let them sample and bid (see Fig. 110). These buyers are usually either local men that have business connections with a firm of cotton buyers in a larger market or traveling representatives of a large firm. Both classes of buyers commonly buy on a "limit," a certain price specified by the firm they are buying for and beyond which they cannot go. The firm bases this limit on the current market and notifies the local buyers one or more times during the day.

Relation between Primary Market Prices and Values.—It frequently happens in primary markets that, if competition for cotton is not keen and the grower is not well informed as to the value of his cotton, he will be given a price considerably below its actual worth. Taylor[2] investigated the variations in price paid for Middling cotton in the same market the same day in the different states of the Cotton Belt. It was found that there was a difference of at least $2.50 a bale in every market listed. In one market the difference was $15.60. The average difference was $8.91.

In discussing the results of the investigation, Taylor says:

> When such variations in prices as shown by this investigation, involving two or more bales of Middling grade, exist in any market, it is apparent that gross injustices may occur in many individual cases and in the aggregate assume enormous proportions. These variations show a condition that is unfair to the producer, for it is the farmer, ignorant of the value of his crop and knowing least about marketing his product, who as a rule is called upon to submit to such practices.

Large Interior Markets.—The cotton assembled and bought in the primary markets is mostly passed on to larger markets for classing and grouping bales in even-running, or like, lots. Of course, these markets vary in size. There is no well-defined line separating them from large primary markets, but some common points of difference may be pointed out. The larger market has a larger warehouse and a compress; does a large volume of business, a goodly portion of which is in cotton collected in other markets; and has firms that sell either directly to mills or to consuming and export markets. Some of the firms in such markets may have connection or business relation with foreign firms and do some export business also.

A cotton firm in a large interior market must have considerable financial backing. It maintains an office in which market quotations are received several times during the day and a force of classers that staple and grade the samples of cotton coming in and make up lots of even-running cotton (see Fig. 111). It has branch offices over its tributary territory and business relations with a number of merchants who buy for the firm. The head office, on the basis of market reports, determines the price to be paid, makes up lots of cotton for sale, and makes sales. As a rule, too, the head office, by "hedging," protects the business from loss that might come from a fall in the market. This is a

FIG. 111.—Well-lighted room for classing cotton and the display of samples for the inspection of buyers. (*Courtesy U. S. Department of Agriculture, Bureau Agricultural Economics.*)

sort of price insurance secured by buying or selling cotton on the future exchange market, the details of which will be explained later.

Charlotte, N. C.; Spartanburg, S. C.; Macon, Ga.; Montgomery, Ala.; Greenwood, Miss.; Pine Bluff, Ark.; Shreveport, La.; and Fort Worth, Tex., are illustrative of the type of markets just described. There are several such markets in almost all of the cotton-producing states.

Consuming Markets.—Certain cities in which or near which there are a number of cotton mills are known as "consuming markets," because much cotton is brought there for actual consumption. Boston, New Bedford, Fall River, New York, and Philadelphia are the leading consuming markets in the Northeast; Charlotte, Greenville, Spartanburg, and Gastonia in the Carolinas, and Atlanta, Ga., are important markets in the South. In these markets are a number of spinners' brokers, men who know the particular kind of cotton desired by certain mills and make a specialty of finding and buying for the mills. There are also to be found in such markets selling brokers, who represent large cotton-buying firms in the larger markets of the South.

Export Markets.—The leading export markets of the United States, in order of rank, are Galveston, New Orleans, Savannah, and New York. Much of the cotton handled in these cities is bought and consigned by interior markets but must be routed through export markets.

Foreign Markets.—Liverpool is the leading cotton market of the world. It handles not only a large part of the American cotton that is exported but also cotton from nearly every producing country in the world. Manchester is another English market of importance. Havre is an importing market in France, Antwerp and Ghent in Belgium, Rotterdam in Holland, Bremen in Germany, Barcelona in Spain, and Genoa in Italy. Japan takes a large quantity of American cotton.

Future Markets.—There are future markets at Liverpool, Osaka, Bremen, Havre, New York, New Orleans, and Chicago. Trading in futures on the Chicago market began in 1924. These markets deal in contracts for the delivery of cotton at a specified future date. Most of the sales or purchases are for the purpose of hedging, or for speculation, and only a small percentage of the cotton sold is actually delivered. The figures for the New York Exchange, as given by Hubbard,[1] show that in some years as small a quantity as 85,400 bales is delivered, whereas in other years the quantity ranges as high as 1,005,300 bales.

Value of the Future Markets to the Cotton Trade.—Although the future exchanges are frequently criticized, and a part of the trading done there is purely speculative, they are of great value to the cotton industry. They establish market prices that are largely based on the supply or prospective supply and demand for cotton. By means of the quotations sent out, the market of the whole world is unified. The exchanges serve to regulate prices and afford legitimate business an insurance against loss from market changes. How the protection is secured will now be described.

Suppose that a spinner sells a large bill of cotton goods to be delivered in 6 months. He at once buys of a cotton merchant a sufficient number of bales of cotton to make the goods. The merchant, to protect himself against a change in the price of cotton prior to the time for delivery, buys an equivalent amount of cotton on the future exchange, or hedges, at a price in keeping with the terms of the present sale. If, when the time for delivery arrives, the price of cotton has advanced so that the actual bales he buys in the market to fill the order cost him more than the price stipulated in the sale, there will be a loss; but if he bought "future" cotton, the price of that will have advanced so that there will be a gain on his "futures" sufficient to balance the loss on the other, or approximately balance it. In actual deals, the balance may go one way in one trade and the other in the next. The original sale price was, of course, fixed so that there would be some net profit. If, in the illustration just given, the market had declined, the merchant would have been able to buy cotton in the open market at a lower price than the one stipulated in his sale. This would give him profit, but there would be in that case a loss on the "future" cotton which would about balance the extra gain. As in the case of other forms of insurance, there is considerable cost attached to hedging. Brokerage fees have to be paid both when the futures are bought and when they are sold, and a margin guarantee must be deposited.

Dealers on the future exchanges keep close watch on the consumption of cotton in all parts of the world and the growth conditions in the producing countries. By this means they hope to anticipate the market and buy or sell to their advantage. This study, however, helps keep the market prices based on supply and demand. The quotations of sales as they are made are studied closely by dealers in all parts of the Cotton Belt.

Market Prices.—The basis for price quotations in all markets is the quotation for Middling cotton on the future exchanges. Exchange quotations are for short-staple cotton of a length not below ⅞ inch. The values of grades above and below Middling fluctuate somewhat, relatively, the fluctuation depending on supply or demand for particular grades and staple lengths. The values quoted for the different grades are determined largely by the values given for actual sales made in the designated spot-cotton markets of the country. These markets, which were selected by the Secretary of Agriculture, are Norfolk, Augusta, Savannah, Montgomery, New Orleans, Memphis, Little Rock, Dallas, Houston, and Galveston. In each city there is a committee, approved by the U. S. Department of Agriculture, which makes a daily survey of the market and reports to the Department of Agriculture and the cotton exchanges of the country market values of different grades of cotton in that market on the day specified. Some illustrative figures will be given in a later paragraph.

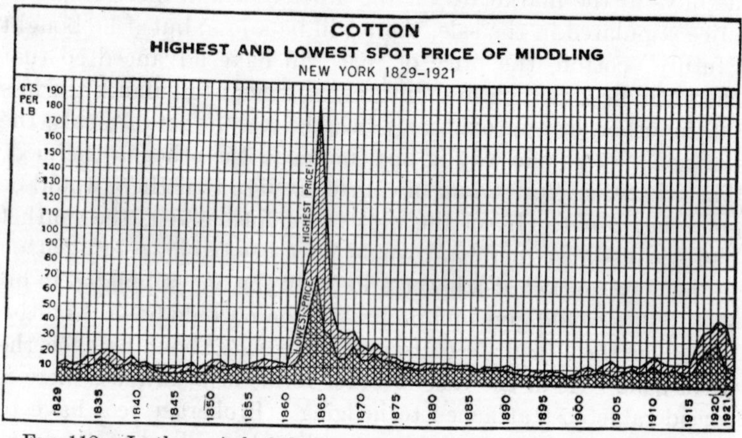

Fig. 112.—In the period of the recent war the price did not rise as high as in the Civil War period, one reason being that production continued and there was always available a good supply, whereas in the earlier period very little was produced and almost no cotton was available. (*Year Book, U. S. Department Agriculture.*)

The buyer in a particular locality, in determining the price he can afford for cotton that he buys, considers first the future market price for cotton, the relation of spots and futures existing at the time, then the relative value of the different grades and staple lengths, the freight charges if the cotton has to be moved at his expense, insurance and storage charges, and some profit in return for his labor and capital used.

The range of prices paid for spot cotton in the New York market for a long series of years (1829–1921) is shown by Fig. 112.

Marketing Period.—A large percentage of the small farmers and tenant farmers who grow cotton have debts to settle and must sell their cotton soon after it is ginned to pay them. The negro tenants, even if they have no debts to settle, almost all want to sell at once so that they may have money to spend. Many of the growers have no satisfactory place to store their cotton after ginning. On account of these conditions and for other reasons, the bulk of the cotton grown passes out of the hands of the growers a few weeks after ginning. Figure 113 shows graphically the percentage of cotton sold each month in the year. It will be noticed that 75 per cent of the cotton is sold by the end of January.

Some of the larger growers, especially the ones who are well conditioned financially, make it a practice to hold their cotton for

COTTON

NUMBER OF BALES MARKETED BY MONTHS AND AVERAGE FARM PRICE FIRST OF EACH MONTH

AVERAGE, 1910-11 TO 1914-15

PRICE	MONTH	PER CENT
11.6	SEPTEMBER	9.14
11.2	OCTOBER	18.28
10.6	NOVEMBER	19.03
10.8	DECEMBER	17.03
10.7	JANUARY	11.52
10.9	FEBRUARY	8.39
11.1	MARCH	5.89
11.2	APRIL	4.11
11.6	MAY	2.42
11.6	JUNE	1.29
11.6	JULY	.98
11.4	AUGUST	1.92

FIG. 113.—The producers sell most of their cotton in the heaviest ginning season, September to January. The monthly average price for the five years, September, 1910, to August, 1914, was lowest in November and highest in May, June, and July. (*After Taylor.*)

a while, or at least a part of it, if the market is not satisfactory at ginning time.

The placing of a large amount of cotton on the market at one time has a tendency, as may be readily seen, to glut the market and to depress the price. Figure 113 shows the relation of farm prices and number of bales sold.

Although most of the cotton soon passes out of the hands of the producers, it is, as a rule, taken by the mills gradually throughout the year. In the meantime, it is stored in warehouses and large public storage places and owned by cotton dealers, traders, etc. (see Fig. 114).

Orderly Marketing.—The present trend of the cotton producers is away from dumping the crop on the market during or soon after the ginning season and toward the goal of orderly marketing, which means the sale of cotton throughout the year as needed by the consumers. This change is being brought about by the diversification in crops grown, which makes the growers less dependent on merchants for supplies, and by the development of cooperative marketing associations among the growers.

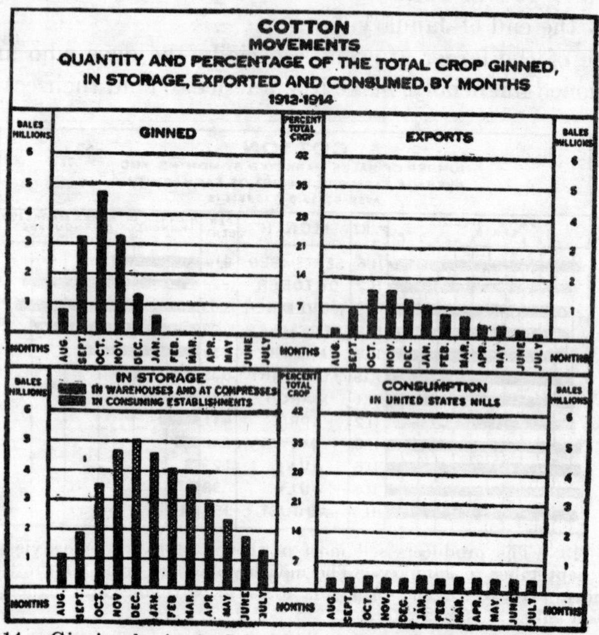

FIG. 114.—Ginning begins in July and ends in February or March. October is the busiest month. Averaging 1913–1914 and 1914–1915, by months, the amount in storage increased from August to December and then gradually declined, not reaching above 36 per cent of the total crop. The heaviest export movements were from September to January, during the heaviest ginning season. Mills consumed regularly, and had on hand at the end of the year only enough cotton to supply them for about two months running. (*After Taylor.*)

Cooperative Marketing Associations.—A good many cooperative marketing associations for the marketing of cotton have been organized in the past and tried, but most of them have not been successful. Organizations formed more recently not only for selling cotton but for selling other commodities cooperatively are meeting with greater success. Probably something has been gained from the former failures, so that pitfalls are being avoided. For a cooperative association to be a success, the individual must make some sacrifice at times for the benefit of the whole. To do this requires a high standard of ethics. Americans are so used to personal independence and to doing just as they please that they are naturally not well fitted for cooperative movements.

One of the most successful of the cotton cooperative marketing associations that have been formed is the Staple Cotton Cooperative Association of Mississippi. This organization has as members about 2,000 growers of upland staple cotton. The organization was formed by growers of staple cotton in the Yazoo-Mississippi Delta of the state of Mississippi, in 1921 but membership is open to growers of staple cotton in other states and at present includes those in some of the neighboring states. It is a nonstock, nonprofit organization.

It was the purpose of the founders of the association to secure a more efficient and satisfactory system for marketing their cotton. Speculative features were not contemplated, and there was no intention of trying to "bull the market" or raise the price of staple cotton by holding cotton. It was the intent simply to supply the consumer with well-classed cotton at a fair market price.

The Staple Cotton Cooperative Association has a board of directors consisting of 21 members chosen from among the leading growers of staple cotton in the district, a general manager, an assistant general manager, a secretary-treasurer, a force of expert classers, and managers of branch offices at principal concentration points in the Delta. The head business office is situated at Greenwood, Miss.

The general manager spends much of his time studying the cotton market and visiting the leading consuming markets of the world, where he secures business agreements and makes sales.

In the usual workings of the organization, the member delivers his cotton at the nearest concentration point where there is a good warehouse. The bales are sampled, and samples forwarded to the nearest branch office. The branch office issues a receipt for the tentative classing on each bale. The grower can borrow through the association, on the basis of this tentative classing sheet, money on his cotton to the extent of about 80 per cent of its value. A comparatively low rate of interest is paid for this money during the time it is charged against the borrower. The samples are then sent to the main office, where they are classed carefully by the expert classers there, and a classing sheet is rendered the grower. This sheet shows the grade, staple length, and character of each bale received. The bales are grouped in even-running lots in respect to staple length, grade, character, etc., and pools are made up for sale.

Different pools are made to meet market demands and to serve the needs and wishes of the members in regard to the sale of the cotton. The member may choose the option of the seasonal pool, in which case he turns the cotton over to the association without pricing and leaves it to the discretion of the officials of the association to sell at any time during the year. The member may choose the optional pool, in which he will have the right to say when the cotton is to be sold. Under this option, he can order the cotton sold at a time that he feels is most advantageous to him. During the years 1935 and 1936, cotton placed in the seasonal pool brought, on the average, $2 more per bale than that in the seller's option pool. Additional selling options are available at present, and new ones are devised from time to time to meet the needs of the members of the association.

Cotton futures may be bought or sold for the purpose of hedging cotton prices or closing out remnant lots of cotton at the end of the season. Futures are not bought and sold by the association for speculative purposes. Statistical reports, showing sales made, quantity of cotton on hand, etc., are made to the members each month. Checks are also sent to them each month covering their pro-rata shares of the sales made during the month.

It appears that the Staple Cotton Cooperative Association has been able to render a definite service to its members. Their cotton is stored in a good warehouse at a reasonable cost and is insured at a rate lower than the individual can secure. It is carefully stapled and graded, and a classing sheet is rendered, so that the owner may know within narrow limits just what the cotton is worth. It is pooled in large even-running lots, so that the best possible market price can be obtained. The individual with a small lot cannot command so high a price. The average price received is higher than that which could be obtained by individuals. Only the actual cost of handling and selling is charged against the cotton. The middleman's profit is eliminated.

It is claimed that the association has a tendency to stabilize staple cotton prices and make better prices available both for members of the association and for growers without. During the season of 1936–1937, the association handled 248,317 bales of cotton, and the 10-year average 1928–1937 was 207,865 bales.

The Staple Cotton Cooperative Association sells cotton chiefly to domestic mills, partly direct and partly through brokers at Charlotte, Greenville, Boston, New Bedford, Fall River, Akron, and New York. It also has foreign selling agents in Montreal, Canada; Liverpool, England; Enschede, Holland; and Milan, Italy.

Other Cooperative Cotton Associations.—Since 1920, cooperative cotton associations have been formed in nearly all the cotton-growing states. At present, there are 14. These state or regional associations operate in somewhat different ways, but the general principles are much the same, and the operations are similar to those of the Staple Cotton Cooperative Association described above. The chief difference, perhaps, lies in the fact that the state or regional associations all belong to a central organization known as the "American Cotton Cooperative Association." This association is chartered as a stock corporation for the purpose of performing the services of classing, storing, financing, and selling for the regional associations, who are the stockholders. The main office of the American Association is in New Orleans. It has an office force of about 300 men and women, who occupy six floors in a large building. More than a million bales of cotton are sold annually.

In a conference of the officials of the cotton cooperatives at Memphis, Tenn., in 1928, it was agreed that cooperatives should attempt to render the following economic services to growers:[13]

Grade and staple cotton accurately through classers licensed and supervised by Government.

Make returns to growers on basis of grade and staple.

Sell direct to mills.

Provide an efficient selling agency for members using short-time pools.

Obtain highest possible average seasonal pool prices through use of trained sales force and sales based on scientific analysis of market conditions.

Reduce market risks of individual growers by pooling.

Store and insure at minimum rates.

Obtain funds for commodity financing at low rates of interest.
Stimulate interest in better ginning.
Encourage the production of better staple.

It is no doubt very evident to all persons interested in cotton production that the aims mentioned were laudable. Fortunately, most of them have been realized.

Warehousing Cotton.—The U. S. Department of Agriculture has estimated that the annual loss due to exposure of cotton bales to the weather is more than $30,000,000. The storage of cotton, immediately after ginning, in a good warehouse prevents weather damage practically altogether. There are also other advantages to be gained from warehouse storage. If the warehouse is a first-class one, insurance rates will be moderate, and the warehouse receipt may be used as security in negotiating loans. Receipts from warehouses licensed in accordance with the Federal Warehouse Act of 1916 are accepted by banks everywhere as negotiable paper.

Selling Cotton in the Seed.—In the extreme northern part of the Cotton Belt and in other parts where gins are not numerous, a considerable quantity of cotton is sold without ginning, or "sold in the seed," as it is termed. A higher percentage is marketed in this way in Virginia, Missouri, Oklahoma, Florida, and Tennessee than in other states. In 1915, 8.5 per cent of the crop of the country was sold in the seed, the total amounting to 937,000 bales.

Selling cotton in the seed usually works to the disadvantage of the producer. Before the cotton is ginned, no one can grade it accurately, and consequently the ginner in his estimate of the grade is careful to give himself some margin. Frequently, several lots of cotton are put together in one bin, with the consequent mixture of different grades and staple lengths. The grower of cotton with good staple length, or of high grade, is apt to get little or no premium over that obtained by the grower of the most inferior cotton. The net result is that the growers are encouraged to grow inferior varieties that will yield plenty of seed cotton.

In communities where most of the cotton is sold in the seed, the grower who tries to grow a good variety and market lint and seed separately is placed at a disadvantage, because, since there is not much cotton sold in the bale in that market, regular buyers do not visit the market. The buyers of the gin-run cotton may not grade the cotton at all before making the purchase, and, of course, the price is low.

Damaged and Inferior Cotton.—In the course of marketing 10,000,000 to 15,000,000 bales of cotton, a considerable quantity of damaged or inferior cotton is found. This is the result of leaving the cotton in the field too long before picking, faulty ginning, improper care of bales after ginning, carelessness in the transportation of bales, and several other factors.

Country Damage.—A bale of cotton is considered a non-perishable product and, if stored properly, will last for years with but little deterioration. Bales may be left out in the rain or exposed to the weather without protection for a longer period than most agricultural products without serious damage. But even cotton bales may be injured by the weather, and the total loss to growers in consequence amounts to millions annually.

It has been, and still is to some extent, the practice of some growers to haul bales of cotton home from the gin and throw them on the ground or perhaps put them on poles in some convenient place. Often a considerable quantity is stored in a similar way around gins or about concentration points in towns during the rush of the ginning season. If cotton is left exposed in this way for a few weeks, the bagging becomes bleached and the exposed cotton discolored, the bale appearing to be damaged even if it is not. If the bale rests on the ground, the lint coming in contact with the soil will be discolored; and if there are rains moisture from the ground is apt to penetrate the bale for a considerable distance, causing a weakening, discoloration, or even rotting of the fibers. Half of the bale may be ruined in this way. When the bales are put on poles, water is apt to run under them and wet them where they come in contact with the poles. The cotton may mildew in these places if the bales are not turned after rains so that they will have a chance to dry out.

Transportation Damage.—Cotton is sometimes damaged to some extent through getting wet while stored on railway platforms and in leaky cars and by the water from rivers when the bales are piled on boats near the water line. A goodly portion of the apparent transportation damage is really not the result of injury during transit but is due to wet cotton being accepted for carriage with the resultant damage appearing during transit or at the end of the journey.

Pickings.—A bale of cotton that has become damaged is not in a marketable condition. Before it can be marketed satisfactorily, it must be worked over, or "conditioned," as it is termed. The bale is opened and all damaged cotton is carefully separated from the rest by hand. The lint of the bale may be sorted out into two or three lots, the number depending on quality, and later rebaled with cotton from other similar bales. Cotton from "pickeries" is usually irregular and of low grade. Its value is not quoted in the market.

Repacked Cotton.—Repacked cotton consists of cotton from parts of two or more bales that has been worked over and rebaled. Much of the repacked cotton is samples from cotton-classing rooms and brokers' offices which have been collected and rebaled. Bales from such samples are sometimes called "loose cotton," or "the city crop."

Since every bale of cotton is sampled one or more times, frequently several times, the aggregate of lint cotton in these samples is considerable. Allowing 1 pound to the bale, which appears to be a reasonable estimate, the whole amount for a 14,000,000-bale crop is equivalent to 28,000 bales.

*Cotton of Perished Staple.**—Cotton of perished staple is defined by the U. S. Department of Agriculture[3] as cotton that has had the strength of fiber, as ordinarily found in cotton, destroyed or unduly reduced through exposure to the weather either before picking or after baling, or through heating by fire, or on account of water packing, or by other causes.

Cotton of Immature Staple.—Cotton of immature staple is cotton that has been picked and baled before the fiber has reached a normal state of maturity. This results in a weakened staple of inferior value.

Gin-cut Cotton.—Gin-cut cotton is defined as cotton that shows damage in ginning, through cutting by the saws, to an extent that reduces its value more than two grades.

Reginned Cotton.—Reginned cotton is cotton that has passed through the ginning process more than once and, after having been ginned, has been subjected to a cleaning process and then baled.

* This type of cotton and the following kinds of irregular cotton are defined by the Secretary of Agriculture in regulations under the United States Cotton Futures Act.

False-packed Cotton.—False-packed cotton is defined as cotton in a bale (1) containing substances entirely foreign to cotton; (2) containing damaged cotton in the interior, with or without any indication of such damage upon the exterior; (3) composed of good cotton upon the exterior and decidedly inferior cotton in the interior, arranged in such manner as not to be detected by customary examination, called a "plated bale"; or (4) containing pickings or linters worked into the bale.

Mixed-packed Cotton.—Under the regulations of the Secretary of Agriculture, cotton is considered to be mixed-packed if in the samples drawn from a given bale (1) there is a difference of more than two grades if of the same color; (2) they show a difference of more than two color gradations if of the same grade but of different color; (3) they show a difference of two or more grades and two or more color gradations; or (4) they show a difference in length of staple exceeding $\frac{3}{32}$ inch.

Water-packed Cotton.—Cotton is considered to be water-packed if the cotton in the bale has been penetrated by water during the baling process, causing damage to the fiber or if, through exposure to the weather or by other means, while apparently dry on the exterior, the cotton on the interior of the bale has been damaged by water.

Cotton Not Tenderable on a Future Contract.—The different types of inferior cotton mentioned in the preceding paragraphs, cotton with a staple length less than $\frac{7}{8}$ inch, and the lower grades of cotton cannot be offered in settlement of a future contract. The following paragraph from the Cotton Futures Act, as amended Mar. 4, 1919, specifies the types of cotton that may not be delivered.

Fifth. Provided that cotton that, because of the presence of extraneous matter of any character, or irregularities or defects, is reduced in value below that of Low Middling, or cotton that is below the grade of Low Middling, or, if tinged, cotton that is below the grade of Strict Middling, or, if yellow-stained, cotton that is below the grade of Good Middling, the grades mentioned being of the Official Cotton Standards of the United States, or cotton that is less than $\frac{7}{8}$ inch in length of staple, or cotton of perished staple, or of immature staple, or cotton that is "gin cut" or "reginned," or cotton that is "repacked" or "false-packed" or "mixed-packed," or "water-packed," shall not be delivered on, under, or in settlement of such contract.

TABLE XXXVII.—GRADE DIFFERENCES REPORTED BY 10 IMPORTANT SPOT MARKETS[a] ON JULY 31, 1922
(After Palmer)

Market	White standards									Yellow-tinged						Yellow-stained			Blue-stained		
	Middling fair, on	Strict good middling, on	Good middling, on	Strict middling, on	Middling, cents	Strict low middling, off	Low middling, off	Strict good ordinary, off	Good ordinary, off	Strict good middling on	Good middling	Strict middling, off	Middling, off	Strict low middling, off	Low middling, off	Good middling, off	Strict middling, off	Middling, off	Good middling, off	Strict middling, off	Middling, off
Norfolk	200	150	100	50	21.38	50	100	175	250	50	Even	50	100	175	275	100	200	275	150	225	300
Augusta	125	100	75	38	21.38	37	100	175	275	50	Even	37	100	175	275	100	175	300	100	200	300
Savannah	125	100	75	50	21.22	50	100	150	200	50	Even	50	150	225	300	100	200	300	150	225	300
Montgomery	163	125	88	50	21.19	50	125	200	275	50	Even	75	175	250	325	125	200	275	150	225	300
New Orleans	175	150	100	50	21.25	50	125	225	325	50	Even	50	200	250	325	125	250	300	125	175	300
Memphis	225	150	100	50	22.50	50	125	225	325	50	Even	50	150	225	325	125	225	275	100	150	200
Little Rock	200	150	100	75	21.50	50	125	225	325	75	On 25	50	150	225	300	125	200	275	125	225	325
Dallas	200	150	100	75	21.20	75	150	250	350	50	Even	75	175	250	325	150	250	350	175	250	325
Houston	150	125	100	50	21.70	75	150	250	320	50	Even	50	175	250	325	150	250	350	150	225	300
Galveston	250	175	100	50	21.75	75	150	250	320	50	Even	50	175	250	325	150	225	325	150	225	300
Average	181	138	94	54	21.51	56	125	213	303	53	On 3	54	155	228	310	125	218	303	138	213	295

[a] The average differences of these 10 markets are used in settlements for cotton delivered on future contracts of the New York Cotton Exchange, the markets being designated for the purpose by the Secretary of Agriculture under authority of the United States Cotton Futures Act.

Relative Value of Different Grades of Cotton.—As was mentioned previously, the price of Middling cotton is the basic price. The values of other grades are usually expressed as so many points on or off Middling price. A point is 0.01 cent. "On" means above the price of Middling, whereas "off" means below. The 10 designated spot markets quote the value of the different grades, their quotations being based on values in actual sales. Averages made from their quotations are used for the values of the different grades in future contract settlements when deliveries are made.

Table XXXVII shows the values of all of the different grades of both White and discolored cotton in the 10 designated spot markets on a certain day.

Value of Various Staple Lengths of Upland Cotton.—Cotton shorter in staple length than 7/8 inch is not acceptable in the settlement of a future contract and is not quoted on the market regularly. Cotton 7/8 to inch in length is considered short cotton, and very little if any difference in price is made for different lengths between these limits. Above 1 inch, there is an extra premium paid for each 1/16-inch increase in length.

TABLE XXXVIII.—A COMPARISON OF STAPLE PREMIUMS QUOTED IN TWO LARGE MARKETS FOR STAPLE COTTON

	Apr. 30, 1920		Apr. 30, 1921		Apr. 29, 1922		May 5, 1923		Apr. 30, 1937	
	New Orleans	Memphis	New Orleans	Memphis	New Orleans	Memphis	New Orleans	Memphis	New Orleans	Memphis
Middling upland short cotton, cents[a]	41.00	42.00	11.25	11.00	17.00	17.25	26.50	27.50	13.31	13.01
	Points on	Points on	Points on	Points on	Points on	Points on	Points on	Points on	Points on	Points on
1 1/16 inches....	625	600	100	150	150	175	50	Even	240	240
1 1/8 inches....	2,425	2,600	200	400	375	375	100	120	325	400
1 3/16 inches....	3,550	4,000	525	700	550	575	150	150	...	500
1 1/4 inches....	4,200	4,600	800	[b]	800	875	225	250	...	600
1 5/16 inches....	[b]	5,000	[b]	[b]	[b]	[b]	300	[b]		
1 3/8 inches....	[b]	5,850	[b]	[b]	[b]	[b]	400	[b]		

[a] Cotton Exchange spot quotation.
[b] Nominal.

Table XXXVIII gives premium paid over short cotton for different staple lengths in the leading staple-cotton markets, New Orleans and Memphis, on a certain date each year for 5 years. The premiums for 1920 were abnormally high, whereas those of 1923 were unusually low.

Regional Names and Values.—The varieties grown and the soil and climatic conditions affect the quality of cotton produced in different regions. This has led buyers and merchants to use certain trade names for cotton from particular regions, as, for instance, Mississippi Deltas, Texas-Oklahoma cotton, Red River staples, Carolina staples, Mississippi Hill cotton, or North Georgias, etc. In each of these regions more than one variety is grown, and the cotton is not altogether uniform, but there is enough similarity to give it some uniformity of character, sufficient to cause the trade to take notice of it. A premium of a cent or more a pound may be paid for cotton from a certain region over that from another region that has cotton of the same grade and staple length. In the regular sales, cotton is, of course, sold on the basis of the grade and staple length. The regional name is used in addition to give some notion of the character of the cotton.

The following paragraph from Palmer[4] gives something of the history and meaning of special names that have been used by the cotton trade, some of which are still in use:

> In early days in the New Orleans and Memphis markets, cotton was known as "Staples" or "Peelers," "Benders," "Rivers," and "Creeks." "Staples," or "Peelers," were the long-staple varieties from the Yazoo and Mississippi Delta plantations. "Benders" referred to cotton of extra length and character which the river boats took up from the plantations in the bends of the river where the soil was extremely rich. "Rivers" were the hard-bodied cottons from the plantations along the Mississippi and its tributaries, and the "Creeks" the slightly softer cottons from along the smaller streams where there was less bottom land. "Uplands" was originally the term used to describe short cotton from the hill section, but gradually it has come to mean in this country all of the crop of the United States except Sea Island and American-Egyptian varieties.

Relation of the Percentage of Spinnable Cotton in the Different Grades and Their Market Value.—The question frequently arises as to whether or not premiums paid for cotton better than

Middling or discounts made for cotton of lower grade than Middling are just or are in proportion to their spinning values. Table XXXIX, from Palmer,[4] shows relation of market values of differ-

TABLE XXXIX.—COMPARISON OF THE RELATIVE PERCENTAGE OF SPINNABLE COTTON REMAINING AFTER ELIMINATING THE VISIBLE WASTE CONTENT IN CERTAIN STANDARDIZED GRADES WITH THEIR RELATIVE VALUE ACCORDING TO COMMERCIAL DIFFERENCES

(After Palmer)

(Average of 10 Important Spot Markets)

Grade	Relative percentage of spinnable cotton[a]	Relative value according to commercial differences, per cent			
		May 15, 1916	July 17, 1920	July 31, 1922	July 31, 1923
Middling Fair	102.4	107.6	101.0	108.4	104.2
Good Middling	101.6	103.8	106.3	104.4	103.1
Middling	100.0	100.0	100.0	100.0	100.0
Low Middling	98.6	94.1	78.8	94.2	96.9
Good Ordinary	95.9	86.2	61.6	85.9	92.0
Good Middling yellow-tinged	99.8	99.8	94.8	100.1	98.3
Middling yellow-tinged	98.5	96.3	86.9	92.8	95.8
Low Middling yellow-tinged	94.5	90.2	72.2	85.6	91.3
Good Middling yellow-stained	100.0	96.0	88.1	94.2	97.0
Middling yellow-stained	95.8	92.2	79.1	85.9	92.2
Good Middling blue-stained	99.6	95.5	85.1	93.6	95.7
Middling blue-stained	96.4	90.5	77.0	86.3	92.1

[a] Computed from percentages of waste as determined in spinning tests conducted by the department. See Dean, William S. and Taylor, Fred, "Manufacturing Tests of the Official Cotton Standards for Grade," *Bulletin* 591; and Meadows, W. R. and Blair, W. G., "Preliminary Manufacturing Tests of the Official Cotton Standards of the United States for Color for Upland Tinged and Stained Cotton," *Bulletin* 990.

ent grades of cotton in 10 important spot markets and their spinning values as shown by results from spinning tests conducted by officials of the U. S. Department of Agriculture. Both premiums and discounts are apparently somewhat in excess. However, spinning values are determined in part by qualities other than grade of cotton (see Chap. VII for further discussion of spinning qualities).

Relation of the Market Price of a Pound of Raw Cotton and the Retail Price of the Goods Made from It.—Several yards of cotton goods may be made from 1 pound of raw cotton. Under the prevailing high retail prices of cotton goods, it may appear at

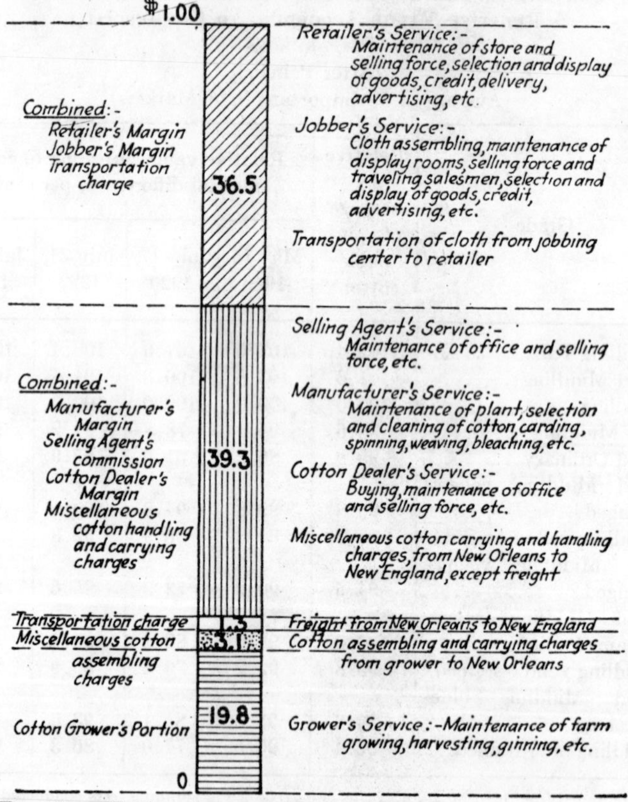

FIG. 115.—Division of consumer's dollar spent for sheeting in 1922.

first thought that the grower is not getting his share of the money that the consumer pays for goods, but a study of the many items of expense between the bale of cotton and sale of a dress pattern to the consumer convinces one that the grower is not so badly mistreated as it at first appeared. The consumer does pay several times as much for the pound of cotton as the grower receives, but a good deal of the extra price is for fancy showrooms, expenses of traveling men, losses due to changes in fashions, advertising, etc. Items of expense must be added to the retail price of the goods and paid by the consumer; yet they add very little or nothing to the real value of the goods.

L. A. Adams, marketing specialist of the U. S. Department of Agriculture, has made an analysis of the difference between the retail price of cotton cloth and the price of raw cotton. Figure 115, taken from his report, shows the distribution of the dollar that the consumer spends for sheeting.

Recently, Whitaker and Monachino, of the Bureau of Agricultural Economics, U. S. Department of Agriculture, and others of the bureau have made a study of the relation between the price that the producer receives for a pound of his raw cotton and the value that cotton mills receive for cloth. The period from 1925–1926 to 1936–1937 was used in the study, and 17 constructions of standard unfinished cloths including print cloth, sheeting, three-leaf twill, drill, sateen, and duck were considered. Allowance was made for waste, waste sold, and the noncotton content of the goods. It was found that, on the average, raw-cotton costs constitute a little more and mill margins a little less than one-half of the wholesale price of unfinished cotton cloth of the types considered. The average price of raw cotton during the period used was 14.57 cents per pound; the average mill margin or charge for manufacturing the goods was 13.26 cents per pound; and the wholesale value of the goods 27.83 cents per pound.

Mill margins on fine cloths are considerably greater than those mentioned above, and the raw-cotton values constitute a much lower percentage of the retail price of finished cotton goods.

References

1. Hubbard, W. Hustace: "Cotton and the Cotton Market," D. Appleton-Century Company, Inc., New York, 1923.
2. Taylor, Fred: "Relation between Primary Market Prices and Qualities of Cotton," *U. S. Dept. Agr. Bull.* 457, 1916.
3. Wallace, Henry C.: "Regulations of the Secretary of Agriculture under the United States Cotton Futures Act," *U. S. Dept. Agr. Cir.* 159, 1922.
4. Palmer, Arthur W.: "The Commercial Classification of American Cotton," *U. S. Dept. Agr. Cir.* 278, 1924.
5. Miller, T. S.: "Cotton Trade Guide and Student's Manual," Steck Austin, Tex., 1915.

6. TODD, JOHN A.: "World's Cotton Crops," A. & C. Black, Ltd., London, 1923.
7. WORKMAN, J. M. and NEWTON, R. V.: "Benefits of an Adequate Warehouse System," *Year Book, U. S. Dept. Agr.*, 1918.
8. MCCONNELL, O. J.: "Suggested Improvements in Methods of Selling Cotton by Farmers," *U. S. Dept. Agr. Cir.* 56, 1919.
9. COOK, O. F.: "The Relation of Cotton Buying to Cotton Growing," *U. S. Dept. Agr. Bull.* 60, 1914.
10. CRESWELL, CHARLES F.: "Disadvantages of Selling Cotton in the Seed," *U. S. Dept. Agr. Bull.* 375, 1916.
11. MCCONNELL, O. J. and CAMP, W. R.: "A Study of Cotton Market Conditions in North Carolina with a View to Their Improvement," *U. S. Dept. Agr. Bull.* 476, 1917.
12. MARTIN, J. G.: "The Handling and Marketing of the Arizona-Egyptian Cotton of the Salt River Valley," *U. S. Dept. Agr. Bull.* 311, 1915.
13. CHRISTENSEN, CHRIS L.: "Farmers' Cooperative Associations in the United States, 1929," *U. S. Dept. Agr. Cir.* 94, 1929.

CHAPTER XX

COTTON EXCHANGES

Exchange means to give or trade one value or commodity for another. The word "exchange" is also used as a name for the place where exchanging or trading is taking place. Nearly every town or city has a "cotton exchange," or a place where men meet to buy and sell cotton. Many of these trading places are not called exchanges but are simply a street or part of a street where traders, through habit or for some other reason, have become accustomed to meet. In some instances, the trading place or exchange is located by a municipal ordinance. Trading is open to everyone in these simple markets. There are no regular members and no rules except the rules of custom. As the business increases, it becomes necessary to have certain definite rules to protect the rights of individuals and help make the business run smoothly. For the convenience of the traders it is also helpful to have a limited membership—have as traders only men who have established a reputation of moral integrity and are responsible financially.

Apparently it is natural for most men to like to trade. The small boy trades tops and marbles largely for the pleasure of the exchange. David Harum traded horses for the fun of the game. There is a fascination about buying and selling cotton which draws many to it. Many fortunes have been made in the cotton trade, but, unfortunately, a good many have been lost. For a cotton trader to succeed he must not only "know" cotton but must be a good businessman.

The Evolution of Cotton Exchanges Where Future Contracts Are Bought and Sold.—Prior to 1818, there were no telegraph or cable lines and no railroads, and no steamboat had crossed the Atlantic. Means of communication, measured by modern standards, were very poor. Under such conditions, trading in cotton as practiced today was not possible. In 1819, the first steamboat crossed the Atlantic Ocean. From this date on during the next half century, means of communication were gradually improved. A steam-railway line was put in operation in 1829, the telegraph in 1844, and the Atlantic cable in 1866.

With the improvement of means of communication came the development of trading in future contracts in the cotton exchanges. It is always the desire of cotton merchants, farmers, and millmen to play safe, or to insure their investments when they can conveniently. When the opportunity to do so by means of futures arose, they made use of it. According to Hubbard,[1] the first step toward our present system of futures was the selling of "cotton afloat," which was practiced a decade or two prior to 1866. The cotton bales were sampled before being stored on board a ship in the American port. The samples were then placed on fast-sailing ships, which reached Liverpool considerably sooner than freighters bearing the cotton bales. Brokers in Liverpool examined these samples when they arrived and bought the cotton afloat, or "to arrive," on the basis of the samples. With the completion of the Atlantic cable in 1866, market reports from America were received in Liverpool daily or more frequently. This made the old plan of selling by sample cotton to arrive unsatisfactory. When market reports are received often, so that buyers can keep posted on market prices at all times, immediate sales contracts are desired. A new scheme was devised to take the place of the old cotton-afloat sales. According to the new plan, cotton from the United States was sold on Middling basis, no samples being used, no grade specified except that the cotton was not to be below Middling, and no particular ship on which the cotton was to arrive being mentioned. It was to be delivered within a certain 2-month period. This form of sale was, as may be seen, of a more general nature than the former and was better liked by both merchant and spinner, since it gave greater latitude. The plan resembles rather closely our modern system of futures.

During the Civil War in America, "cotton to arrive" was bought and sold in New York. This was probably the first step toward the development of future exchanges in America.

The origin of future exchanges was discussed by Arthur R. Marsh, a former vice-president of the New York Exchange, in his address before the National Association of Cotton Manufacturers at Washington, D. C., in 1907. Portions of this address, as quoted by Miller,[2] are as follows:

Before the Civil War the cotton business in New York was simply one form of the old-fashioned commission business, exactly like the business of handling molasses, sugar, hides, wool, country produce, and many other similar agricultural commodities.

The Civil War completely upset the conduct of the cotton business in New York, as just described. While it lasted, there were, of course, no regular shipments of cotton to New York from the South, and the only source of supply consisted of lots of cotton which the government from time to time got hold of through the capture of blockade runners or through confiscation in the South. Naturally, such lots of cotton could not be handled on a commission basis, but had to be bought outright as a speculation. The huge profit made by some of those who bought this Government cotton was the real beginning of general speculation in cotton in this country. And the same thing happened across the water in Liverpool. The fierce demand and the uncertain and inadequate supply gave opportunity for vast and sudden profits, such as have never been seen before or since in connection with any commodity. And, curiously enough, it was out of this wild speculation of the time of the Civil War that the entire modern method of handling the cotton business was evolved; for, in their eagerness to get hold of cotton, speculators began to buy not only actual cotton on the spot in New York or Liverpool, but "cotton to arrive," when they got wind of a lot of cotton on some ship destined for one or the other of those ports. Here was the beginning of the system of trading in cotton futures, which has gradually revolutionized the whole cotton business in every root and branch, for very clever men, whose business was that of cotton merchants and not speculators, saw a way to make use of the extensive trading in contracts for "cotton to arrive" as a protection to themselves in their legitimate buying and selling actual cotton.

It was two or three years after the Civil War that this new conception of the cotton business took shape in the mind of one of the most brilliant cotton merchants the world has ever known, the late Mr. Rew, of Liverpool, whose firm is still in existence. In 1868 or 1869, Mr. Rew saw that the newly laid Atlantic cable made it possible for a cotton merchant in Liverpool to ascertain with unheard-of quickness the price at which actual cotton could be bought in the Southern States, and the approximate date at which it could be shipped to England. He saw also that, if the price that was being bid in Liverpool for "cotton to arrive" was high enough to enable him to buy the cotton in the South and sell contracts for this same "cotton to arrive" in Liverpool, two or three months later, he could enter into the transaction with entire safety, as when his cotton reached Liverpool he could either deliver it to the parties to whom he had sold the contract, or if some spinner was willing to pay a higher relative price than the holder of the contracts had agreed to pay, he could back his contracts and sell the cotton to the spinner with the larger profit to himself.

Following the successful venture of Mr. Rew, other buyers in Liverpool and elsewhere began to deal in contracts for future delivery of cotton. The New York Cotton Exchange was organized and began trading in futures in 1870. The New Orleans Cotton Exchange was organized in 1871, and the Liverpool Cotton Association adopted rules governing future trading the same year.

Spot Cotton and Future Contracts.—Before proceeding further in the discussion of cotton exchanges, it may be well to explain more fully the distinction between "spot cotton," or "spots," and "future contracts," or "futures."

Spot cotton, or spots, which is simply a contraction of the term, includes all cotton that is bought or sold except that sold on future contracts in cotton exchanges. The latter is futures, or future-contract cotton.

The term "spot cotton" was at first applied to cotton actually in a warehouse, piled up at a port, or on the spot in the market. The term is also used for cotton being shipped, or en route, to the market or to cotton that will be shipped at some future date. As brought out by Hubbard,[1] it is not altogether correct to designate spot cotton as simply actual cotton and futures as fictitious cotton. A dealer may sell spot cotton for future delivery—cotton that he does not have but expects to purchase and deliver. On the other hand, a trader who sells futures may deliver the actual bales of cotton at the expiration of the contract period. This may be somewhat confusing to a person beginning the study of the subject, but the matter may be clarified if it is remembered that all cotton sold is spot cotton except that sold on future contracts in exchanges.

New York Cotton Exchange.—At present, there are seven exchanges that deal in future contracts for American cotton. They are situated in New York, New Orleans, Chicago, Liverpool, Bremen, Havre, and Osaka. Other future markets for foreign cotton are in Liverpool, England; Bombay, India; Alexandria, Egypt; São Paulo, Brazil; and Shanghai, China. The operation of all the futures exchanges dealing in American cotton is similar. A discussion of the New York Exchange will suffice, in a general way, for all.

The New York Cotton Exchange was organized at first simply as a means of convenience for the people in New York who were engaged in buying and selling cotton. There were many cotton traders scattered about over the city who dealt in both spot cotton and futures. To reach these traders, or for them to reach each other, it was necessary to do much traveling about over the city and much waiting for men who happened to be out of their offices. As the trading business increased, the trouble of looking up men wanted became so burdensome that a number of the traders established the rule of meeting at a certain office at regular periods. It was soon realized that rules were needed by means of which the trading might be regulated. To make and enforce rules required an organization. This led to the organization of the New York Cotton Exchange in 1870. It was incorporated the following year.

According to its charter:

> The purposes of said corporation shall be to provide, regulate, and maintain a suitable building, room or rooms, for a Cotton Exchange, in the City of New York, to adjust controversies between its members, to establish just and equitable principles in the trade, to maintain uniformity in its rules, regulations, and usages, to adopt standards of classification, to acquire, preserve, and disseminate useful information connected with the cotton interests throughout all markets, to decrease the local risks attendant upon the business, and generally to promote the cotton trade of the city of New York, etc.

The New York Cotton Exchange is limited to 450 members. Applicants for membership must be over twenty-one years of age and of good character and financial standing. The initiation fee is $1,000 and there are certain annual dues. Membership rights may be bought, either from a member or at auction in the exchange room. In 1934, the value of a membership was about $18,000. The exchange includes many prominent cotton men, some living in parts of the country distant from New York City, cotton dealers, exporters, manufacturers, etc., who buy and sell both spot and future cotton. About 150 members, or their representatives, are on the floor of the exchange daily transacting business.

The New York Cotton Exchange has as officers a president, a vice-president, a treasurer, and a board of 15 directors. All the officers are elected annually. The chief officer in active management is the secretary, who is paid a salary. He is the superintendent of the exchange, collects and publishes statistics, manages the office, and acts as chairman when calls are made on the floor.

Fig. 116.—New Orleans Cotton Exchange Building. (*Courtesy H. G. Hester.*)

The New York Cotton Exchange is a corporation acting under a charter given by the state of New York, but in the trading in cotton the members act individually.

Daily Routine on the Floor of the New York Exchange.—At ten o'clock the superintendent of the exchange appears on the rostrum at one side of the trading ring, rings a gong, and announces trading open for the current month. Then all traders wishing to buy or sell for that month make their offers. The superintendent, acting as chairman, repeats the offers and announces sales, which are recorded on a blackboard at the opposite side of the ring (see Fig. 117). After the trading for the current

Fig. 117.—Trading ring, New Orleans Cotton Exchange. (*Courtesy H. G. Hester.*)

month has stopped, the chairman announces the next month. Most of the trading is for March, May, July, October, December, and January. These are known as "active" months. After the trading for the second month has closed, another is announced and so on around the year. After the round has been made, the chairman announces the "call adjourned, market open." Trading is then carried on promiscuously around the ring by members, any month being dealt in. Offers and bids must be announced aloud, and sales reported.

The quotations on sales are recorded on the blackboard. The pit, or spot where the traders stand, is surrounded by a ring of steps, three up and three down (see Fig. 117). Another call of months is made by the chairman at 11:45 A.M. and another at 2:45 P.M. On Saturday there is but one call. After the close of the exchange at 3 P.M., no further trades are allowed.

The traders carry purchase and sale slips, on which they keep a temporary record of their sales and purchases. These are later recorded in their offices. Final settlement is made in the clearinghouse.

The New York Futures Contract.—The New York futures contract is as follows:

New York,................192....

....................of the County and State of New York, have this day {sold to / or bought from} ...., and agree to {deliver to / or receive from} ....of the same place, 50,000 pounds in about 100 square bales of cotton, growth of the United States, at the price of.....cents per pound for Middling, with additions or deductions for other grades in accordance with the provisions of the United States Cotton Futures Act, Sec. 6, and allowance for staple premium, deliverable from licensed warehouse in (....................) between the first and last delivery days of ....
Point of Delivery
inclusive, the delivery within such time to be at seller's option, in one warehouse, upon notice to the buyer, as provided by the By-laws and Rules of the New York Cotton Exchange. The cotton dealt with herein or delivered hereunder shall be of, or within, the grades for which standards are established by the Secretary of Agriculture, except cotton prohibited being delivered on a contract by the fifth subdivision of Sec. 5 of the United States Cotton Futures Act, and shall be of no other grade or grades, and shall be subject to the New York Cotton Exchange inspection..

Either party may call for a margin, as the variations of the market for like deliveries may warrant, which margin shall be kept good.

This contract is made in view of, and in all respects subject to, the regulations made pursuant to said United States Cotton Futures Act and to the By-laws and Rules of the New York Cotton Exchange not in conflict with the said Act or said regulations. Subject to the United States Cotton Futures Act, Sec. 5.

How a Nonmember May Trade on a Cotton Exchange.—Some members of exchanges dealing in future contracts buy and sell for their own firm only. Others do a brokerage business in addition to their own firm business. They will buy or sell futures for responsible parties living anywhere in the country. One hundred bales, or about 50,000 pounds, is the minimum amount that may be bought or sold on the New York Exchange. Fifty bales is the minimum at New Orleans and Chicago. The brokerage charge, which varies somewhat at different times, is about $15 for buying or selling 100 bales. In addition to paying the brokerage, the purchaser must make a deposit, for each 100 bales bought, of $100 to $500, which is known as a "margin." This sum is deposited with the clearinghouse or a trust company to guarantee the broker making the purchase against loss due to market fluctuations. If the market is declining sharply, the broker may call for an additional margin. If this is not supplied, the broker may sell and close out the transaction. Brokers may buy for firms of known financial responsibility without requiring a deposit.

Hedging.—Hedging, as noted in Chap. XIX, is a sort of trade insurance which a cotton grower, merchant, or manufacturer may secure from a cotton exchange by selling or purchasing future contracts for cotton. It is defined by Hubbard[1] as follows: "Hedging is trade insurance carried on in such a manner as to make a world-wide market for instantaneous buying and selling." The use of hedging by a cotton grower may be illustrated as follows: We shall suppose that a grower has 300 acres of cotton planted and that in July it gives promise of a good yield. The future market for December is 25 cents a pound. At that price per pound, the grower will make a fair profit in his business. Since his neighbor's crops are good and crop-condition reports indicate good crops over a considerable part of the Cotton Belt, it looks as if the market price might be lower during October and November, when his cotton would be ready to go on the market. There is, of course, a possibility that there will be mishaps to the crop and that the price will be better than 25 cents a pound in the fall. To the grower, however, a price of 25 cents a pound with a sure profit is better than an uncertain one, even if there is a possibility of a better price. Consequently, he sells 100 bales of December futures at 25 cents a pound. If, when December comes, the market price is better than 25 cents a pound, he can sell his cotton on the market and buy futures to balance the ones he has sold or to deliver on his contract. If the market price for spot cotton is much higher than 25 cents a pound the future market will likely be higher than 25 cents also.

If the grower buys futures to deliver for those sold, he will lose on his future deals, but he has gained on the spot cotton. If the market has declined as he expected, he can deliver his spot cotton on the contract and realize 25 cents a pound less brokerage fees.

A manufacturer sells goods several months in advance and protects himself by buying future contracts for cotton to be delivered in the future when the goods are made. He bases the price of his goods on the price of the future cotton, so his profits are secure.

A cotton buyer or merchant protects himself continually against a fall in prices by selling future contracts to balance the spot cotton that he buys.

It occurs to the writer that cotton growers might profitably make more extensive use of hedging than they do. Probably many do not understand the benefits that may be obtained. Many others do not have control of enough cotton to make a future sale. But it seems as if two or more might combine interests and make a sale.

Delivery on a Future Contract.—Theoretically, all the cotton that is sold on a future contract is to be delivered, but actually only a small percentage is delivered. The amount varies somewhat in different years, the variation depending on market conditions. In the New York Exchange it is rarely more than 1 or 2 per cent of the amount sold and is usually only a small fraction of 1 per cent. Deliveries may be stopped by ring settlements. Delivery on contracts sold on the New York Cotton Exchange may be made at any one of the eight ports: New York, Norfolk, Charleston, Savannah, Mobile, New Orleans, Houston, and Galveston. As in the spot-cotton markets, delivery is effected by the transfer of warehouse receipts, together with the necessary accompanying documents.

The cotton that is delivered is good cotton. It must have a staple length of ⅞ inch or more, be of Low Middling grade or better, and meet other government specifications mentioned in Chap. XIX. The bales delivered may vary in grade and staple; that is, they are not even-running. This makes them less desirable for the spinner but does make delivery on contract easier. If a certain grade and staple length are specified, it may be difficult to procure the cotton in the market when wanted. The cotton delivered being of different grades, it is settled for on a

Middling basis, the relative values being determined by an average of the values for various grades in the 10 designated spot markets of the country which quote prices daily. New Orleans is a designated spot market and uses prices of that particular market in settling future-contract business.

The cotton-futures law now in force requires that the person or firm selling a future contract must give notice in writing of intention to deliver cotton on the future contract. This notice must be given 5 days before date of delivery. It is probable that the person to whom the contract was first sold has sold contracts himself and will transfer the notice of delivery to another. The notice may be transferred by several firms. The last to receive the notice within the time limit must either take the cotton or arrange some form of settlement.

Preparation of Bales for Delivery.—Prior to delivery, bales of cotton in a warehouse are inspected, weighed, sampled, and classed. The classing is done by the government Board of Classers, and the inspecting, weighing, and sampling are done by exchange officials.

When the person delivering the cotton has received the classification certificates, these, with warehouse receipts, are turned over to the buyer, and a check in payment of full amount of the bill is rendered in return.

Value of Exchanges to the Cotton Industry.—A great deal is said in favor of cotton exchanges, but, on the other hand, they are subject to much criticism. A considerable part of the criticism comes, however, from people who either do not understand their workings or fail to realize the benefits. Almost all well-informed men connected with the cotton industry and all marketing specialists have come to the conclusion that cotton exchanges, when operating under proper government regulation, are of assistance to the cotton trade. They broaden the market; stabilize prices; spread abroad information in regard to market values; and afford a good, convenient means of trade insurance.

Perhaps it will be well to quote from authorities who have made a careful study of the subject of cotton exchanges. Here is what the Committee on Agriculture on the Smith-Lever bill, Sixty-third Congress, Second Session, had to say in their report about the functions and economic uses of cotton exchanges:

It is the opinion of the committee that the abolition of cotton exchanges would result inevitably in the monopolizing of the entire cotton crop into the hands of a very few powerful interests, with the force and means to fix the price at which the farmer would be compelled to sell his cotton. Fully 75 per cent of American-produced cotton leaves the hands of the producer during the four months of September, October, November, and December. It takes no stretch of the imagination to foresee how utterly helpless the farmer, as a class, would be in his present disorganized condition as a factor in fixing the price of his own products, as against the organized genius and money of the spinners and powerful spot-cotton dealers. Any legislation, therefore, which eliminates from the cotton trade the element of a legitimate speculation and legitimate speculators must, in the opinion of the committee, result disastrously to the producers.

We have also a copy of a letter by W. B. Thompson, of New Orleans, recently read before the United States Senate when the Comer Amendment to the United States Futures Act was being considered. Mr. Thompson was formerly president of the Louisiana division of the American Cotton Association, an extensive cotton grower, a cotton merchant, and at one time president of the New Orleans Cotton Exchange. He was thus in a position to view the subject of cotton exchanges from various angles. He says in part:

In times past I have made a rather close study of the future contract, and I have long since come to the conclusion that its chief, and indeed its only, value lies in the protection it affords buyers and sellers of cotton and manufacturers of cotton goods. When the future-contract cotton functions normally, the owner of the cotton, pending such time as he can market the same, or, if he be a farmer, before he has made the same, can insure himself against loss by selling a future-contract hedge; the cotton buyer can make forward contracts with the spinner before he has any cotton in hand and protect himself against loss by buying a future-contract hedge; and the manufacturer may make forward contracts for the sale of his product before he has cotton in hand or a specific grade contract with the spot merchant or broker and insure himself against loss by buying a future-contract hedge.

It is the hedge, or price-insurance function of the future contract, which makes it valuable to the trade. It was never intended nor will it ever be possible to make the future contracts for cotton bought and sold around the rings of the exchanges take the place of direct dealings for specific grades and staples between the manufacturer and the spot merchant.

These contracts, which provide the protection aforesaid to those who do not want to speculate, are supplied primarily by those who want to speculate and whose business is to speculate. These latter are the underwriters, who assume the risk. From this postulate it is obvious, therefore, that the form of the contract or the character and the extent of the obligation to which the speculative seller or buyer is called upon to commit himself becomes the determining factor in the existence of a future-contract market. If the contract is too narrow or if it contains conditions which place the seller at a disadvantage, then there will always be more buyers than sellers, and, in consequence, the contract will go to an artificial premium, and if these disadvantages are exaggerated, the price of contracts, irrespective of the value of cotton, will go through the roof, and you will no longer have any contract market, because nobody will be willing to take the short side of the contract. If, on the other hand, the contract be too broad and the advantages all be with the seller, buyers will be driven from the market, and the whole system of the future trading will be destroyed.

In order, therefore, that supply of contracts may be forthcoming and a future-contract market possible, it is necessary that the authorized contract shall be one to which both the speculative buyer and seller will be willing to commit himself. It must be a basic contract, because no speculative seller of future contracts would commit himself to an obligation to deliver a specific grade which it might be impossible for him to obtain, and the range of deliverable grades should be sufficiently wide to give him a reasonable assurance that he could secure the cotton necessary to liquidate his contract should he be called upon to do so. On the other hand, the buyer must have assurance that he will not be called upon to take delivery of unmarketable cotton or of any cotton at artificial price differences.

I am very much in favor of the future-contract market properly regulated. In fact, it is my opinion that it would be a great calamity to the producer if the future markets were destroyed or impaired.

The following extract from a letter written by Henry Wallace, Secretary of Agriculture, to the chairman of the Committee on Agriculture in the House of Representatives, gives his views on limiting future trading. After mentioning the fact that the House committee had at different times thoroughly investigated the subject of cotton futures, he says:

In each case, as we understand it, the committee reached the conclusion that, while the regulation of the future exchanges in some form was desirable, it was not prepared to recommend even a limitation upon speculative trading.

Objections to Exchanges.—Criticism is often made against exchanges to the effect that the members manipulate buying or selling in such a way as to cause a rise or fall in the markets as suits their financial interests. It is true that a number who are interested in seeing the price advance—"bulls" they are called—can act together in buying and cause some immediate change in prices, but the movement will be of short duration unless there is some real market condition to back it up, as very adverse weather conditions, unusual demand for cotton, or something of the sort. Some of the members of the exchange are always interested in seeing the price of cotton fall—"bears" so called, who will rally and struggle against the advances fostered by the "bulls."

Two attempts at "corners," that is, attempts to buy up all available cotton that could be delivered on contract at a certain market at a certain time, have been made by New York firms and others since 1900. These corners make available cotton so scarce that those who have sold future contracts may pay extreme prices for the cotton to deliver on their contracts when they come due.

Engineering a corner requires much capital and is attended by a large financial risk. Much cotton must be bought at a high price as the market advances. A part of this must be sold later at a lower price on a declining market. It is questionable whether either of the corners mentioned was a success financially. In the second corner in 1910, the leaders were indicted and fined for violation of the Sherman Antitrust Act. The government charged that there was an agreement between the parties concerned which resulted in restraint of trade.

During the period of high prices incident to a corner, the growers, merchants, or anyone else who has cotton to sell may profit by the high prices, but usually the corner is engineered at a time when there is a scarcity of cotton and after it has passed out of the hands of the growers and primary buyers. The high prices are artificial in a sense and are not just. All that one party gains someone else loses. Corners are illegal. It is the intent of rules adopted by exchanges and of our federal laws to prevent them.

Effect of Closing the Future Exchanges.—During the summer of 1914, the future exchanges of the country were closed for 3 months. Extreme market fluctuations due to the outbreak of the World War and the failure of firms dealing on the future market made this action necessary. While the exchanges were closed, cotton trading was largely at a standstill. There was no definite world market. Prices differed in different towns, the variation depending on the sales in sight. Merchants could not safely buy cotton, because they could not hedge to secure themselves. The same was true of manufacturers. It was not long after the exchanges had been closed before there were calls for them to reopen. Since that time there have been fewer demands for the abolition of future exchanges than previously. It was clearly shown that they have a legitimate function.

Legislation to Control Exchanges.—As was stated previously, there has been a clamor for the abolition of future exchanges or the adoption of such stringent control measures that trading on them would be hindered. A number of bills dealing with the matter have been introduced in Congress. When the bills were up for consideration, the Senate and House committees gave public hearings, inviting prominent men in various branches of the cotton industry. They made an earnest effort to secure all possible information from all interests.

The United States Cotton Futures Act was passed in 1914 but was later declared unconstitutional. It was reenacted, with certain changes, in 1916 and amended in 1919 and also in 1927. Some of the principal provisions of the law as it now stands are:

1. The cotton bales traded in shall be considered as weighing 500 pounds.

2. Middling shall be the basic grade unless some other is specified in the contract.

3. The cotton delivered shall be of the grades for which official standards have been prepared.

4. The relative value of different grades shall be determined by commercial differences existing in the 10 designated spot markets.

5. Cotton shall not be delivered if the grade is below Low Middling; or, if tinged, below the grade of Strict Middling; or, if yellow-stained, below the grade of Good Middling; or if the staple is less than ⅞ inch in length; or if of perished staple, gin-cut staple, reginned staple, or immature staple; or if the cotton is repacked, false-packed, mixed-packed, or water-packed.

6. Notice of intended delivery must be given on the fifth business day prior to the date of delivery.

7. When the cotton is tendered, there must accompany each tender a written statement of each and every bale by tag number, showing the grade.

8. All cotton tendered shall be classed by government classers.

9. A tax of 2 cents a pound shall be levied on cotton sold for future delivery if the provisions of the act are not obeyed.

In addition to the federal laws for the control of cotton exchanges, each exchange has a number of restrictive rules of its own. Some of these, as, for instance, the one limiting market fluctuations to 200 points for any one day, have been adopted at the suggestion of government authorities.

Speculation in Cotton.—The buying or selling of cotton futures with the sole idea of making money by a rise or fall of the market price is termed "speculation," or "gambling," in cotton. Speculation is defined in Webster's International Dictionary as:

> Engaging in business out of the ordinary, or by dealing with a view to making a profit from conjectural fluctuations in the price rather than from earnings of the ordinary profit of trade, or by entering into a business venture involving unusual risks for a chance of an unusually large gain or profit.
>
> Speculation, while confined within moderate limits, is the agent for equalizing supply and demand, and rendering the fluctuations of price less sudden and abrupt than they would otherwise be.

There is speculation in futures, in spot cotton, and in various other commodities. A merchant who buys spot cotton and holds it for a higher price is, in a sense, speculating. He is taking a chance on making a profit by a rise in price. While speculation is harmful to the public in some ways—has a tendency to create artificial or unnatural market conditions—it is not without value. As was brought out in the second paragraph of the definition quoted above, it assists in equalizing supply and demand and in reducing price fluctuations.

Many of the sales and purchases made on future exchanges are for the purpose of hedging and are considered legitimate trading. Hubbard[1] says that, when a careful survey was made of the New York Exchange when it closed in 1914, it was found that 85 per cent of the business was for hedging. A few years later, President Marsh estimated that 75 per cent of the business then being carried on was for hedging.

Market Statistics.—A trader on a future exchange, to be successful, must be wide-awake, do much thinking, and make a careful study of crop statistics. To assist the trader, the exchange collects and posts or publishes all available data on crop conditions in all parts of the world: weather conditions; insect depredations; cotton ginned, exported, stored in warehouses, consumed; etc.

References

1. HUBBARD, W. HUSTACE: "Cotton and the Cotton Market," D. Appleton-Century Company, Inc., New York, 1923.
2. MILLER, T. S.: "Cotton Trade Guide and Student's Manual," Steck, Austin, Tex., 1915.
3. HAMMOND, M. B.: "The Cotton Industry," American Economic Association, New York, 1897.
4. Anonymous: "Cotton Futures," A Compilation of Certain Public Documents Dealing with the Enactment, Amendment, and Administration of the Act to Regulate Trading in Cotton for Future Delivery, Government Printing Office, Washington, D. C., 1922.

CHAPTER XXI

ECONOMICS OF COTTON PRODUCTION

Conditions obtaining in the American Cotton Belt, both natural and man made, are very favorable for the production of cotton. In fact, it is doubtful if there is in the world any other territory of such wide expanse so well suited. The climate, as has been said, is exceptionally favorable; the summers are warm and long; the rainfall is rather plentiful during the spring and early summer months, when the young cotton plants need rain for good growth, but scanty during the fall, when dry weather is desirable for picking and for the protection of the open cotton bolls. The soil is naturally rich. Where it has been run down by continuous cropping, it may be made productive again by the use of proper fertilizers. The labor supply is fairly plentiful. The cotton fields are not far removed from cotton mills and consuming markets. Numerous good roads and adequate railroad facilities make transportation easy.

Approximately 3,000,000 people work in the cotton fields of the South. In 1929, cotton was the major crop on 62 per cent of the farms in the leading cotton-growing states and yielded 56 per cent of the income of farmers in these states. The cotton crop, including lint and seed, normally has a greater cash value than any other crop grown in the United States except corn. During the period 1924–1928, its annual value averaged about one and one-half billion dollars.

Foreign Competition.—For a good many years, the United States has had virtually a monopoly of the cotton market, due to the fact that it produced one-half to two-thirds of the cotton supply of the world. But this monopoly cannot be maintained unless the country produces cotton economically, so that it may be sold at a moderate price and yet yield profit to the producer. If it is not sold at a moderate price, foreign nations will be stimulated to activity in growing cotton. During the period of high prices from 1860 to 1870, England and some other European countries made special efforts to foster cotton growing in their colonies. More recently during the period of high prices incident to the World War, renewed interest was taken in cotton growing in several countries where cotton had not been grown extensively.

There are large areas of fertile land suited to the growing of cotton in the Sudan region and in the Congo district of Africa; in Brazil and Argentina in South America; in Mexico; and in other parts of the world. The land in much of these regions is very well suited to cotton, and the climate, while probably not so well adapted as that of the American Cotton Belt, does serve very well. The main hindrances to the industry in these new territories are lack of labor and inadequate means of transportation. But if prices are high enough to make the growing of cotton very profitable, unemployed labor in Europe will probably be drawn to the regions; and if there is a chance for good profit on the investment, capital for building railroads and securing other means of conveyance will not be hard to obtain. If large quantities of cotton are produced in other countries, America cannot produce her usual amount without there being an overproduction. Overproduction brings a glut on the market and reduced prices.

Sanders,[12] in discussing foreign competition in cotton production, says:

> All the more important producers of foreign cotton cultivate a very low acreage per man compared with America. If mechanized cotton harvesting is made practical in the near future, America undoubtedly would have a distinct advantage over nearly all foreign producers of cotton. In this case we would be certain to increase radically our proportion of the world's cotton production. Furthermore, the problem of increasing cotton production involves, in nearly all producing areas of the world, the most skillful control of insects, pests, and diseases, and the most skillful selection of seed and quality improvements. It is believed that America, with its elaborate experiment station work and its superior agricultural extension activities, has distinct advantages in these respects over all foreign producers.

Cost of Cotton Production.—A good many estimates of the cost of cotton production have been issued. These show a wide variation in opinion. The variation is largely due to different ideas in respect to the size of the wage the laborers should receive and whether or not all the time of laborers and mules should be charged to the cotton crop. Some, of course, express opinions that are mere impressions and are not based on facts.

Some estimators fail to consider that the cotton-field laborers should have the opportunity of living in some degree of comfort and have some educational advantages. During slavery days and for a good many years afterward, they lived largely in squalid cabins and had very limited rations. Under such conditions, cotton could be produced at a low cost. Plantation owners figured that they could make money even if the cotton did not sell for more than 8 or 10 cents a pound. With impressions gained from viewing such conditions, some have come to feel that nothing more is needed by the laborers and that cotton can be produced very cheaply.

Figures expressing the cost of cotton production at any particular time may not be applicable to other times because of fluctuating values in the purchasing power of money. Costs are best expressed in terms of hours of man and mule labor necessary for production, but even these will vary considerably in different regions, the amount depending on the productivity of the soil, the prevalence of troublesome weeds, insect pests, etc. If the date at which certain particular prices prevail is given, it assists in estimating values.

In figuring the cost of cotton production, some authorities consider only the land rent and the hours of man and mule labor actually spent in plowing, hoeing, picking, etc., and allow the workmen a moderate wage for that time. This is not quite fair or right, because there are overhead expenses, such as manager's salary, taxes, and interest on money invested in machinery and in mules. Mules and laborers must be fed the whole year in order that they may be available during the months when needed in the cotton fields. It does not seem as if *all* the expense of support for the year should be charged to the cotton crop alone. A large part of it should be so charged, because, if the cotton crop has the first call for labor at all times, no great amount of other profitable or productive work can be done on the farm. To be sure, it should be the effort of the management to keep all labor employed at all times, but during the winter and during periods of rainy weather at other times in the year this is not always possible. Consequently, the income from the cotton crop must be drawn upon for support during this off season

Other authorities are inclined to charge the entire expense for the year to the cotton crop, even if no special attempt is made to find other employment for laborers during the spare time. Man labor is estimated as a regular monthly wage, at a fairly high figure, throughout the year. Extra labor hired for hoeing and picking is also allowed a comparatively high wage. This, it occurs to the writer, is going to the extreme in the other direction and is not fair. In so far as is possible, the work on the farm should be so organized as to keep man and mule labor employed continually during good weather and so that soil fertility will be maintained. Overhead expenses should be prorated between the different incomes according to receipts. The proportion for cotton plus the cost per acre of the actual labor, rent, fertilizers, seed, etc., will give the real cost of cotton production per acre. This total divided by the pounds produced will give the cost of production per pound.

Low Estimates on the Cost of Production.—Hammond[1] reports that, about 1840, certain American planters became alarmed lest India should become a rival in the production of cotton. They held a meeting to consider measures to counteract the work of the East India Company. After a lengthy discussion, they decided that so long as the American planters could get 8 cents a pound for their cotton, delivered at the nearest market, they could afford to produce it. In 1849, several of the large planters of the Mississippi bottoms estimated that they could grow cotton for 6 cents a pound. In 1893, the Senate Committee on Agriculture and Forestry made a report to Congress on the condition of the cotton growers in the United States. This committee stated:

> It is the general consensus of opinion that cotton cannot, except under the most favorable circumstances, be raised profitably at less than 8 cents a pound, nor without loss under 7 cents.

During most of the time since 1850, cotton has been worth 10 cents a pound or more on the New York market. Up to the World War period, a price of 10 cents a pound was considered sufficient to pay cost of production and marketing expenses and perhaps to yield the producer a small margin of profit.

Estimate of the Cost of Cotton Production in Alabama in 1928 *and* 1929.—The following estimate of the cost of producing cotton in Alabama[13] covers farms in but two counties. The average yield of lint was 358 pounds, which is a good yield. The cotton was grown during a period of high prices, but the estimates appear to be moderate. No allowance is made for supervision or support of labor and mules when not being used in the cotton.

COSTS AND RETURNS PER ACRE FROM PRODUCING COTTON IN MARSHALL AND DEKALB COUNTIES, ALABAMA, 1928 AND 1929

Item	1928	1929	2-year average
Number of farms	111	102	106
Acres cotton per farm	14.3	14.8	14.6
Pounds lint per acre	362	355	358
Costs per acre:			
Seed	$ 0.90	$ 0.98	$ 0.94
Fertilizer	8.86	10.41	9.63
Manure or compost	0.27	0.58	0.42
Ginning	3.05	3.03	3.04
Use of land	5.29	5.62	5.46
Use of equipment	1.39	1.28	1.34
Man labor, preharvest	9.94	7.55	8.74
Man labor, hauling to gin and marketing	0.51	0.60	0.56
Picking labor	9.37	8.92	9.15
Mule labor	3.81	3.71	3.76
Truck cost	0.04	0.13	0.08
Share of automobile cost	1.07	1.24	1.16
Interest on seed and fertilizer	0.78	0.91	0.84
Total cost per acre	$45.28	$44.96	$45.12
Value of cotton seed	11.29	6.97	9.13
Cost lint per acre	33.99	37.99	35.99
Cost lint per pound	0.094	0.107	0.100
Returns per acre:			
Lint cotton	69.85	58.58	64.21
Cotton seed	11.29	6.97	9.13
Total returns per acre	$81.14	$65.55	$73.34
Profit per acre	35.86	20.59	28.22
Return for labor per acre	55.68	37.66	46.67
Return for hour of man labor	0.46	0.37	0.42

Estimates by U. S. Department of Agriculture Investigators.—Several research men connected with the U. S. Department of Agriculture have made studies on the cost of producing cotton. In 1918, Moorhouse and Cooper[2] obtained records of labor and material requirements for 842 crops grown in 10 different counties in four southern states. The laborers were allowed a fair wage, and charges were made for overhead expenses, but no allowance was made for time that the labor was not actually employed in the crop. The range in cost was from 8 cents to $1.07 per pound, but most of the cotton was produced at 28 cents a pound or less. The average cost was 23 cents.

Moorhouse and Cooper,[2] in 1919, obtained additional cost records from 821 crops grown in 12 counties in six different

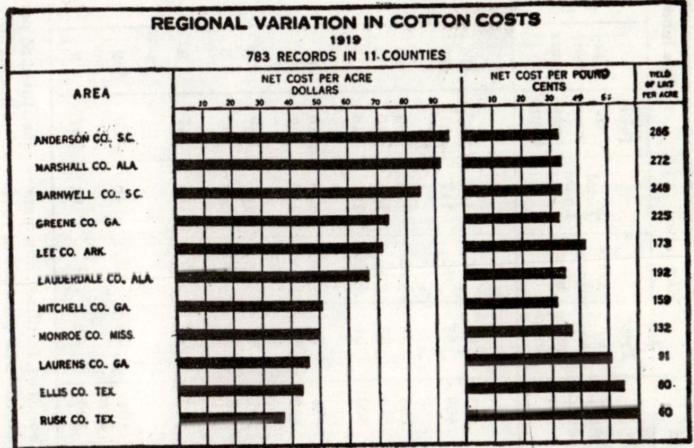

FIG. 118.—Variations both in the cost per acre and in the yield per acre cause variations in the net cost per pound of lint. The average acre in Anderson County cultivated at the highest cost in 1919 produced the highest average yield at the lowest cost per pound. (*After Agelasto and others.*)

southern states. The bases used were very similar to those used in 1918. Table XL lists the counties and states in which data were collected and gives the main results of the survey. These results are expressed in terms of material and labor requirements and not, with the exception of ginning charges, in dollars and cents. An effort was made to express the costs in a form that would be permanent, that is, usable any year and independent of fluctuating values. By considering prices paid for seed, labor, fertilizer, etc., during any particular year, this table may be used in determining the cost of production that year.

In a study of Table XL, it will be observed that the man-labor requirements are low in Ellis and Rush counties, Texas, and relatively low in Laurens County, Georgia. Comparatively low yields were obtained in these regions. The man-labor require-

TABLE XL.—LABOR AND MATERIAL REQUIREMENTS PER ACRE
(821 Records, 1919 Crop)

Region	Number of records	Yield		Man labor			Mule labor			Seed, pounds	Fertilizer, pounds	Ginning charge per hundredweight
		Lint, pounds	Seed, pounds	Prior to harvest, hours	Harvest, hours	Total, hours	Prior to harvest, hours	Harvest, hours	Total, hours			
South Carolina:												
Anderson County[a]	74	286	495	80	60	140	45	14	59	35	449	$1.00
Barnwell County	76	248	408	65	52	117	41	12	53	28	699	1.04
Georgia:												
Laurens County	77	93	168	55	23	78	39	3	42	26	254	1.24
Greene County	74	225	413	53	45	108	40	8	48	37	295	1.11
Mitchell County	50	159	300	61	39	100	43	5	48	30	277	1.07
Alabama:												
Marshall County	79	272	473	70	58	128	46	11	57	31	369	1.02
Lauderdale County	84	192	345	69	51	120	47	7	54	29	168	1.10
Mississippi:												
Washington County	29	171	391	87	54	141	47	5	52	35	...	1.69
Monroe County	49	132	238	54	34	88	35	6	41	34	[b]	1.39
Arkansas:												
Lee County	83	174	363	109	55	164	47	8	55	34	[b]	1.35
Texas:												
Ellis	71	50[c] 29[d] 24[e]	134	31	15	46	29	2	31	22	...	1.80
Rusk	75	61	106	48	16	64	37	3	40	22	105	1.87

[a] On 34 owned farms producing wage cotton, man labor, mule labor, seed, fertilizer, and manure constituted 85 per cent of the total operating expense. By adding ginning to the foregoing list, the operating expense amounted to 89 per cent of total cost, excluding interest on land.

[b] In Monroe County, Mississippi, fertilizer was applied on only 13 farms; in Lee County, Arkansas, on only 1.

[c] Picked cotton.

[d] Bolly cotton.

[e] Unginned seed cotton.

TABLE XLI.—COST OF PRODUCTION BY YIELD GROUPS, 1923

Yield, pounds of lint per acre	Number of reports	Acres in cotton per farm	Yield per acre, pounds of lint	Gross cost per acre										Credit per acre (cotton seed)	Net cost of lint		Value of lint	
				Prepare and plant	Cultivate	Harvest and market	Miscellaneous labor[a]	Fertilizer and manure	Seed	Ginning	Land rent	Miscellaneous costs[b]	Total		Per acre	Per pound	Per acre	Per pound
20 and under	32	55	14	$3.69	$5.25	$2.11	$0.70	$2.94	$1.14	$0.22	$3.52	$1.52	$21.09	$0.74	$20.35	$1.45	$ 4.28	$0.30
21–60.......	249	69	44	3.84	5.24	3.98	0.79	4.25	1.25	0.59	4.33	2.69	26.96	2.10	24.86	0.56	13.32	0.30
61-100......	451	55	89	3.91	5.73	5.57	0.79	3.97	1.16	1.33	4.88	2.57	29.91	3.60	26.31	0.30	26.55	0.30
101–140.....	407	54	124	4.25	6.12	6.69	1.03	3.39	1.24	1.76	4.98	3.06	32.52	5.13	27.39	0.22	36.48	0.30
141–180.....	394	70	161	4.18	5.92	7.60	1.11	3.55	1.22	1.99	5.99	2.74	34.30	6.79	27.51	0.17	47.73	0.30
181–220.....	279	51	200	4.37	6.20	8.74	1.36	4.48	1.18	2.23	6.90	3.31	38.77	7.64	31.13	0.16	60.29	0.30
221–260.....	257	63	245	4.71	6.59	9.85	1.56	5.04	1.40	2.77	7.08	3.33	42.33	9.71	32.62	0.13	72.60	0.30
261–300.....	165	30	290	5.01	7.08	11.13	1.64	6.27	1.54	3.08	8.05	3.44	47.24	11.00	36.24	0.12	87.77	0.30
301–340.....	34	56	324	5.26	8.51	13.39	2.46	9.03	1.64	3.32	9.59	5.09	58.29	12.64	45.65	0.14	95.34	0.30
341–380.....	54	33	356	5.50	7.18	12.13	2.83	8.75	1.48	3.90	9.53	4.10	55.40	12.62	42.78	0.12	107.14	0.31
381–420.....	94	31	401	5.71	8.62	13.74	2.37	10.41	1.51	3.71	8.91	4.37	59.35	14.98	44.37	0.11	124.22	0.31
421–460.....	27	31	444	5.82	8.14	14.87	2.91	9.99	1.59	4.46	12.42	4.56	64.76	14.53	50.23	0.11	132.63	0.31
461–500.....	60	26	495	6.09	7.75	17.71	3.04	9.73	1.55	5.17	10.43	5.54	67.01	17.94	49.07	0.10	150.42	0.30
501 and over	16	27	618	6.34	7.51	28.28	2.79	13.86	1.44	6.92	10.79	5.07	83.00	26.17	56.83	0.09	197.45	0.32

[a] Includes miscellaneous labor, irrigating and water, dusting and dusting material.

[b] Includes picking sacks and sheets, crop insurance, use of implements, use of storage buildings, and overhead.

ments in Washington County, Mississippi, and in Lee County, Arkansas, are especially high because much hoeing was necessary to keep down grass and weeds. Heavy applications of fertilizer were made in South Carolina, and little or none in Mississippi, Arkansas, and Texas. It is apparent that the yields increase greatly with an increase in the use of fertilizers.

As has been mentioned elsewhere, the cost of producing cotton varies greatly in different regions and during different seasons in the same region. Much depends on the natural productiveness of the soil, the weather, the fertilizers used, and other factors. Figure 118 shows in graphic form regional variation in cost per acre and cost per pound for the year 1919.

In 1923, Cooper and Hawley[4] collected data on the cost of cotton production from 2,519 farmers living in various parts of the Cotton Belt. At that time, the price of labor and supplies had fallen from wartime levels but was still considerably above prewar figures. The main results of their investigations are shown in Table XLI. In discussing their results, the authors say:

Cotton reports were received from 2,519 farmers, but the greater number were growers having yields considerably above the average. For this reason the costs are shown by yield groups, rather than by average costs by states and by the entire Cotton Belt.

The average yield of lint cotton in the United States in 1923 was 130.6 pounds per acre, according to the Division of Crop and Livestock Estimates, U. S. Department of Agriculture. Of the 2,519 reports received regarding the cost of producing cotton, 407 showed yields of 101 to 140 pounds per acre, averaging 124 pounds. This group appears most nearly to represent general conditions in the cotton states during 1925; the average net cost of production on these 407 farms was 22 cents per pound of lint, and the average price received was 30 cents.

Fifty-five per cent of all farmers reporting had yields of more than 140 pounds per acre and, on an average, produced their cotton at considerably less cost per pound than did those who had yields of 101 to 140 pounds of lint per acre.

Seven hundred and thirty-two, or 29 per cent, of the farmers reporting produced 100 pounds of lint or less per acre. Of these, 281 produced at a cost above the price received.

Financial Condition of Cotton Farmers.—Much has been said and written about the financial status of the cotton growers in the South. For several years after the Civil War, conditions were indeed bad. The farms were run down; tools and equipment were scarce or lacking altogether; and there was practically nothing upon which to obtain loans. Even as late as 1893, J. Washington Watts, of Laurens, S. C., writing for a United States Senate committee report (Hammond[1]), said:

> The actual financial and material condition of the cotton raisers is very bad; generally in debt. In my opinion, the majority are making their crop with supplies bought on credit, and are in debt to that extent at least. I estimate the proportion of insolvent to be at least three-fourths of our farmers. Those who raise even a portion of their supplies are doing fairly well, but very few do this, preferring to risk the cotton, which sinks them deeper in debt each year.

A good many croppers and a small number of tenants become insolvent in bad crop years, but this does not worry some of them, least of all many of the colored croppers. They know that the farm owner will supply them with rations and some clothing, even though they do not make enough to pay all their debts. They can grow a part of their needed supplies in the garden. A house is furnished them free of charge, and they can get wood to burn by going to the forest after it. Although the living is meager, there is very little real want. Most years, they make enough to pay their debts and a few to a few hundred dollars besides. This extra money is spent very freely. But extremely little is saved for a "rainy day." On one plantation that the writer has visited frequently, 26 Ford cars were bought by the tenants and the croppers one good crop year.

In the seven southeastern cotton-growing states, North Carolina, South Carolina, Georgia, Alabama, Mississippi, Louisiana, and Arkansas, covered in *Research Monograph* V of the Works Progress Administration of the U. S. Government, the aggregate farm-mortgage debt on plantations and small farms combined amounted to less than $166,000,000 in 1910 but rose to $637,597,000 in 1928. In 1910, 33 per cent of all the farms in the United States were mortgaged; in 1920, 37 per cent; in 1925, 35 per cent; and in 1928, 36 per cent. From 1925 to 1928, the increase in the number of mortgages in the cotton states was 8.4 per cent, whereas that of the whole United States was only 1.2 per cent.

Financing Cotton Growers.—Most of the share-croppers, renters, and small farmers who grow cotton require financial aid while working in their crop. With the share-croppers, the aid is usually furnished by the landowner, who charges about 10 per cent for the use of his money. The renters may likewise be furnished money by the landowner, or they may open an account with a local merchant, who agrees to sell them, on credit, food, feed, and other supplies needed while growing the crop. Since the goods are sold on time and frequently to people who have little regard for their obligations, the time prices charged are considerably higher than the cash prices. The merchant is given the right to buy the cotton grown and can, of course, collect on the account if sufficient cotton is made. It sometimes happens that, if crop prospects look unfavorable during the summer or fall, the borrower runs away and leaves his crop. The loss occasioned by his departure must be made up by the more honest ones who remain. Consequently, time prices are high—in many instances unreasonably high. The poor, plodding farmer may get in debt to the merchant and, if his family does not practice strict economy, may find it almost impossible to get out again. Hammond,[1] writing in 1897, said:

> The use of the credit system is not confined to a small number of the farmers of the Cotton Belt. Ninety per cent of the cotton growers of Alabama, it is stated on high authority, make their purchases in this way, and pay prices on an average 25 per cent higher than do their neighbors who buy for cash. Throughout the Cotton Belt it is probably no exaggeration to say that three-fourths of the cotton planters and their tenants, white and black, buy on time, and pay usury to factors, merchants, or others who have advanced money or supplies on the prospective cotton crops of the borrowers.

While the percentage of cotton growers who buy on time from local merchants is not so large as it was in 1897, there are still a good many. An increasingly large number borrow money from banks or other money-lending concerns and then pay cash for their supplies.

According to studies made by the U. S. Department of Agriculture[3] in 1921, the average prevailing rate of interest charged farmers in the leading cotton-growing states on personal and collateral loans was:

	Per Cent		Per Cent
North Carolina	6.23	Alabama	8.46
Tennessee	7.88	Georgia	8.94
South Carolina	8.06	Texas	9.68
Mississippi	8.11	Oklahoma	9.84
Louisiana	8.34	Arkansas	9.70

The actual charges were somewhat higher than the foregoing figures indicate, as the interest was usually collected in advance, and in some cases the borrower was required to maintain a deposit at the bank.

Table XLII from Agelasto[3] furnishes data on the forms of security given banks by cotton farmers.

TABLE XLII.—FORM OF SECURITY GIVEN FOR PERSONAL AND COLLATERAL BANK LOANS TO FARMERS IN 10 LEADING COTTON STATES; PERCENTAGE OF LOANS SECURED BY VARIOUS FORMS OF SECURITY

State	Note without indorsement	Note with one or more indorsements	Mortgage on livestock	Crop lien	Warehouse receipt	Stocks and bonds	Other ways
North Carolina	10.5	68.6	1.7	5.2	2.1	7.5	4.4
South Carolina	9.1	41.0	13.6	20.2	9.7	4.8	1.6
Georgia	12.5	50.1	14.5	4.9	10.0	3.5	4.5
Tennessee	18.1	67.2	5.0	1.5	0.8	5.8	1.6
Alabama	10.4	20.1	31.5	26.1	7.5	2.4	2.0
Mississippi	12.7	27.0	20.2	15.1	8.0	9.1	7.9
Arkansas	12.1	37.9	22.7	19.9	3.0	2.2	2.2
Louisiana	15.5	52.7	12.4	5.2	2.7	9.0	2.5
Oklahoma	17.2	12.5	49.3	18.1	0.7	1.2	0.6
Texas	21.9	18.0	38.1	18.3	1.6	1.1	0.0

Recently certain cotton cooperative marketing associations have been able to borrow large sums of money for use in making advances on cotton and for other expenses. By joining together, the members of the association need no outside financial backing and are able to get plenty of money at a low rate of interest.

Other growers are financed by loan agencies fostered by the government. One of these, the Production Credit Corporation, makes the majority of the cotton-production loans in some areas. This organization has a local representative in each county or parish, who assists the borrower in making application to the district office for a loan. Money may be borrowed for a limited time, say three to nine months or longer, at an interest rate of 5½ to 6 per cent. A mortgage on livestock and equipment and a lien on the crop are taken as security.

Some small farmers who are unable to secure credit elsewhere may obtain crop-production aid from the Emergency Crop and Feed Loan Fund. This, however, is an emergency fund which is available only in years when Congress makes an appropriation for this particular purpose.

One-crop System.—A one-crop system of farming is used on a large number of the farms of the Cotton Belt. The reason for this is as follows: Growing cotton in the South is a well-established custom. The labor knows how to grow it, and the farm owners and managers feel that they can superintend a cotton farm better than any other. To change a custom requires considerable effort. Cotton can be grown and harvested with the use of very little machinery. The less machinery the average colored laborer has to use the better he gets along. The negroes like to grow cotton because they are used to it and because they have a good time working together in the fields, in groups much of the time, especially while hoeing and picking. Cotton in the form of bales keeps indefinitely without deterioration if kept dry. It is easily marketed.

Probably the chief reason for growing cotton largely to the exclusion of other crops is to be found in the belief of the landowners that cotton will pay better than any other crop. This is probably true if the practice is not carried to the extreme in growing too much. But many growers do go to the extreme and neglect to grow supplies needed for home consumption. They see an acre of cotton and have visions of a bale to the acre and the bale selling for a large sum, say $100. They compare the returns from an acre of corn with the visionary returns from the acre of cotton. The acre of corn is not apt to yield more than 25 bushels, which at $1 a bushel will bring but $25. The cotton appears to be a far better crop. But there are several things that the grower fails to consider. First, the yield is apt to be much less than a bale. The average yield is not more than one-third of a bale, and less than that may be made. The price is very apt to be low if a good crop is made. As a rule, the price is nearer 10 than 20 cents. A third of a bale at 10 cents a pound does not make a very large sum, only $16.66. Cotton is an expensive crop to produce in that it requires much hand labor. The landowner will have to give at least half of the cotton produced to pay for the man labor used in making the crop. This leaves him only $8.33 for his acre of cotton. With this sum he must pay for feed for his work animals for a year (feed bought at a high price if it has to be shipped in from a distance); buy harness, plows, seed, and fertilizer; pay taxes; provide for the upkeep on houses and buildings; and meet losses sustained when croppers, who have been "furnished," or given their supplies for making the crop, fail to make enough to pay their debts.

But this is not all. The one-crop cotton farmer is continually using up the fertility of his soil. The land is becoming poorer and poorer. In a corn crop may be grown soybeans or cowpeas, which will improve the soil and also yield a fair money return in hay or seed. Sudan grass, sweet potatoes, peanuts, sorghum, sugar cane for syrup, and oats are other crops that may be grown to advantage on a cotton farm. In many instances, cotton growers who have attempted to diversify or grow other crops in addition to their cotton have failed because no market was provided in advance for the product, and the labor was inexperienced in handling the new crop. Owners and tenants operating small acreages have an advantage in planting a part of their land to foodstuffs. They can use the produce themselves. The large planters may have trouble in selling profitably all the food that is produced if a large acreage is grown.

There is considerable business hazard in the one-crop system of cotton farming. If the crop is a failure, considerable loss is sure to be involved, because there is not only the loss of the rent on the land, but the labor and work animals have to be supported a whole year to make one crop. In most instances, the labor is furnished with supplies without any security. If the yield is very large, there is danger of overproduction and a low price; yet it costs much to harvest a large crop.

The family of the one-crop farmer usually suffers from lack of good food. If it all has to be bought at the store at a high price, the variety is not apt to be great or satisfactory; fruits and fresh vegetables may be lacking altogether.

Also the one-crop system is not economical, as it does not provide sufficient employment for the labor and work animals. In a survey made by Willard[5] in Ellis County, Texas, in 1914, it was found that on small farms the mules worked at productive labor but 60.3 days during the year. On the larger farms they worked 79.1 days. The men on the all-cotton farms do not work in the fields more than half the working days of the year, and the women not more than one-third.

Share-crop System.—A large part of the cotton grown in the United States is produced on a share-crop basis. The growers who furnish only their own labor, or labor and a part of the fertilizer if fertilizer is used, are known as "share-croppers." They get half of the cotton that is produced. The growers that furnish labor and teams are commonly known as "renters," or "tenants." They furnish the planting seed and a part of the fertilizer and pay for their share of the ginning. They commonly get three-fourths of the cotton and seed produced.

Although the share-crop system is sometimes criticized adversely, it has some good features. In case of crop failure, the share-crop farmer is not subject to so much loss as the cash renter. The tenant takes more interest in the crop than he would if he were working for wages, and he requires less supervision. The negro laborer likes the system, because he is his own boss to some extent, and he has a chance to make more money than he would have if he were working by the day for wages. He is usually given rations or supplies by the landowner. He is thus guaranteed a living whether he makes anything or not, and has a chance to live off the future crop.

Growth of Tenancy.—Figures published by the Census Bureau show a gradual increase in tenancy in the United States up to the year 1930. From 1930 to 1935, there was a slight decrease. In 1930, tenants operated 42.4 per cent of all the farms in the United States, whereas in 1935 they operated 42.1 per cent. Declines in proportion of tenants were registered only in the southern states, North Carolina, South Carolina, Georgia, Florida, Alabama, Mississippi, Arkansas, Louisiana, Oklahoma, Texas, and New Mexico. The other 32 states showed an increase. Tenancy is most prevalent in the South and lowest in the Northeast. In 1935, the percentages ranged from 69.8 in Mississippi and 65.6 in Georgia to 6.2 in Massachusetts and 6.9 in Maine.

The change in the relative number of farm owners and tenants is not caused necessarily by the landowners' becoming bankrupt and losing their farms. During the depression of 1930–1934, when mortgage foreclosures were most frequent, there was a decline in percentage of tenants. That the proportion of tenant farmers is becoming greater is probably due, in part at least, to the fact that with the increase in value of lands and the increase in population, it is becoming more and more difficult for men who do not have land to obtain it. Consequently, more of them, if they become farmers, rent land to farm.

The increase of tenancy is viewed with some degree of displeasure by the public, because tenants move from place to place and cannot be expected to take so much interest in building up the community as do the more permanent residents. A hopeful sign is to be seen, however, in the working out, and adoption in some sections, of long-time leases. These are of mutual advantage and make the tenant a more permanent citizen. With the usual short-time lease under which land is usually rented, the tenant takes no interest in improving the land or building up the property. He feels that it is to his interest to get as much out of the place as he can during the short time he has it and to spend as little as possible in improvements. As a result, the farms that are rented deteriorate rapidly to the loss of both owner and renter.

Overproduction.—It frequently happens that more cotton is produced than can be marketed advantageously. This results in what is termed "overproduction." It does not mean necessarily that the world does not need the cotton but that conditions are such that it cannot be marketed satisfactorily and low prices prevail. The period from 1891 to 1895 was one of low prices, the average value of lint cotton being 7.81 cents. Hammond and others attribute these low prices to overproduction. About that time, extensive new areas were opened west of the Mississippi River, especially in Texas. During the season of 1914, a fairly large crop was produced. World War hostilities so interfered with the exporting and marketing of this crop that cotton became a glut on the market, and there was apparently great overproduction.

With the beginning of the depression of 1929–1930, there was a decrease in purchasing power on the part of the buyers of cotton goods in most parts of the world. This resulted at once in a decrease of cotton consumption and in an increase of the cotton surplus. The crop of 1931, plus the carry-over from the previous year, resulted in a world supply of 41,000,000 bales of cotton. The world crop of cotton in 1932 was almost 4,000,000 bales smaller than that of 1931, but with reduced consumption the total world supply remained abnormally high—almost 40,700,000 bales. The carry-over of American cotton on Aug. 1, 1933, was 11,600,000 bales—still an abnormally large supply. Under these conditions of supply and demand, cotton prices dropped to low levels. Spot-cotton prices fluctuated from 5 to 7 cents per pound between June, 1931, and April, 1933. The income from cotton and cotton seed slumped from $1,389,000,000 in 1929–1930 to $464,000,000 in 1932–1933. This decline in income, together with the general effects of the economic depression, caused widespread distress.

High prices for a year or two may stimulate cotton growing and result in the production of more cotton than can be sold at a good price. The one-crop system in vogue in much of the South is probably a contributory factor to overproduction. There is no other crop to which the farmers are prepared to turn or that they know how to grow successfully. They are forced into a goodly cotton acreage, although the outlook is not bright, and they feel that they should reduce acreage. The credit system in use in many parts of the Cotton Belt is another contributory factor to frequent overproduction. Credit merchants and banks have usually considered a cotton crop better security than any other the farmer could plant and have, consequently, encouraged him to plant more and more cotton.

The production of more cotton than can be marketed advantageously is, of course, not sound business practice and may result in considerable financial loss to the grower. As a rule, when there is a prospect of overproduction the following year, considerable effort is made by newspapers, the extension force of the U. S. Department of Agriculture, chambers of commerce, and other public-spirited organizations and individuals to persuade the farmers to curtail their acreage. Their efforts may have some effect, especially on the farmers that are in a position to diversify their crops.

Ever Normal Granary Plan.—During good crop years and during years following periods of high prices, there is a tendency for a greater production of agricultural crops than the public can consume and, consequently, more than can be marketed at a fair price. On the other hand, during other years when crops are partial failures, there is a scarcity of certain products, and prices become so high that the consumer suffers. The purpose of the Ever Normal Granary Plan advocated by Henry A. Wallace, Secretary of Agriculture, is to provide a means by which some of the commodity can be held in reserve and kept off the market until there is need for it. This surplus that has been held in reserve can be used to advantage after bad crop years.

Secretary Wallace has been accused by some of his opponents of trying to create a scarcity of certain agricultural commodities in order to raise prices. In reply he says:

> As a matter of fact, our policy has been steadily exerted, not toward scarcity but in the direction of greater abundance controlled and balanced.
>
> The corn and cotton loans constituted an approach to the ever-normal granary plan. Farmers received a large measure of benefit from these loans. Quantities of corn and cotton which would otherwise have acted as price-depressing surpluses remained in the hands of the producers until the market was able to absorb them. The corn loans were also effective in establishing a reserve of feed which became valuable in carrying corn-hog farmers through the drouth of 1934.
>
> This concept of balanced abundance looks beyond the immediate present and the near future in its design to provide a margin of supply of food and fiber and a fair price for farmers. It looks to the safety of the land itself.

Lack of Thrift in the Cotton Belt.—Many families living on farms in the Cotton Belt have little or no garden, although with slight effort they might have an abundance of fresh vegetables nearly the whole year. An orchard is a rarity, yet peaches, pears, grapes, figs, plums, blackberries, and some other fruits can be grown in abundance. It is the practice of many of the tenant farmers to spend money freely if they have it and to use poor management in planning their business.

Efficiency in Farm Management.—Experience on cotton farms the last few years has demonstrated that it is best to provide for growing needed home supplies first and then plant the rest of the land in cotton. The most successful farmers are the ones who grow the feed needed for animals on the farm; keep plenty of cows, pigs, and hens; and grow corn, potatoes, cowpeas, and garden vegetables. Such farmers have well-fed families; and although only a moderate amount of cotton is grown, the money it brings is clear money—it need not be paid out for supplies.

On a well-organized farm where diversification is practiced, living conditions are better than on a one-crop farm; expenses may be kept down to a minimum; labor and work animals may be employed a large part of the time on the different crops; and the productiveness of the soil may be maintained. Where all-cotton farming is practiced, the soil is sure to decrease in richness after a time.

That the advantage of diversification is not mere theory is shown by many practical illustrations. An interesting one is given in some detail by Goodrich.[6] He describes the work and gives figures on the operations of a farm in South Carolina. The owner bought a farm of 131¾ acres which had been rented for 8 or 10 years previously and was badly run down. The fields were irregular from bushes along their boundaries and contained some bad gullies. The land was so poor that it had been making only 300 pounds of seed cotton or 5 to 8 bushels of corn to the acre.

The new owner of the farm filled the gullies, straightened the fields by removing brush along the edges, plowed deep, bought some barnyard manure and kept some hogs and cattle to make more, used commercial fertilizers freely, cultivated the crops well, and used the following rotation:

First year, corn, with cowpeas planted between the rows at last cultivation.

Second year, winter oats, followed by cowpeas for hay.

Third year, cotton.

The cultivated land was in three fields of 22 acres each.

From 5 to 7 tons of manure per acre was put in furrows upon which cotton beds were made, and from 700 to 1,200 pounds of commercial fertilizer was used at planting time. This fertilizer consisted of acid phosphate, cottonseed meal, and muriate of potash in the proportion of 5–5–2. About 150 to 200 pounds of nitrate of soda was also applied to the cotton about July 1. Corn was fertilized with commercial fertilizers also, about 500 pounds of a complete fertilizer being used. The oats were fertilized

TABLE XLIII.—CROPPING SYSTEM AND CROP YIELDS, 1902–1908 INCLUSIVE

Year	Field A		Field B		Field C	
	Crop	Yield	Crop	Yield	Crop	Yield
1902	Cotton	1.6 bales	Corn[a]	37 bushels	Oats[b]	10 bushels
1903	Corn[a]	50 bushels	Cotton	1.8 bales	Oats[b]	45 bushels
1904	Oats[b]	60 bushels	Corn[a]	65 bushels	Cotton	1.68 bales
1905	Cotton	2.14 bales	Oats[b]	75 bushels	Corn[a]	62 bushels
1906	Corn[a]	75 bushels	Cotton	2 bales	Oats[b]	60 bushels
1907	Oats[b]	75 bushels	Corn[a]	85 bushels	Cotton	1.7 bales
1908	Cotton	2.27 bales	Oats[b]	80 bushels	Corn	62 bushels

[a] Corn with cowpeas. [b] Oats followed by cowpeas.

differently different years, but the most common treatment was 200 pounds of nitrate of soda applied in the spring.

That the cropping system used was a success is well shown by Table XLIII, which gives acre yields for 7 years. Prior to 1902,

TABLE XLIV.—TOTAL COST OF PRODUCTION AND VALUE OF PRODUCTS, 1908

Total cost of farm operation in 1908	$2,855.04
Value of Products	
60 bales cotton, at $42.50	2,125.00
25 tons cotton seed exchanged for 18¾ tons cottonseed meal, at $24	450.00
1,800 bushels oats, at 75 cents a bushel	1,350.00
54 tons oat straw, at $10	540.00
44 tons cowpea hay, at $15	660.00
1,364 bushels corn, at $1	1,364.00
2¾ tons pulled fodder, at $25	68.75
Total value crops produced	$6,557.75
4,000 pounds pork was produced having a market value of	$ 320.00
Estimated cost of production of pork	160.00
Net proceeds on pork	160.00
Net proceeds from farm	$3,862.71

the yields were about 100 pounds of lint cotton and 5 to 8 bushels of corn.

Table XLIV shows the value of products in 1908 and the total cost of production. In the costs are figured labor, fertilizer, manure, seed, interest on money invested in land and machinery, and miscellaneous articles.

References

1. HAMMOND, M. B.: "The Cotton Industry," The Macmillan Company, New York, 1897.
2. MOORHOUSE, L. A. and COOPER, M. R.: "The Cost of Producing Cotton," *U. S. Dept. Agr. Bull.* 896, 1920.
3. AGELASTO, A. M. *et al.*: "The Cotton Situation," *Year Book, U. S. Dept. Agr.*, 1921.
4. COOPER, M. R. and HAWLEY, C. R.: "Cost of Producing Field Crops, 1923," *U. S. Dept. Agr. Cir.* 340, 1925.
5. WILLARD, REX E.: "A Farm Management Study of Cotton Farms of Ellis County, Texas," *U. S. Dept. Agr. Bull.* 659, 1918.
6. GOODRICH, C. L.: "A Profitable Cotton Farm," *U. S. Dept. Agr. Farmers' Bull.* 364, 1915.
7. MOORHOUSE, L. A. and JUVE, O. A.: "Labor and Material Requirements of Field Crops," *U. S. Dept. Agr. Bull.* 1000, 1921.
8. SOULE, A. M.: "Some Factors Affecting the Economic Production of Cotton," *Ga. State Coll. Agr. Bull.* 247.
9. SOULE, A. M.: "A Cotton and Live Stock Production Program for 1923," *Ga. State Coll. Agr. Bull.* 278, 1923.
10. NEWMAN, C. L.: "Some Conditions Influencing Cotton Production," *S. C. Agr. Exp. Sta. Bull.* 140, 1908.
11. BENNETT, R. L.: "Cotton Crop Mortgage Credit," *Texas A. & M. Coll. Ext. Service Bull.* 4.
12. SANDERS, J. T.: "How Much Cotton Should America Plant?" *Current Farm Economics*, Oklahoma Agricultural Experiment Station, vol. 8, No. 5, 1935.
13. GARMAN, C. G.: "Factors Related to Income and Costs of Production on Farms in Marshall and DeKalb Counties, Alabama, 1927–1929," *Ala. Agr. Exp. Sta. Bull.* 236, 1932.

CHAPTER XXII

COTTONSEED PRODUCTS AND OIL-MILL PROCESSES

During the year 1935–1936, the cottonseed-oil mills in the United States crushed 4,775,000 tons of cotton seed. The average price received for these seed was $31.60 per ton, and the total farm value $150,877,000. The average annual receipts for the ten years 1927–1936 were 6,120,600 tons. This does not represent the total production of cotton seed, since about 20 per cent of the seed produced on the farms each year is reserved for planting purposes.

During the year 1935, the oil mills of the country produced approximately 554,000 tons of crude oil, 1,614,000 tons of cake and cottonseed meal, 805,000 bales of linters, and 912,000 tons of cottonseed hulls. In addition to the products just mentioned, the oil mills remove from cotton seed each year some 60,000 bales of hull fiber and 20,000 bales of inferior lint known as "grabbots." All these products have an annual value in excess of $200,000,000.

Prior to the advent of the cottonseed-oil mill, or before the Civil War, cotton seed were considered of little value and were largely thrown away. Aside from the seed reserved for planting and the small amount for fertilizers and cattle feed, no use was made of them. They were piled up and allowed to rot or dumped into a creek. They were often considered a nuisance around gins, and some state laws were enacted prohibiting gins from allowing them to accumulate if the gin was near a town.

All this has changed now. Cotton seed are all saved with care and marketed. Cotton seed and cottonseed products are among the main assets of the South. Lint, fuzz, hulls, oil—in fact, every part of the seed except the sand and dirt clinging to it—are used.

Parts of a Cotton Seed.—The seeds of nearly all the varieties of American upland cotton have feltlike covering, the fuzz, which consists of numerous short epidermal hairs. Very commonly there are also small tufts of the longer fibers, lint which

Table XLV.—Oil and Protein Content of Seed of Varieties of Cotton, Grown at College Station, Mississippi 1917

(*After Brown and Anders*)

Variety	Per cent of meats or kernels	Per cent of hulls	Per cent of moisture in seed	Per cent of oil in meats	Per cent of oil in seed	Total gallons oil per ton	Per cent of ammonia in meats	Per cent of ammonia in seed	Available oil per ton in clean seed, gallons	Pounds of cake per ton from clean seed	Pounds of hulls and linters per ton from clean seed	Length of lint	Per cent of lint
Dodd's Prolific	59.06	40.94	7.89	35.19	20.78	55.40	6.63	3.95	47.98	1,128.52	511.63	1 1/16	30.4
Trice-271–43	55.50	44.50	7.45	34.00	18.87	50.31	6.64	3.69	42.45	1,054.23	627.40	1 1/16	29.0
Trice-270–41	57.00	43.00	7.79	33.90	19.32	51.51	6.85	3.90	43.20	1,114.23	561.77	1 1/8	30.6
Wannamaker Cleveland	51.74	48.26	7.94	33.91	17.55	46.79	6.98	3.61	39.10	1,031.38	675.37	1	35.6
Cleveland Big Boll	54.64	45.36	7.92	32.80	17.92	47.77	7.25	3.96	39.33	1,130.37	573.65	1 1/16	33.2
Cleveland-641	56.50	43.50	7.21	34.29	19.37	51.64	6.99	3.95	43.22	1,128.52	547.33	1	34.3
Cleveland-43	56.73	43.27	8.16	35.14	19.93	53.13	6.71	3.81	45.00	1,089.50	572.86	1 1/8	32.6
Cook-919	55.74	44.26	7.67	35.06	19.54	52.09	6.72	3.75	44.10	1,071.38	597.87	1	38.4
Miller	53.31	46.69	7.23	31.91	17.01	45.35	6.93	3.69	37.49	1,054.23	664.58	1 1/8	33.4
Rowden	55.50	44.50	6.91	35.43	19.66	52.41	6.54	3.63	44.67	1,037.09	627.88	1 1/8	33.4
Lone Star-15	54.50	45.50	7.83	33.93	18.49	49.29	7.19	3.92	40.94	1,119.94	573.01	1 1/8	35.8
Lone Star-132	57.50	42.50	7.78	34.90	20.07	53.51	6.69	3.85	45.30	1,099.95	560.30	1 1/8	28.4
Lone Star × Express	54.00	46.00	7.57	36.06	19.47	51.91	6.25	3.38	44.71	965.67	699.00	1 1/8	31.5
Express-412	53.60	46.40	7.93	33.81	18.12	48.31	6.89	3.69	40.45	1,054.23	642.39	1 3/16	28.3
Express-122–433	56.00	44.00	7.39	35.11	19.66	52.41	6.71	3.75	44.42	1,071.38	595.47	1 3/16	28.9
Express-314	55.50	44.50	7.74	34.04	18.89	50.36	6.83	3.79	42.28	1,082.80	600.10	1 1/8	29.0
Express-432	55.66	44.34	7.70	34.19	19.03	50.73	6.62	3.68	42.89	1,051.38	626.94	1 3/16	31.7
Express-350	57.50	42.50	7.30	33.84	19.46	51.88	6.81	3.92	43.53	1,119.94	553.58	1 3/16	28.0
Polk	59.17	40.83	7.75	33.63	21.67	57.77	6.07	3.59	50.12	1,025.66	598.44	1 3/16	29.4
Kekchi	55.00	45.00	6.76	36.07	19.84	52.89	6.84	3.76	44.88	1,074.23	589.17	1 3/16	31.8
Foster-120	55.80	44.20	7.52	35.02	19.54	52.09	6.61	3.69	44.23	1,054.23	614.04	1 3/16	30.9
Foster-120–449	57.50	42.50	7.67	34.69	19.95	53.19	6.88	3.96	44.75	1,131.37	533.00	1 3/16	31.7
Columbia (U. S. Department of Agriculture)	56.00	44.00	6.89	34.94	19.57	52.17	6.69	3.75	44.19	1,071.38	597.19	1 1/4	28.1
Columbia (Sherard's)	57.50	42.50	7.56	35.84	20.61	54.95	6.28	3.61	47.26	1,031.38	614.17	1 3/16	28.2
Sunflower	58.00	42.00	7.38	36.04	20.90	55.72	6.46	3.75	47.73	1,071.38	570.64	1 3/8	28.0

the gin failed to remove, clinging to the seeds. Beneath the fuzz is the tough, leathery seed coat, which forms most of the commercial product known as "hulls." The kernel inside the hull consists almost entirely of plant embryo. In the large much-folded cotyledons of the embryo are stored oil and other food materials which the seed contains. The reader is referred to Chap. IV for more detailed discussion of the structure of cotton seeds.

Oil Content of Cotton Seed.—The oil content of cotton seed varies with the variety, locality in which the cotton is grown, soil fertility, fertilizers used, and climatic conditions. Probably the factor causing the greatest difference is the variety. Rast[1] in laboratory studies found a difference of 16.5 gallons of oil per ton between different varieties of cotton grown in Georgia. The difference in quantity that could be extracted in regular mill operation would be somewhat less probably, but still the difference would be large. Brown and Anders[2] in laboratory tests found a difference of 12.42 gallons of oil per ton between different varieties grown on the same plots in a variety test at State College, Mississippi, in 1917. Table XLV, from Brown and Anders,[2] gives considerable data on the content and composition of seed of a number of common varieties of cotton.

In discussing the foregoing table, the authors say:

> Table XLV gives the moisture, oil, and protein content of the seed of twenty-five different strains of cotton grown at the College Station in 1917. The seed of Polk ranked highest in oil content and that of Miller lowest. Polk is a variety with small seed, low lint percentage, long staple, and rather small bolls, while Miller (a strain of Rowden) on the other hand, has large seeds, medium-high lint percentage, and large bolls. It seems that, in general, varieties having the characteristics of Polk tend to have a high oil content, while the shorter, big-boll cottons with higher lint percentage tend to have low oil content.

In further studies made by Brown and Anders,[2] it is shown that there appears to be a positive correlation between length of lint and oil content. In general, the varieties with longest staple have the highest oil content. When lint percentage and oil content are considered, the correlation appears to be negative. As a rule, varieties with a high lint percentage have low oil content. Apparently there is also a negative correlation between the oil content and the size of the seed. The larger seed have a comparatively lower oil content. Varieties with a high percentage of protein rank comparatively low in oil content.

Effect of Environmental Conditions on Oil Content of Cotton Seed.—A great many analyses of cotton seed are made by or for oil mills each year, but, as a rule, the samples of seed tested are from mixed lots or from lots of seed whose history is uncertain or unknown. This being the case, most of the numerous analyses made are of little value in determining relationship between oil content of seed and particular environmental factors.

Creswell and Bidwell,[3] of the U. S. Department of Agriculture, collected data from more than 50,000 analyses of cotton seed. This mass of data reveals several interesting points. It is shown that the average production of oil per ton in the United States for the period 1916–1919 was 306 pounds. The amount ranged from 284 pounds for Texas and Oklahoma to 322 pounds for Mississippi. All the states except Oklahoma and Texas produced more than 300 pounds of oil per ton. It is probable that the more arid climate of these two states had some influence on the oil percentage, but the varieties grown doubtless had greater effect. The average oil production varied considerably in different years, ranging from 296 pounds in 1918–1919 to 314 pounds in 1916–1917. This probably shows some seasonal effect. The average loss in weight per ton due to dirt, trash, and inpurities was 118 pounds. The average loss was highest in South Carolina with 139 pounds and lowest in Texas with 105 pounds.

The quality of cotton seed varies somewhat in different parts of the season. During August and September, they are green and contain much moisture. Late in the season, seed are sometimes damaged by rains or bad weather. The best seed come from November deliveries.

White,[4] of Georgia, in studies to determine the effect of soil fertilizers on the oil content of cotton seed, found that nitrogen and potassium had no effect apparently but that phosphorus increased the percentage of oil considerably, the increase being 3.6 per cent in 1911 and 5 per cent in 1912.

Protein Content of Cotton Seed.—Table XLV, prepared by Brown and Anders,[2] shows percentage of ammonia in the seed and also in the meats of the seed of a number of varieties of cotton where the varieties were grown under practically identical conditions. The percentage of ammonia in the meats ranges from 6.07 for the Polk variety to 7.25 for Cleveland Big Boll. Wells and Smith,[5] of Georgia, report a variation of 1.06 per cent in nitrogen between certain short-staple varieties that they studied in Georgia. Mexican Big Boll was next to lowest with a nitrogen percentage of 5.21, while Wannamaker Cleveland was near the top with a percentage of 6.15.

Cottonseed-oil Mills in the United States.—In 1783, a society in London known as the "Society of Arts" came to the conclusion that the oil contained in cotton seed is of commercial value and voted to encourage an invention by means of which it might be extracted.

In 1826, Benjamin Waring of Columbia, S. C., expressed oil from cotton with crude machinery. This was probably the first instance in America. In 1829, Gen. David R. Williams of Society Hill, S. C., also built a machine for the manufacture of cottonseed oil. From that time until the period of the Civil War, the industry was largely in an experimental stage. In European countries, considerable advancement had been made, but there the problem of oil extraction was not such a difficult one, since the millers crushed seed of Egyptian and Sea Island cotton—varieties with smooth seeds. The seed of most American varieties were covered with a heavy coating of fuzz which absorbed much oil when the seed were compressed. At the outbreak of the Civil War, there were but seven oil mills in the United States. By 1870, the number had increased to 26.

From 1870 to 1890, improvements in the extraction of crude oil as well as in its refinement came along steadily. More efficient delinting machines were invented, and edible oils were discovered, which found a ready market in foreign countries. By 1890, the number of oil mills had increased to 119; the production of crude oil increased from 547,165 gallons in 1872 to 13,384,385 gallons in 1890. The amount of the cotton seed crushed increased from 4 to 25 per cent. The value of a ton of seed increased from 40 cents to $4.80. The total value of the products obtained from the seed increased from $293,546 to $5,291,178.

From 1890 to 1900, several further improvements in the methods of extracting oil were made, including better methods of cooking the meats and the use of the hydraulic press for expressing the oil. The combining of cottonseed oil with beef fat to make lard compound and artificial butter and its further use for packing sardines increased the demand for refined oil. The number of mills increased to 357, and their combined output at the close of the period represented 46,902,390 gallons of oil annually. The value of seed had risen to $11.60 per ton, and 53 per cent of the production was crushed. By 1900, other cottonseed products had become of much importance. In that year, the oil mills produced 884,000 tons of cottonseed meal, 1,169,000 tons of hulls, and 114,544 bales of linters.

From 1900 to the present time, still other improvements in manufacturing have been made, and new uses found for the products. Processes have been discovered by means of which the oil can be hardened and deodorized, thus increasing its use for human consumption. It has been found that the meal and hulls, if properly fed, afford excellent food for animals. Quantities of linters have been used in making explosives, and a great deal is used in making rayon, or artificial silk. There are now about 500 oil mills in the South, which crush about 80 per cent of the cotton seed grown. The other 20 per cent is reserved for planting. The total value of the product from these mills, as was shown, represents an annual value of more than $200,000,000. This is an enormous sum, when the fact is considered that a few decades ago cotton seed was considered practically worthless.

Marketing Cotton Seed.—The price paid by oil mills for cotton seed is determined largely by the value of oil and other seed products in the larger markets of the country. The mill knows within a narrow margin just what it can get out of a ton of seed, and it knows, too, the cost of milling and so fixes the price per ton accordingly. Competition for seed is keen, and the margin of profit is narrow.

Some seed is hauled in wagons from gins near by, but the greater part received by larger mills comes in carlots. Larger plantations near railroads usually so locate their gins that they can blow seed from the gin into a car and sell by the carload. Most of the larger mills have buyers at railway stations remote from oil mills who buy small lots of seed and ship by the carload. The price paid for wagon lots is about $3 a ton less than for carlots.

When a car of seed arrives at the mill, it is sampled in different parts. These samples are put together and a composite sample taken from them to represent the car. This sample is tested by a disinterested party for oil, foreign matter, moisture, and damaged and immature seed.

Official Standards for Cotton Seed.—The Secretary of Agriculture, by virtue of the authority vested in him by an act of Congress in 1931, established the following rules and specifications for cotton seed bought or sold for crushing purposes:

Section 1. The grade of cottonseed shall be determined from the analysis of samples, and it shall be the result, stated as the nearest whole number without fractions obtained by multiplying a Quantity Index by a Quality Index as hereinafter provided.

a. The Basis Grade of cottonseed shall be Grade 100.

b. High grades of cottonseed shall be those grades above 100.

c. Low grades of cottonseed shall be those grades below 100.

Section 2. The following equations shall be used in determining the Quality Index of cottonseed.

a. For cottonseed that by analysis contain not less than 17 percent of oil the Quality Index shall equal four times the percentage of oil, plus six times the percentage of ammonia, plus five.

b. For cottonseed that by analysis contain less than 17 percent oil the Quality Index shall equal five times the percentage of oil, plus six times the percentage of ammonia, minus twelve.

Section 3. The Quality Index of cottonseed shall be percentage of purity and soundness, and shall be determined as follows:

a. Superior Quality Cottonseed.—Cottonseed that by analysis contain less than one-half percent foreign matter, and more than 8 percent but less than 10 percent moisture, and less than three-quarter percent free fatty acids in the oil in the seed, shall be known as Superior Quality Cottonseed and shall have a Quality Index of 102 percent.

b. Prime Quality Cottonseed.—Cottonseed that by analysis contain not more than 3 percent foreign matter, not more than 12 percent moisture, and not more than 1.8 percent free fatty acids in the oil in the seed, shall be known as Prime Quality Cottonseed and shall have a Quality Index of 100 per cent.

c. Sub-quality Cottonseed.—Cottonseed that by analysis contain foreign matter, moisture, and/or free fatty acids in the oil in the seed in excess of the percentages shown in Section 3-*b*, or are seed from seed cotton that has been processed in a boll breaker or other device for preparing snapped cotton or bollies for ginning, shall be known as sub-quality cottonseed; and the Quality Index of such cottonseed shall be found by reducing the Quality Index of Prime Quality Cottonseed as follows:

1. Not to exceed five-tenths of a unit for each 0.1 percent of free fatty acids in the oil in the seed in excess of 1.8 percent; provided that this reduction shall not exceed 50 units of the Quality Index of Prime Quality Cottonseed.

2. Not to exceed one unit for each one percent of foreign matter in excess of 3 percent.

3. Not to exceed one unit for each one percent of moisture in excess of 12 percent.

4. Not to exceed eight units when the seed cotton has been processed as snapped cotton or bollies before ginning.

d. Off-quality Cottonseed.—Cottonseed that have been treated by either 'mechanical or chemical process other than the usual cleaning, drying, and ginning (except such sterilization as may be required by the United States Department of Agriculture for quarantine purposes) or that are hot or fermented, or that upon analysis are found to contain more than 25 percent foreign matter, or more than 25 percent moisture, or more than 40 percent combined moisture and foreign matter, shall be known as Off-quality Cottonseed and may not be graded.

Section 4. Sampling of Cottonseed.—In the application of these standards the following methods shall be observed in the drawing and handling of cottonseed.

a. Sampling before Unloading.—Portions shall be drawn at different points in each end and in the middle of the car with a suitable cottonseed trier or sampling device. In drawing samples with a trier, cross sections shall be taken from the top to the bottom of the car, if possible. In the absence of a trier, holes shall be dug at various points at least 30 inches deep with a small (8-tine) fork and portions taken from the bottom and sides of these holes.

b. Sampling during Unloading.—For this purpose the sampler shall be provided with suitable receptacles, which he shall place in the center of the unloading chute at regular intervals, as the seed are being ejected from the car, to receive portions of the seed.

Whether drawn before or during unloading the several portions drawn from car-lots shall total not less than 50 pounds in weight.

c. Sampling of Truck or Wagon Seed.—In drawing samples of truck or wagon loads of cottonseed, the same methods shall be used as in sampling car-lots before unloading. The total weight of the portion drawn shall be not less than 2½ pounds for each ton of seed in the load.

d. Handling Samples.—Samplers shall be provided with metal containers with close-fitting covers large enough to hold 60 or 70 pounds of cottonseed. Each portion of a sample as drawn shall be immediately placed in such a receptacle and the cover promptly replaced. As soon as the full sample has been taken, it shall be carefully weighed, then cleaned of foreign matter, and carefully reweighed. The loss in weight shall be calculated as foreign matter. After the sample is cleaned, the seed shall be mixed either by means of a suitable mechanical mixer or by heaping together and mixing by passing the hands or a small shovel up through the heap, repiling and spreading by pressing. Finally, not less than two quarts shall be packed in an air-tight tin can or Mason jar and sent to the laboratory for analysis and grading. All cleaning, mixing, and handling of samples shall be done expeditiously and without undue exposure.

Section 5. Analysis.—The methods for analyzing cottonseed recommended from time to time by the inter-bureau committee of this department on standard methods of sampling and analyzing cottonseed shall be used.

According to the rules of the National Cottonseed Products Association, deductions may be made in the price to be paid for the seed if they grade below Prime Quality cottonseed as specified in the Official Standards for Cottonseed.

Storage of Cotton Seed.—The storage of seed is a problem of considerable importance to oil mills. Large quantities of seed are needed to keep the mill going, but if large amounts are bought and stored in bulk or loose, they are apt to heat and deteriorate. This is especially true if the seed are green or contain an undue amount of moisture. The heating results in a decided increase in the free fatty acid in the oil. Rather[6] found that seed in storage which heated to a temperature of 111°F. increased in free fatty acid content from 2.26 to 11.89 per cent. Free fatty acid in cottonseed oil is objectionable and must be removed in refining processes. If present in the oil in quantity, there is considerable loss. If the heating of the seed is considerable, there may be a further chemical reaction which will result in loss of nitrogen and thus lower the value of the meal that is to be made from the seed.

If the seed cotton is allowed to remain in the small cotton houses on the plantations for 2 or 3 weeks after picking, or until thoroughly dry, there is not so much danger of the seeds' heating in the milling processes later. Or if the green seed from the gin are stored in small lots, say a ton or less, for a few weeks, there is not much heating later. Mills usually have large seed houses, so that the seed may be scattered throughout considerable space, and air forced through them. As far as possible, bulk seed received by the mills are crushed soon after they arrive to avoid deterioration. If a lot of seed is beginning to heat, it is thrown into the conveyors and used at once.

Oil-mill Processes and Machines.—The following discussion of the various machines used in oil mills must be considered to apply in general. Nearly every mill will have one or more machines of a type different from the one described, and the process followed may be somewhat different from the one outlined.

FIG. 119.—Sand-and-boll screen. (*Courtesy Atlanta Utility Works.*)

In the limited space available, an exhaustive discussion is, of course, not possible.

Cleaning Seed.—When a lot of cotton seed arrives at the oil mill, it contains considerable foreign matter, which must be removed before the seed are put through the milling processes. These impurities include dirt, sand, cotton burs, locks, unginned seed, pieces of metal, and other foreign matter. To remove these, the seed are put through different machines which will be described briefly.

The seed are transferred from the seed house to the first cleaning machine, the sand-and-boll screen, by an auger conveyor, which runs the whole length of the house, and by bucket elevators. The sand-and-boll screen consists essentially of an inclined reel covered with revolving screens of various-sized mesh (Fig. 119). The seed are fed into the higher end of the reel, and as they are moved toward the lower end there is a selective separation of seed and foreign matter according to size. The unginned cotton locks and burs pass on through the reel. This miscellaneous mass of material is collected and saved for ginning. The lint obtained is, when ginned, of low grade and inferior but is in sufficient quantity to be of commercial importance. This inferior lint, as was mentioned previously, is known as "grabbots." Dirt and sand pass through fine openings in the screen in the upper part of the reel, while the seed pass through larger meshes in the lower part. From the sand-and-boll screen the seed are commonly passed over a shaker screen, which removes other foreign matter. In this machine, or in another adjacent to it, the seed are subjected to the action of a blower, a strong current of air, which separates out other foreign matter yet remaining. The size and the pattern of cleaner used vary considerably, of course, in different mills, depending on the capacity of the mill and whether or not the most recent types of machines are used.

Delinting.—After the seed are cleaned as thoroughly as possible, they are passed on to machines known as "delinters" for the removal of a part of the fuzz clinging to the seed. These machines work on the same principle as cotton gins and are similar in structure. They differ from the cotton gin in that the saws have finer teeth and are set closer together.

Without the linters the hulls are of higher value as livestock feed. Not all the fuzz is removed from the seed; some of it is left to cause the hulls to cling together more or less when they are separated from the meats in the hulling process. From 40 to about 80 pounds of linters per ton are removed from seed in ordinary commercial operations, but it is possible to remove as much as 200 pounds, as was sometimes done during the World War, when there was a great demand for linters for making explosives. The average cut is about 65 pounds. As linters are worth from 3 to 6 cents a pound, their commercial value is another reason for their removal from the seed.

All American upland varieties of cotton grown at present, or at least all that are grown extensively, have seed covered with fuzz, which makes delinting necessary. Sea Island cotton and certain foreign varieties have smooth seeds, which require no delinting, but because of this lack of fuzz it is difficult to remove the hulls from the meats. The seed of these varieties are frequently compressed without removing the hulls. This method gives a lower grade of oil, since it is discolored by coloring matter from the hulls.

Hulling.—After the seed have had the required amount of fuzz removed by the delinting machines, they are next carried to a machine known as a "huller" for removal of the hull from the kernel, or meat. This is accomplished by subjecting them

Fig. 120.—Disk cotton-seed huller.

to the action of a machine containing two sets of knives, one stationary and the other on a rapidly revolving plate or cylinder. The knives are so adjusted that when the plate revolves they almost touch (Fig. 120). The huller cracks, breaks, or cuts the hull and kernel so that the kernels may be separated from the hulls. It must not mash or compress the seed, for that would result in a loss of oil from the kernel. If the seed is dry, the kernel is loose inside and is easily separated from the hull, which is then brittle and easily cut or broken. If the seed is damp, the hull is tough, and the seed is liable to be crushed before the hull can be cut.

Separating Meats and Hulls.—The seed that come from the huller are cut into rather small pieces. The next step in the milling process is to separate these pieces of hulls and meats.

This is accomplished by passing the material over a shaker screen, which has perforations large enough for the pieces of meats and but little of the hulls to pass through.

The hulls are next carried to a beater, where they are pounded up to loosen any additional pieces of meats that may be clinging inside. These remaining pieces of kernel are separated from the hulls by perforated screens over which the hulls pass on their way to the hull bin. In the bin the hulls are next sacked in 100-pound bags for the market.

FIG. 121.—Cotton-seed crusher. (*Courtesy French Oil Mill Machinery Company.*)

The meats, in which there is still a small amount of little pieces of hulls, are next passed under one or two suction blasts, which may be set so as to remove just the quantity of hulls desired. By this means the ammonia content of the meal is regulated. The meal made from pure kernels is too rich in nitrogen to meet trade demands.

Crushing.—From the hull and meat separator the cleaned kernels are transferred to a crusher. This consists essentially of a series of large iron rollers placed one above another and so arranged that they revolve in contact (Fig. 121). The kernels are fed between the two upper rollers and pass in succession between every pair. The object of this crushing is to rupture every cell in the kernel structure, so that the cell contents may cook evenly and the oil within the cell may find easy egress.

Fig. 122.—Cooker. (*Courtesy French Oil Mill Machinery Company.*)

Cooking.—From the crusher the pulverized kernels are carried to a heater, or cooker. This consists of a set of four or five steam-jacketed iron drums having a diameter of 52 to 72 inches and a depth of 14 to 18 inches. These drums are stacked one above the other and resemble a large iron boiler built in sections (Fig. 122). The bottom of each drum contains a trap door, through which the contents of an upper drum may pass to the one below. Each drum is also equipped with a pressure gauge to govern the steam or temperature, a steam trap for the removal of moisture, and an agitator to stir the meats.

The meats are first placed in the top drum where they are heated to a temperature of about 140°F. The temperature is gradually increased in successive drums until a temperature of 220°F. is reached in the bottom drum. The meats are allowed to remain in each drum from 15 to 40 minutes, the length of time varying with the moisture in the seed which must be expelled. If the seed are very dry, it may be necessary to moisten them slightly with steam to facilitate cooking. Cooking is a particular

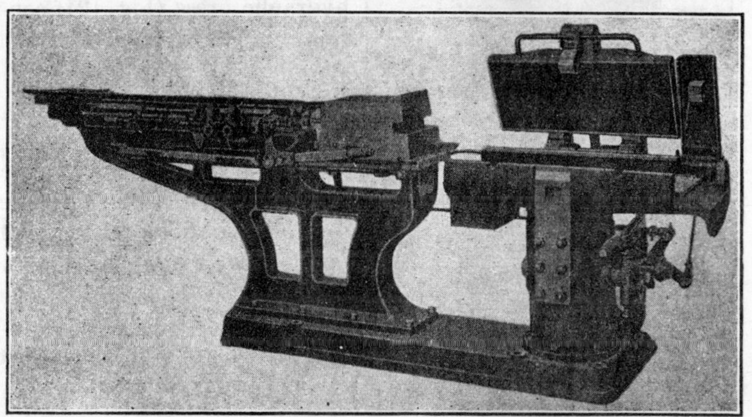

Fig. 123.—Cake former. (*Courtesy French Oil Mill Machinery Company.*)

process. Success depends largely on the judgment of the cook. There are no definite rules that can be followed.

Cooking the meat pulp serves several purposes. It drives out excess moisture, modifies the consistency of the pump so that the maximum amount of oil may be extracted, increases the fluidity of the oil, and coagulates the albumen of the seed.

Cake Forming.—From the cooker, the seed pulp is transferred to a receptacle known as the "subheater," which holds the material and preserves its temperature until it is placed in the cake former (Fig. 123). The cake former consists of a carriage which takes a measured quantity of the cooked meats from the subheater to a cast-iron mold, usually 14 by 32 inches in breadth and length. This mold is lined with a strong press cloth made of animal hair. After the pulp is put into the mold, the cloth is folded over its upper surface so that it is entirely surrounded by the cloth. A limited amount of pressure is then applied to the wrapped mass, sufficient to set the cake but not enough to start the oil within the pulp to flowing. After the pressure is removed, the iron tray containing the wrapped pulp is lifted by hand and placed in an empty compartment of the hydraulic press. The tray is then withdrawn, the wrapped pulp being left for compression.

FIG. 124.—Hydraulic press. (*Courtesy French Oil Mill Machinery Company.*)

Hydraulic Press.—The hydraulic press (Fig. 124) contains 15 compartments, one on top of another, into which the trays containing cooked meats are inserted. Pressure is applied from below by means of a hydraulic ram. The pressure is applied gradually, about 400 pounds per square inch until the slack is taken up and the cakes set, and then it is increased to about 4,000 pounds per square inch. The length of time in the press varies with the relation of the capacity of the cooker and presses but is usually 24 to 30 minutes. A large mill has several presses like the one shown in Fig. 124. Figure 125 shows a view in a mill in Japan where soybeans are crushed for oil. The equipment is the same as for cotton seed. At the extreme left is a cooker; between the workmen is a cake former; and at the right are several presses.

The crude oil from the presses is carried by gravity to a tank below. From here it passes through a filter press or settling tanks into storage tanks or to large tanks on cars for shipment to refineries.

Cake Grinding.—After the oil is extracted, the pulp left is removed from the press and unwrapped. It is now in the form of firm cakes weighing 12 to 14 pounds each. These cakes are stacked to dry for at least 24 hours. Some of the cake is sold on the market and used for feed for livestock in this form. As a rule, the cakes are ground to form a meal known as "cottonseed meal." The cakes are first broken into small pieces by a machine

FIG. 125.—Press room, Nisshin Oil Mills, Yokohama, Japan. (*Courtesy French Oil Mill Machinery Company.*)

known as a "cake breaker" and next ground into a fine meal by a grinding mill. The meal is sacked in 100-pound bags for the market.

Products from a Ton of Cotton Seed.—Sheets and Thompson[8] give the following figures for the quantity of raw products obtained from a ton of cotton seed. They are based on Bureau of Census figures for the years 1914–1915 to 1918–1919 inclusive.

	Pounds
Linters, or short fiber	110
Hulls	514
Cake or meal	954
Crude oil	303
Loss in manufacture	119
Total	2,000

Since the figures cover the war period, during which a heavy cut of linters was made, the amount of linters given is considerably above the average obtained, which is about 65 pounds. This extra weight of linters, about 45 pounds, ordinarily goes to increase the weight of the hulls. The average hull weight at present is about 559 pounds.

Linters.—Of the several crude products from cotton seed, linters is one of the most important. The origin and development of cotton fuzz, or linters, were discussed in Chap. VII, and methods used in delinting seed were given on a preceding page of this chapter. The character, uses, and marketing of linters will now be considered.

The fiber of linters is very short, ranging from ⅛ to ½ inch in length, but there is also material mixed with the fibers that is too short to measure. Linters vary in color, ranging from gray, yellow, and green to brown. The color depends largely on the closeness of the cut. If much fuzz is removed from the seed, the color is darker, because more coloring matter is obtained from seed coat, and there is more dust and other foreign matter mixed with the linters. As brought out by Agelasto,[7] the oil mill can vary the grade of linters at any time by varying the number of pounds cut from a ton of seed.

In 1900, the United States produced 114,000 bales of linters. This was 1.2 per cent of the cotton crop. The production gradually increased until in 1910 it was 313,000 bales, which was 2.9 per cent of the crop. Following this, the increase was gradual until the period of the World War, when the production increased greatly; in 1917, it was 1,331,000 bales, which was 10.9 per cent of the cotton crop for that year. Since the war, the production has decreased somewhat, but it is still considerable. From Aug. 1, 1934, to Aug. 1, 1935, 805,000 bales was ginned.

Linters are packed in 600-pound bales very similar to cotton bales. Samples are drawn from the bales in much the same way as they are taken from ordinary bales except that with linters they are taken from all four sides of the bale. Bales of linters are bought and sold by certain cotton factors. Trading is mostly done on the basis of types or samples presented.

There are numerous uses to which linters are applied, and the number is increasing. The following outline from Agelasto[7] gives in brief most of the uses. Most of the product is used, however, in making explosives and mattresses.

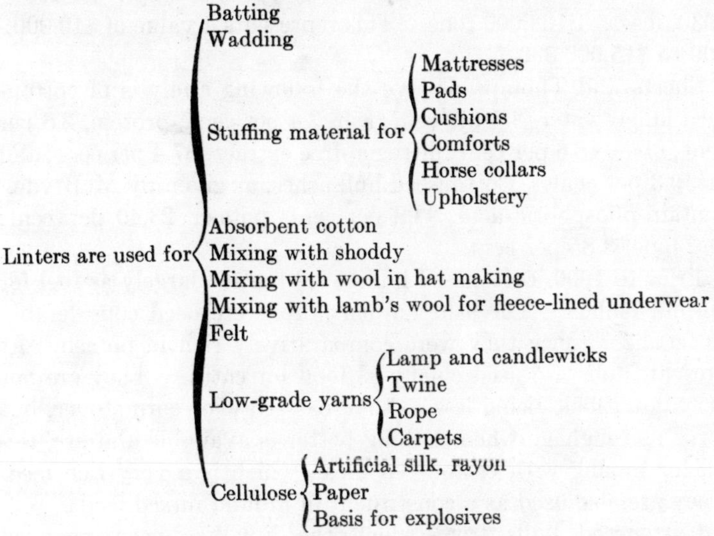

Grades of Linters.—In 1918, a committee from the U. S. Department of Agriculture, headed by G. L. Meloy,[11] began the study of cotton linters with the idea of establishing grades for their classification. Not much was accomplished at that time, but in July, 1924, the studies were resumed and, with the aid of consumers of linters and various other commercial interests, were carried to completion. Seven grades were devised and designated as U. S. Grade No. 1, U. S. Grade No. 2, U. S. Grade No. 3, etc., and illustrative boxes of practical standards prepared for sale. Each of these boxes contains 12 samples which illustrate the range in character, color, foreign matter, neps, staple length, etc., for linters produced in three divisions of the Cotton Belt—the eastern states, the valley section, and the western states. Acting under the Cotton Standards Act of Congress, the Secretary of Agriculture, on July 7, 1925, issued an order that on Aug. 1, 1926, the seven grades mentioned above would become the official standard grades of the United States for American cotton linters.

Cottonseed Hulls.—As was mentioned on a preceding page, under the present milling system a ton of cotton seed yields about

559 pounds of hulls. The amount or weight varies with the cut of linters and the grade of meal made. The total production in the United States is about 1,000,000 tons annually. In 1934–1935, it was 1,103,000 tons. This represents a value of $10,000,000 to $15,000,000.

Sheets and Thompson[8] give the following analysis of cottonseed hulls: water, 8.5 per cent; ash, 2.4 per cent; protein, 2.8 per cent; fiber, 48.6 per cent; nitrogen-free extract, 37.4 per cent; and fat, 0.3 per cent. Cottonseed hull ashes, as given by McBryde,[9] contain phosphoric acid, 9.08 per cent; potash, 23.40 per cent; and lime, 8.85 per cent.

Prior to 1900, cottonseed hulls were burned largely as fuel for oil-mill boilers. The ashes, when pure, were used considerably as fertilizer, since they were comparatively rich in potash. At present, hulls are used chiefly as food for cattle. They are not very nutritious, being less so than oat straw or corn stover, but serve as roughage when nothing better is available and are useful for mixing with cottonseed meal, which is a very rich feed. They are also used as a constituent of ground mixed feeds.

Cottonseed hulls from which the lint has been removed are sometimes ground to make what is known as "cottonseed-hull bran." The feeding value of the bran is similar to that of hulls.

Cottonseed Meal.—In a preceding part of this chapter, the manufacture of cottonseed cake and the grinding of cake to make meal were explained.

Cottonseed meal, the grades of meal, and certain other cottonseed products are defined and classified as follows by the Association of Feed Control Officials of the United States:

Cottonseed Meal.—Cottonseed meal is a product of the cotton seed only, composed principally of the kernel, with such portions of the fiber or hull and oil as may be left in the course of manufacture, and shall be graded and classed as follows:

Cottonseed Meal, Prime Quality.—Cottonseed meal, Prime Quality, must be finely ground, not necessarily bolted, of sweet odor, reasonably bright in color, yellowish, not brown or reddish, not artificially colored, free from excess of lint, and shall contain not less than 36 per cent of protein, 7 per cent of ammonia, or 5.76 per cent of nitrogen. It shall be designated and sold according to its protein, its ammonia, or its nitrogen content.

Cottonseed meal with 36 per cent of protein, 7 per cent of ammonia, or 5.76 per cent of nitrogen shall be termed "36 per cent Protein Cottonseed Meal, Prime Quality," or "7 per cent Ammonia Cottonseed Meal, Prime Quality"; or "5.76 per cent Nitrogen Cottonseed Meal, Prime Quality"; and higher grades similarly designated.

Munsell Color Standard for Prime Cottonseed Cake and Meal.—Whenever in these rules reference is made to color of prime cottonseed cake or prime cottonseed meal, 41 per cent or higher protein cottonseed cake or cottonseed meal shall not be darker in color than Munsell Color Standard "1½ yellow, ⅘"; and 36 per cent protein prime cottonseed cake or cottonseed meal shall not be darker in color than Munsell Color Standard "10 yellow-red, ⅘."

Cottonseed Meal, Off Quality.—Cottonseed meal not fulfilling the preceding requirements as to color, odor, and texture, or weevily cottonseed meal shall be graded Cottonseed Meal, Off Quality.

Whole-pressed Cottonseed.—Whole-pressed cottonseed is the product resulting from subjecting the whole undecorticated cotton seed to the expeller process for the extraction of oil and includes the entire cotton seed less the oil extracted and the lint removed.

In most oil mills, the cottonseed cake, or meal, is tested on the ammonia basis. To determine the percentage of protein in the cake, the ammonia content given is multiplied by 5.15. To illustrate, if the analysis shows 7.5 per cent ammonia, the protein percentage is 7.5×5.15, which is 38.62, or 38.62 pounds of protein in 100 pounds of cake. If the analysis is given in terms of nitrogen content, the protein percentage is determined by multiplying the nitrogen figures by 6.25.

Sheets and Thompson[8] give the table on page 520 showing nutrients in cotton seed and cottonseed products.

Lamborn[10] gives the fertilizing constituents of cottonseed meal as follows: nitrogen, 6.79 per cent; phosphoric acid, 2.88 per cent; potash, 1.77 per cent. These figures are based on 204 analyses.

The annual production of cottonseed meal and cake in the United States for the 9-year period 1926–1935 was 2,179,000 tons. This represents an enormous value, since the average market value of meal is about $30 a ton and has ranged as high as $67 a ton within the past 20 years.

Use of Cottonseed Meal and Cake.—The chief use of cottonseed meal and cake is for feeding livestock. Because of their rich protein content, they are very nutritious and are valuable foods if properly fed. The meal is used in the United States chiefly, but for feeding on the ground or for feeding on the range when the wind is blowing the cake is more desirable. Much cake is used in foreign countries.

TABLE XLVI.—PERCENTAGE COMPOSITION OF COTTONSEED PRODUCTS
(Pounds of nutrients in 100 pounds)

Product	Water	Ash	Crude protein	Carbohydrates		Fat (ether extract)
				Fiber	Nitrogen-free extract	
Cotton seed	6.6	3.7	21.7	19.7	26.9	21.4
Cotton seed meal:						
Choice	6.4	5.9	43.3	10.0	25.7	8.7
Prime	6.6	6.1	38.9	13.2	28.5	6.7
Good	6.5	5.8	37.8	13.6	30.6	5.7
Cold-pressed cotton seed cake	6.2	4.7	25.8	24.0	31.5	7.8
Cotton seed hulls	8.5	2.4	2.8	48.6	37.4	0.3

Cottonseed meal is especially good for cattle, being excellent feed for both dairy and beef animals. Best results are obtained when the meal is given with other feeds, such as corn silage and bran, which are not so rich in protein. It may be fed to mules in limited quantities to good advantage. It may also be fed to horses, sheep, and hogs if fed cautiously. Meal contains a toxic substance, or poison, known as "gossypol." This is injurious to all animals if taken in sufficient quantity for a period of time, but cattle can eat a moderate amount of meal indefinitely without injury. Other animals are more susceptible to the injurious effects. This is especially true of hogs. If meal is fed to them for more than 5 or 6 weeks, harmful results are very likely to follow.

Cottonseed meal was formerly used considerably as a fertilizer, but its present value for feeding purposes is such as practically to prohibit such use.

Crude Cottonseed Oil.—Cottonseed oil as it flows from the press contains some impurities—water, a little meal, albumen, free fatty acids, coloring matter, etc. Most of these impurities may be eliminated or lessened by proper management in the cooking process and by means of settling tanks or filtration. The coloring matter and free fatty acids can be removed at the refinery only.

Cottonseed oil is fluid at ordinary temperatures, but the fatty acids require a temperature of 95 to 122°F. for melting. The freezing point is 53°F. to several degrees lower. When the freezing-point temperature is reached, the stearin in the oil separates from the rest. This is helpful in certain manufacturing processes but is not desirable when the oil is on the way to the refinery. To avoid this separation, tank cars are provided with steam pipes for heating the oil if necessary during cold weather.

Crude cottonseed oil is classified by the National Cottonseed Products Association as "Prime," "Basis Prime," "Off," "Reddish Off," "Low Grade," "Cold Pressed," and "Extracted." These different grades are defined in detail in the book of rules issued by the association. The classes and definitions for the classes are changed somewhat from time to time. As defined at present, Prime Crude Cottonseed Oil must be sweet in flavor and odor and must produce, when refined. Prime Summer Yellow Oil. The other grades mentioned above are of somewhat lower value.

The average annual production of oil in the United States for the years 1926–1936, as given by the U. S. Department of Commerce, was 1,469,776,000 pounds. The market value per pound during the past 10 years has ranged from 3.27 to 8.96 cents. The price recently has been 8 to 10 cents a pound.

Refined Cottonseed Oil.—Some oil mills have refineries in connection with the mill, but, as a rule, the oil is shipped to refineries elsewhere. It is transported in tank cars, 60,000 pounds making a carload.

Refined cottonseed oil differs from crude oil in that most of the free fatty acids, coloring matter, and other impurities have been removed from it. In the refining process, the crude oil is

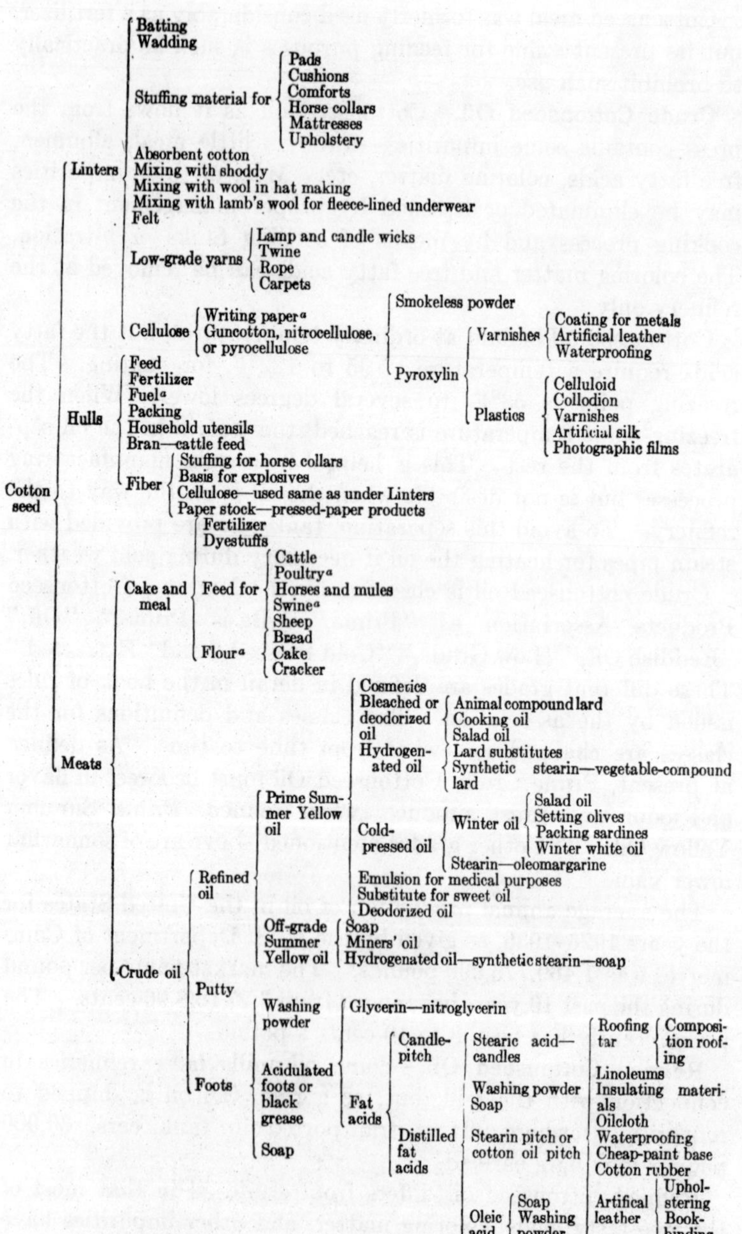

[a] Possible uses to which small quantities only are devoted.

placed in a large iron tank, a solution of caustic soda is poured on it, and the two are stirred and mixed by agitator. After they are mixed thoroughly, the temperature is raised to 120 to 140°F. The impurities go to the bottom as settlings. The purified oil has a yellow appearance and is drawn off. The settlings are sold as soap stock. The yellow oil contains some of the caustic soda. To free it of this alkali, it is run into a finishing tank, where the soda is washed out with water.

Summer Yellow Oil is classified into several different grades: Choice Summer Yellow, Prime Summer Yellow, Good Off Summer Yellow, Off Summer Yellow, Reddish Off Summer Yellow, etc. The classification depends on the color, flavor, odor, and percentage of free fatty acid.

According to the rules of the Interstate Cotton Seed Crushers' Association, Choice Summer Yellow Cottonseed Oil must be sweet in flavor and odor, prime in color, clear and brilliant in appearance, and free from water and settlings, and shall contain not more than one-eighth of 1 per cent free fatty acid.

Use of Cottonseed Oil.—Large quantities of cottonseed oil are mixed with animal fats to make a lard compound which is used as a substitute for hog lard. Much is also used in making oleomargarine, a butter substitute. Low grades of oil, and better grades also if the market price of oil is low, are used in making soap. Some oil is mixed with petroleum to make miners' oil, which miners use in their lamps in the mine. Some is used for packing sardines. A considerable quantity is used also as a table or salad oil.

Numerous Uses of Cottonseed Products.—The number of uses to which different cottonseed products are put is now legion; yet new uses are being discovered yearly. A full discussion of this subject would require a volume. The outline shown on p. 522 published by the U. S. Department of Agriculture will give the reader some notion of the numerous uses.

References

1. Rast, Loy E.: "Oil Content of Cotton Seed as Influenced by Variety and Selection," *Ga. State College Agr. Cir.* 70, 1918.
2. Brown, H. B. and Anders, C. B.: "Cotton Experiments 1919 and 1920," *Miss. Agr. Exp. Sta. Bull.* 187, 1920.
3. Creswell, Charles F. and Bidwell, George L.: "Composition of Cotton Seed," *U. S. Dept. Agr. Bull.* 948, 1921.

4. White, H. C.: "The Feeding of Cotton II," *Ga. Agr. Exp. Sta. Bull.* 114, 1915.
5. Wells, C. A. and Smith, F. H.: "The Nitrogen and Fat in Short Staple Cotton Seed," *Jour. Ind. Eng. Chem.*, vol. 7, No. 3, 1915.
6. Rather, J. B.: "Biochemical Changes in Cotton Seed in Storage," *Ark. Agr. Exp. Sta. Bull.* 125, 1916.
7. Agelasto, A. M.: "Linters," *U. S. Dept. Agr. Cir.* 175, 1921.
8. Sheets, E. W. and Thompson, E. H.: "Feeding Cotton Seed Products to Livestock," *U. S. Dept. Agr. Farmers' Bull.* 1179, 1920.
9. McBryde, J. B.: "Chemistry of Cotton," in "The Cotton Plant," *U. S. Dept. Agr. Bull.* 33, 1896.
10. Lamborn, L. L.: "Cotton Seed Products," D. Van Nostrand Company, New York, 1904.
11. Meloy, G. S.: "Development and Use of Standards for Grade, Color, and Character of American Cotton Linters," *U. S. Dept. Agr. Misc. Pub.* 242, 1936.

CHAPTER XXIII

USES AND SPINNING QUALITIES OF VARIOUS KINDS OF COTTON

The subject of this chapter is a rather difficult one to discuss, because there is much variation in the staple lengths and grades used by different mills in making one kind of goods. The quality of the goods made varies considerably with the grade and the staple of the cotton spun. Changes in price of cotton or in economic conditions bring about some changes in products turned out. It is also difficult to get reliable data on just what is used by different mills in making their products. They feel that this information is a trade secret and are loath to give it out. Also, the methods used in practical work are frequently improved or changed.

How Cotton Yarns Are Measured.—Before cotton fibers can be woven into cloth, they must be spun into threads, or yarns. Yarns differ considerably in size or fineness, the finer grades being used in making laces and delicate fabrics. The size of a yarn is indicated or measured by counts. The count is determined by the number, or count, of hanks of any particular yarn that are required to weigh a pound. A hank of yarn contains 840 yards. To illustrate, a yarn designated as 20's requires 20 hanks of 840 yards each to weigh a pound. The counts range from 1's to 200's in practical work. In rare instances, they may run as high as 700's.

Staple Lengths Used in Making Different Counts of Yarn.—Howarth,[1] on the basis of the size of the yarn, groups yarns into five classes: thick, coarse, medium, medium-fine, and fine. The thick yarns are of counts up to 10's. Short-stapled cottons and cotton waste, that is, short fibers combed from longer cottons by mill machines, are used chiefly in making the thick yarns. The coarse yarns range up to 22's. Short-staple varieties from India and ordinary American short-staple uplands are used in spinning these yarns. The medium yarns have counts that run up to 46's, and staples from $1\frac{1}{16}$ to $1\frac{1}{4}$ inches in length are used in making them. These come from American Benders and Long Staples, Upper Egyptians, and similar cottons. The medium-fine yarns range approximately from 46's to 80's, Egyptian and Sea Island cottons being used in making them. The longest and most uniform staples grown are used in making the fine yarns, whose counts range from 80's up to 400's or higher.

Fabrics Made from Different Counts of Yarn.—In general, coarse fabrics are made from the thicker or heavier yarns, and finer fabrics from the higher counts. The quality of cloth desired determines what is to be used. One kind of goods may vary greatly in quality, ranging from coarse to fine. Some goods made from yarns ranging in counts from 12's to 20's are sheeting, some grades of calico, some domestic, toweling, cords, fustians, corduroys, etc. Shirting and gingham are made from various yarns, the class depending on the fineness of the goods wanted. Sateens are made from yarns ranging from 36's to 80's. Mulls range from 60's to 100's; and cambrics, from 80's to 160's. The yarns of highest counts are used for making laces and the finest fabrics.

Character of Cotton Desired by Spinners.—Although it is true that there is great variation in the kind or quality of cotton wanted by different mills on account of their varied product or the many kinds of goods made, there are certain qualities that are generally sought. These are summarized by Howarth[1] as follows:

First, and most important of all: The fibers must be even in length.

Secondly: The cotton must be ripe.

Thirdly: The grade must be even.

Fourthly: The cotton must be as free as possible from dead cotton, bearded motes, seed, fibers other than cotton, sand, iron, stones, excessive moisture, and other foreign substances.

If long-staple and short-staple cottons are mixed and an attempt is made to spin the mixture, the long fibers will wrap and clog the machines if they are set for the short fibers. If the machines are set for the long fibers, a part of the short fibers will be discarded as waste; others will ride on the longer fibers without being drafted at all. The result will be an irregular thread with thick and thin parts.

Unripe fibers lack strength and twist. A part of the weak fibers are broken in the spinning process. Some of the pieces go into the waste; others get rolled together and make neps, or knots on the threads. Some of the weak or unripe fibers pass through unbroken, but the resultant yarn is less strong because of their presence, and it is more difficult to dye uniformly fabrics made from such yarns. If the fibers lack twist, they will not cling well, and consequently the machines cannot be run at maximum speed. The millworkers will not be able to turn out the usual number of pieces; so the total output of the mill will be lowered.

As a rule, evenness of grade in a lot of cotton is not so important as evenness in length of staple or the ripeness of fibers; yet it is a matter of considerable importance to mills that produce yarns used in making a standardized cloth, a cloth finished in a particular color, or a cloth that retains the natural color of the cotton.

Improvement in Quality of American Upland Cotton.—About 1926, there was considerable complaint from spinners that American upland cotton was deteriorating. Probably the chief reason for this complaint was to be found in an increased acreage, in certain localities, of varieties with a staple shorter than 7/8 inch.

Since 1928, the U. S. Department of Agriculture has collected and classed each year lint samples from every bale ginned by hundreds of gins scattered over the Cotton Belt. These samples show a steady improvement in staple length (see Table XLVI*A*).

TABLE XLVI*A*.—STAPLE LENGTHS OF AMERICAN UPLAND COTTON PRODUCED IN THE UNITED STATES, 1928–1934[a]
(Percentage of Total)

Year	All lengths, inches	Under 7/8	7/8 and 29/32	15/16 and 31/32	1 and 1 1/32	1 1/16 and 1 3/32	1 1/8 and longer
1928	100	14.5	41.5	22.6	11.0	5.6	4.8
1929	100	20.1	38.1	18.9	11.7	6.5	4.7
1930	100	13.3	38.8	24.9	12.6	7.1	3.3
1931	100	6.1	39.7	27.2	15.4	6.5	5.1
1932	100	6.6	37.7	28.9	14.3	6.9	5.6
1933	100	4.2	35.5	31.6	16.0	6.5	6.2
1934	100	8.3	36.8	21.9	15.0	9.4	8.8

[a] From "Preliminary Reports on Grade and Staple Length and Tenderability of Cotton Ginned in the United States, 1928–1934," issued by Bureau of Agricultural Economics U. S. Department of Agriculture, Washington, D. C.

The production of cotton shorter than $\frac{7}{8}$-inch staple has definitely decreased, both in number of bales and in the proportion that they constitute of the annual crop. Ginnings of these lengths from the crop of 1929 amounted to 2,921,500 bales, or 20.1 per cent of the total crop, whereas only 539,100 bales of these lengths, 4.3 per cent of the total, was ginned from the 1933 crop. The staple has also improved in uniformity because of the fact that a greater acreage of better bred varieties with a more uniform staple is now produced, and more uniform cotton is being grown in the numerous one-variety communities that have been established.

The grades of cotton obtained in different years varies considerably in those years, the variation depending on the weather prevailing during harvest season, but there has been some improvement in grade. This is reflected in the increase in the amount of tenderable cotton produced. During recent years, considerable attention has been given to conditioning and drying cotton before ginning and to the regulating of gins. All of this improves the quality of the cotton ginned.

A number of other cotton-growing countries of the world are making an effort to improve the quality of their cotton by trying to breed or to introduce better varieties and to do better ginning.

Spinning Quality of Cotton in Relation to Seed Purity and Care of Seed Stocks.—Moore and Stutts,[12] of North Carolina, made a study of the effect of care of seed stock of cotton varieties on their spinning qualities. Cotton from plants grown from registered seed direct from the breeder was compared with that from plants grown from seed one to several years removed from the breeder and handled in various ways, some of which permitted a considerable degree of mixing. The combed samples of seed cotton from pure and mixed seed showed considerable difference in variability of staple length, and these differences showed a correlation with yarn strength. Cotton from registered seed made the strongest yarns in all counts. Yarns spun from cottons grown from mixed seed gave the lowest breaking strength; they were 12 to 13.5 per cent weaker than the same counts spun from registered seed of the same variety.

Spinning Qualities of Snapped Cotton as Compared with Those of Picked Cotton.—Tests were made by Willis[14] on cottons grown in Texas and Oklahoma as to the effect on spinning value of the two different methods of harvesting cotton. The picked cotton and the snapped cotton used for comparison were in each instance of the same variety, grown in the same field and harvested on the same day.

The test indicated that snapping lowers the classing about two grades; that there is an increase in waste, as is usual for lower grades; that there is no appreciable difference in the strength of the yarns spun from picked and snapped cotton of the same variety grown under the same conditions; that there is no significant difference in the uniformity of the yarns spun from picked and snapped cotton. Results obtained in the two states were very similar, except that yarns spun from snapped cotton from one section of the Texas test field appeared to be consistently somewhat weaker than those from picked cotton from this section of the field.

Comparative Waste, Tensile Strength, and Bleaching Qualities of Cotton Belonging to Different Grades of Upland White Cotton. Dean and Taylor[2] made manufacturing tests of cotton representing Middling Fair, Good Middling, Middling, Low Middling, and Good Ordinary of the Official Cotton Standards to determine the relative values of the different grades. Cotton from 109 bales collected from different parts of the Cotton Belt was tested both in a cotton mill and in a textile school. By this means, comparison was made of the spinning qualities of cotton from different sections of the Cotton Belt. It was found in the mill tests that the percentage of combined visible and invisible waste of cotton from east of the Mississippi River ranged from 8.63 for Middling Fair grade to 15.64 for Good Ordinary. The waste from the intermediate grades was intermediate, as may be seen from Table XLVII. The cotton grown west of the Mississippi River was about 1 per cent more wasty than the other. The strength of the yarns of the eastern cotton was found to decrease from 74.6 pounds per skein for Middling Fair grade to 68.0 pounds for Good Ordinary. The western cotton was slightly stronger, ranging from 83.9 to 74.9 pounds per skein.

After allowance has been made for difference in moisture conditions prevailing where the tests were made, the results from the textile-school tests agreed very closely with those from the mill tests.

Bleaching tests made by Dean and Taylor[2] showed that goods made from Middling Fair and Good Middling grades, when

TABLE XLVII.—COMBINED VISIBLE- AND INVISIBLE-WASTE PERCENTAGES OF FIVE GRADES OF BOTH EASTERN UPLAND AND WESTERN UPLAND COTTON (MILL TESTS)

(From Dean and Taylor)

	Middling Fair	Good Middling	Middling	Low Middling	Good Ordinary
Eastern Upland.....	8.63	8.75	10.66	12.23	15.64
Western Upland.....	8.22	9.35	11.47	13.37	16.89

TABLE XLVIII.—COMPARATIVE TENSILE STRENGTH OF 22'S YARNS IN POUNDS PER SKEIN OF 120 YARDS (MILL TESTS)

(From Dean and Taylor)

	Middling Fair	Good Middling	Middling	Low Middling	Good Ordinary
Eastern Upland.....	74.6	74.5	72.5	71.5	68.0
Western Upland.....	83.9	81.4	80.9	71.1	74.9

bleached under the same conditions, were practically identical in color. The goods from Middling were not so pure white as those from Middling Fair and Good Middling, but for commercial purposes the results were satisfactory. The goods from Low Middling appeared to have a slightly slaty color when closely compared with the other grades, while the goods made from Good Ordinary were easily distinguished by a slaty, bluish cast when compared with goods made from a higher grade.

As a conclusion of their results, Dean and Taylor[2] say:

The tests based on the Official Cotton Standards of the United States show that, after making allowances for the losses due to the cleaning processes, there is comparatively little difference between the grades above and those below Middling in the price paid by the manufacturer for each pound of the usable cotton obtained from bales of the different grades, but there is a difference in the intrinsic value per pound of the manufactured product. Accordingly, on the basis of quotations and values at the time of the tests, the inducement in the price paid to the farmer for the production of high-grade cotton was not commensurate with the greater value to the manufacturer of the product derived from such cotton.

Waste, Tensile Strength, Bleaching and Dyeing Properties of Cottons Belonging to Different Grades of Upland Tinged and Stained Cotton.—Meadows and Blair[3] report results of tests made to determine the relative values of the following grades of cotton: Low Middling, Good Middling Yellow-Tinged, Middling Yellow-Tinged, Low Middling Yellow-Tinged, Good Middling Yellow-Stained, Middling Yellow-Stained, Good Middling Blue-Stained, and Middling Blue-Stained.* The cotton was bought throughout the Cotton Belt by expert cotton classers connected with the Bureau of Markets and, in the tests, was subjected to the same mechanical conditions.

The percentages of visible waste made by the different grades were as follows:

Grade	White	Yellow-Tinged	Yellow-Stained	Blue-Stained
Good Middling	....	7.04	6.85	7.24
Middling		8.21	10.75	10.24
Low Middling	8.23	11.99		

From the foregoing percentages of waste it appears that the waste follows the grade of cotton.

The Middling Yellow-Stained and Low Middling Yellow-Tinged gave off considerable fly. These grades were followed in order by Good Middling Yellow-Stained, Middling Blue-Stained, and Middling Yellow-Tinged. The other three grades gave off the usual amount of fly.

The breaking strength, in pounds per skein of 120 yards of 22's yarn spun from the different grades, was as follows:

Grade	White	Yellow-Tinged, pounds	Yellow-Stained, pounds	Blue-Stained, pounds
Good Middling	....	85.7	87.2	76.6
Middling	....	88.5	82.9	79.5
Low Middling	81.3	81.2		

* Some changes have been made in the Official Standards since this research was conducted. See p. 416 for grades now recognized.

The results given show that the highest grade did not always give the strongest yarn.

The bleaching and dyeing tests showed that Low Middling, Good Middling Yellow-Tinged, and Good Middling Yellow-Stained could be bleached satisfactorily for white yarns. The Middling Yellow-Tinged and Low Middling Yellow-Tinged when bleached can be used for dyeing both light and dark shades. The Good Middling Blue-Stained and Middling Blue-Stained can be bleached satisfactorily for dyeing dark shades only.

The tensile-strength tests of single strands showed that the bleaching and dyeing processes did not materially affect the strength of the yarn, as shown by the average of all tests. The average strength of the gray yarn was 10.73 ounces; bleached, 11.05 ounces; pink, 10.83 ounces; blue, 10.60 ounces.

Comparative Spinning Values of Certain Varieties of Long-staple Upland Cotton.—Many spinners and cotton-mill men have held to the belief that long-staple cotton grown in the Mississippi Delta was superior in spinning qualities to long-staple cotton grown on the uplands of the eastern states. Experiments carried on by Taylor and Sherman of the U. S. Department of Agriculture to determine whether or not these views were well founded showed that Webber, Lewis, and Columbia, when grown in the Southeast, were no more wasty than average Delta cottons but that they had less strength. The three varieties just mentioned, however, were grown from well-bred seed, but the Delta cotton used was a mixture of five bales from different sources in the Delta and was all grown from unselected seed. This doubtless favored the eastern-grown varieties.

TABLE XLIX.—STRENGTH OF CARDED AND COMBED YARNS

Variety	Grade	Staple	Carded yarn	Combed yarn
Durango	G.M.	1 3/16	24.03	33.07
Columbia	S. G. M.	1 3/16	25.56	34.50
Lewis	G. M.	1 5/16	34.93	41.92
Webber No. I	M.	1 5/16	25.88	32.12
Webber No. II	M.	1 5/16	26.13	33.47
Delta	Blended grades	1 1/4	32.98	42.05

Table XLIX lists the varieties used in the tests made by Taylor and Sherman[4] and gives their staple length, the grade of each used, and the breaking strength in pounds per skein of carded yarns and of combed yarns resulting from the 13 to 15 per cent comber settings.

R. W. Boys of the Farr Alpaca Mills, Holyoke, Mass., reports a waste of 18 to 20 per cent for carded cotton and 30 to 35 per cent for combed. Strict Middling, 1⅛ inch full, Delfos cotton grown in the Mississippi Delta was used.

Comparative Spinning Values of Certain Prominent Varieties with a Staple 1 1/16 Inches in Length.—Work carried on by Meadows and Blair[5] in 1922 gave evidence to the effect that several well-bred strains of 1 1/16-inch cotton possessed qualities that made them more attractive to spinners than cotton of that length from "North Georgia," a region with an established reputation for good cotton of such a length. The following summary by Meadows and Blair gives in brief the results from their experiment:

The cottons tested were from the crop of 1921, and consisted of the fiber of the following varieties: Acala, Lone Star, Mexican Big Boll, Rowden, and of typical cotton of the kind commercially known as "North Georgia." The Acala was grown in Alabama, the Lone Star, Mexican Big Boll, and Rowden were grown at different points in North Carolina, and the typical North Georgia cotton was grown in "North Georgia."

The grades, lengths of staple, percentages of visible waste, strengths of the yarns, and percentages of average deviation or irregularity of the sizings and strengths, as shown in Table L, indicate that for hard-twisted or warp yarns the varieties tested if placed in order of their merit and attractiveness from a spinner's viewpoint would fall in the following rank:

1. Acala, Mexican Big Boll — equal
2. Lone Star, Rowden — equal
3. Typical "North Georgia."

Willis and McNamara[13] grew several of the leading Texas varieties in adjacent four-row plots in fields at four different places in Texas for a 3-year period and made spinning tests of the cotton produced. The object of the experiment was to determine just how much difference there is between the spinning qualities of different varieties when grown under identical or very nearly identical conditions. The following different varieties, listed in the order of breaking strength of their yarns, the strongest being first, were used: Acala, Kekchi, Lone Star, Mebane, Rowden, "Hoground," and Kasch. These covered fairly well the range of the principal varietal types grown in Texas.

TABLE L.—GRADES, LENGTHS OF STAPLES, PERCENTAGE OF VISIBLE WASTE STRENGTHS OF THE YARN, AND PERCENTAGES OF AVERAGE DEVIATION OF THE SIZINGS AND STRENGTHS OF THE YARN

	Acala	Lone Star	Mexican Big Boll	Rowden	Typical "North Georgia"
Grade	Mid.	S. M.	G. M.	G. M.	S. M.
Length of staple, inches	1⅙	1⅙	1 full	1⅙	1⅙
Percentage of visible waste	8.28	7.54	6.71	5.97	6.29
Strength of yarn, in pounds per skein of 120 yards:					
28's	70.5	62.3	67.2	63.5	56.4
36's	49.9	44.4	47.3	43.7	40.1
44's	38.2	33.5	34.4	33.9	25.6
Percentage of average deviation or irregularity of sizing of the yarn:					
28's	2.02	2.02	2.11	2.01	1.94
36's	2.06	2.06	2.18	1.79	2.10
44's	2.40	1.87	2.16	2.17	1.81
Percentage of average deviation or irregularity of strength of the yarn:					
28's	4.54	3.72	4.72	4.72	3.74
36's	4.51	3.86	5.09	3.98	3.86
44's	4.58	5.30	5.27	5.70	6.29

It was noted that there was a difference both in waste content and in strength of yarn spun from a particular variety when grown in different crop seasons and when grown in different parts of an experiemental field the same year, although the soil of the field used was apparently uniform. Kasch, Mebane, Lone Star, Acala, and Kekchi ranked in lint percentage in the order listed. Their rank in staple length and yarn strength was in reverse order, except that Acala ranked first in strength. No significant difference in the yarn spun from the several varieties was evident.

Comparative Spinning Qualities of Sea Island, Meade, Egyptian, and American-Egyptian Cottons.—Within the last decade or two, American-grown Egyptian cotton has come to be a commodity of commercial importance. A considerable quantity of this cotton is grown, and there is need for it to take the place of Sea Island, which can no longer be grown profitably on account of the boll weevil. Meade, a long-staple upland variety which was introduced in the Sea Island district to take the place of Sea Island, has not been grown extensively. It has a staple similar to Sea Island but apparently is not so profitable in regions with heavy boll-weevil infestation as are short-staple varieties.

Several tests of the spinning qualities of the three cottons mentioned above and of Egyptian cotton have been made to see how they rank comparatively and to learn whether or not values placed on them were in accordance with their real merits. Taylor and Dean[6] found that the relative waste in the manufacturing processes of the four grades of Arizona-Egyptian cotton tested was as follows: Extra, 17.69 per cent; Choice, 18.56 per cent; Standard, 20 per cent; Medium, 20.90 per cent. The four grades of Arizona-Egyptian were, with respect to grade, proportionately less wasteful than the Sea Island in their tests, and the Sea Island was proportionately less wasteful than the imported Egyptian cotton used. There seemed to be no relation between the price values put on the different kinds of cotton and the percentage of waste in the manufacturing processes. Arizona-Egyptian, which was considered as having the lowest commercial value, was found to be the least wasteful.

Taylor and Dean[6] found no significant relationship between tensile strength and grade in the different grades of Arizona-Egyptian cotton. The difference in the tensile strengths of the three cottons, Sea Island, Egyptian, and Arizona-Egyptian, was slight. On the whole, Egyptian Sakellaridis averaged slightly stronger, but Sea Island excelled in strength of fine yarns.

In respect to the bleaching and dyeing properties of the cottons under investigation, Taylor and Dean[6] say:

> The laboratory test indicated that, after bleaching, dyeing, and mercerizing, the Arizona-Egyptian and the Sea Island cottons were practi-

cally equal to each other and were slightly superior to the Sakellaridis in their bleaching and mercerizing properties; that they were fully equal to each other in dyeing properties; and in tensile strength the advantage was slightly in favor of the Sea Island and Sakellaridis. The finished gray and mercerized yarns were comparatively equal in luster; however, the yellow color was a little more evident in the Arizona-Egyptian than in the Sakellaridis, which, in turn, was somewhat more yellow than the Sea Island. The difference in color was more apparent between the

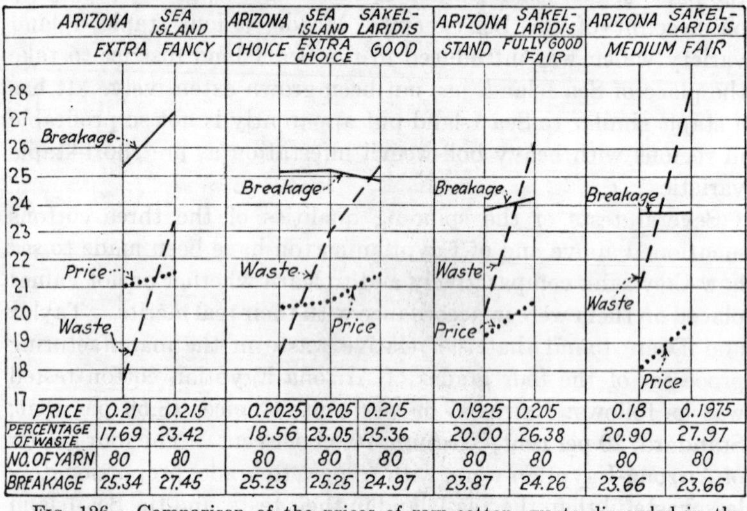

	ARIZONA	SEA ISLAND	ARIZONA	SEA ISLAND	SAKELLARIDIS	ARIZONA	SAKELLARIDIS	ARIZONA	SAKELLARIDIS
	EXTRA FANCY		CHOICE	EXTRA CHOICE	GOOD	STAND	FULLY GOOD FAIR	MEDIUM FAIR	
PRICE	0.21	0.215	0.2025	0.205	0.215	0.1925	0.205	0.18	0.1975
PERCENTAGE OF WASTE	17.69	23.42	18.56	23.05	25.36	20.00	26.38	20.90	27.97
NO. OF YARN	80	80	80	80	80	80	80	80	80
BREAKAGE	25.34	27.45	25.23	25.25	24.97	23.87	24.26	23.66	23.66

Fig. 126.—Comparison of the prices of raw cotton, waste discarded in the manufacturing processes, and the tensile strength of the yarn in pounds per skein of 120 yards each for Arizona-Egyptian, Sea Island, and Sakellaridis Egyptian cottons. (The figures at the left indicate the cents per pound for the price, percentage for the waste, and pounds per skein for the breaking strength.) (*After Taylor and Dean.*)

Arizona and the Sakellaridis than between the Sea Island and Sakellaridis.

Figure 126 from Taylor and Dean[6] shows graphically the comparative prices of the three kinds of cotton mentioned above, percentages of waste discarded in the manufacturing process, and the tensile strength of the yarn made from them. Equivalent grades are placed side by side. The figures on the left-hand margin indicate cents per pound for the price, percentage for the waste, and pounds per skein for the breaking strength. Number 80's yarn was used in the comparison.

Tests made by Meadows and Blair[7] on the comparative spinning values of Meade and Sea Island cottons showed that, on an average of results from three seasons, the Meade cotton was 3.5 per cent more wasteful than Sea Island. In the 3-year average, Sea Island showed a breaking strength 17.2 pounds greater than Meade when 23's yarns were compared and 1.68 pounds greater when 100's were compared. The grades used were almost the same, and the staple length used was all 1⅝, except one lot of 1$\frac{7}{16}$ Meade.

Effect of Bale Compression on Spinning Qualities.—Much of the cotton fiber used in making cloth is compressed to make high-density bales before it is used, in order to economize space in boxcars, warehouses, etc. The question often arises as to whether or not the extra compression that the lint in the bales receives reduces its spinning value.

Spinning tests were made by Meadows and Blair[7] to secure information on the question just mentioned. Cotton from Cleveland, Rowden, Delta, and Webber-49 varieties was used as examples of the different types of cotton with staples, respectively, of $\frac{15}{16}$, 1, 1⅛, and 1¼ inches. Lint was taken from the regular uncompressed, or "flat," bale, as it it is called; from standard or railroad compressed bales with a density ranging from 22 to 28 pounds per cubic foot; from high-density bales, the density ranging from 28 to 40 pounds per cubic foot; from high-density bales that were wet when compressed (from two of the varieties); and from a round bale (for one variety).

All the tests made showed that compressing cotton to standard or high density when dry is not injurious to its spinning value but that compressing it to high density when wet may increase the percentage of waste, or it may decrease the strength of the yarn, or it may both decrease the strength and increase waste. It was found that compressing cotton into a round bale with a hard core reduced the strength of yarn about 7 per cent.

Effect of Fumigation with Hydrocyanic Acid Gas on the Spinning Qualities of Cotton.—Spinning and dyeing experiments were carried on by Dean[8] with Egyptian, Peruvian, and Chinese cottons that had been subjected to the fumes of hydrocyanic acid gas. Some of the samples were subjected to the gas for 30 minutes to an hour from one to five times. The volume of gas that penetrated some of the bales was ten times as great as is ordinarily used in fumigating cotton.

As a result of the tests, Dean[8] says:

These tests indicate that the fumigation of cotton with hydrocyanic acid gas does not affect, to any material extent, the percentage of waste, spinning qualities, tensile strength, bleaching, dyeing, or mercerizing properties of the cotton.

Special Value of Egyptian and Peruvian Cottons to American Spinner.—Although America exports annually several million bales of cotton, she imports each year about a quarter million bales. The imported cottons are chiefly Egyptian and Peruvian and inferior cottons from China and India.

The Egyptian cotton is especially good for making sewing thread, fine underwear, hosiery, etc., and for goods requiring a smooth finish or a high luster. Since it has a good luster, it is used to mix with silk in the manufacture of certain silk goods. It is also used extensively in the manufacture of mercerized goods.

Much of the Peruvian cotton imported into this country is the "rough Peruvian," a cotton with a stiff, rough fiber which resembles wool. This is used by manufacturers of woolen goods to mix with wool in making mixed or shoddy woolen goods, underwear, hosiery, etc. None of the American cottons is so well suited to this purpose.

Utilization of Cotton Waste from Spinning Mills.—Cotton waste from spinning mills consists principally of short fibers that have been rejected by machines in the process of combing and carding; and also floor sweepings, odds and ends from weaving, and various scraps.

According to the estimate by Lomax,[10] 25 per cent of the raw cotton that enters spinning and weaving mills in England is resold in the form of cotton waste as a by-product. This gives some idea of the quantity of waste produced in cotton mills.

The better grades of waste are used in making quilts, blankets domestic, sheets, towels, and flannelettes. The flat cylinder strips made in the process of carding are frequently used as raw material for products in which strength is required, such as warps, twine, ropes, and nets. Large quantities of strips from Egyptian cotton are mixed with wool, on account of the strength of their fibers, and used in making mixed woolen goods. Lower qualities of waste are used in making sponge cloth, carpet yarns, and low-quality mixed woolen goods. Floor sweepings and fibers unfit for spinning are bleached and used in making guncotton, cellulose, and artificial silk. Thready material that cannot be broken up or spun again is made into cleaning waste for engines and machines.

Use of Linters.—Cotton linters have a variety of uses other than for spinning purposes. The reader is referred to Chap. XXII for further discussion. In cotton mills, linters are mixed with lamb's wool in making fleece-lined underwear and are used in making low-grade yarns for lampwicks and candlewicks, twine, carpets, and rope.

References

1. Howarth, William: "Characteristics of Cotton Required by the Spinner and Present Defects in the Raw Material," Report World Cotton Conference, Liverpool, 1921.
2. Dean, William S. and Taylor, Fred: "Manufacturing Tests of the Official Cotton Standards for Grade," *U. S. Dept. Agr. Bull.* 591, 1917.
3. Meadows, W. R. and Blair, W. G.: "Preliminary Manufacturing Tests of the Official Cotton Standards of the United States for Color for Upland Tinged and Stained Cotton," *U. S. Dept. Agr. Bull.* 990, 1921.
4. Taylor, Fred and Sherman, Wells A.: "Spinning Tests of Upland Staple Cottons," *U. S. Dept. Agr. Bull.* 121, 1914.
5. Meadows, W. R. and Blair, W. G.: "Comparative Spinning Tests of Superior Varieties of Cotton (Grown under Weevil Conditions in the Southeastern States; Crop of 1921)," *U. S. Dept. Agr. Bull.* 1148, 1923.
6. Taylor, Fred and Dean, William S.: "Comparative Tests of the Different Grades of Arizona-Egyptian with Sea Island and Sakellaridis Egyptian Cottons," *U. S. Dept. Agr. Bull.* 359, 1916.
7. Meadows, William and Blair, W. G.: "Comparative Spinning Tests of Meade and Sea Island Cottons," *U. S. Dept. Agr. Bull.* 946, 1921.
8. Dean, William S.: "Manufacturing Tests of Cotton Fumigated with Hydrocyanic Acid Gas," *U. S. Dept. Agr. Bull.* 336, 1916.
9. Shepperson, A. B.: "Cotton Facts," Shepperson Publishing Company, New York, 1913.
10. Lomax, J. Barber: "Cotton Waste and Its Value as a Raw Material." Report World Cotton Conference, Liverpool, 1921.
11. Lanham, W. B. and others: "Grade, Staple Length and Tenderability of Cotton in the United States, 1928–29 to 1934–35," *U. S. Dept. Agr. Bull.* 56, 1937.
12. Moore, J. H. and Stutts, R. T.: "Spinning Quality of Cotton in Relation to Seed Purity and Care of Seed-stocks," *N. C. Agr. Exp. Sta. Tech. Bull.* 45, 1934.
13. Willis, Horace H. and McNamara, Homer C.: "Spinning Tests of Some Texas-grown Varieties of Cotton (Crops of 1923, 1924 and 1925)," *U. S. Dept. Agr. Prelim. Rept.*, May, 1928.
14. Willis, Horace H.: "Effects of Methods of Harvesting and Weather Exposure on Spinning Quality of Cotton (Texas and Oklahoma Crop of 1926)," *U. S. Dept. Agr. Bur. Agr. Econ. Prelim. Rept.*, July, 1928.

CHAPTER XXIV

MAKING COTTON CLOTH

It does not fall within the province of this text to treat in detail the many phases of the cotton-manufacturing industry. That is a large subject within itself, and the student seeking details may find special works on the subject or on its different phases which will give him the desired information. The aim here is to trace cotton from the field to its finished product, and consequently the discussion will not be complete unless some mention is made of spinning, weaving, and other manufacturing processes. Information on these subjects should be of some interest and value to cotton growers, many of whom have never been inside a cotton mill and have but a vague notion of what takes place within its walls.

History of Cotton Manufacturing.—When cotton fiber was first used in making cotton cloth is not definitely known, although it is certain that it was used in India prior to the Christian era. At that early date, a primitive distaff was employed in spinning, and the loom was little more than a few sticks or reeds. Many centuries elapsed before the cotton plant or cotton manufactures were introduced in Europe. According to Handy,[1] the first authentic record of the making of cotton cloth in England was in 1641. The industry grew slowly. Raw material was scarce, and the machinery used in spinning and weaving was crude. Because of the influence of people interested in the wool industry or the manufacture of woolen goods, laws were passed prohibiting the wearing of cotton goods, and a fine was imposed on anyone who should weave or sell a piece of calico. This, of course, retarded the advancement of the cotton-weaving industry in England but could not keep it down. Spinning and weaving were at first done in the home, and the worker came to have the idea that, if improved machinery, which turned out a greater amount of goods, should be introduced, not so many people would be needed to make the required amount of goods and some would be deprived of work. Consequently, the workers were opposed to new inventions and did everything possible to prevent their introduction. But several of the inventors worked with the spirit of martyrs. Gradually, the machines were improved until they became real mechanical wonders. It is almost beyond belief that machines, with almost no assistance by man, can take the tangled mass of cotton fibers and produce the fine, evenly woven cotton fabrics.

Until the latter part of the eighteenth century, spinning was done with the distaff or spinning wheel, and the weaving on hand looms. In America, these machines were in use in homes all over the country prior to the American Revolution, and in the southern states they were used considerably later—even after cotton mills were well established. Women and children did the spinning and weaving. A generation or so ago, spinning wheels were frequently seen stored in the attics of houses in America. The writer remembers having seen, when a boy, one that his grandmother had.

Inventions in England.—England early took an active interest in the textile industry. Being a commercial country, merchants found a sale for much cotton goods, and a heavy demand was created, greater, in fact, than could be met easily with the equipment at hand. The heavy demand stimulated inventors in an endeavor to devise machines that would turn out goods more rapidly. Several important inventions were the result.

The first in this series was the fly shuttle invented by John Kay in 1738. Prior to his invention, the shuttle was thrown back and forth across the loom by hand. The new contrivance moved it back and forth by mechanical means and more rapidly. This invention improved the efficiency of the loom greatly, but it brought the inventor no reward. In fact, it brought him into disfavor, and weavers even threatened his life. Manufacturers made use of his invention but refused to remunerate him, and he died in poverty.

The success of the loom improvement just mentioned induced others of inventive mind to study spinning machines and led to the invention of James Hargreaves's "spinning jenny" in 1767. The first spinning jenny made contained 8 spindles, but the number was later increased to 80. This was a big improvement over the one spindle of the former spinning wheel. Its use alarmed neighboring spinners, who broke into Hargreaves's house and destroyed his machine. He later rebuilt the machine and had it patented but never succeeded in getting just returns from his invention. He died in comparative poverty.

Richard Arkwright in 1767 improved the rollers for drawing out cotton strands, or sliver, in the spinning operation. The idea was originated by John C. Wyatt about 1730, but it was not put into successful operation. Arkwright was more successful in a business way than the previous inventors mentioned. He amassed a fortune and in 1786 was knighted by King George III.

Samuel Crompton in 1779 made an improved spinning machine which would spin fine yarns. This was known as the "mule" spinning jenny, so called because it was a sort of cross between Arkwright's machine and Hargreaves's jenny. This machine brought cotton-spinning machinery to a high state of perfection. The next marked improvement in textile machinery was the invention of a power loom by Edmund Cartwright in 1785–1787.

Cotton Mills.—With the development and improvement of power-driven machinery for spinning and weaving came other changes. When the work was done by hand, or with simple machines run by hand, it was done in the home. But with the invention of power machinery it was moved to factories. These were small at first, of course, but were the beginning of our modern factories. The first mills were located on streams where the advantage of water power could be secured.

The first cotton mill in the United States was erected at Beverly, Mass., in 1787. Several others were established within the next few years, but progress was slow for several years. With the invention of Whitney's cotton gin in 1793, Lowell's power loom in 1814, and the ring spindle in 1830 came considerable improvement.

Table LI.—Increase in Cotton Spindles in the United States

Year	Number spindles	Year	Number spindles
1805	4,500	1860	5,200,000
1810	87,000	1870	7,100,000
1820	220,000	1880	10,700,000
1825	820,000	1890	14,200,000
1831	1,200,000	1900	19,000,000
1840	2,300,000	1910	28,500,000
1850	3,600,000	1920	34,500,000
		1931	32,673,000

From Table LI it may be seen that there was a rather steady increase in the number of cotton spindles in the United States during the first hundred years of the cotton-manufacturing industry. The New England states first took the lead in the industry, Providence, Fall River, New Bedford, and Lowell being important centers. But since 1860, the southern states have gained rapidly in comparative number of spindles. In 1891, the northern states had 12,900,000 spindles, and the southern states had 1,740,000. In 1923, the northern states had 20,946,538 spindles; and the southern, 16,483,657. In 1936, the northern states had but 9,125,000, whereas the southern states had 19,032,000. Copeland[2] considers that this phenomenal increase of mills in the South is due to four causes, namely, proximity to source of raw material, water power, lower taxation, and cheaper labor.

The increase in cotton mills in the South has had an effect on social and industrial conditions there. The increase in factories brings an increase in wealth, more taxes, more money for public improvements, schools, etc.; and higher standards of social life are possible.

Cotton Mills in Foreign Countries.—England has long been the leader in the manufacture of cotton goods and is still in the lead, but her margin is decreasing, as may be seen from Table LII. More than 90 per cent of her spindles are in the one county of Lancashire. The relative rank in spindleage of several countries of the world is shown by Table LII.

How Cotton Cloth Is Made.—To a person who has never seen a large cotton mill a trip through one is a most interesting experience. It is a revelation to see what the various machines can do. The theory of cloth manufacture and a brief description of the most important processes will now be given.

Cotton that is to be made into cloth must be thoroughly cleansed of foreign matter; then it must be opened so that each fiber is well separated from every other fiber; then the fibers must be straightened so that they will lie parallel, and then be twisted together to make a yarn. Such, in general, are the steps in getting the cotton ready for the loom. The whole manufacturing process may be divided into five parts: (1) preparatory processes; (2) spinning; (3) spooling, warping, and sizing; (4) weaving; and (5) finishing.

The type of machine used varies, of course, to some extent in different mills, and the processes that the cotton is put through vary, depending on the quality of goods to be made. In the following discussion, a brief sketch will be given of the processes commonly used in making an ordinary type of goods.

TABLE LII.—COTTON SPINDLES OF THE WORLD

Location	1936	1920	1907
United States, North	9,125,000	19,600,000	16,850,000
United States, South	19,032,000	15,100,000	9,500,000
Great Britain	41,391,000	57,300,000	50,700,000
Germany[a]	10,109,000	9,400,000	9,300,000
France	9,932,000	9,400,000	6,800,000
Russia and Poland	11,507,000	8,000,000	8,100,000
India	9,705,000	6,700,000	5,300,000
China	5,010,000		
Austria	773,000	5,000,000	3,600,000
Italy[a]	5,483,000	4,800,000	3,500,000
Japan	10,867,000	3,400,000	1,500,000
Spain	2,070,000	2,000,000	1,900,000
Brazil	2,712,000	1,600,000	800,000
Switzerland	1,241,000	1,500,000	1,400,000
Belgium	2,009,000	1,538,000	1,140,000
Other countries	10,140,000	5,975,000	4,730,000
Total	151,698,000	151,313,000	124,320,000

[a] The figures for Germany and Italy are for the year 1934 instead of 1936.

Preparatory Processes.—Before spinning proper can begin, the raw cotton must pass through several machines—bale breaker, picker, carding machine, combing machine, and drawing frame.

Bale Breaker.—When bales of cotton are first opened in the cotton mills, the lint cotton is found to be in more or less compact masses, or laps, due to the pressure that has been applied to it in baling or compressing bales. These laps are passed into a bale breaker, which contains three pairs of rollers, two of them spiked and one fluted, which disentangle and spread apart the fibers to some extent and deliver a more fluffy mass of cotton. From the bale-breaker room, the cotton is usually blown through large pipes to the room in which the pickers are situated.

Picker.—The purpose of the picker, sometimes called "scutcher," is to separate fibers from each other, or spread them apart so that they will appear to touch only in places, to remove bits of leaf, dirt, and other foreign matter, and to roll the cotton into a more or less regular layer or "lap."

The cotton is carried to the picker, or opener, by a movable lattice apron, from which it is taken up by two feed rollers (Fig.

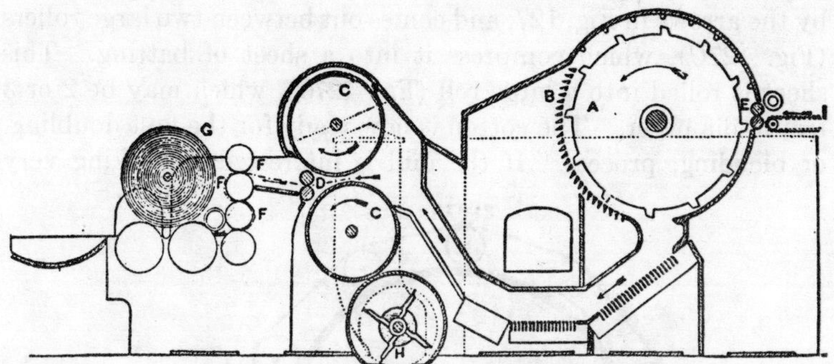

FIG. 127.—Cross-section diagram of picker.

FIG. 128.—Picker or opener machine. (*Courtesy Whitin Machine Company.*)

127*E*). In the picker, it is first subjected to the action of a beater (Fig. 127*A*), which consists of a large rapidly revolving roller with teeth over the surface. These teeth seize the cotton and fling it about, a part being flung by centrifugal force against an iron grid (Fig. 127*B*), through which dirt and foreign matter fall. Other teeth seize the cotton again and whirl it about. It is subsequently carried around an indefinite number of times and thrown against the grating until a large part of the dirt and foreign matter is removed. The cotton gradually works its way along a channel in the lower part of the machine, as shown by the arrows in Fig. 127, and comes out between two large rollers (Fig. 127*C*), which compress it into a sheet of batting. This sheet is rolled into a large roll (Fig. 127*G*), which may be 2 or 3 feet in diameter. The cotton is now ready for the first doubling, or blending, process. If the mill is interested in making very

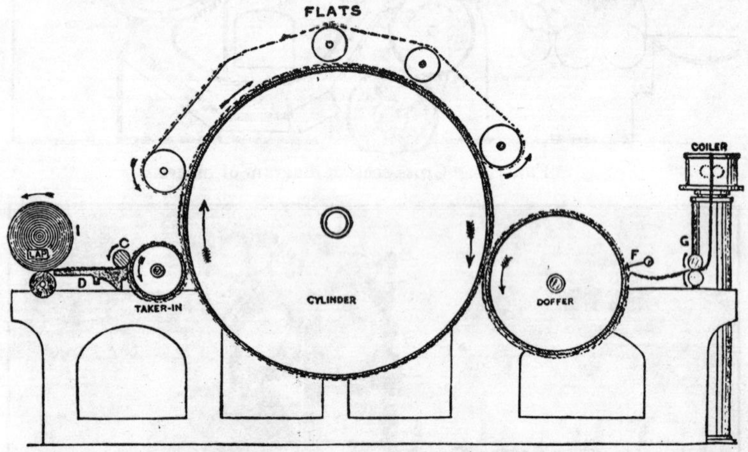

Fig. 129.—Cross-section diagram of revolving flat card.

strong or even yarn, four sheets, or laps, from the picker are placed on top of one another and fed into another picker. This machine breaks up the laps, thoroughly mixes their fibers, takes out more foreign matter, and makes a more even lap. When the cotton comes from this picker, or scutcher, it is much whiter and cleaner looking than when first taken from the bale.

Carding Machine.—When the cotton comes from the last picker, or finisher picker, the fibers are tangled and point in every direction just as they happened to be arranged as they come out. It is the function of the carding machine to place the fibers parallel, to remove remaining impurities and immature fibers, and to form the mature fibers into a porous rope or band called a "sliver."

The fibers are straightened out and made to lie parallel by combing or brushing them with wire brushes or cards. In the revolving flat card, which is in common use at present, there are, as a rule, three principal cylinders. The lap passes first under the smallest of the three, called the "taker-in," or "licker," which is covered with very fine saw teeth all in one long strip of steel, wound and fixed spirally in the surface of the cylinder. The taker-in receives the cotton from a feed roller (Fig. 129*C*) which turns above a smooth iron plate (Fig. 129*D*) called the "feed plate." The saw teeth comb the fibers which are embedded in the lap, remove some impurities, and deliver the loose fibers

FIG. 130.—Carding machine. (*Courtesy Whitin Machine Company.*)

to the second cylinder, which is the largest of the group. This main cylinder is covered with wire teeth all bent at exactly the same angle. The cotton clings to them and is carried around to the top of the cylinder, where it is engaged by teeth on the revolving flat card—teeth bent in the opposite direction. This flat card, which is arranged on a traveling lattice, moves in the same direction as the large cylinder but moves very slowly. It cards the fibers between two sets of wire points. The short and immature fibers tend to remain on the card wires of the lattice, while the perfect fibers, which are now almost entirely parallel, are carried over on the main cylinder to the doffer cylinder, the third of the three cylinders mentioned above (Fig. 129). From this cylinder the fibers are removed by an oscillating comb (Fig. 129*F*), coming off in a light, fleecy lap, which is condensed through a funnel into a soft, untwisted roping, or sliver, about half an inch in diameter. The sliver is coiled and deposited in a deep, narrow can. These cans may be carried to other machines, and the sliver removed without getting it tangled or damaged.

Combing Machine.—When very fine or very uniform yarns are sought, the cotton is subjected to the process of combing. This

FIG. 131.—Drawing frame. (*Courtesy Whitin Machine Company.*)

is an actual combing, as the name implies. The cotton is raked by a fine-tooth comb with needle-like steel teeth ranging from 16 to 90 per inch. This process removes short fibers left in the cotton. The slivers from 14 to 20 cans which are placed near together are drawn along side by side and passed between drawing rollers, which lengthen them and reduce their diameter. The slivers next pass between two pairs of calender rollers and emerge as a narrow lap. These laps are drawn and doubled by being superimposed one or more times, the number depending on the quality of yarn to be made. The lap then passes the combs, where it is raked and combined with other combed laps to make a sliver which is again deposited in a can.

Drawing Frame.—Slivers from the card, or comber, are next taken to the drawing frame (Fig. 131). The masses of fibers in the sliver are still irregularly distributed. It is the function of the drawing frame to eliminate irregularities in the sliver and make a smooth, uniform yarn. This is accomplished by doubling or uniting laterally several slivers and then drawing them out to the dimension of one. For finer yarns, 216 or more doublings may be made.

The attenuation or drawing out of the sliver is effected by passing it between successive pairs of rollers, each succeeding pair

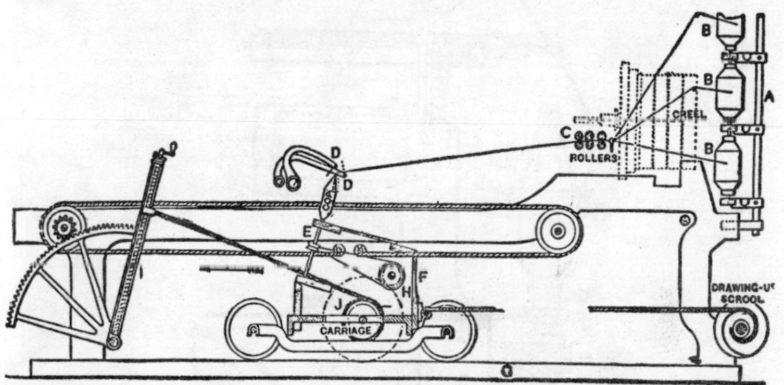

FIG. 132.—Diagram of mule spinning machine.

revolving at a greater rate. The last pair in a series revolves six or eight times as rapidly as the first. This produces a constant pull on the roping of cotton and lengthens it.

From the drawing frame the sliver is taken to the slubbing frame, where it is attenuated, slightly twisted, and wound on spools. The sliver becomes the slubber roving, on being twisted.

The slubber roving is next passed to the intermediate frame, where it is doubled, drawn out to less diameter than the original roving, twisted, and wound on a spindle.

The next drawing frame, or "speeder," as it is called, receives the rovings from the intermediate frame, combines two of them into one, twists them a little more, and winds them on tubes. Still another speeder frame, known as a "jack frame," is used for very fine yarns spun from Sea Island or Egyptian cotton.

Spinning.—In the spinning process, the roving is further attenuated and twisted into a firm, coherent yarn of the required strength and fineness. There are two types of spinning machines in use, the "mule" and the "ring" spindle. The mule, which was invented by Crompton, is the older type.. About 75 per cent of the spindles in use in England are of this type. The ring

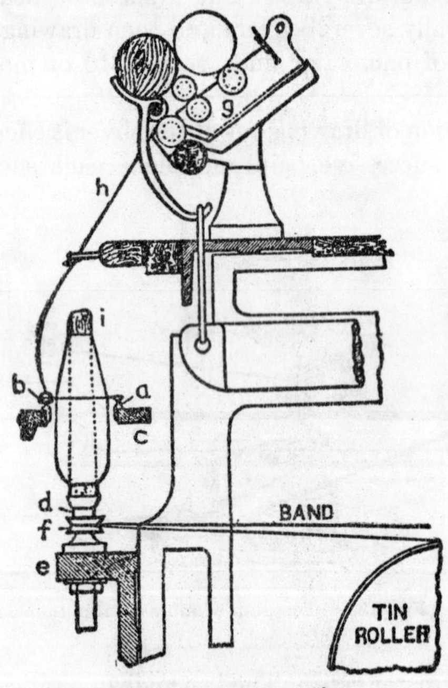

Diagram of ring spindle

FIG. 133.—Diagram of ring spinning machine.

spindle, which was invented by an American, James Thorpe, in 1828, is used in America almost altogether. It is claimed that the mule spindle will spin smoother and finer yarns, but the ring type is simpler and more rapid in its work.

The mule spinning machine is a large machine, carrying in some instances as many as 1,300 spindles. In this machine, the drawing and twisting are intermittent. The rovings (Fig. 132*B*), which are held on a creel at the back of the machine, run from the spools through three pairs of rollers (Fig. 132*C*) to the spindles on the carriage. The carriage (Fig. 132) at intervals moves away from the creel and draws the rovings somewhat, since it moves slightly faster than the rovings are unwound. As the carriage recedes, the spindles, which are revolving rapidly, give twist to the yarn. Reaching the end of its course, which is about 63 inches, the carriage reverses and returns. The spindles reverse their course on the return trip of the carriage, and the yarn on

FIG. 134.—Ring spindles in operation. (*Courtesy Whitin Machine Company.*)

them is unwound and then rewound in the opposite direction on the spindles.

The ring spindle is more simple than the mule and has no movable carriage (Fig. 133). The roving (Fig. 133*H*) runs from its bobbin between drawing rollers (Fig. 133*G*) and then is twisted, hooked into the traveler (Fig. 133*B*), and fastened to a spool (Fig. 133*I*) which is placed on the spindle. The spindle revolves with great rapidity, 11,000 revolutions per minute being used in some cases. As it revolves, the traveler (Fig. 133*B*) is pulled round the ring by the thread. Since the traveler revolves more slowly than the spindle, a drag is put on the yarn, and it is given a twist. The frame to which the rings are attached (Fig. 133*C*) moves up and down slowly, so that the yarn may be properly distributed over the spool on which it is wound. Figure 134 shows a machine with ring spindles in operation.

From Spinning Machines to Loom.—The spinning machine leaves its yarn, or finished product, wound on a bobbin known as a "cop." The cops that are used for weft, or filling in weaving, that is, the threads that run across the goods, are placed in shuttles and used in weaving without going through any further processes. The yarn from cops that is to be used for warp, the threads that run

Fig. 135.—Warping being wound on beam (the large roller). Creel containing numerous spools in the rear.

the long way of the goods, must be wound on large spools before it can be warped. A simple machine known as a "spooler" is used in this work. The spool revolves at a rapid rate, and the thread is wound on it evenly. The yarn used for warp is usually twisted a little more than the weft yarn.

The spools from the spooler are placed on a large frame called a creel (Fig. 135). The creels have an average capacity of about 600 spools. Threads from the spools are unwound to make the warping for the loom. When the threads leave the spools, they are drawn between the dents of an adjustable reed and then under and over a series of rollers. From here they are led down to the beam, upon which they are wound. The revolving of the beam unwinds the yarn from the spools and winds it regularly and evenly on the beam itself. There is a device for measuring the length of the warp wound, and a stop mechanism for arresting the operation should a thread break or an accident occur.

The yarn of the warp must usually be impregnated with a sizing which will smooth out and stick down its furry surface and also add to its tensile strength, so that the strain of weaving may be withstood. For the sizing process, a machine known as the "slasher" is most generally used. The chief feature of this is a roller whose lower side is immersed in the sizing solution. Threads from the warp beam are run around this roller through

FIG. 136.—Plain cotton loom.

the solution and then dried by passing them over a heated roller. The threads are next wound on another beam for the loom. More than one loom beam can be filled from the one set of warper beams mounted in the slasher. The warp is next taken to the loom.

Weaving.—In the process known as "weaving," the warp and the weft are interlaced with each other as seen in woven goods. This work is done by a machine known as a "loom" (Fig. 136). The following are the principal steps in weaving:

1. Raising or lowering alternate threads of the warp so that the weft thread may be passed between them by means of the shuttle.

2. Pushing each new thread of the weft into position against the preceding thread.

3. Unwinding the warp from its beam as fast as it is woven and winding up the woven cloth.

Finishing.—When cotton goods leave the loom, they are ready for the finishing operations. These include bleaching, printing, mercerizing, dying, starching, dampening, pressing, and other treatments.

Bleaching.—When a piece of cotton goods leaves the loom, it is more or less soiled; it has some colored motes from leaf fragments and other foreign matter which were in the lint cotton; and it is colored somewhat by the natural yellowish or brownish color of the lint cotton. The bleaching renders the goods a pure white, and, if it is properly done, the fabric is not injured.

In the bleaching process, the goods are first boiled in plain water, which removes soluble material. They are next boiled in a strong alkaline solution to remove fatty and waxy substances in the fibers. Such substances interfere with bleaching and dyeing if left in the goods. Next the goods are steeped in a bleaching solution, chloride of lime being commonly used for this purpose. These chemicals must be neutralized and washed out thoroughly, or the goods will be damaged. In the first washing following the bleacher, a dilute solution of sulphuric acid is used. This is followed by another washing in which the water may be made slightly alkaline.

Dyeing.—In the dyeing process, the goods are passed through dyeing vats and washing tanks until the desired effect is obtained. Much skill is required in making up the dyeing solution in order that it may give a good even shade of the color desired.

Mercerizing.—The finishing process known as "mercerizing" was invented by John Mercer, an English calico printer, in 1844. Mercerized goods are characterized by a beautiful luster which makes them resemble silk. Yarn, warp, skein, or goods may be mercerized by treating them with strong caustic soda solution or certain other reagents. With this treatment the goods are more attractive in appearance, stronger in fiber, and take dyes better. Principally long-staple cottons, especially Sea Island and Egyptian, are used for mercerizing.

Printing.—Certain goods, such as calicoes, are printed or stamped on one side with dyes of various colors, so as to give them a figured appearance. The printing is done on a press, which has a large cylinder and smaller engraved rollers. Under each engraved roller is a long copper trough containing a particular coloring matter. Between the engraved roller and the trough of coloring matter and dipping into the latter is a second roller known as a "furnisher," which transfers dye from the trough to the engraved roller. As the cloth moves by the different engraved rollers, each adds its particular color until the complete figure containing several colors is made.

The goods next pass into a drying room and are afterward steamed under slight pressure. The steaming causes the colors to penetrate fibers more thoroughly. Better classes of goods are

Fig. 137.—Woodside Cotton Mills, Greenville, S. C.

also washed to rid them of thickening matters mixed with the coloring.

There are many other details connected with the printing process, and there are other methods of making figured goods, but lack of space prevents further discussion.

References

1. Handy, R. B.: "History and General Statistics of Cotton," in "The Cotton Plant," *U. S. Dept. Agr. Bull.* 33, 1896.
2. Copeland, M. T.: "The Cotton Manufacturing Industry in the United States," Harvard University Press, Cambridge, Mass., 1917.
3. Anonymous: "The Fabric of Civilization," Guaranty Trust Company, New York, 1919.
4. Bigwood, George: "Cotton," Henry Holt & Company, New York, 1919.
5. Anonymous: "Cotton Spinning Machinery," article on Cotton Manufacture, Encyclopaedia Britannica, 11th ed., 1910.
6. Todd, John A.: "The World's Cotton Crops," A. & C. Black, Ltd., London, 1923.

CHAPTER XXV

COTTON STATISTICS

Statistics on any subject are commonly considered to be dry reading matter. They are devoid of interest except to the person who wishes to use them in interpreting a trend. By studying carefully figures on the cotton acreage, production, and resultant prices in different countries of the world for periods in the past, it is possible for a cotton grower, dealer, or manufacturer of cotton goods to work out certain laws which he may fit to his immediate conditions and thus be able to forecast, to some extent, the trend of future demand or future prices.

Accurate statistics furnish a convenient and reliable means of studying the history of the cotton industry in different countries and in different parts of the same country. It is only by means of statistical data that the industries of different countries can be compared.

The available statistics on various phases of the cotton industry are not very satisfactory. In the past, many countries had no regular or reliable system for collecting data. The figures given out were largely based on estimates. Even at present, some of the large cotton-producing countries, China, for instance, have no good system for collecting data. The United States probably has the best system in use, yet it is open to some objections. Some of the figures are based on estimates; in the past, careful distinction was not made between gross and net weights of bales or, at times, among crop year, calendar year, and fiscal year; nor was weight of bales given.

The size of bale, or package of lint cotton, in use in different countries varies from 100 to 750 pounds. The weights of tare, bagging, and ties around the bale vary considerably in different countries. In some countries the cotton goods produced are measured by the pound; in others, by the linear yard; and in still others, by the square yard. This lack of uniformity in different countries in methods of recording data makes comparison and study difficult.

Statistical System in the United States.—Since the United States has the most elaborate and probably the most satisfactory system for collecting cotton statistics and data on other crops, it may be well to discuss it briefly.

The Census Bureau of the Department of Commerce has local agents scattered over the cotton-growing states. These agents canvass the ginners twice a month during the ginning season and obtain figures on the actual number of bales ginned. They also visit cotton mills and other consuming establishments and obtain information on the amount of cotton and linters consumed and of stocks held in consuming establishments, compresses, and public storage places. Statistics of exports, imports, and stocks held at ports are collected by the Bureau of Foreign and Domestic Commerce.

The Division of Crop and Livestock Estimates, Bureau of Agricultural Economics, U. S. Department of Agriculture, has a large corps of field agents and local crop reporters, who make estimates of acreage, crop condition, probable yield, etc. This force is grouped into five divisions:

1. Regional cotton statisticians.
2. State agricultural statisticians, each with a corps of helpers.
3. Township reporters.
4. Ginners.
5. Farmers and others interested in the cotton industry.

The cotton statisticians are especially qualified for this work. They have had training in statistics and possess practical knowledge of crops. Each agent is assigned to a group of states, over which he travels systematically, rating acreage, noting general crop conditions, and consulting with the best informed authorities. The cotton statisticians meet with the Crop Reporting Board when a report on cotton is to be issued.

In each of the cotton-growing states there is a state agricultural statistician, who is an employee of the government. He has qualifications similar to those of the cotton statisticians mentioned above. He spends much time traveling over the state studying crop conditions, acreage, etc. He has a large force of crop reporters, scattered over the state, who report to him. From these reports and from his own observations, the state agricultural statistician makes his report to the bureau at Washington at regular periods.

There is also a large force of township and special reporters, who regularly make reports to the bureau at Washington a short time before each government report. They report conditions in

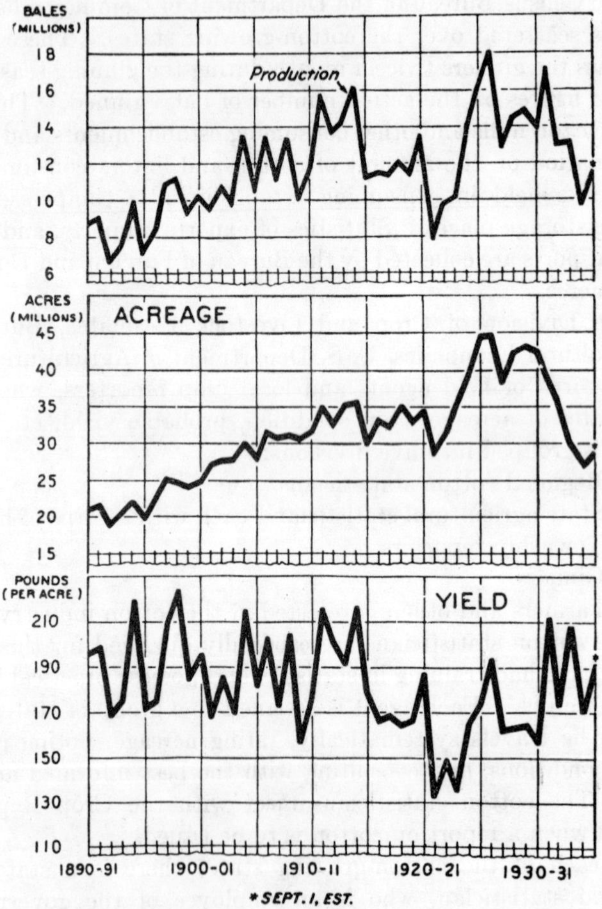

FIG. 138.—Cotton: United States production, acreage, and yield, 1890–1891 to 1935–1936.

their immediate vicinity or in the local area with which they are familiar. These reporters are either farmers or men interested in and acquainted with local crop conditions.

There are still other reporters, mostly farmers and merchants, who make reports at irregular periods or when called on.

TABLE LIII.—COTTON: ACREAGE, PRODUCTION, VALUE, AND FOREIGN TRADE, UNITED STATES, 1866–1935

Year	Acreage cultivation July 1,[a] 1,000 acres	Acreage harvested, 1,000 acres	Average yield per acre, pounds	Production,[b] 1,000 bales	Season average price per pound received by producers,[c] cents	Farm value, $1,000	Market price per pound, year beginning August[d]		Foreign trade, year beginning August		
							New York, cents	New Orleans, cents	Domestic exports,[e,f,g] 1,000 bales	Imports,[f,h] 1,000 bales	Net exports,[e,f,g,i] 1,000 bales
1866	......	7,666	121.5	2,097	.....	....	32.16	.....	1,323[j]	2	1,324
1867	......	7,864	142.6	2,520	.....	....	24.54	.....	1,511[k]	2	1,510
1868	......	6,973	150.7	2,366	.....	....	28.64	.....	1,288	6	1,284
1869	......	......	.....	*3,012*							
1869	......	7,751	155.4	3,011	.....	....	25.31	.....	1,980	4	1,977
1870	......	9,238	208.2	4,352	.....	....	17.04	.....	2,894	3	2,893
1871	......	8,285	159.0	2,974	.....	....	21.88	.....	1,851	7	1,844
1872	......	9,580	182.3	3,933	.....	....	20.22	.....	2,437	11	2,426
1873	......	10,998	168.3	4,168	.....	....	17.29	.....	2,706	5	2,702
1874	......	10,753	157.0	3,836	.....	....	15.67	.....	2,523	5	2,520
1875	......	11,348	181.2	4,631	.....	....	13.10	.....	3,003	5	2,999
1876	......	11,747	167.6	4,474	9.71	200,791	11.89	.....	2,869	6	2,864
1877	......	12,606	170.4	4,773	8.53	192,217	11.17	.....	3,198	7	3,194
1878	......	13,539	167.5	5,074	8.16	194,224	10.82	.....	3,265	6	3,259
1879	......	*14,480*	.....	*5,755*							
1879	......	14,474	180.5	5,756	10.28	281,784	12.13	.....	3,711	7	3,705
1880	......	15,921	190.9	6,606	9.83	312,925	11.36	.....	4,409	9	4,403
1881	......	16,483	149.0	5,456	10.66	274,440	12.09	.....	3,430	9	3,426
1882	......	15,638	208.9	6,949	9.12	311,644	10.81	.....	4,582	9	4,577
1883	......	16,295	162.0	5,713	9.13	252,501	10.87	.....	3,745	15	3,734
1884	......	16,849	155.1	5,682	9.19	251,581	10.74	.....	3,740	10	3,733
1885	......	17,922	169.9	6,576	8.39	267,481	9.47	.....	4,193	11	4,185
1886	......	18,370	164.3	6,505	8.06	254,733	9.91	.....	4,274	9	4,266
1887	......	18,703	175.1	7,047	8.55	294,527	10.15	.....	4,557	11	4,547
1888	......	19,520	169.5	6,938	8.50	294,183	10.44	.....	4,720	17	4,704
1889	......	*20,175*	.....	*7,473*							
1889	......	20,191	176.9	7,473	8.55	319,334	11.27	10.69	4,934	19	4,915
1890	......	20,937	195.5	8,653	8.59	368,108	9.48	9.08	5,859	45	5,815
1891	......	21,503	198.7	9,035	7.24	323,943	7.68	7.28	5,888	61	5,827
1892	......	18,869	168.7	6,700	8.34	277,556	8.45	8.15	4,456	90	4,363
1893	......	20,256	175.3	7,493	7.00	260,096	7.75	7.30	5,309	58	5,258
1894	......	21,886	219.0	9,901	4.59	230,071	6.38	5.86	7,010	104	6,908
1895	......	19,839	172.2	7,162	7.62	272,378	8.10	7.68	4,710	115	4,595
1896	......	23,230	175.2	8,533	6.66	283,463	7.71	7.28	6,172	119	6,057
1897	......	25,131	209.0	10,899	6.68	367,065	6.40	5.84	7,757	102	7,656
1898	......	24,715	223.1	11,278	5.73	330,282	6.00	5.46	7,662	105	7,557
1899	......	*24,275*	.....	*9,535*							
1899	......	24,163	185.0	9,346	6.98	326,208	8.36	8.03	6,228	140	6,091
1900	......	24,886	194.7	10,124	9.15	463,295	9.38	.....	6,800	109	6,692
1901	......	27,050	168.2	9,508	7.03	334,075	8.73	8.40	6,949	202	6,750
1902	......	27,561	184.7	10,630	7.60	403,717	9.96	9.64	7,084	151	6,936
1903	......	27,762	169.9	9,851	10.49	516,764	12.84	12.49	6,207	103	6,107
1904	......	30,077	213.7	13,438	8.98	603,433	9.09	8.70	8,908	129	8,781
1905	......	27,753	182.3	10,576	10.78	569,788	11.30	10.97	7,118	144	6,980
1906	......	31,404	202.3	13,274	9.58	635,537	11.24	10.92	8,943	227	8,741
1907	......	30,729	172.9	11,106	10.36	575,207	11.53	11.41	7,666	153	7,518
1908	......	31,091	203.8	13,241	9.01	596,608	10.23	9.80	8,955	181	8,778
1909	......	*32,044*	.....	*10,649*							
1909	31,744	30,555	156.5	10,005	13.60	680,246	14.66	14.33	6,353	170	6,194
1910	32,480	31,508	176.2	11,609	13.95	809,724	14.87	14.65	8,027	245	7,787
1911	35,634	34,916	215.0	15,694	9.60	752,925	10.85	10.85	11,116	233	10,885
1912	33,199	32,557	201.4	13,703	11.49	787,232	12.29	12.20	9,146	249	8,899
1913	35,721	35,206	192.3	14,153	12.50	884,926	13.21	13.12	9,508	273	9,251
1914	36,197	35,615	216.4	16,112	7.36	592,830	8.89[l]	.....	8,702	400	8,322
1915	30,544	29,951	178.5	11,172	11.22	626,774	11.98	11.68	6,113	458	5,673
1916	33,977	33,071	165.6	11,448	17.34	992,304	19.28	18.84	5,525	311	5,219
1917	33,064	32,245	167.4	11,284	27.12	1,529,862	29.68	28.96	4,402	231	4,175
1918	36,123	35,038	164.1	12,018	28.93	1,738,071	31.01	29.87	5,774	211	5,568
1919	......	*33,740*	.....	*11,376*							
1919	34,573	32,906	165.9	11,411	35.41	2,020,398	38.29	38.21	6,707	732	5,993

TABLE LIII.—COTTON: ACREAGE, PRODUCTION, VALUE, AND FOREIGN TRADE, UNITED STATES, 1866–1935.—(*Continued*)

Year	Acreage cultivation July 1,[a] 1,000 acres	Acreage harvested, 1,000 acres	Average yield per acre, pounds	Production,[b] 1,000 bales	Season average price per pound received by producers,[c] cents	Farm value, $1,000	Market price per pound, year beginning August[d]		Foreign trade, year beginning August		
							New York, cents	New Orleans, cents	Domestic exports,[e,f,g] 1,000 bales	Imports,[f,h] 1,000 bales	Net exports,[e,f,g,i] 1,000 bales
1920	35,872	34,408	186.7	13,429	15.92	1,069,257	17.89	16.55	5,973	237	5,753
1921	29,716	28,678	132.5	7,945	17.01	675,773	18.92	17.92	6,348	380	5,980
1922	32,176	31,361	148.8	9,755	22.87	1,115,578	26.24	25.94	5,007	492	4,536
1923	37,000	35,550	136.4	10,140	28.69	1,454,320	31.11	30.33	5,815	306	5,530
1924	......	*39,204*	.....	*13,683*							
1924	40,692	39,503	165.0	13,630	22.91	1,561,022	24.74	24.21	8,240	328	7,923
1925	45,972	44,390	173.5	16,105	19.59	1,577,091	20.53	19.71	8,267	340	7,939
1926	45,847	44,616	192.8	17,978	12.47	1,121,185	15.15	14.74	11,299	419	10,900
1927	39,479	38,349	161.7	12,956	20.19	1,308,088	20.42	19.98	7,857	354	7,522
1928	43,735	42,432	163.3	14,477	17.99	1,302,036	19.73	18.98	8,419	479	7,957
1929	......	*43,227*	.....	*14,574*							
1929	44,458	43,242	164.1	14,825	16.79	1,244,846	16.60	16.16	7,035	396	6,650
1930	43,339	42,454	157.0	13,932	9.46	659,041	10.38	10.08	7,133	112	7,029
1931	39,109	38,705	211.5	17,096	5.66	483,627	6.34	6.20	9,193	138	9,081
1932	36,542	35,939	173.3	13,002	6.52	424,006	7.37	7.26	8,895	136	8,766
1933	40,852	29,978[m]	208.5	13,047	10.17	663,507	11.09	10.92	7,964	156[n]	7,815
1934	27,883	26,987	170.9	9,636	12.36	595,602	12.44	12.44	5,037	112	4,925
1935[o]	27,872	27,331	188.0	10,734	11.10[p]	593,677					

[a] For 1909–1926, inclusive, the acreage figures relate to June 25 instead of July 1.
[b] Department figures are in running bales for all years prior to 1899 and in 500-pound gross-weight bales 1899–1935. Agricultural census figures for all periods are in running bales.
[c] Prior to 1908 prices are as of Dec. 1.
[d] New York prices 1866–1867 to August, 1871, Donnell, E. J., "Chronological and Statistical History of Cotton"; 1871–1872 to August, 1900, Commercial and Financial Chronicle, average of daily quotations, beginning 1900 from reports of the New York Cotton Exchange except Sept. 23–Nov. 16, 1914, when the exchange was closed (prices for this period from the Commercial and Financial Chronicle). New Orleans prices were from same sources prior to Aug. 16, 1915; since this date, from reports of the New Orleans Cotton Exchange direct to this bureau. These central market prices are for Middling grade, ⅞-inch staple, only.
[e] Excluding linters from 1914 to 1934.
[f] Compiled from "Commerce and Navigation of the United States," 1866–1917; "Foreign Commerce and Navigation of the United States," 1918; Monthly Summary of Foreign Commerce of the United States, June and July, 1919–1934, and January, 1927–1935.
[g] Bales of 500 pounds gross weight.
[h] Bales of 478 pounds net, which are equivalent to bales of 500 pounds gross weight.
[i] Total exports (domestic plus foreign) minus imports.
[j] Year beginning July
[k] 13 months, July-July.
[l] Average for 9 months only. Exchange closed Aug. 1–Nov. 17, on account of war.
[m] Area in cultivation July 1 less removal of acreage reported by the Agricultural Adjustment Administration, less abandonment on area not under contract.
[n] Includes imports for consumption, January–June, 1934; reexports not considered beginning this date.
[o] Preliminary.
[p] Season average price to Dec. 1.

NOTE: From Bureau of Agricultural Economics. Agricultural census figures in italics. Production figures conform with census annual ginning enumerations, with allowance for cross-state ginnings; state figures rounded to thousands and added for United States total. Cotton grown in Lower California, Old Mexico, ginned in California, 1913–1924 has been excluded.

The date and hour when each report shall be issued are fixed by law. These official reports are prepared and released by the Crop Reporting Board at Washington. The information from various sources is carefully guarded until the board meets on the day of the report, when, behind locked doors, it is all brought together for the first time. The utmost precautions are taken to prevent any possibility of the premature release of any of the official report, owing to the effect that it frequently has on the speculative market.

TABLE LIV.—COTTON: ACREAGE, YIELD, PRODUCTION OF LINT IN 500-POUND GROSS-WEIGHT BALES, AND SEASON AVERAGE PRICE PER POUND RECEIVED BY PRODUCERS, BY STATES, AVERAGES, AND ANNUAL 1934 AND 1935

(*Bureau of Agricultural Economics*)

State	Acreage harvested			Yield per acre			Production[a]			Price for crop of—	
	Average, 1928–1932, 1,000 acres	1934, 1,000 acres	1935,[b] 1,000 acres	Average, 1923–1932, pounds	1934, pounds	1935,[b] pounds	Average, 1928–1932, 1,000 bales	1934, 1,000 bales	1935,[b] 1,000 bales	1934, cents	1935,[c] cents
Missouri	374	316	307	256	366	311	229	242	200	12.37	11.20
Virginia	79	58	53	270	290	271	45	35	30	12.08	11.30
North Carolina	1,432	951	932	269	316	300	752	629	585	12.31	11.40
South Carolina	1,879	1,302	1,341	208	250	266	856	681	745	12.35	11.20
Georgia	3,166	2,103	2,177	176	220	233	1,241	968	1,060	12.12	11.00
Florida	124	91	86	124	146	161	35	28	29	12.24	10.30
Tennessee	1,065	744	729	197	260	210	479	404	320	12.05	10.90
Alabama	3,373	2,133	2,240	172	213	226	1,255	950	1,060	12.10	10.60
Mississippi	3,977	2,485	2,622	191	220	230	1,559	1,143	1,260	12.50	11.30
Arkansas	3,382	2,162	2,140	188	192	199	1,351	867	890	12.23	11.20
Louisiana	1,847	1,160	1,230	192	200	216	745	485	555	12.46	11.00
Oklahoma	3,707	2,708	2,383	149	56	108	1,109	317	535	11.77	10.70
Texas	15,598	10,297	10,606	139	112	138	4,580	2,406	3,050	12.51	10.90
New Mexico	122	90	89	318	474	420	90	89	78	12.99	11.90
Arizona	186	136	156	322	410	383	128	117	125	13.97	12.40
California	222	223	219	386	556	448	200	259	205	12.98	12.10
All other	20	28	21	227	282	161	11	16	7	12.24	10.90
United States	40,554	26,987	27,331	169.9	170.9	188.0	14,666	9,636	10,734	12.36	11.10
Lower California (old Mexico)	101	55	113	242	190	297	48	22	70		

[a] Compiled from reports of the Bureau of the Census. Slight differences from census figures on ginnings are due to ginnings in one state of cotton grown in another.

[b] Preliminary estimate of the U. S. Department of Agriculture.

[c] Preliminary. Season average price to Dec. 1.

Sources of Information.—Much of the data given in Tables LIII to LXXII is taken from *Statistical Bulletin* 62, U. S. Department of Agriculture. Where the data are from other sources, the source is stated.

Table LV.—Acreage, Yield, and Production of American-Egyptian Cotton, 1911–1930
(Bureau of Agricultural Economics)

Year	Acreage harvested			Yield of lint per acre		Production		
	Arizona, acres	California, acres	Total, acres	Arizona, pounds	California, pounds	Arizona, bales[a]	California, bales[a]	Total, bales[a]
1911	......	30	30	...	500	......	30	30
1912	400	150	550	300	387	240	116	356
1913	4,000	62	4,062	275	315	2,200	39	2,239
1914	12,000	550[b]	12,550	258	100	6,200	110	6,310
1915	2,600	......	2,600	222	...	1,150	.....	1,150
1916	7,300	......	7,300	226	...	3,300	.....	3,300
1917	33,000	2,400	35,400	230	206	15,200	990	16,190
1918	72,000	6,600	78,600	238	228	34,227	3,007	37,234
1919	87,000	1,500	88,500	229	225	39,817	675	40,492
1920	200,000	43,000	243,000	205	112	82,041	9,650	91,691
1921	75,000	9,100	84,100	234	181	35,032	3,300	38,332
1922	77,000	100	77,100	220	325	33,907	65	33,972
1923	40,000	......	40,000	287	...	22,960	.....	22,960
1924	8,000	......	8,000	274	...	4,374	.....	4,374
1925	40,000	......	40,000	257	...	20,586	.....	20,586
1926	28,000	......	28,000	277	...	15,512	.....	15,512
1927	44,000	......	44,000	275	...	25,000	.....	25,000
1928	50,000	......	50,000	284	...	30,000	.....	30,000
1929	67,000	......	67,000	211	...	30,000	.....	30,000
1930	46,000	......	46,000	251	...	24,000	.....	24,000

[a] Bales of 478 pounds net weight.

[b] Includes 100 acres of American-Egyptian grown in Lower California, Old Mexico, which yielded 250 pounds per acre, equal to 50 bales. All ginned in California.

TABLE LVI.—PRODUCTION OF SEA ISLAND COTTON IN THE UNITED STATES, 1899–1930
(*Bureau of the Census*)

Growth year	Production, bales[a]	Growth year	Production, bales[a]
1899	97,279	1915	91,844
1900	88,294	1916	117,559
1901	77,879	1917	92,619
1902	104,953	1918	52,208
1903	75,393	1919	6,916
1904	104,317	1920	1,868
1905	112,539	1921	5,327
1906	57,550	1922	5,125
1907	86,895	1923	785
1908	93,858	1924	11
1909	94,791	1925	18
1910	90,368	1926	23
1911	119,293	1927	179
1912	73,777	1928	22
1913	77,563	1929	7
1914	81,654	1930	20

[a] Running bales.

TABLE LVII.—NUMBER OF ACTIVE AND IDLE GINNERIES, AND AVERAGE NUMBER OF RUNNING BALES, EXCLUDING LINTERS, GINNED PER ACTIVE ESTABLISHMENT, BY STATES, 1920–1923

State	Growth year	Number of ginneries			Average number of running bales ginned per active establishment
		Total	Active	Idle	
United States	1923	19,195	15,298	3,897	665
	1922	19,939	15,420	4,519	631
	1921	20,938	16,192	4,746	493
	1920	21,876	18,440	3,436	720
Alabama	1923	1,750	1,364	386	439
	1922	1,840	1,427	413	575
	1921	1,922	1,415	507	415
	1920	2,040	1,585	455	423
Arizona	1923	63	46	17	1,689
	1922	54	36	18	1,226
	1921	51	36	15	1,192
	1920	52	50	2	2,104
Arkansas	1923	1,540	1,353	187	476
	1922	1,582	1,400	182	722
	1921	1,653	1,462	191	539
	1920	1,702	1,531	171	772
California	1923	41	26	15	2,127
	1922	46	26	20	1,095
	1921	49	30	19	1,160
	1920	51	45	6	1,731
Florida	1923	135	68	67	200
	1922	149	84	65	327
	1921	165	73	92	167
	1920	198	87	111	223
Georgia	1923	2,677	1,715	962	357
	1922	2,914	1,890	1,024	389
	1921	3,152	2,218	934	371
	1920	3,313	2,639	674	548
Louisiana	1923	962	720	242	519
	1922	1,042	748	294	462
	1921	1,095	793	302	359
	1920	1,149	884	265	441
Mississippi	1923	1,751	1,433	318	434
	1922	1,853	1,513	340	652
	1921	1,927	1,535	392	532
	1920	1,988	1,643	345	548
Missouri	1923	144	136	8	917
	1922	101	94	7	1,488
	1921	103	78	25	874
	1920	110	94	16	812
North Carolina	1923	2,070	1,690	380	623
	1922	2,140	1,730	410	508
	1921	2,237	1,807	430	445
	1920	2,358	1,961	397	484
Oklahoma	1923	931	853	78	781
	1922	933	770	163	827
	1921	961	737	224	648
	1920	1,018	963	55	1,353
South Carolina	1923	2,648	1,779	869	446
	1922	2,850	1,788	1,062	289
	1921	3,029	2,258	771	348
	1920	3,127	2,757	370	599
Tennessee	1923	518	449	69	524
	1922	509	448	61	861
	1921	539	454	85	655
	1920	572	480	92	656
Texas	1923	3,791	3,511	280	1,200
	1922	3,772	3,340	432	936
	1921	3.905	3,176	729	671
	1920	4,047	3,592	455	1,155
Virginia	1923	142	124	18	419
	1922	136	110	26	246
	1921	134	106	28	157
	1920	136	115	21	190
All other states[a]	1923	32	31	1	1,118
	1922	18	16	2	1,222
	1921	16	14	2	624
	1920	15	14	1	950

[a] Includes Illinois, Kansas, Kentucky, and New Mexico.

TABLE LVIII.—PERCENTAGE OF TOTAL COTTON GINNED TO SPECIFIED DATES, BY STATES, 1920–1923

State	Growth year	Percentage total cotton ginned to								
		Sept. 1	Sept. 25	Oct. 18	Nov. 1	Nov. 14	Dec. 1	Dec. 13	Jan. 1	Jan. 16
United States.......	1935	7.9	30.2	63.2	74.3	81.0	88.8	93.6	....	98.3
	1923	11.2	31.8	36.0	74.3	82.3	90.9	93.9	96.4	97.8
	1922	8.3	39.7	71.7	83.7	91.2	95.8	97.5	98.6	99.2
	1921	6.1	36.6	68.9	83.3	91.2	95.8	97.7	98.8	99.2
	1920	2.6	17.0	43.4	56.6	67.2	76.4	82.0	87.1	90.5
Alabama..........	1923	0.8	26.9	66.6	82.1	91.6	97.4	98.7	99.3	99.5
	1922	6.8	40.2	74.4	86.8	94.3	98.1	99.1	99.6	99.7
	1921	2.2	39.2	72.7	87.2	94.3	97.7	98.8	99.4	99.5
	1920	0.2	12.5	42.1	61.5	75.0	85.0	90.2	94.3	96.2
Arizona............	1923	1.0	9.8	28.9	43.5	52.4	67.2	74.3	80.3	88.4
	1922	a	4.6	23.2	37.4	50.3	64.5	71.3	83.1	90.9
	1921	0.3	6.9	20.2	31.9	43.7	60.1	70.6	82.6	89.6
	1920	0.3	5.3	18.7	28.6	38.4	54.2	63.5	73.7	83.3
Arkansas...........	1923	0.7	7.6	47.3	60.9	73.1	87.4	91.0	94.5	96.4
	1922	0.5	27.8	64.5	81.8	91.6	96.6	97.0	98.8	99.4
	1921	a	20.0	59.4	79.4	90.6	95.9	98.1	99.2	99.6
	1920	a	4.8	29.7	44.2	57.0	69.1	75.3	81.2	86.0
California..........	1923	1.8	7.8	21.9	37.5	46.5	59.2	64.6	72.2	82.0
	1922	0.2	3.7	16.6	29.9	45.1	60.9	74.1	83.3	86.9
	1921	0.5	4.3	11.3	19.7	30.6	45.2	56.8	66.7	75.5
	1920	2.3	8.7	17.0	21.8	32.0	42.0	49.0	60.5	72.3
Florida............	1923	5.9	33.9	75.8	83.8	90.2	96.0	98.1	98.6	98.9
	1922	19.6	46.2	74.0	82.0	93.0	95.6	98.9	99.6	99.7
	1921	3.2	35.1	67.1	76.9	91.8	96.2	98.6	99.1	99.4
	1920	0.6	16.9	50.4	60.6	77.0	82.4	91.5	93.7	95.4
Georgia............	1923	3.2	30.4	67.5	79.8	88.6	95.0	97.9	98.9	99.2
	1922	19.2	50.7	77.3	85.9	92.6	96.8	98.4	99.2	99.4
	1921	5.8	47.7	77.4	89.5	94.9	97.6	98.7	99.0	99.5
	1920	1.0	19.3	50.2	68.8	80.5	86.8	91.2	94.3	95.7
Louisiana..........	1923	3.3	29.0	66.4	79.3	88.0	95.3	97.1	98.1	98.6
	1922	4.2	46.1	79.9	90.3	95.2	98.2	99.0	99.6	99.7
	1921	1.0	35.7	68.6	83.3	91.7	97.0	98.2	99.1	99.3
	1920	0.7	22.3	52.5	69.5	76.8	85.8	91.0	95.0	96.7
Mississippi.........	1923	0.2	17.6	56.3	73.5	85.6	95.5	97.8	98.5	98.9
	1922	1.1	35.8	70.6	85.2	93.4	97.7	99.0	99.5	99.6
	1921	0.5	30.7	62.5	78.6	89.7	96.6	98.5	99.4	99.5
	1920	0.1	10.7	38.8	55.3	67.8	80.3	86.2	91.2	93.4
Missouri...........	1923	....	0.6	29.3	43.1	53.8	70.7	75.5	82.7	88.8
	1922	a	14.9	45.1	64.6	80.1	90.9	93.5	96.2	98.4
	1921	a	20.9	59.4	81.2	94.0	97.1	98.9	99.7	99.7
	1920	....	0.3	17.8	35.1	46.6	57.8	64.7	72.2	80.3

[a] Less than one-tenth of 1 per cent.

TABLE LVIII.—PERCENTAGE OF TOTAL COTTON GINNED TO SPECIFIED DATES, BY STATES, 1920–1923.—(*Continued*)

State	Growth year	Percentage total cotton ginned to								
		Sept. 1	Sept. 25	Oct. 18	Nov. 1	Nov. 14	Dec. 1	Dec. 13	Jan. 1	Jan. 16
North Carolina	1923	[a]	14.9	53.5	69.3	79.0	89.3	93.7	96.5	97.6
	1922	0.1	17.0	50.5	65.6	78.7	90.0	94.4	96.7	98.0
	1921	[a]	17.6	55.2	72.4	83.6	91.9	95.5	97.5	98.4
	1920	[a]	3.3	20.1	37.5	52.8	64.4	71.7	79.4	84.5
Oklahoma	1923	0.9	9.6	32.9	44.4	55.3	76.3	84.9	93.2	96.6
	1922	0.5	29.1	69.0	84.8	92.6	97.0	98.3	99.3	99.5
	1921	0.3	29.4	73.8	91.1	96.7	98.5	99.2	99.7	99.8
	1920	[a]	5.1	29.0	36.5	47.1	59.2	66.1	74.2	80.9
South Carolina	1923	0.3	24.2	64.0	79.0	87.7	94.5	97.1	98.4	98.8
	1922	0.9	28.6	65.0	79.8	89.4	95.3	97.1	98.2	98.8
	1921	0.1	27.4	62.7	79.2	87.1	93.5	96.4	98.0	98.7
	1920	[a]	10.4	34.0	52.5	66.5	76.2	82.6	88.0	91.2
Tennessee	1923	....	1.4	35.1	52.0	66.8	85.0	89.9	93.9	95.7
	1922	[a]	13.9	52.8	72.8	85.9	94.7	96.7	98.1	99.0
	1921	[a]	14.2	53.8	75.0	87.5	94.4	98.1	99.2	99.5
	1920	....	0.3	17.0	36.6	52.6	66.6	74.8	80.0	88.1
Texas	1923	25.9	51.8	76.4	83.1	87.8	93.0	94.9	97.0	98.0
	1922	18.1	57.4	84.5	91.1	94.7	96.6	97.9	98.9	99.3
	1921	19.5	57.4	81.6	90.6	95.2	97.5	98.5	99.4	99.6
	1920	7.9	32.8	62.6	69.9	76.1	82.5	86.7	90.3	93.1
Virginia	1923	....	1.7	27.4	42.3	57.8	72.8	83.1	89.3	92.3
	1922	....	2.5	29.4	50.4	72.3	86.3	92.5	95.2	96.2
	1921	....	6.8	45.1	63.8	79.8	90.2	94.5	96.5	98.0
	1920	....	....	5.0	14.1	29.4	45.1	56.5	62.8	66.6
All other states[b]	1923	....	3.0	22.8	37.4	48.0	65.3	71.3	75.3	83.6
	1922	....	6.6	24.5	37.3	52.3	74.6	82.0	90.3	95.1
	1921	....	4.3	41.4	58.7	74.2	84.7	94.7	97.5	99.6
	1920	....	....	15.3	23.4	34.2	49.8	62.9	72.9	85.6

[a] Less than one-tenth of 1 per cent.

[b] Includes Illinois, Kansas, Kentucky, and New Mexico.

Table LIX.—Percentage Reduction from Full Yield per Acre of Cotton from Stated Causes, as Reported by Crop Correspondents, 1909–1924

Year	Deficient moisture	Excessive moisture	Other climatic	Total climatic	Plant diseases	Insect pests	Other or unknown	Total
1924	14.0	5.0	2.3	21.3	0.8	12.0	0.6	34.7
1923	7.2	8.0	2.8	18.0	0.7	26.6	0.2	45.5
1922	10.3	4.9	2.3	17.5	0.8	26.7	0.2	45.2
1921	8.6	4.3	3.1	16.0	1.0	35.4	0.5	52.9
1920	2.2	8.8	2.1	13.1	1.1	24.0	0.8	39.0
1919	2.7	15.3	3.2	21.2	1.4	18.8	0.5	41.9
1918	23.8	0.9	4.5	29.2	2.0	7.9	1.2	40.3
1917	15.1	1.7	8.7	25.5	1.3	12.3	0.8	39.9
1916	9.2	9.1	6.9	25.2	0.9	15.7	0.6	42.4
1915	6.8	5.7	6.8	19.3	1.9	12.2	3.4	36.8
1914	7.9	2.9	3.0	13.8	0.2	9.8	1.6	25.4
1913	15.2	2.0	5.9	23.1	0.5	8.9	1.2	33.7
1912	8.1	7.6	5.0	20.7	4.3	6.5	1.2	32.7
1911	9.8	2.6	3.0	15.4	0.4	7.9	2.4	26.1
1910	12.2	5.1	5.3	22.6	0.4	7.5	5.1	35.6
1909	14.9	6.0	7.7	28.6	4.2	7.9	1.3	42.0

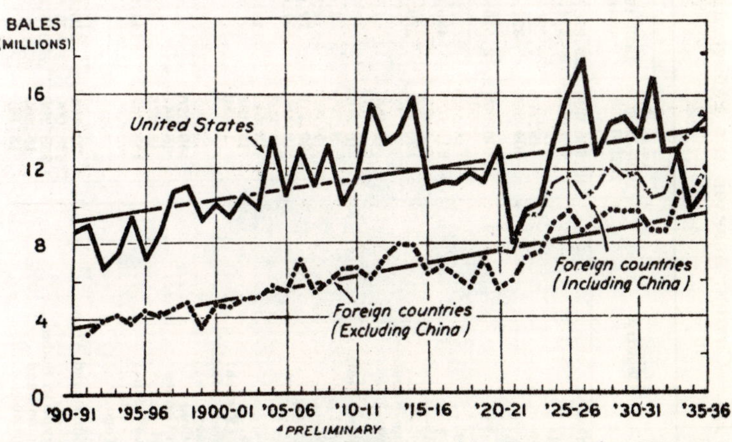

Fig. 139.—Cotton production in United States and foreign countries, 1890–1891 to 1935–1936.

TABLE LX.—COTTON: ACREAGE AND PRODUCTION IN SPECIFIED COUNTRIES, AVERAGE 1925–1926 TO 1929–1930, ANNUAL 1933–1934 TO 1935–1936

Country	Acreage				Production			
	Average, 1925–1926 to 1929–1930, acres	1933–1934, acres	1934–1935, acres	1935–1936,[a] acres	Average, 1925–1926 to 1929–1930, bales[b]	1933–1934, bales[b]	1934–1935, bales[b]	1935–1936,[a] bales[b]
United States	42,606,000	29,978,000	26,987,000	27,331,000	15,268,000	13,047,000	9,636,000	10,635,000
Mexico	471,632	424,288	418,000	585,000	252,805	260,426	222,973	234,569
Venezuela	....	....	....	....	33,095[c]	10,000	7,245	
Colombia	49,273	....	....	....	14,305			
Peru	304,302	322,000	367,000	....	245,705	277,781	341,970	
Ecuador	....	....	....	....	5,776	5,188	7,782	
Brazil	1,306,000	2,519,000	....	....	525,140	1,011,000	1 324,000	1,743,000
Bolivia	5,601[d]	....	....	....	2,139[c]			
Paraguay	25,163[c]	....	....	....	12,328[c]			
Argentina	241,073	480,000	707,000	909,000	115,392	200,000	295,352	
Guatemala	697	....	....	....	397			
Haiti	130,269	270,080	....	....	22,324[e]	30,000		
Dominican Republic	....	....	....	....	351[c]			
Porto Rico	10,020	....	....	....	2,030			
Salvador	....	....	....	....	774[c,e]			
British West Indies	16,807	....	....	....	4,500			
Italy	8,772[c]	3,620	7,000	....	3,300[f]	1,183	4,000	
Yugoslavia	1,763	2,026	....	....	392	315		
Greece	39,819	72,000	91,000	133,000	15,016	32,000	36,000	62,000
Bulgaria	10,869	51,000	48,000	89,000	3,046	11,000	18,000	39,000
Malta	993	62	....	....	427			
Spain	13,643	18,533	....	....	2,974	4,179		
Algeria	15,138	50	....	....	6,200	21		
Morocco (French)	1,480	....	....	....	448			
French West Africa:								
Dahomey	....	....	....	....	6,344	3,920	4,100	
Ivory Coast	149,376[g]	....	....	....	7,646[e]			
French Guinea	18,841[c]	....	....	....	2,406[c]			
Senegal	48,975	9,884	....	....	1,695	369		
French Sudan	158,267[g]	144,974	....	....	9,486	8,071		
Upper Volta	....	....	....	....	13,268			
French Togo	....	....	....	....	7,732			
Nigeria	....	....	....	....	28,846[e]	21,757	42,000	
French Equatorial Africa	22,164[d]	210,035	....	....	1,698[g]	20,708	32,300	
Egypt	1,828,000	1,873,000	1,798,000	1,733,000	1,587,000	1,777,000	1,566,000	1,750,000
Anglo-Egyptian Sudan	269,200	333,000	365,000	....	125,547	135,000	227,474	
Italian Somaliland	15,862	....	....	....	4,005	....	3,099	
Niger Territory	18,162[c]	....	....	....	1,764	1,476		

Eritrea	6,487[c]	12,000	...	...	1,660	872	1,207	1,000
Gold Coast	...	...	...	...	133[e]			
Belgian Congo	26,156	...	...	...	26,202			
Kenya	...	...	...	...	1,299	5,648		
Uganda	615,441	1,091,000	1,186,000	...	131,257	239,031	205,742	
Angola	...	...	...	...	3,022[e]			
Tanganyika	...	...	...	...	20,537	23,841		
Nyasaland	33,374	30,240	...	...	4,759	4,490	8,100	15,000
Northern Rhodesia[h]	2,566	...	...	...	121			
Southern Rhodesia	16,706	...	...	...	1,495			
Mozambique	36,634	...	...	...	9,040			
Union of South Africa	64,491	...	...	...	11,318[i]	1,952	2,652	
Cyprus	11,342	4,757	...	...	2,532	764		
Ceylon	1,631	...	...	...	192			
Turkey (Asiatic)	334,230	400,000	486,000	...	93,938	127,710	162,000	203,000
Syria and Lebanon	54,977	19,276	31,000	...	9,886	5,277	4,262	
Union of Soviet Socialist Republics	2,017,000	5,070,000	4,787,000	...	1,032,000	1,887,000	1,738,000	2,250,000
Iraq	15,000[f]	...	...	...	2,977	428	2,000	
Iran	...	...	...	...	95,160	100,000[j]		
India	26,192,000	24,136,000	24,023,000	...	4,724,000	4,241,000	4,065,000	4,793,000
China[k]	4,519,000	6,142,000	6,827,000	...	2,060,000	2,726,000	3,125,000	2,500,000
Japan	2,857	1,826	...	...	1,090	814		
Chosen	495,232	433,000	474,000	514,000	149,593	140,000	136,000	175,261
Manchuria	...	141,000	228,000	160,000	...	70,000	86,471	55,788
French Indo-China	36,984	38,140	...	...	5,664	5,797		
Netherland India	21,708	20,000	...	...	4,708	5,649		
Siam	8,951	...	...	...	3,244			
Australia	18,332	50,000	...	...	7,311	18,533		
New Hebrides	...	...	...	...	2,374[f]			
Estimated world total, including China	83,120,000	74,600,000	73,600,000	...	26,763,000	26,569,000	23,622,000	26,000,000

[a] Preliminary. [b] Bales of 478 pounds net. [c] Average for 4 years. [d] Average for 2 years. [e] Exports. [f] Estimates for 1 year.
[g] Average for 3 years.
[h] Production has been discontinued with the exception of a few experimental plots under government supervision.
[i] Includes Swaziland.
[j] From an unofficial source.
[k] From reports of the Chinese Cotton Statistics Association. Figures represent the crop in the most important cotton provinces where the commercial crop is grown.

NOTE: From Bureau of Agricultural Economics; from official sources, International Institute of Agriculture, and estimates of the Bureau of Agricultural Economics except as noted.

Data for crop year as given at the head of table are for crops harvested between Aug. 1 and July 31.

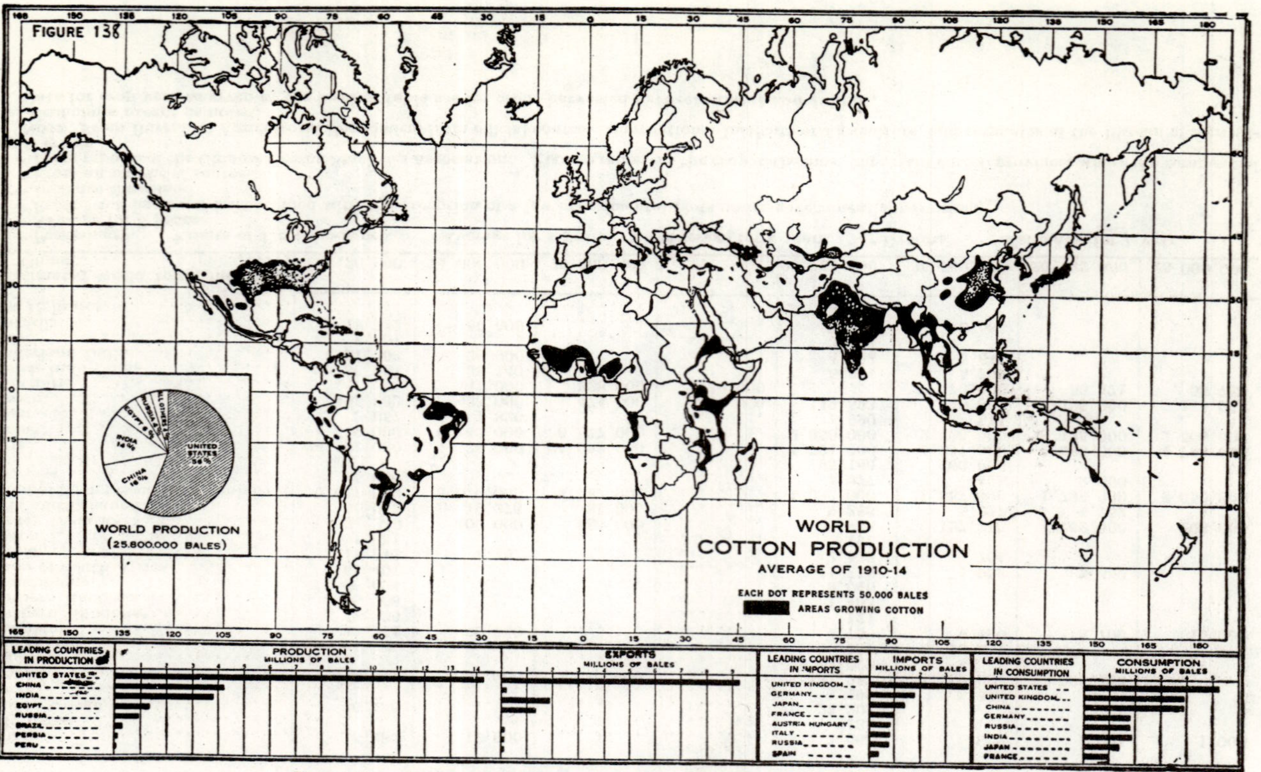

Fig. 140.—Cotton-producing areas of the world. (*After U. S. D. A.*)

World's Production and Consumption of Cotton.—The sources of information in regard to the production, consumption, and stocks of cotton have never been adequate or altogether satisfactory for many countries. The United States is the only country that has provided an efficient means of determining the supply and distribution of cotton within its borders. Other governments in the principal cotton-producing countries have been providing estimates, approximately correct and fairly reliable. The Indian government publishes a memorandum from time to time during the cotton year based upon estimates received from the several producing provinces and the statistics of yarn production compiled from returns of the spinners. The Egyptian government publishes monthly returns showing the number and weight of steam-pressed bales from Sept. 1 of each season to the date to which the report relates, but complete data are not available until the end of the year and do not necessarily represent the actual growth of the season.

Information as to production, consumption, and stocks in all countries except the United States is compiled from reports transmitted by the representatives of the government in foreign countries and from special correspondence, trade reports, and such accounts of the movements of cotton as are available.

While the average weight of the American bale approximates 500 pounds, the weights of the packages into which the several foreign cottons are packed vary from 200 pounds or less to 750 pounds or more. For this reason, the statistics for all these countries have been expressed in bales of 478 pounds, which is practically equivalent to a 500-pound gross-weight bale, in order to provide a practical unit for comparative purposes that will at the same time convey the conventional idea of actual quantity.

TABLE LXI.—COTTON: GRADE, STAPLE LENGTH, AND TENDERABILITY OF CROP AND CARRY-OVER, UNITED STATES, 1931–1932 TO 1934–1935

Item	Crop, 1,000 bales				Carry-over Aug. 1,[a] 1,000 bales				
	1931–1932	1932–1933	1933–1934	1934–1935	1931	1932	1933	1934	1935
Total[b]	16,628 9	12,709.6	12,664.0	9,472.0	6,262.7	9,576.8	8,079.5	7,645.1	7,137.5
Total American upland	16,615.2	12,701.3	12,654.3	9,458.0	6,246.0	9,560.3	8,069.7	7,638.1	7,128.9
Total American-Egyptian	13.7	8.3	9.7	14.0	16.7	16.5	9.8	7.0	8.6
Grade (American upland):									
Extra White:									
Above Good Middling	0.6	1.8	16.0	10.3	......	......	0.1	0.3	0.2
Good Middling	76.7	106.9	363.3	914.1	30.4	33.3	21.7	157.7	50 8
Strict Middling	174.2	132.8	623.6	2,259.3	24.6	40.4	32.1	821.0	519.4
Middling	88.1	88.2	409.5	1,284.2	16.8	19.0	34.2	1,128.1	926.2
Strict Low Middling	62.2	99.6	189.5	281.9	6.6	8.9	32.2	460.7	383.0
Low Middling	29.7	28.9	44.7	63.0	0.9	2.5	14.1	81.4	74.2
Below Low Middling	4.3	1.8	0.9	1.6	0.1	0.1	0.9	4.1	0.7
White:									
Middling Fair									
Strict Good Middling	10.9	1.2	2.0	0.6	3.2	3.2	2.2	1.8	0.9
Good Middling	940.0	251.3	264.7	110.0	219.9	454.7	202.1	125.2	119.3
Strict Middling	5,873.4	3,147.0	2,409.5	1,187.2	1,536.3	3,183.5	1,931.7	1,079.4	1,141.7
Middling	5,233.2	4,474.5	2,899.1	1,424.6	2,077.8	3,292.2	2,801.6	1,438.7	1,770.3
Strict Low Middling	1,759.2	1,569.2	1,143.3	468.2	928.3	1,083.3	1,210.1	686.5	848.1
Low Middling	640.3	330.3	238.0	107.2	273.9	243.1	255.0	171.9	160.6
Strict Good Ordinary	421.9	116.3	52.6	33.4	71.4	148.6	144.7	72.5	40.0
Good Ordinary	160.8	55.5	11.1	11.4	21.3	98.5	82.6	51.8	16.0
Spotted:									
Good Middling	115.3	193.6	479.4	130.2	93.1	102.4	102.6	111.9	45.5
Strict Middling	428.5	1,054.0	2,130.1	664.4	383.0	392.3	547.3	478.1	445.3
Middling	247.9	673.0	1,041.6	345.4	348.2	244.3	385.9	378.8	354.6
Strict Low Middling	185.2	217.5	224.5	96.3	95.3	59.0	101.3	112.5	76.8
Low Middling	71.3	78.8	57.9	28.2	27.1	31.4	56.8	57.5	19.5
Yellow-tinged:									
Strict Good Middling	......	......	......	0.1	0.1	0.1	......	0.2	0.3
Good Middling	1.6	2.7	4.0	0.9	4.9	3.2	2.7	3.6	1.3
Strict Middling	4.5	10.7	7.7	3.2	18.3	16.5	14.2	22.9	17.4
Middling	7.3	9.0	6.2	5.5	21.2	19.5	16.0	27.5	28.0
Strict Low Middling	8.1	9.7	2.3	3.0	11.5	11.3	8.3	45.4	36.3
Low Middling	2.4	1.7	1.2	1.3	5.5	4.9	4.0	35.3	14.3
Light Yellow-stained:									
Good Middling	0.1	0.1	0.2	......	0.1	0.1	0.1	0.2	0.1
Strict Middling	0.2	0.1	0.1	0.2	0.4	0.2	0.2	0.1	0.2
Middling	0.4	0.1	0.2	0.1	1.1	0.5	0.2	0.2	0.3
Yellow-stained:									
Good Middling	......	......	......	......	......	......	......	0.1	0.1
Strict Middling	......	......	......	......	0.4	0.4	0.1	0.4	0.3
Middling	0.1	0.1	......	......	1.6	0.9	0.1	1.8	1.7
Gray:									
Good Middling	0.5	1.2	......	0.1	0.1	0.4	0.2		•
Strict Middling	6.9	5.9	1.8	7.4	0.7	2.1	2.1	2.7	2.6
Middling	5.1	3.0	1.1	2.6	0.6	2.1	1.7	2.7	1.4
Blue-stained:									
Good Middling	......	......	......	......	0.1				
Strict Middling	0.1	......	......	......				0.2	
Middling	......	......	......	......	0.2	0.2	......	0.3	0.1
No grade[c]	54.2	34.8	28.2	12.1	21.0	57.2	60.6	74.6	31.4
Staple length (American upland):									
Shorter than 7/8 inch	1,019.5	837.7	539.1	783.0	463.2	298.3	188.4	233.4	528.8
7/8 and 29/32 inch	6,593.3	4,786.5	4,504.4	3,490.9	2,615.7	3,392.6	2,503.6	2,534.1	2,773.6
15/16 and 31/32 inch	4,511.9	3,671.0	3,992.2	2,065.4	1,528.2	2,704.0	2,199.3	2,112.9	1,799.5
1 and 1 1/32 inches	2,557.1	1,822.0	2,004.3	1,415.6	849.2	1,657.6	1,774.6	1,477.4	996.2
1 1/16 and 1 3/32 inches	1,087.8	871.8	824.1	880.1	414.8	754.5	671.7	615.6	533.1
1 1/8 and 1 5/32 inches	590.0	622.1	640.3	680.6	269.5	546.7	562.9	529.2	353.7
1 3/16 and 1 7/32 inches	224.6	84.5	143.8	123.1	89.7	174.0	143.6	111.6	108.9
1 1/4 inches and longer	31.0	5.7	6.1	19.3	15.7	32.6	25.6	23.9	35.1
Tenderability:[d]									
Total tenderable	14,833.9	11,489.1	11,782.8	8,518.7	5,543.3	8,882.7	7,437.4	6,969.8	6,371.0
Total untenderable	1,781.3	1,212.2	871.5	939.3	702.7	677.6	632.3	668.3	757.9

[a] Carry-over of foreign cotton not included.
[b] Report of Bureau of the Census.
[c] Includes bales not otherwise classified above.
[d] According to sec. 5, United States Cotton Futures Act.

NOTE: From Bureau of Agricultural Economics. Adapted from *Statistical Bulletins* 40, 47, 52, and a manuscript now ready for the printer.

TABLE LXII.—PERCENTAGE OF AMERICAN UPLAND COTTON OF EACH STAPLE LENGTH, BY STATES, 1928–1929, 1929–1930, AND 1930–1931

State	Year	Total	Staple length, inches							
			13/16 and under	7/8	15/16	1 and 1 1/32	1 1/16 and 1 3/32	1 1/8 and 1 5/32	1 3/16 and 1 7/32	1 1/4 and longer
		%	%	%	%	%	%	%	%	%
United States	1928–1929	100.0	14.5	41.5	22.6	11.0	5.6	3.4	1.2	0.2
	1929–1930	100.0	20.1	38.1	18.9	11.7	6.5	3.8	0.8	0.1
	1930–1931	100.0	13.4	38.8	24.9	12.7	7.0	2.8	0.4	[a]
Alabama	1928–1929	100.0	24.0	69.7	4.5	1.6	0.2	[a]	[a]	
	1929–1930	100.0	44.8	52.6	2.2	0.3	0.1	[a]		
	1930–1931	100.0	38.6	55.4	5.2	0.6	0.2	[a]		
Arizona	1928–1929	100.0	2.2	9.5	14.7	33.2	37.5	2.9		
	1929–1930	100.0	1.2	3.6	23.6	53.3	16.0	2.2	0.1	
	1930–1931	100.0	0.2	4.2	32.1	37.8	23.4	2.3		
Arkansas	1928–1929	100.0	10.5	29.6	23.8	18.4	10.7	5.4	1.4	0.2
	1929–1930	100.0	12.8	28.2	27.1	18.7	9.2	3.3	0.6	0.1
	1930–1931	100.0	9.0	22.7	29.3	23.9	12.3	2.6	0.2	[a]
California	1928–1929	100.0	1.6	12.0	17.2	27.8	35.1	6.3		
	1929–1930	100.0	0.3	2.0	6.9	47.3	40.6	2.9		
	1930–1931	100.0	....	0.7	5.7	16.2	62.0	15.3	0.1	
Florida	1928–1929	100.0	11.5	72.6	12.4	2.0	1.0	0.5		
	1929–1930	100.0	37.6	57.4	4.0	0.7	0.3			
	1930–1931	100.0	19.8	68.2	11.4	0.6				
Georgia	1928–1929	100.0	12,3	70.6	13.7	2.7	0.5	0.1	0.1	
	1929–1930	100.0	21.1	68.5	8.5	1.6	0.3	[a]		
	1930–1931	100.0	14.9	68.6	13.6	2.3	0.5	0.1	[a]	
Louisiana	1928–1929	100.0	12.9	29.6	26.0	18.6	7.6	4.0	1.2	0.1
	1929–1930	100.0	13.9	25.8	26.6	21.4	8.0	3.7	0.6	[a]
	1930–1931	100.0	12.9	22.7	23.7	26.6	9.6	4.0	0.5	[a]
Mississippi	1928–1929	100.0	17.4	17.7	7.7	8.7	17.8	20.9	8.3	1.5
	1929–1930	100.0	10.1	18.8	12.1	10.8	20.7	22.0	5.2	0.3
	1930–1931	100.0	7.9	20.5	18.1	14.3	21.5	14.9	2.7	0.1
Missouri	1928–1929	100.0	5.8	20.5	35.5	26.9	9.8	1.2	0.3	
	1929–1930	100.0	1.4	26.1	40.1	25.4	6.2	0.8	[a]	
	1930–1931	100.0	7.3	37.3	36.3	13.9	4.5	0.7		
New Mexico	1928–1929	100.0	6.0	7.5	10.5	26.9	41.3	7.8		
	1929–1930	100.0	5.4	3.1	7.2	46.9	32.1	5.3		
	1930–1931	100.0	0.5	1.4	8.1	37.2	44.3	8.4	0.1	
North Carolina	1928–1929	100.0	10.0	69.7	14.9	3.5	1.0	0.6	0.3	[a]
	1929–1930	100.0	10.7	62.6	19.7	5.9	0.9	0.2	[a]	
	1930–1931	100.0	5.2	51.2	29.3	10.2	3.3	0.7	0.1	
Oklahoma	1928–1929	100.0	13.6	35.6	33.4	14.5	2.1	0.7	0.1	[a]
	1929–1930	100.0	28.6	44.3	18.5	6.0	2.0	0.6	[a]	
	1930–1931	100.0	14.5	40.9	34.5	8.4	1.4	0.3	[a]	
South Carolina	1928–1929	100.0	6.9	55.0	19.5	9.2	4.6	3.2	1.3	0.3
	1929–1930	100.0	11.2	52.9	18.9	9.7	5.1	1.7	0.4	0.1
	1930–1931	100.0	2.7	41.7	28.2	13.9	8.4	3.9	1.1	0.1
Tennessee	1928–1929	100.0	20.8	40.8	25.5	10.0	2.5	0.3	0.1	[a]
	1929–1930	100.0	15.9	34.4	32.5	14.9	2.0	0.3	[a]	[a]
	1930–1931	100.0	15.1	32.8	33.7	14.5	3.6	0.3		
Texas	1928–1929	100.0	16.1	37.6	31.5	11.9	2.2	0.6	0.1	[a]
	1929–1930	100.0	25.6	33.2	25.1	12.6	2.7	0.7	0.1	[a]
	1930–1931	100.0	12.3	34.2	35.0	15.1	2.6	0.7	0.1	[a]
Virginia	1928–1929	100.0	6.3	82.8	10.5	0.4				
	1929–1930	100.0	6.9	71.0	20.0	1.9	0.2			
	1930–1931	100.0	9.4	69.1	19.4	1.9	0.2			
All other states	1928–1929	100.0	9.7	3.2	8.1	16.1	29.1	16.1	14.5	3.2
	1929–1930	100.0	3.4	1.1	12.4	25.8	22.5	23.6	11.2	
	1930–1931	100.0	7.8	15.6	21.9	28.1	21.9	4.7		

[a] Less than one-tenth of 1 per cent.
NOTE: From Division of Cotton Marketing.

TABLE LXIII.—COTTON: AVERAGE DISCOUNTS AND PREMIUMS FOR STAPLES SHORTER AND LONGER THAN 7/8-INCH FOR MIDDLING SPOT COTTON, 1925–1926 TO 1934–1935
(*Bureau of Agricultural Economics*)

Year beginning August	Discount for 13/16 inch,[a] points[d]	7/8-inch, average price per pound,[b] cents	Premiums for[c]					
			15/16 inch, points[d]	1 inch, points[d]	1 1/16 inches, points[d]	1 1/8 inches, points[d]	1 3/16 inches, points[d]	1 1/4 inches, points[d]
1925–1926	125	19.68	76	106	202	396	635	935
1926–1927	100	14.40	66	106	159	266	480	860
1927–1928	94	19.72	37	93	166	275	409	631
1928–1929	67	18.67	33	96	177	237	332	587
1929–1930	108	15.79	45	118	182	232	347	630
1930–1931	95	9.61	41	91	154	192	317	670
1931–1932	36	5.89	21	51	93	154	244	425[e]
1932–1933	21	7.15	14	39	75	106	201[e]	425[e]
1933–1934	23	10.81	19	53	110	161	270[e]	453[e]
1934–1935	36	12.36	34	78	120	158	236[e]	479[e]

[a] Average of New Orleans, Houston, and Galveston, calculated from actual sales and partly estimated.

[b] Average for the 10 designated spot markets.

[c] Average of New Orleans and Memphis for 1 1/16 inches and longer and for 15/16 inch and 1 inch from 1925–1926 to 1926–1927, inclusive. Average of the six designated markets (New Orleans, Memphis, Houston, Galveston, Dallas, and Little Rock) for 15/16 inch and 1 inch from 1927–1928 to 1934–1935, inclusive.

[d] Hundredths of a cent a pound.

[e] Memphis only.

TABLE LXIV.—COTTON: AVERAGE PREMIUMS AND DISCOUNTS FOR GRADES[a] ABOVE AND BELOW MIDDLING FOR THE 10 DESIGNATED SPOT MARKETS, 1925–1926 TO 1934–1935

Year beginning August	Premiums for				Middling,[a] average price per pound, cents	Discounts for			
	Middling Fair, points[c]	Strict Good Middling, points[c]	Good Middling, points[c]	Strict Middling, points[c]		Strict Low Middling, points[c]	Low Middling, points[c]	Strict Good Ordinary,[b] points[c]	Good Ordinary,[b] points[c]
1925–1926	124	98	73	50	19.68	110	268	432	563
1926–1927	129	106	82	58	14.40	104	238	381	501
1927–1928	100	76	51	33	19.72	51	114	197	284
1928–1929	81	60	42	28	18.67	73	153	236	322
1929–1930	92	76	61	41	15.79	74	170	278	376
1930–1931	88	70	52	31	9.61	59	138	226	305
1931–1932	70	56	41	24	5.89	29	64	101	138
1932–1933	62	50	39	25	7.15	27	55	89	123
1933–1934	71	56	44	30	10.81	35	75	123	165
1934–1935	71	58	47	32	12.36	38	81	131	177

[a] White standards and ⅞-inch staple.
[b] These grades untenderable according to sec. 5 of the United States Cotton Futures Act.
[c] Hundredths of a cent a pound.
NOTE: From Bureau of Agricultural Economics. Data for earlier years in 1934 "Yearbook," Table 126.

TABLE LXV.—COTTON: AVERAGE PRICE PER POUND RECEIVED BY PRODUCERS, UNITED STATES, 1926–1927 TO 1935–1936
(In cents)

Year	Aug. 15	Sept. 15	Oct. 15	Nov. 15	Dec. 15	Jan. 15	Feb. 15	Mar. 15	Apr. 15	May 15	June 15	July 15	Weighted average
1926–1927	16.1	16.8	11.7	11.0	10.0	10.6	11.5	12.5	12.3	13.9	14.8	15.5	12.5
1927–1928	17.1	22.5	21.0	20.0	18.7	18.6	17.0	17.8	18.7	20.1	19.7	21.0	20.2
1928–1929	18.8	17.6	18.1	17.8	18.0	17.9	18.0	18.8	18.5	18.0	17.9	17.8	18.0
1929–1930	18.0	18.2	17.5	16.2	16.0	15.8	14.8	13.8	14.7	14.5	14.0	11.9	16.8
1930–1931	11.4	9.9	9.2	9.6	8.7	8.6	9.1	9.6	9.3	8.8	7.7	8.5	9.5
1931–1932	6.3	5.9	5.3	6.1	5.5	5.6	5.8	6.2	5.7	5.2	4.6	5.1	5.7
1932–1933	6.5	7.2	6.4	5.9	5.4	5.6	5.5	6.1	6.1	8.2	8.7	10.6	6.5
1933–1934	8.8	8.8	9.0	9.6	9.6	10.3	11.7	11.7	11.6	11.0	11.6	12.3	10.2
1934–1935	13.1	13.1	12.5	12.3	12.4	12.3	12.2	11.5	11.7	12.0	11.8	11.9	12.4
1935–1936	11.5	10.6	10.9	11.5	11.4	....	....	....	....	....	....	....	11.1[a]

[a] Preliminary.
NOTE: From Bureau of Agricultural Economics. Based upon returns from special price reporters. Monthly prices, by states, weighted by production to obtain a price for the United States; average for the year obtained by weighting state price averages for the crop-marketing season. Data for earlier years in 1928 "Yearbook," Table 266; only monthly prices are comparable.

TABLE LXVI.—COTTON: INTERNATIONAL TRADE, AVERAGE 1925–1926 TO 1929–1930, ANNUAL 1931–1932 TO 1934–1935
(In thousands of bales)

Country	Year beginning July									
	Average 1925–1926 to 1929–1930		1931–1932		1932–1933		1933–1934		1934–1935[a]	
	Exports	Imports	Exports	Imports	Exports	Imports	Exports	Imports	Exports	Imports
Principal Exporting countries:										
United States	8,579	399	8,989	139	8,647	133	8,366	157[b]	5,066	116
British India	2,938	176	1,565	476	2,126	193	2,585	202	2,733	379
Egypt	1,484	0	1,652	0	1,274	0	1,875	0	1,669	0
Brazil	119	0	40	0	5	0	235	0	725	0
Argentina	88	1[c]	123	0	122	0	92	0	142	0
Total	13,208	576	12,369	615	12,174	326	13,153	359	10,335	495
Principal Importing countries:										
United Kingdom	0	3,070	0	2,475	0	2,460	0	2,955	0	2,230
Japan	0	3,061	0	3,628	0	3,089	0	3,563	0	3,658
Germany	325	1,900	350	1,666	270	1,771	253	1,923	203	1,154
France	100	1,640	47	787	24	1,402	12	1,473	31	951
Italy	1	1,053	0	856	0	898	2	1,009	0	747
China[d]	289	636	220	1,298	185[e]	1,036[e]	202	556	97	536
Czechoslovakia	4	567	1	395	0	340	1	349	16[f]	328
Belgium	14	400	73	300	61	368	114	389	122	342
Poland	0	283	0	218	0	241	0	314	0	286
Canada	0	271	0	202	0	191	0	317	0	255
Netherlands	2	192	2	189	1	152	2	207	1	181
Austria	1	149	0	116	0	88	1	138	1	142
Switzerland	0	141	2	109	0	117	0	123	0	122
Sweden	0	106	0	121	0	109	0	137	0	129
Total	736	13,469	695	12,360	541	12,262	587	13,453	471	11,061

[a] Preliminary.
[b] Beginning 1933–1934, imports for consumption.
[c] 3-year average.
[d] Calendar year.
[e] Beginning July 1, 1932, figures do not include Manchuria.
[f] Monthly Crop Report and Agricultural Statistics, International Institute of Agriculture.

NOTE: From Bureau of Agricultural Economics. Official sources except where otherwise noted. Bales of 500 pounds gross weight or 478 pounds net. The figures for cotton refer to ginned and unginned cotton but do not include linters, mill waste, cotton batting, scarto (Egypt and Sudan), when separately stated. Wherever unginned cotton has been separately stated in the original reports, it has been reduced to ginned cotton in this statement at the ratio of 3 pounds unginned to 1 pound ginned.

TABLE LXVII.—NET RECEIPTS OF RAW COTTON AT PRINCIPAL COTTON PORTS, FOR SPECIFIED YEARS, 1875–1924

(Compiled from reports of New Orleans Cotton Exchange. The statistics for 1915 to 1924 relate to the 12 months ending July 31, and those for prior years to the 12 months ending Aug. 31)

Port	Net receipts of cotton, running bales							
	1924	1923	1922	1921	1920	1919	1918	1915
Galveston	2,875,783	2,345,674	2,551,935	3,144,361	2,101,119	1,933,092	1,621,886	4,001,710
Port Arthur and Texas City	16,852	69,917	34,273	46,900	329,637	124,276	80,728	560,103
New Orleans	1,372,664	1,365,382	1,277,802	1,504,751	1,366,735	1,635,444	1,664,267	1,810,184
Mobile	86,344	90,377	166,317	111,690	265,176	155,516	107,290	166,997
Pensacola	6,014	6,049	2,108	....	13,298	4,038	15,024	87,236
Jacksonville and Fernandina	6,986	9,584	5,407	5,805	15,617	36,082	38,585	34,770
Brunswick	469	25,391	30,224	12,424	168,366	186,114	158,791	215,504
Savannah	456,474	454,027	761,971	727,506	1,306,361	1,150,618	1,140,591	1,762,418
Charleston	146,605	138,964	156,862	91,269	445,123	217,226	203,879	405,504
Wilmington	134,244	107,117	108,019	104,306	142,758	151,882	98,913	279,097
Norfolk and Newport News	426,448	287,659	352,764	296,037	351,137	304,012	299,882	829,683
Baltimore	31,594	21,547	61,443	51,982	92,263	23,157	80,014	83,114
Philadelphia	1,361	4,942	29,800	16,707	21,284	8,065	24,752	11,134
New York	25,053	9,541	28,207	37,027	30,229	12,970	149,970	30,022
Boston	44,907	77,464	50,046	40,111	44,362	30,175	112,367	88,043
San Francisco	84,406	68,842	71,132	81,367	112,390	127,233	158,890	189,561
Seattle and Tacoma	53,676	18,465	82,226	191,384	354,965	608,935	416,978	277,269
Houston	1,071,798	723,622	480,294	466,185	[a]	[a]	[a]	[a]
Laredo, Eagle Pass, etc.	2,082	8,150	796	44,271	71,655	....	2,677	9,707
Minor points	107,248	102,931	151,359	114,409	67,192	27,063	30,645	52,996

Port	1910	1905	1900	1895	1890	1885	1880	1875
Galveston	2,501,412	2,879,336	1,710,263	1,659,999	860,112	463,463	480,352	354,927
Port Arthur and Texas City	163,778	[a]	[a]	[a]	[a]	[a]	[a]	[a]
New Orleans	1,315,328	2,689,520	1,867,153	2,584,115	1,973,571	1,529,592	1,504,654	993,481
Mobile	255,665	329,556	340,646[b]	253,287	261,957	237,071	358,971	320,822
Pensacola	138,234	195,151	[c]	[a]	[a]	[a]	[a]	[a]
Brunswick	227,301	199,193	94,278	[a]	[a]	[a]	[a]	[a]
Savannah	1,365,825	1,877,343	1,088,807	944,410	956,517	728,087	741,018	606,727
Charleston	228,728	225,366	265,523	425,487	327,079	507,802	464,332	412,931
Wilmington	312,511	375,383	282,360	234,621	134,916	94,054	78,876	76,601
Norfolk and Newport News	587,363	841,174	432,727	472,540	404,056	545,418	590,032	387,279
Baltimore	85,526	72,427	101,648	[a]	[a]	[a]	[a]	[a]
Philadelphia	2,581	13,645	36,238	[a]	[a]	[a]	[a]	[a]
New York	40,706	33,798	119,215	187,794	176,502	99,200	229,426	179,163
Boston	14,792	83,644	118,891	[a]	[a]	[a]	[a]	[a]

[a] Not shown separately.
[b] Includes receipts of Pensacola.
[c] Included in receipts of Mobile.

TABLE LXVIII.—CONSUMPTION OF AMERICAN COTTON IN THE WORLD, BY COUNTRIES

(In thousands of running bales, counting round as half bales, linters not included)

(*From "Yearbook, New York Cotton Exchange"*)

Consumed in	1928–1929	1929–1930	1930–1931	1931–1932	1932–1933	1933–1934	1934–1935	1935–1936
United States	6,778	5,803	5,084	4,744	6,004	5,553	5,241	6,221
Great Britain	1,936	1,390	944	1,323	1,365	1,403	941	1,295
Austria and Hungary	103[a]	76[a]	111	124	134	155	128	150
Belgium	217	184	141	152	172	155	122	150
Czechoslovakia	375	329	284	279	247	227	180	242
Denmark	21	20	23	24	28	34	32	30
Finland	34	30	33	33	32	46	49	49
France	820	746	738	599	808	781	535	660
Germany	1,020	947	748	906	973	1,099	376	390
Holland	151	156	148	128	133	130	79	81
Italy	741	681	495	587	714	663	460	425
Norway and Sweden	101	106	85	121	116	133	127	126
Poland	208	189	183	174	240	237	204	217
Portugal	56	56	52	40	49	51	46	41
Russia	389	290	80	......	......	80	35	89
Spain	289	259	242	299	308	314	248	200
Switzerland	53	47	40	43	53	50	36	28
Other countries in Europe	129	111	37	47	72	75	82	85
Total for European continent	4,707	4,227	3,440	3,556	4,079	4,230	2,739	2,963
China	304	292	362	883	748	423	256	104
India	33	10	60	190	106	41	39	34
Japan	1,198	1,095	962	1,623	1,847	1,857	1,737	1,655
Total for the Orient	1,535	1,397	1,384	2,696	2,701	2,321	2,032	1,793
Brazil								
Canada	243	186	179	198	187	238	226	244
Other countries[b]	27	18	25	11	49	35	27	23
Total various	270	204	204	209	236	273	253	267
Total world	15,226	13,021	11,056	12,528	14,385	13,780	11,206	12,539

[a] Covers Austria only. Hungary included in other countries.

[b] Includes Mexico and numerous countries of Central and South America.

TABLE LXIX.—COTTON: MILL CONSUMPTION OF AMERICAN AND OTHER GROWTHS IN THE WORLD, UNITED STATES, AND FOREIGN COUNTRIES, 1915–1916 TO 1934–1935
(In thousands of bales[b])

Year beginning Aug. 1	World			United States			Foreign countries		
	All growths	American[a]	Other growths	All growths	American[a]	Other growths	All growths	American[a]	Other growths
1915–1916	21,978	13,039	8,939	6,398	6,081	317	15,580	6,958	8,622
1916–1917	21,109	12,561	8,548	6,789	6,470	319	14,320	6,091	8,229
1917–1918	18,516	10,871	7,645	6,566	6,382	184	11,950	4,489	7,461
1918–1919	16,705	9,909	6,796	5,766	5,590	176	10,939	4,319	6,620
1919–1920	19,300	11,898	7,402	6,420	6,003	417	12,880	5,895	6,985
1920–1921	16,905	10,268	6,637	4,893	4,677	216	12,012	5,591	6,421
1921–1922	19,990	12,209	7,781	5,910	5,613	297	14,080	6,596	7,484
1922–1923	21,325	12,449	8,876	6,666	6,322	344	14,659	6,124	8,535
1923–1924	19,982	10,917	9,065	5,681	5,353	328	14,301	5,564	8,737
1924–1925	22,642	13,311	9,331	6,193	5,917	276	16,449	7,394	9,055
1925–1926	23,930	14,010	9,920	6,456	6,176	280	17,474	7,834	9,640
1926–1927	25,869	15,748	10,121	7,190	6,880	310	18,679	8,868	9,811
1927–1928	25,285	15,576	9,709	6,834	6,535	299	18,451	9,041	9,410
1928–1929	25,782	15,226	10,556	7,091	6,778	313	18,691	8,448	10,243
1929–1930	24,878	13,021	11,857	6,106	5,803	303	18,772	7,218	11,554
1930–1931	22,402	11,056	11,346	5,263	5,084	179	17,139	5,972	11,167
1931–1932	22,896	12,528	10,368	4,866	4,744	122	18,030	7,784	10,246
1932–1933	24,986	14,385	10,601	6,137	6,004	133	18,849	8,381	10,468
1933–1934	25,324	13,780	11,544	5,700	5,553	147	19,624	8,227	11,397
1934–1935	25,283	11,206	14,077	5,361	5,241	120	19,922	5,965	13,957

[a] "American" cotton means cotton grown in the United States.

[b] American in running bales and other growths in bales of 478 pounds net. Prior to 1919–1920, the quantities given for world consumption of all growths were reported in bales of 500 pounds net and have been converted to equivalent 478-pound bales.

NOTE: From Bureau of Agricultural Economics; compiled from reports of the Bureau of the Census except consumption figures for American cotton in foreign countries, which are compiled from the *Yearbook* of the New York Cotton Exchange, p. 127, 1935.

The figures for the consumption of "other growths" in the world and in foreign countries were computed by deduction.

TABLE LXX.—COTTON SEED: PRODUCTION OF COTTON SEED, CRUDE COTTONSEED OIL, AND BY-PRODUCTS; PRICE PER TON, FARM VALUE, AND CRUSHINGS OF COTTON SEED, 1926–1927 TO 1935–1936

Year beginning August	Cotton seed				Cottonseed products[b]			
	Production,[a] 1,000 short tons	Season average price per ton received by producers, dollars	Farm value, $1,000	Quantity crushed,[b] 1,000 short tons	Crude oil, 1,000 short tons	Cake and meal, 1,000 short tons	Linters, 1,000 running bales	Hulls, 1,000 short tons
1926–1927	7,989	21.55	172,131	6,306	944	2,840	1,042	1,854
1927–1928	5,758	35.94	206,940	4,654	738	2,093	875	1,320
1928–1929	6,435	35.26	226,874	5,061	802	2,282	1,086	1,368
1929–1930	6,590	30.43	200,521	5,016	786	2,232	1,038	1,384
1930–1931	6,190	21.93	135,753	4,715	721	2,165	824	1,304
1931–1932	7,604	9.52	72,412	5,328	847	2,402	876	1,511
1932–1933	5,783	10.35	59,881	4,621	723	2,093	741	1,312
1933–1934	5,804	14.21	82,474	4,157	652	1,889	801	1,103
1934–1935	4,282	34.79	148,986	3,549	554	1,614	805	912
1935–1936[c]	4,775	31.60	150,877					

[a] Estimated from the production of lint cotton, assuming 65 pounds of seed for each 35 pounds of lint. Refers to the cotton crop of the year stated.

[b] Crushings and products are not limited to the crop specified.

[c] Preliminary.

NOTE: From Bureau of Agricultural Economics. Quantity crushed and products from Cotton Production and Distribution," annual reports of the Bureau of the Census.

TABLE LXXI.—QUANTITIES OF THE SEVERAL PRODUCTS OBTAINED PER TON OF SEED CRUSHED, BY STATES, FOR THE YEARS ENDING JULY 31, 1922, 1923, AND 1924

State	Products obtained, pounds														
	Total			Crude oil			Cake and meal			Hulls			Linters		
	1924	1923	1922	1924	1923	1922	1924	1923	1922	1924	1923	1922	1924	1923	1922
United States	1,884	1,903	1,900	296	309	309	918	918	901	569	582	623	101	94	67
Alabama	1,900	1,884	1,878	321	311	308	911	921	889	571	568	620	97	84	61
Arkansas	1,864	1,923	1,889	284	312	305	860	896	856	618	623	663	102	92	65
Georgia	1,896	1,882	1,900	316	317	321	958	957	901	512	504	609	110	104	69
Louisiana	1,869	1,874	1,882	285	310	314	917	922	916	565	554	589	102	88	63
Mississippi	1,896	1,901	1,896	329	330	329	850	880	880	606	595	617	111	96	70
North Carolina	1,868	1,875	1,884	322	333	320	941	905	943	513	549	555	92	88	66
Oklahoma	1,850	1,879	1,911	277	273	290	948	943	928	522	552	624	103	111	69
South Carolina	1,876	1,893	1,894	311	322	315	959	928	912	490	523	580	116	120	87
Tennessee	1,891	1,894	1,879	276	315	306	826	853	838	677	632	671	112	94	64
Texas	1,889	1,930	1,917	282	295	293	927	947	916	582	597	648	98	91	60
All other states	1,893	1,906	1,933	309	310	324	888	892	909	606	624	636	90	90	64

TABLE LXXII.—NUMBER COTTONSEED-OIL MILLS AND THEIR DISTRIBUTION

State	Total number of cotton seed oil mills	Number of cotton seed oil mills crushing—					
		Less than 1,000 tons	1,000 but less than 2,000 tons	2,000 but less than 5,000 tons	5,000 but less than 10,000 tons	10,000 but less than 20,000 tons	20,000 tons and over
United States	532	53	56	167	164	75	17
Alabama	33	6	6	14	4	3	
Arkansas	29	4	2	9	9	5	
Georgia	52	13	8	13	12	5	1
Louisiana	20	2	3	6	5	4	
Mississippi	41	3	2	16	13	6	1
North Carolina	58	1	7	28	15	6	1
Oklahoma	42	7	5	12	12	6	
South Carolina	47	7	12	16	7	5	
Tennessee	19	. . .	4	2	9	1	3
Texas	174	9	5	48	72	32	8
All other states	17	1	2	3	6	2	3

References

1. Austin, William L. and Zimmerman, Harvey J.: "Cotton Production and Distribution," *U. S. Dept. Commerce, Bur. Census, Bull.* 156, 1925.
2. Anonymous: *Yearbook, U. S. Dept. Agr.*, Washington, D. C., 1923 and other years.
3. Shepperson, A. B. and Shepperson, C. W.: "Cotton Facts," Shepperson Publishing Company, New York, 1924 and other years.
4. Todd, John A.: "The World's Cotton Crops," A. & C. Black, Ltd., London, 1923.
5. Anonymous: "Book of the World's Cotton Conference, 1921," Executive Committee, Manchester, England, 1921.
6. Anonymous: "Agricultural Statistics," *U. S. Dept. Agr.*, 1936.
7. Garside, A. H.: *Yearbook, New York Cotton Exchange*, 1936, New York N. Y.

INDEX

T

U

V